Information Systems Engineering and Management 49

The book series "Information Systems Engineering and Management" (ISEM) publishes innovative and original works in the various areas of planning, development, implementation, and management of information systems and technologies by enterprises, citizens, and society for the improvement of the socio-economic environment.

The series is multidisciplinary, focusing on technological, organizational, and social domains of information systems engineering and management. Manuscripts published in this book series focus on relevant problems and research in the planning, analysis, design, implementation, exploration, and management of all types of information systems and technologies. The series contains monographs, lecture notes, edited volumes, pedagogical and technical books as well as proceedings volumes.

Some topics/keywords to be considered in the ISEM book series are, but not limited to: Information Systems Planning; Information Systems Development; Exploration of Information Systems; Management of Information Systems; Blockchain Technology; Cloud Computing; Artificial Intelligence (AI) and Machine Learning; Big Data Analytics; Multimedia Systems; Computer Networks, Mobility and Pervasive Systems; IT Security, Ethics and Privacy; Cybersecurity; Digital Platforms and Services; Requirements Engineering; Software Engineering; Process and Knowledge Engineering; Security and Privacy Engineering, Autonomous Robotics; Human-Computer Interaction; Marketing and Information; Tourism and Information; Finance and Value; Decisions and Risk; Innovation and Projects; Strategy and People.

Indexed by Google Scholar. All books published in the series are submitted for consideration in the Web of Science.

For book or proceedings proposals please contact Alvaro Rocha (amrrocha@gmail.com).

Zahurin Mat Aji · Rohaida Romli ·
Shafinah Farvin Packeer Mohamed ·
Mohamed Ali Saip · Tutut Herawan
Editors

Digital Innovation in Knowledge Management

Proceedings of the 11th Knowledge Management International Conference (KMICe 2024)

Springer

Editors
Zahurin Mat Aji
School of Computing (SOC)
Universiti Utara Malaysia
Sintok, Kedah, Malaysia

Rohaida Romli
School of Computing (SOC)
Universiti Utara Malaysia
Sintok, Malaysia

Shafinah Farvin Packeer Mohamed
School of Computing (SOC)
Universiti Utara Malaysia
Sintok, Malaysia

Mohamed Ali Saip
School of Computing (SOC)
Universiti Utara Malaysia
Sintok, Malaysia

Tutut Herawan
Department of Information System
Faculty of Computer Science and Information
Technology
Universiti Malaya
Kuala Lumpur, Malaysia

ISSN 3004-958X ISSN 3004-9598 (electronic)
Information Systems Engineering and Management
ISBN 978-3-031-91484-3 ISBN 978-3-031-91485-0 (eBook)
https://doi.org/10.1007/978-3-031-91485-0

This Springer imprint is published by the registered company Springer Nature Switzerland AG
The registered company address is: Gewerbestrasse 11, 6330 Cham, Switzerland

Preface

The 11th Knowledge Management International Conference (KMICe 2024) took place from October 2nd to 3rd, 2024, in Putrajaya, Malaysia, with the theme "Embracing Digital Acceleration & Social Innovation in Knowledge Management". The conference served as a platform for presenting international research findings, fostering discussions, and sharing recent advancements in the field. It brought together leading researchers, academia, and practitioners from various fields, united by a common interest in advancing the field of Knowledge Management (KM). The conference received approximately 69 articles, each of which underwent a double-blind review process. Reviewers were assigned a maximum of three articles, with each article being reviewed by at least two reviewers. A total of 42 articles have been included in this book.

This book explores a broad spectrum of research that underscores the importance of technology and knowledge management in addressing modern challenges. In the dynamic and rapidly evolving landscape of digital transformation, knowledge management, blockchain technologies, artificial intelligence, and big data solutions have emerged as fundamental pillars in addressing contemporary global challenges. This book presents a carefully curated compilation of scholarly research and case studies, showcasing innovative strategies, technological advancements, and practical applications across a range of disciplines. Structured into four distinct thematic sections, the work reflects the collaborative efforts of academics and practitioners dedicated to advancing knowledge and fostering meaningful progress.

The first section examines the critical role of knowledge management in fostering innovation, sustainability, and organizational efficiency. It highlights diverse approaches to leveraging knowledge systems as tools for achieving improved outcomes. The second section delves into the convergence of blockchain technology, cybersecurity, and knowledge management, offering insightful analyses on blockchain's transformative applications, emerging trends in trust management, and the implementation of cybersecurity solutions for smart contracts, as well as the broader implications of digital behaviors.

The third section focuses on big data systems, software solutions, and their applications, emphasizing the enhancement of user experiences, optimization of organizational processes, and promotion of sustainability. The studies in this section cover a wide array of topics, including heterogeneous data integration, mobile usability, and the gamification of learning environments. The final section highlights state-of-the-art research in artificial intelligence, deep learning, and pattern recognition, with a focus on real-world applications and forward-thinking solutions. These studies address key areas such as social innovation, employment challenges, agricultural development, and the optimization of advanced computational models.

This book offers a comprehensive and critical exploration of technology's transformative potential in addressing today's global challenges. By examining the synergies between knowledge management, blockchain, artificial intelligence, and big data, this book presents forward-thinking perspectives on innovation, usability, and sustainability.

Each theme invites readers to engage deeply with emerging technological paradigms and to contribute to a future marked by equity, efficiency, and technological excellence.

We extend our sincere appreciation to all contributors, reviewers, and collaborators for their dedication to the advancement of knowledge and their commitment to fostering positive change. It is our aspiration that this book will serve as a foundational resource, inspiring further research, inquiry, and innovation in these critical and rapidly evolving fields.

We would like to express our gratitude to everyone who contributed to the KMICe 2024, including the authors, committees, and sponsors. Their involvement and support were greatly appreciated.

January 2025

Zahurin Mat Aji
Rohaida Romli
Shafinah Farvin Packeer Mohamed
Mohamed Ali Saip
Tutut Herawan

KMICe 2024 Organization

Patron

Mohd Fo'ad Sakdan (Vice Chancellor)	Universiti Utara Malaysia (UUM)

Advisor

Osman Ghazali (Dean)	School of Computing (SOC), UUM

Conference Chair

Haslina Mohd	School of Computing (SOC), UUM

Conference Co-Chair

Yuhanis Yusoff	School of Computing (SOC), UUM

Organizing Committee

Fauziah Baharom	SOC, UUM, Malaysia
Aniza Mohamed Din	SOC, UUM, Malaysia
Musyrifah Mahmod	SOC, UUM, Malaysia
Azizi Ab Aziz	SOC, UUM, Malaysia
Nor Iadah Yusop	SOC, UUM, Malaysia
Nor Hazlyna Harun	SOC, UUM, Malaysia
Mohd Samsu Sajat	SOC, UUM, Malaysia
Baharudin Osman	SOC, UUM, Malaysia
Fazli Azzali	SOC, UUM, Malaysia
Ahmad Hanis Mohd Shabli	SOC, UUM, Malaysia
Zahurin Mat Aji	SOC, UUM, Malaysia
Rohaida Romli	SOC, UUM, Malaysia
Shafinah Farvin Packeer Mohamed	SOC, UUM, Malaysia

Mohamed Ali Saip	SOC, UUM, Malaysia
Rahayu Ahmad	SOC, UUM, Malaysia
Wan Hussain Wan Ishak	SOC, UUM, Malaysia
Mohamad Farhan Mohamad Mohsin	SOC, UUM, Malaysia
Tutut Herawan	FCSIT, UM, Malaysia
Najia Saher	BJC, IUB, Pakistan

Technical Program Committee

Fauziah Baharom	SOC, UUM, Malaysia
Najia Saher	BJC, IUB, Pakistan
Nor Iadah Yusop	SOC, UUM, Malaysia
Noraziah Che Pa	SOC, UUM, Malaysia
Wan Hussain Wan Ishak	SOC, UUM, Malaysia
Mohd Hasbullah Omar	SOC, UUM, Malaysia
Alawiyah Abd Wahab	SOC, UUM, Malaysia
Azizi Ab Aziz	SOC, UUM, Malaysia
Farzana Kabir Ahmad	SOC, UUM, Malaysia
Haslina Mohd	SOC, UUM, Malaysia
Mazida Ahmad	SOC, UUM, Malaysia
Mazni Omar	SOC, UUM, Malaysia
Mustafa Ali Abuzaraida	Misurata University, Libya
Nooraini Yusoff	FSD, UMK, Malaysia
Nor Hazlyna Harun	SOC, UUM, Malaysia
Nur Haryani Zakaria	SOC, UUM, Malaysia
Omar Husain Altarawneh	Amman Arab University, Jordan
Rahayu Ahmad	SOC, UUM, Malaysia
Rohaida Romli	SOC, UUM, Malaysia
Shuzlina Abdul Rahman	FKSM, UiTM, Malaysia
Sobihatun Nur Abdul Salam	SMMTC, UUM, Malaysia
Yuhanis Yusof	SOC, UUM, Malaysia
Zahurin Mat Aji	SOC, UUM, Malaysia
Azlin Binti Nordin	Kulliyyah of ICT, IIUM, Malaysia
Hamzah Alaidaros	Al-Ahgaff University, Yemen
Juhaida Abu Bakar	SOC, UUM, Malaysia
Mohamed Ali Saip	SOC, UUM, Malaysia
Mohd. Nor Akmal Khalid	FTSM, UKM, Malaysia
Aisha Siddiqa	SOC, UUM, Malaysia
Atyaf Sami Noori	University of Information and Technology, Iraq
Azliza Othman	SMMTC, UUM, Malaysia

Bakr Salim Ba-Quttayyan	SOC, UUM, Malaysia
Fariza Hanim Rusly	TISSA, UUM, Malaysia
Fazli Azzali	SOC, UUM, Malaysia
Hapini Awang	SOC, UUM, Malaysia
Juliana Wahid	SOC, UUM, Malaysia
Khuzairi Mohd Zaini	SOC, UUM, Malaysia
Moath Husni	WISE, Jordan
Mohamad Farhan Mohamad Mohsin	SOC, UUM, Malaysia
Mohammed Gamal Ahmed Alsamman	SOC, UUM, Malaysia
Mohd Fitri Mansor	FPK, UniMAP, Malaysia
Mohd Shamrie Sainin	FCI, UMS, Malaysia
Munya Saleh Saeed Ba Matraf	Hadhramaut University, Yemen
Ruziana Binti Mohamad Rasli	SMMTC, UUM, Malaysia
Samera Obaid Barraood	Hadhramaut University, Yemen
Shafinah Farvin Packeer Mohamed	SOC, UUM, Malaysia
Syahida Hassan	SOC, UUM, Malaysia
Wan Aida Nadia Wan Abdullah	SOC, UUM, Malaysia
Wiwied Virgiyanti	FSKM, UMT, Malaysia
Zhamri Che Ani	SOC, UUM, Malaysia
Ahmad Hanis Mohd Shabli	SOC, UUM, Malaysia
Inusa Sani Maijama'a	HAFEDPOLY, Nigeria
Suwannit Chareen Chit	SOC, UUM, Malaysia
Aniza Mohamed Din	SOC, UUM, Malaysia
Cik Fazilah Hibadullah	SOC, UUM, Malaysia
Musyrifah Mahmod	SOC, UUM, Malaysia
Nur Azzah Abu Bakar	SOC, UUM, Malaysia
Nur Fadziana Faisal Mohamed	MMTC, UUM, Malaysia
Sharifah Nawaitul Masdinar	UPTM, Malaysia

Contents

Part 1: Knowledge Management and Practices

About the Editors

Zahurin Mat Aji is an Associate Professor at the Information Systems Department, School of Computing (SOC), Universiti Utara Malaysia (UUM). She started her academic career as a Tutor in 1992 and has been a faculty member since then. She holds a BSc. in Business Administration with a concentration in Computer Information Systems from California State University, Dominguez Hills, an MSc. Information Systems from Leeds University, and a PhD. in Information Technology from UUM. Her research interests include community informatics, rural ICT development, and management information systems. Her research often integrates practical applications with theoretical insights, aiming to bridge the digital divide and promote e-inclusion. In addition to her research contributions, Dr. Zahurin is also committed to teaching and mentoring. The mentoring of undergraduate and graduate students plays a crucial role in shaping the next generation of IT professionals and researchers. She is also a member of the Malaysia Chapter of the Association for Information Systems (MyAIS), Internet Society (ISOC), and Internet Society (ISOC) Malaysia Chapter. These affiliations underscore her engagement with professional communities dedicated to advancing information technology and its applications. Overall, her work has significantly advanced the understanding of the socio-technical dimensions of ICT. Her research, teaching, and community engagement activities demonstrate her commitment to using technology as a tool for social good, particularly, in ensuring that its benefits are accessible to all, including the underserved and marginalized communities. Her contributions continue to shape the field of information technology, providing valuable insights and practical solutions for leveraging ICT in ways that promote inclusive and sustainable development.

Rohaida Romli is an Associate Professor at the School of Computing, UUM College of Arts and Sciences. Her research primarily focuses on advancing software quality and testing, particularly on automated assessment techniques. In addition, her work explores software engineering practices, with a particular emphasis on Agile methodologies. Her publications span a wide range of topics, including automated programming assessment, Agile software development, and the integration of augmented reality in object-oriented programming education. She also contributes extensively to the academic community through her role as a reviewer for several renowned journals, including *Computer Applications in Engineering Education*, *Journal of King Saud University*, *Information & Software Technology*, *Journal of Information and Communication Technology (JICT)*, *Journal of Behavior and Information Technology*, *Journal of Open Innovation – Technology, Market & Complexity*, *Malaysian Journal of Computer Science (MJCS)*, and *ComSIS Journal*. She is dedicated to fostering the development of the next generation of scholars and professionals in the fields of software engineering and related disciplines.

Shafinah Farvin Packeer Mohamed is a lecturer at the School of Computing (SOC), Universiti Utara Malaysia (UUM). She holds a Ph.D. in Information Technology from the same university. Currently serving as the program chair for the Bachelor of Science with Honours (Information Technology), she oversees curriculum reviews and program management. Regarding research, she has led and contributed to various research grants and published over 20 publications. Her research focuses on software process quality, Agile software development, and multi-criteria decision making. She has supervised students at all levels and remains dedicated to advancing computing education and research. As academic recognition, she has been involved in reviewing quite a few journals, including the Journal of Information and Communication Technology, the Journal of Analytic Hierarchy Process and the Arabian Journal of Geosciences. Besides, she was also invited as a reviewer for conference proceedings like the Knowledge Management International Conference (KMICe), International Conference on Computing and Informatics (ICOCI), International Conference on Fundamental, Applied Sciences and Technology (ICoFAST) and International Soft Science Conference (ISSC). She is committed to nurturing the growth of future scholars and professionals in related fields.

Mohamed Ali Saip is a senior lecturer at Universiti Utara Malaysia and has significantly contributed to Computer Science with a PhD from the University of Bradford, UK. His research focuses on e-government, big social data analytics, sentiment analysis, and social network analysis, often bridging theoretical insights with practical applications. A notable example is his collaboration with British local governments, where he applied data-driven approaches to enhance public services based on citizen needs. He is a prolific researcher with numerous publications in prestigious journals and conference proceedings. His work covers diverse areas, such as the impact of cybersecurity risks on social media and the development of assessment tools for preschool educators. His studies on sentiment analysis and bibliometric analyses have provided valuable insights into current trends, influencing future research and policymaking. He has also contributed to book chapters and engaged in impactful research projects. His recent projects include collaborations with Malaysia's Ministry of Higher Education on data analytics for graduate employability and industry-based initiatives such as developing an image recognition system for rabbit farm monitoring and a yield prediction model for paddy. He actively supervises master's and PhD students and serves as a reviewer for numerous academic conferences and journals, reflecting his commitment to both research and academic mentorship.

Tutut Herawan is an associate professor at the Department of Information Systems, Faculty of Computer Science and Information Technology, Universiti Malaya, Malaysia. He was named in the Top 2% World Scientists Ranking by Stanford University and Elsevier BV, since 2019-present. He received a PhD degree in information technology from Universiti Tun Hussein Onn Malaysia in 2010. He has more than 20 years experiences as academic staff and has supervised many Master & PhD students. He is the associate editor of many international journals of Scopus and Clarivate Web of Science indexes. He has edited many books and published more than 330 articles in various book chapters, international journals, and conference proceedings. He is an active reviewer for many reputable international journals. He delivered more than 50 keynote addresses, invited workshop and seminars, and has been actively served as a chair, co-chair, program committee member and co-organizer for numerous international conferences/workshops. His research area includes data engineering, information systems, business data analytics.

Part 1: Knowledge Management and Practices

Integrating Digital Innovation in Knowledge Management for Water Quality and Sustainability

Nur Suhaili Mansor[1](✉), Hapini Awang[1], Nor Hazlyna Harun[2], Mohamad Ghozali Hassan[3], Sarkin Tudu Shehu Malami[4], Lariyah Mohd Sidek[5], Juhaida Abu Bakar[2], and İsmail Rakıp Karaş[6]

[1] Institute for Advanced and Smart Digital Opportunities, School of Computing, Universiti Utara Malaysia, 06010 Sintok, Kedah, Malaysia
nursuhaili@uum.edu.my

[2] Data Science Research Lab, School of Computing, Universiti Utara Malaysia, 06010 Sintok, Kedah, Malaysia

[3] Disaster Management Institute (DMI), School of Technology Management and Logistics, Universiti Utara Malaysia, 06010 Sintok, Kedah, Malaysia

[4] Department of Computer Science and Information Technology, School of Technology, Management and Logistics, Faculty of Science, Sokoto State University, Sokoto, Nigeria

[5] Institute of Energy Infrastructure (HiCOE), Universiti Tenaga Nasional, Kajang, Malaysia

[6] Computer Engineering Department, Faculty of Engineering, DemirCelik Campus, Karabuk University, Karabuk, Turkey

Abstract. The development of industrial areas is frequently thought to positively impact society because it helps meet the population's requirements. However, it also negatively impacts the environment, such as river pollution. If a country experiences rapid economic growth without considering the effects of industrial development on the environment, then human well-being will be affected. In this context, advanced development is attained by a developed nation while ensuring the well-being of society. Uncontrolled and unsustainable development increases environmental pollution, adversely affecting human health. As water is essential for human life, water resources, such as rivers, must be appropriately managed and protected to ensure human well-being. Thus, the study conducted on the changes in the water quality of the Kim Kim River provides insights into how unintended consequences of urban and industrial development can be identified and addressed using three sets of satellite imagery data from 2015, 2019, and 2020. The study was led to identify research problems and achieve the study objectives to obtain research results. The results of image classification analyses indicate that the development area near Sungai Kim Kim increased to 50%, with 38% of industrial regions. The RMSE for the image classification is 1.51, significantly better than the minimum allowed value of 2.00. Integrating digital innovation in knowledge management for water quality and sustainability, this study provides an understanding of the change in river water quality due to urban and industrial development. By highlighting the importance of sustainable development and the need to balance economic growth with environmental protection, this study contributes to the development of effective policies and strategies for the sustainable development of industrial areas.

Z. Mat Aji et al. (Eds.): KMICe 2024, ISEM 49, pp. 3–15, 2025.
https://doi.org/10.1007/978-3-031-91485-0_1

Keywords: Water quality · GIS · Remote Sensing · Environment · Sustainability

1 Introduction

Water quality refers to water's physical, chemical, thermal, and biological properties [1]. Pollution from point or non-point sources can deteriorate water quality in most river ecosystems worldwide. Surface water quality monitoring and evaluation, such as in rivers, are essential to improving water quality. In-situ measurements and laboratory analyses of water samples are commonly used to evaluate water quality [2, 3]. In-situ measurements can indicate a clean condition of the study location, as seen in various studies conducted on the Selangor River's water quality. However, it isn't easy to establish a single water quality standard for all consumer uses and needs. The parameters suitable for irrigating crops may differ from those suitable for human consumption, thus requiring a method to create an information system for the study problem. In the context of Malaysia, water quality characteristics vary based on the region and water bodies under study [4]. However, the attributes of water quality are frequently of concern in Malaysia [5]. Biological Demand for Oxygen (BOD) measures how much oxygen bacteria require when decomposing organic waste [4]. High BOD levels signify organic contamination and can cause water bodies to run out of oxygen [6]. Dissolved oxygen (DO) levels are another important quality attribute for aquatic life [4]. Low DO levels may indicate pollution or organic matter decomposition, harming aquatic ecosystems and water quality [7].

Technological advances in this era have helped handle spatial data and attributes from various sources and specific times. Several studies have used technological techniques to assess water quality [8, 9], such as using MultiRAE and Gasmet Detectors to detect chemical waste dumped into the Kim Kim River. Remote sensing and GIS techniques have been reported [10, 11] to be valuable tools in monitoring water quality changes and producing thematic maps, as seen in the study of the Kim Kim River.

GIS methods can facilitate spatial analysis, and GIS databases can help avoid wasting time updating data and enable quick decision-making processes. In this study, the researcher will use remote sensing techniques and GIS to identify changes in the water quality of the Kim Kim River caused by suspended sediments, chemicals from factories, algae, metals, oil, household waste, and other factors that affect water quality.

With appropriate in-situ validation, remote-sensing satellite images can provide high-accuracy interpolation and produce spatial water quality maps more effectively [12] Thematic map applications are a common choice in remote sensing of river water quality, and they can provide a valuable tool for monitoring water quality. Therefore, remote sensing and GIS technologies have actual and potential applications for assessing and monitoring river water quality.

The evolution of technology has facilitated the processing of spatial data and attributes derived from diverse sources and specific time frames. Consequently, researchers can achieve satisfactory outcomes when examining water quality using various cutting-edge technological methods.

This study will use remote sensing and GIS techniques to identify changes in the water quality of the Kim Kim River. Materials in surface water can significantly change the characteristics of surface water. Remote sensing techniques rely on measuring spectral changes in river water and relating these measurement changes by empirical or analytical models to water quality parameters. The main factors that affect water quality in river water and water bodies throughout the landscape are suspended sediments, chemicals from factories, algae, metals, oil, household waste, and so on [13]. These factors can change river water quality, which remote sensing techniques can measure.

Integrating remote sensing and GIS technology provides a valuable tool for monitoring water quality. Therefore, this study proves that remote sensing techniques have many actual and potential applications for assessing and monitoring river water quality.

2 Related Works

Development is considered a process of change toward progress. In the context of state development, it is generally defined as an effort to achieve progress in community lives. According to [14], development encompasses all spheres of society and is related to change and progress. It is a process of change toward advancing economic well-being, modernization, national development, and improving the quality of human life. Another study [6] assessed spatial water quality in three Malaysian river basins using cluster analysis, principal component analysis, and discriminant analysis. The analysis identified pollution sources mainly originating from industrial waste. [5] analyzed the influence of land use and wastewater treatment plants on water quality in the Johor River Basin, Malaysia. Results emphasize the importance of sustainable land use having a greater influence on water quality during the wet season and effective wastewater management for long-term water resource sustainability.

Development can boost revenue and output while changing social institutions, administrative systems, cultures, habits, and community beliefs. The society undertakes development to reduce societal problems and improve living standards. This includes addressing societal requirements such as providing security, job opportunities, improved educational facilities, emphasizing moral values and attitudes, developing critical thinking abilities, and boosting national revenue. However, economic and industrial development does not necessarily guarantee human well-being and may cause social problems due to unsustainable practices. As per [15], environmental degradation is one of the social problems associated with development that negatively impacts human health.

Over the past few decades, rapid development has caused various environmental changes, such as global warming, water pollution, air pollution, and the extinction of flora and fauna. Even rural development affects natural ecosystems on land and the ocean. In Malaysia, particularly river pollution is not unusual. Human acts prioritizing economic expansion for national development are responsible for this environmental damage. The development raises income and living standards, benefiting the economy and society, but it also has detrimental effects on the environment due to unsustainable behaviors.

According to [16], water pollution refers to changes in water quality, taste, and smell that impact its function as a source of life. River conditions have deteriorated due to development, particularly in areas near rivers. Development has led to controlled landscape changes that cater to population needs, resulting in a severe decline in river water quality. Most rivers in Malaysia face issues such as sedimentation, soil surface pollution, and industrial waste disposal.

Economic and technological developments, particularly infrastructure, have contributed to environmental destruction. In areas like Pasir Gudang, which have undergone urbanization and development since 1977, rapid and uncontrolled development has had a significant impact [17]. To address this, the concept of a sustainable environment has been introduced internationally, emphasizing the importance of environmental hygiene. Sustainable development has become a key approach to achieving a sustainable environment.

Since the development of the industrial area in Pasir Gudang, there have been environmental effects such as pollution of the Kim Kim River. The pollution of the Kim Kim River is one of the largest environmental degradations occurring in Pasir Gudang due to industrial development. The study's focus is the rapid development causing environmental degradation, especially the changes in the Kim Kim River's water quality. This study emphasizes that the changes in the water quality of the Kim Kim River are serious and less desirable for a sustainable development environment. Clean water is essential for various human activities and is a natural habitat for aquatic life. Therefore, current rapid development highlights the need for clean river water as a priority.

Rapid urbanization and development require balancing development and environmental protection [18]. However, the growing problem of water quality changes, especially in the Kim River, indicates that humans prioritize national modernity, improved living standards, and economic growth rather than the importance of water quality. Therefore, humans need to understand the significance of sustainable development to achieve environmental equality. This implies that, apart from the government's actions in the situation of the Kim Kim River in Pasir Gudang, a more sustainable approach to development in the Pasir Gudang area is necessary.

Thus, leading to this study is on the changes in the water quality of the Kim Kim River and the impact of industrial and human activities on the river. Furthermore, this study investigates the application of remote sensing technology to identify changes in the Kim Kim River's water quality. Remote sensing is a technology that uses satellite technology to differentiate information about the Earth's surface and water bodies and then interprets it as an image or picture using electromagnetic radiation in the form of the electromagnetic spectrum [19].

This study examines changes in Kim Kim River water quality from 2015 to 2020 and secondary data to assess Kim Kim River's pollution in 2019 and the variables impacting it. The study also looks at changes in water quality from 2020 to 2022 during the Covid-19 pandemic.

3 Methodology

This research examines land use and water quality changes along the Kim Kim River using Landsat and SPOT-6 satellite imagery. Integrating GIS and remote sensing techniques into research procedures provides a robust tool for processing and interpreting geographical data.

ERDAS Imagine performs geometric and radiometric corrections on the SPOT-6 data provided by the Malaysian Space Agency (MYSA) and the Landsat satellite images obtained from the USGS website. Using a supervised classification technique, the imagery data is categorised according to land-use classes such as water bodies, forests, built-up areas, land, roads, and rivers. The resulting categorised imagery data is used to investigate land use changes in the research area between 2015, 2019, and 2020.

Kim Kim River water quality changes are monitored using Landsat data and SPOT-6 satellite images. SPOT-6 satellite imagery is well-suited for this study because of its high resolution and extensive coverage. Image analysis of water bodies' spectral fingerprints reveals quality shifts in the water they contain.

Fig. 1. Study area.

The Kim Kim River in Fig. 1 is an essential waterway in Malaysia for Johor, used for numerous uses, including drinking, farming, and industry. Additionally, several other river basin systems are located between the Kim Kim River and the Johor River. However, rapid urbanisation and industrialisation have led to severe pollution in the river, resulting in a decline in water quality. Hence, a comprehensive study evaluating water quality changes in the Kim Kim River, Johor, is imperative.

In response, a theoretical investigation was conducted to examine potential modifications in water quality and their relationship to development and environmental conservation. The primary objective of this literature review was to assess the effectiveness of integrating remote sensing with Geographic Information System (GIS) technology to identify alterations in the water quality of the Kim Kim River and produce thematic maps.

SPOT-6 satellite imagery, as in Fig. 2 data from the Malaysian Space Agency (MYSA), was used as it is considered the most suitable data source for implementing remote sensing and GIS techniques to examine changes in water quality within the

Kim Kim River, Johor. Specifically, satellite images of the Kim Kim River were acquired for 2015, 2019, and 2020 and then analysed using remote sensing tools to monitor the river's water quality changes. Every step of the remote sensing process serves a distinct purpose and is essential for subsequent data processing. By combining remote sensing and GIS data, it is possible to identify and visualise changes in water quality within the Kim Kim River and the contributing factors, such as land-use alterations and analysis of adjacent spatial areas.

Fig. 2. SPOT-6 satellite imagery.

This study used the ERDAS software application as the remote sensing technique for data collection, while ArcGIS was applied to ensure accuracy during the final data analysis steps. These two software applications allowed for an accurate assessment of the water quality changes in the Kim Kim River.

Additionally, qualitative data was incorporated into the study to meet the second research objective of elucidating the factors that caused changes in water quality between 2015 and 2020. The qualitative data obtained from various sources, including government agencies and academic publications, will point out the factors contributing to the water quality changes in the Klang River.

Table 1 implies that the SPOT-6 satellite's high resolution and comprehensive coverage make it versatile for various applications, especially those requiring precise and detailed information like disaster management, urban planning, and land use mapping.

The satellite's extensive coverage capacity enables it to capture more thorough imaging of a broader area in a single pass, rendering it appropriate for global and regional environmental monitoring and detecting extensive alterations over extended periods.

Furthermore, it can reduce the resources and time needed for image acquisition and processing, increasing productivity for users and researchers.

Table 1. The characteristics of spot-6 image

Sensor-Satellite	Spatial Resolution	Spectral Resolution	Orbital	Temporal	Swath (km)
SPOT-6	1.5 m	Panchromatic and Multispectral	694 km	1 Day	60 km × 600 km (1 strip)

(Malaysian Space Agency, 2021).

The results of this literature review indicate that remote sensing techniques, when combined with GIS technology, are highly effective in identifying changes in water quality in the Kim Kim River, Johor. The study highlights the importance of integrating different data sources to understand water quality changes in the river comprehensively.

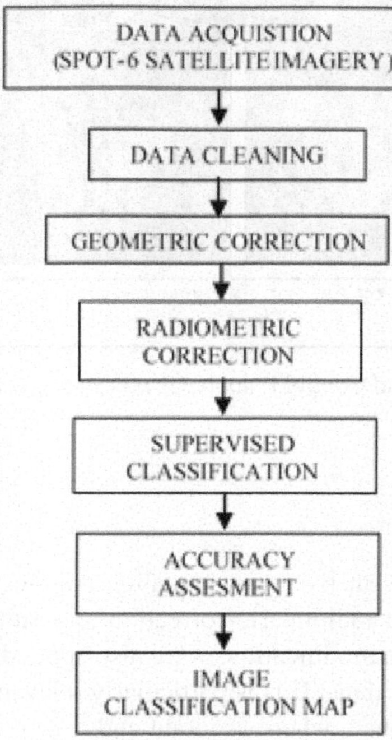

Fig. 3. Research work for water quality changes.

Figure 3 illustrates a detailed depiction of the investigation's data collecting, processing, and analysis phases. This study's findings give compelling evidence that this approach can be utilised to evaluate water quality fluctuations and identify pollutant sources in different river systems. The research demonstrates that GIS and remote sensing are essential for effective water resource management and monitoring.

3.1 Geometric Correction

To address this inaccuracy arising from interference or geometric errors in raw image data obtained from satellite imagery, geometric correction is performed to rectify discrepancies between the image element's location coordinates in the data and its actual location coordinates on the ground.

Image-to-image registration modifies the location and orientation of two or more images to align with one another. The researcher used this method and selected 10 GCPs from one of the photos in Fig. 4 to use as ground truth. These GCPs align the image with other images or maps by providing reference points with known geographic coordinates.

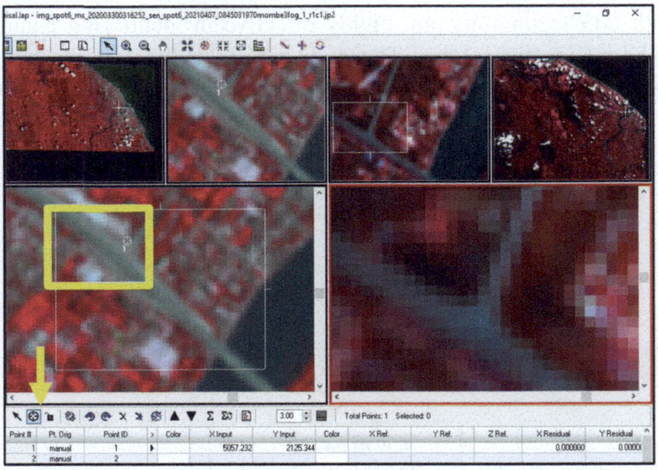

Fig. 4. Ground Control Point (GCP) for image correction.

3.2 Radiometric Correction

To remotely identify changes in river water quality, among other applications, [20]. They emphasised the need for radiometric correction to ensure reliable data collection and avoid errors. Atmospheric modifications were also applied to mitigate their impact, as they can distort raw image data. This is particularly relevant for physical properties of the scene, such as reflectance, where accurate and trustworthy data are crucial for further analysis and interpretation.

Therefore, this study offers a highly effective way to examine geospatial data to investigate changes in land use and water quality along the Kim Kim River using geographic information systems (GIS) and remote sensing techniques. The study's findings can help public authorities and urban planners better comprehend the environmental changes in this area.

4 Analysis and Findings

In the present section, the researcher will analyse and discuss the changes in the water quality of the Kim Kim River that have affected the residents around Pasir Gudang. The Kim Kim River's water quality changes were studied using image classification, which obtained images showing the river's water quality changes in 2015, 2019, and 2020.

Image geometric correction is an essential step in remote sensing analysis to ensure proper alignment with the target area's actual geographic location. The root means square error (RMSE) is one of the most critical metrics for evaluating the success of picture geometric correction. For most applications, an RMSE of less than 2 pixels indicates that the corrected image is within a tolerable range of the natural world's location.

By applying the geometric correction to the images, the requirement was for the RMSE value to be less than 2 pixels, and the researcher's result showed an RMSE value of 1.51, significantly better than the minimum allowed value of 2.00 [21]. Therefore, the image has been successfully rectified and can be used for further study. If the RMSE value is too high, the corrected image may be misaligned with the actual geographic location, which could lead to erroneous analysis and conclusions. Therefore, the fact that the researcher obtained an RMSE value of 1.51 is a significant achievement that underscores the quality of the study.

The remote sensing technique used to process Landsat images for 2015, 2019, and 2020 was a supervised classification technique using the ERDAS Imagine application. This technique classified the Landsat image into six main categories: build-up area, land, forest, river, water bodies, and road. The classification was based on river, land use, and land cover.

In 2015, the water quality of the Kim Kim River showed slight pollution. The change in river water quality was in the colour of the river, as with the dark blue colour on the supervised image indicating there are pollutants detected through remote sensing techniques. The river water quality index (WQI) of Kim Kim River was at level 68, categorised as a slightly polluted river and a class III river. Thus, between 2014 and 2015, the Kim Kim River experienced changes in water quality.

Based on the supervised image in 2019, changes in the water quality of the Kim Kim River were increasingly visible. The increase in land use in Pasir Gudang led to a change in water quality. Built-up areas, including residential, light, medium, and heavy industries, were increasingly developed in Pasir Gudang and along the Kim Kim River. On March 7, 2019, the most considerable Kim Kim River pollution occurred when irresponsible factories illegally dumped toxic and chemical waste into the river. As a result, the WQI readings of the Kim Kim River were recorded as 67, classified as a slightly polluted river and class III river as in Table 2. Hence, the pollution of the Kim Kim River in 2019 caused significant changes in water quality compared to 2015 and was among the worst pollution that occurred in Pasir Gudang, as in Fig. 5.

In 2020, the water quality of the Kim Kim River was getting cleaner from pollutants compared to 2019 due to clean-up operations carried out on a 1.5 km stretch of the river contaminated with hazardous chemicals 2019. Transferring and cleaning toxic waste and chemicals from the Kim Kim River area to a licensed chemical disposal plant was estimated at 1500 metric tons, including river soil and water. According to (Zarzar et al., 2020b), 900 tons of sludge with hazardous chemical waste had been removed from the

Table 2. Water quality status.

River	Year	WQI	Category	Class
Kim Kim River	2015	68	SP	III
	2016	58	P	IV
	2017	57	P	IV
	2018	58	P	III
	2019	67	SP	III

Source: Ministry of Environment & Water, 2018

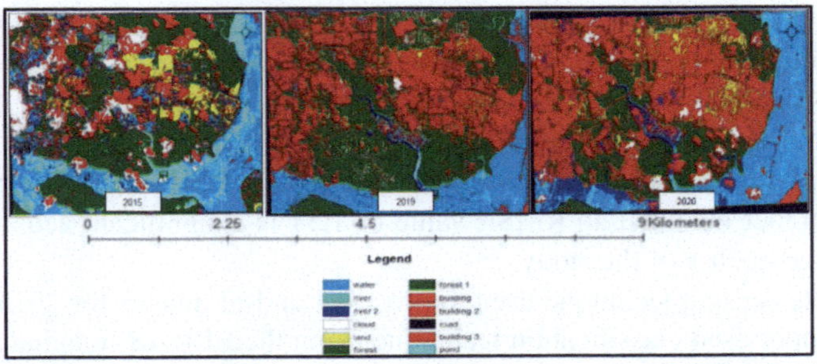

Fig. 5. Image classification for water quality changes in kim kim river for years 2015, 2019 & 2020.

Kim Kim River, making the 1.5 km stretch of the river safe and pollution-free. Until 2020, no cases of chemical waste disposal were detected in the Kim Kim River, consequently reducing water pollution.

The water quality of the Kim Kim River used to be clean and slightly polluted, but after the increase in land use in Pasir Gudang, changes in water quality led to a decline in river water quality. However, the river water quality can still be restored to class I river. The changes in the water quality of the Kim Kim River have been driven by rapid development in the area around the river, especially the development of the industrial area, manufacturing, and settlements. Therefore, the Kim Kim River is at risk of becoming an industrial waste disposal area due to the density of development in the Pasir Gudang district and near the Kim Kim River. Development areas or built-up.

5 Discussion

The use of remote sensing techniques and Geographic information systems (GIS) in conducting studies on changes in the water quality of the Kim Kim River has several significant contributions. One of the primary contributions of this study is to raise awareness and provide general knowledge to the community about the importance of sustainable development in maintaining the quality of the environment, mainly rivers. This study shows that rivers are an essential water source and that water is a basic need for all life on Earth. In addition, it is crucial to ensure the cleanliness and maintenance of river water quality to meet the needs of communities.

Research highlights the current issue of river pollution and emphasises the importance of controlling development to maintain the quality of the environment. It also gives evidence that rapid and uncontrolled development is the leading cause of the deterioration of river water quality, especially the development of industries and manufacturing. They contribute to increasing the community's awareness of environmental problems and emphasise the need for responsible environmental management to maintain the quality of the environment.

Moreover, it emphasises the significance of environmental education, which is easy to understand, particularly about the problems related to river water. This research provides valuable information about using remote sensing and GIS techniques in conducting research. The results contribute to the systematic approach to GIS and remote sensing applications, allowing more realistic analysis results that accurately depict changes in river water quality and land use. A suitable approach facilitates understanding, especially for researchers requiring spatial data management knowledge.

The water quality of the Kim Kim River has undergone significant changes due to the rapid increase in its surroundings. This research indicates that this is a common environmental issue in Malaysia and warns about the potential consequences of rapid development without proper ecological management. All experience highlighted the need to consider responsible environmental management when planning rapid growth, especially in developing and developed countries, to ensure the protection of the environment for future generations.

Henceforth, this study provides valuable insights into the relationship between rapid development and environmental degradation and the use of remote sensing and GIS techniques in research. Subsequently, it contributes to increasing the community's awareness of environmental issues and emphasises the need for responsible environmental management to maintain the quality of the environment. It also contributes to the importance of environmental education and provides valuable knowledge for researchers who require knowledge on spatial data management. Therefore, the findings are crucial in ensuring the sustainable development of the environment for future generations.

6 Conclusion

In essence, the comparison of changes in water quality was accomplished by utilising the supervised images displayed above. These images helped create thematic maps that were instrumental in tracking changes in land use in the Klang area over time. These maps provided insights into the types of land activities that might have contributed to the pollution of the Klang River. Recognising the significance of water for all forms of life, it becomes imperative to take concrete measures to curtail pollution in water sources and waterways. This article has effectively illustrated that combining remote sensing techniques and GIS to generate supervised and monitored images is a powerful method to assess shifts in water quality over a specific region and time frame. Integrating digital innovation in knowledge management for water quality and sustainability further enhances the capability to monitor, analyse, and respond to water quality issues. The findings of the image classification analyses divulge a substantial 50% expansion in the developmental region neighbouring Sungai Kim Kim, with nearly 38% stemming from industrial zones. The root mean square error (RMSE) computed for the image classification stands at a notably low value of 1.51, surpassing the required minimum threshold of 2.00. Hence, there is optimism that this study could serve as a pivotal moment in advocating for preserving sustainable water resources. Furthermore, in the long run, achieving Sustainable Development Goals (SDG) 13: Climate Action, SDG 14: Life Below Water, and SDG 15: Life on Land could become attainable objectives.

Acknowledgment. This research is supported by Universiti Utara Malaysia (UUM) Generation Grant Scheme: S/O Code: 21529.

References

1. Uddin, M.G., Nash, S., Olbert, A.I.: A review of water quality index models and their use for assessing surface water quality. Ecol. Ind. **122**, 1–21 (2021)
2. Zhao, C., Zhang, Y., Guo, W., Baqa, M.F.: Dynamics and drivers of water clarity derived from Landsat and in-situ measurement data in Hulun Lake from 2010 to 2020. Water (Switzerland) **14**(8) (2022)
3. Jamshidi, S., Zand-Parsa, S., Niyogi, D.: Assessing crop water stress index of citrus using in-situ measurements, Landsat, and sentinel-2 data. Int. J. Rem. Sens. **42**(5) (2021)
4. Mohd-Asharuddin, S., Zayadi, N., Rasit, W., Othman, N.: Water quality characteristics of Sembrong dam reservoir, Johor, Malaysia. IOP Conf. Ser.: Mater. Sci. Eng. **136**(1), 6–12 (2016)
5. Pak, H.Y., Chuah, C.J., Yong, E.L., Snyder, S.A.: Effects of land use configuration, seasonality and point source on water quality in a tropical watershed: a case study of the Johor river basin. Sci. Total Environ. **780**, 146661 (2021)
6. Masthurah, A., Juahir, H., Mohd Zanuri, N.B.: Case study Malaysia: Spatial water quality assessment of Juru, Kuantan and Johor river basins using environmetric techniques. J. Surv. Fisher. Sci. **7**(2), 19–40 (2021)
7. Katimon, A., Shahid, S., Mohsenipour, M.: Modeling water quality and hydrological variables using arima: a case study of Johor river Malaysia. Sustain. Water Res. Manage. **4**(4), 991–998 (2018)
8. Yang, X., Gao, J., Liu, X., Huang, A., Sha, W., Zhou, Z.: Analysis of the impact of the Yangtze Huai river diversion project on the hydrodynamics and water quality of the Xi and Zhao rivers. In: 2022 8th International Conference on Hydraulic and Civil Engineering: Deep Space Intelligent Development and Utilization Forum, ICHCE 2022 (2022)
9. Bralewska, K., Rogula-Kozłowska, W., Bralewski, A.: Indoor air quality in sports center: assessment of gaseous pollutants. Build. Environ. **208** (2022)
10. Galinato, R.O., Santillan, M.M.: Using remote sensing and GIS to identify alternative water sources for Butuan City, Philippines. In: IEEE Region 10 Annual International Conference, Proceedings/TENCON (2020)
11. Spyrakos, E., et al.: Moving towards global satellite-based products for monitoring inland and coastal waters. Regional examples from Europe and South America. IEEE Latin American GRSS & ISPRS Remote Sensing Conference (LAGIRS) IV-3/W2-2020, pp. 363–368 (2020)
12. Alsabhan, W., Alotaiby, T.: Automatic building extraction on satellite images using UNet and Resnet50. Comput. Intell. Neurosci. 2022 (2022)

13. Madhav, S., et al.: Water Pollutants: Sources and impact on the environment and human health. In: Pooja, D., Kumar, P., Singh, P., Patil, S. (eds.) Sensors in Water Pollutants Monitoring: Role of Material. Advanced Functional Materials and Sensors. Springer, Singapore (2020)
14. Zhou, K.: The formation mechanism and characteristics of urban multimodal traffic system based on the interaction between urban spatial mode and traffic development. In: Proceedings - 2020 International Conference on Urban Engineering and Management Science, ICUEMS 2020 (2020)
15. Moreno-Hurtado, C., Ochoa-Moreno, W. S., Ochoa-Jiménez, D.: ICT exports and environmental degradation: the case of South America. In: Iberian Conference on Information Systems and Technologies, CISTI (2019)
16. Luo, X., Yang, J.: Problems and challenges in water pollution monitoring and water pollution source localization using sensor networks. In: Proceedings - 2017 Chinese Automation Congress, CAC 2017 (2017)
17. Alkhatib, F. H., Kasim, N., Goh, W. I., Al-Masudi, A. A. H.: Multidisciplinary computational optimization: an integrated approach to achieve sustainability in tall building design at early stage-review. In: 2021 3rd International Sustainability and Resilience Conference: Climate Change (2021)
18. Mansor, N.S., Shafri, H.Z.M., Mansor, S., Paradhan, B.: The influence of urban development and social mobility on socioeconomic level: the application of GIS on urban ecosystems. IOP Conf. Ser.: Earth Environ. Sci. (2014)
19. Reddy, G.P.O.: Satellite remote sensing sensors: principles and applications. In: Reddy, G., Singh, S. (eds.) Geospatial Technologies in Land Resources Mapping, Monitoring and Management. Geotechnologies and the Environment, vol. 21. Springer, Cham (2018). https://doi.org/10.1007/978-3-319-78711-4_2
20. Zarzar, C.M., Dash, P., Dyer, J.L., Moorhead, R., Hathcock, L.: Development of a simplified radiometric calibration framework for water-based and rapid deployment Unmanned Aerial System (UAS) operations. Drones **4**(2) (2020)
21. Johansen, K., Phinn, S., Witte, C.: Mapping of riparian zone attributes using discrete return lidar, quickbird and spot-5 imagery: assessing accuracy and costs. Rem. Sens. Environ. **114**(11) (2010)

Malaysian Undergraduate Students' Acceptance of Discord as a Learning Management System in a HyFlex Course

Liang Jing Teh[1], Su Luan Wong[1(✉)], Mohd Zariat Abdul Rani[2], Mas Nida Md Khambari[1], and Sai Hong Tang[3]

[1] Faculty of Educational Studies, Universiti Putra Malaysia, Seri Kembangan, Malaysia
suluan@upm.edu.my

[2] Faculty of Modern Languages and Communication, Universiti Putra Malaysia, Seri Kembangan, Malaysia

[3] Faculty of Engineering, Universiti Putra Malaysia, Seri Kembangan, Malaysia

Abstract. Managing participation, creating a sense of community, and providing an equitable learning experience between online and in-person students in a HyFlex course is challenging. Previous studies have highlighted a knowledge gap in students' acceptance of the Discord application as a learning management system (LMS). Thus, this research investigated undergraduate students' perceived usefulness and ease of use towards using Discord to manage knowledge and participation in a HyFlex learning environment. Through a convergent mixed methods design, quantitative analyses revealed high perceived usefulness and ease of use among students. Qualitative insights indicated nuanced benefits such as structured channels which enhanced accessibility to information, thread functions which fostered inclusive and deeper learning, and promoted communication between instructor and learners. A few challenges also emerged, including initial confusion due to students' unfamiliarity with Discord and technical issues that hindered student acceptance of Discord. To maximise the potential of Discord as an LMS in a HyFlex learning setting, educational stakeholders and developers should collaborate to provide quality support, such as user onboarding and addressing usability issues. Future studies could compare the effectiveness of Discord with other learning platforms and investigate educators' perspectives on integrating Discord into the teaching and learning process.

Keywords: Discord Application · HyFlex Learning · Learning Management System · Perceived Usefulness · Perceived Ease of Use

1 Introduction

The rapid technological advancements in 21st-century education allowed practitioners and institutions to explore innovative instructional models. Founded in 2006 [1], Hybrid-flexible (HyFlex) learning is one of the modern blended learning models that gained

Z. Mat Aji et al. (Eds.): KMICe 2024, ISEM 49, pp. 16–28, 2025.
https://doi.org/10.1007/978-3-031-91485-0_2

immense attention among scholars and educators since the Coronavirus Disease 2019 (COVID-19) outbreak [2]. In a HyFlex course, students can choose their desired learning mode from any of the three learning modes: physical, online synchronous, and online asynchronous. The description for each learning mode is presented in Table 1.

Table 1. Description of learning modes in a HyFlex learning model.

Learning mode	Description
Physical	Students learn by attending the physical classroom in person
Online synchronous	Students learn via an online conference call platform
Online asynchronous	Students learn by watching the recorded class session on their own time

With the flexibility and accessibility provided by HyFlex learning, the student learning outcomes could be enhanced as learners are allowed to learn according to their learning styles, preferences, and needs [3, 4]. Also, their learning progress would not be halted by physical health issues, the clashing of schedules, and personal commitments such as extra-curricular activities, off-campus events, or family issues [1].

Despite the various inherent benefits of HyFlex learning, creating a sense of community among the in-person and online students in a HyFlex setting is one of the main predicaments instructors face [5]. Fostering active participation among remote students can be challenging as it is relatively more complex for instructors to keep online students accountable during the instruction. Ensuring equitable learning experiences between virtual and physical students is also tricky. Instructors are burdened with redesigning the existing course and identifying digital learning platforms to ensure that online and physical learners have an equally engaging and productive learning environment [1]. Further, technical difficulty is another obstacle in a HyFlex learning setting [4]. The digital platforms must be reliable, accessible, and suitable for students with diverse levels of digital competency.

Discord, an online community platform developed in 2015 that was well-known among gamers [6, 7], has been increasingly used in educational settings [8]. Discord is a free digital application that allows users to create servers. Within a Discord server, separate channels with distinct rules and user permission can be made for specific purposes. In other words, a Discord server is like a house, while the channels on a server are like the house's rooms. Discord combines features of online conference call platforms and instant messaging services, allowing students and instructors to connect synchronously and asynchronously. This enables educators to create various channels for educational purposes, learning activities, and discussion topics.

Given these features within the Discord application, it has the potential to be used as a learning management system (LMS) to mitigate the challenges above of HyFlex learning. However, how undergraduate students in Malaysia perceive the utilisation of Discord as the LMS in a HyFlex course remains unclear. Thus, the authors need to conduct this study to understand students' acceptance of integrating Discord into HyFlex learning.

2 Literature Review

The fundamental functions of LMSs include sharing educational content, fostering communication among instructors and students, managing course materials and learning activities, and assessing learners' progress [9]. In a mixed-methods study, [10] concluded that although LMS is beneficial to students and instructors in a blended learning model, they felt that the traditional LMS was complicated to use and time-consuming. They also faced other main problems, such as the stability of LMS and connectivity issues.

Previous studies have shown that the main advantages of incorporating Discord in education were Discord's perceived user-friendliness and usefulness among students and teachers. The learners found Discord easy to use and accessible as it only requires a low-bandwidth internet speed for non-video communication [11–14]. The perceived user-friendliness among the students was also attributed to the appealing user interface of Discord [11, 12, 14] also suggested that the learning curve of Discord usage was short for students unfamiliar with the application.

Regarding the perceived usefulness of Discord, students indicated that the use of Discord enabled them to connect with their peers and stay updated on the information of the course with ease [13, 15, 16]. This also could promote more interaction between students and instructors [13, 17]. Moreover, Discord also assisted them in achieving their learning objectives as they could also undergo learning activities on Discord [11, 14, 15]. As for the instructors, they also found Discord to be effective in conducting remote learning and creating an online community [12]. Educators could also use Discord to create learning activities and analyze student performance [13]. An experimental study conducted by [18] concluded that students who learnt through Discord performed better in terms of vocabulary acquisition than those who learned via the traditional method.

Nevertheless, Discord has a few inherent drawbacks that could negatively impact teaching and learning. [13] investigated student perspectives using a rubric, the findings signified that Discord offers limited assistance in fostering cognitive engagement among students. Additionally, [16] employed Discord as a learning central hub, revealing that students were easily distracted when using Discord. These findings suggest that Discord may adversely affect online lecturing and the comprehension of complex learning content. Contrary to the findings by [11, 14] found that most of the students without experience using Discord struggled to pick up the application. Also, technical issues such as unstable internet connectivity and unpleasant audio quality may arise when using Discord [14].

Given the mixed conclusions shown by the previous studies, there is a knowledge gap in undergraduate students' acceptance towards Discord as an LMS, particularly in the context of HyFlex learning among Malaysian students. Therefore, this study aims to investigate Malaysian undergraduate students' acceptance towards Discord as an LMS in a HyFlex learning setting. As such, below are the research questions of this study:

1. What is the extent of the perceived usefulness of Discord in a HyFlex course among undergraduate students?
2. What is the extent of Discord's perceived ease of use in a HyFlex course among undergraduate students?
3. How do undergraduate students perceive the use of Discord in a HyFlex course?

3 Methodology

3.1 Overview of the Research

A parallel-database variant of convergent mixed-methods design was employed in this study. As illustrated in Fig. 1, both quantitative and qualitative data were collected simultaneously from the participants using one instrument. The collected quantitative and qualitative data were then analyzed independently, followed by merging and interpreting both results to identify similarities or disparities between them. This study was conducted on a faculty member of a public university in Malaysia that consists of 1,270 undergraduate students. A purposive criterion sampling was adopted to select 50 undergraduate students. The inclusion criteria were undergraduate students who (1) enrolled in the HyFlex course provided by the university faculty and (2) completed the HyFlex course, which utilized Discord.

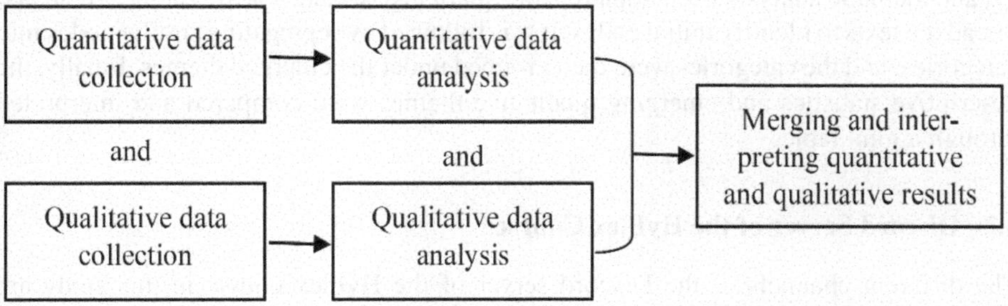

Fig. 1. The parallel-database variant of convergent mixed methods design was adopted from [19].

As shown in Table 2, the instrument in this study comprises 17 items, with four items on participants' demographics, five items each for the perceived usefulness (PU) and perceived ease of use (PEOU) variables, and three open-ended questions. The quantitative questionnaire was developed based on [20] PU and PEOU questionnaires. A five-point Likert scale (1 = strongly disagree to 5 = strongly agree) was used for all 10 PU and PEOU variables items. The three open-ended questions used to collect the qualitative data are as follows:

1. In a few sentences, please describe what you like most about using Discord for this course.
2. In a few sentences, please describe any specific challenges you faced while using Discord in this course.
3. How can the use of Discord in this HyFlex course be improved? Please provide suggestions.

The second and fourth authors evaluated the instrument of this study. A pilot study was also conducted to assess the reliability of the quantitative strand of the questionnaire. It involved 30 undergraduate students. The reliability test indicated Cronbach alpha, *a* Values of.87 and.90 for the PU and PEOU scales respectively, suggesting that the questionnaire used in this study was reliable. Data obtained from the 30 undergraduate

Table 2. Research instrument of this study.

	Section A	Section B	Section C	Section D	
	Demography	PU	PEOU	Open-ended questions	
Number of items	4	5	5	3	Total: 17

students in the pilot study were not included in the data analysis of this study. Both quantitative and qualitative data were collected by using an online survey form. The questionnaire was administered to all 50 undergraduate students via email. Forty-seven of them completed the survey, and 47 samples were analysed in this study. Data was collected after the students had experienced six weeks of using Discord as the LMS for the HyFlex course. The students took about 15 min to complete the questionnaire.

Descriptive analysis was conducted in the quantitative strand using SPSS version 27, and thematic analysis was adopted in the qualitative strand. The researchers read and reread the texts to identify initial codes. It was followed by segregating similar codes into categories, and the categories were then grouped under the emerged themes. Finally, the descriptive statistics and emerging qualitative themes were compared and interpreted through a joint table.

3.2 Discord Server of the HyFlex Course

The different channels of the Discord server of the HyFlex course in this study are illustrated in Fig. 2. The purpose of each channel is described in Table 3.

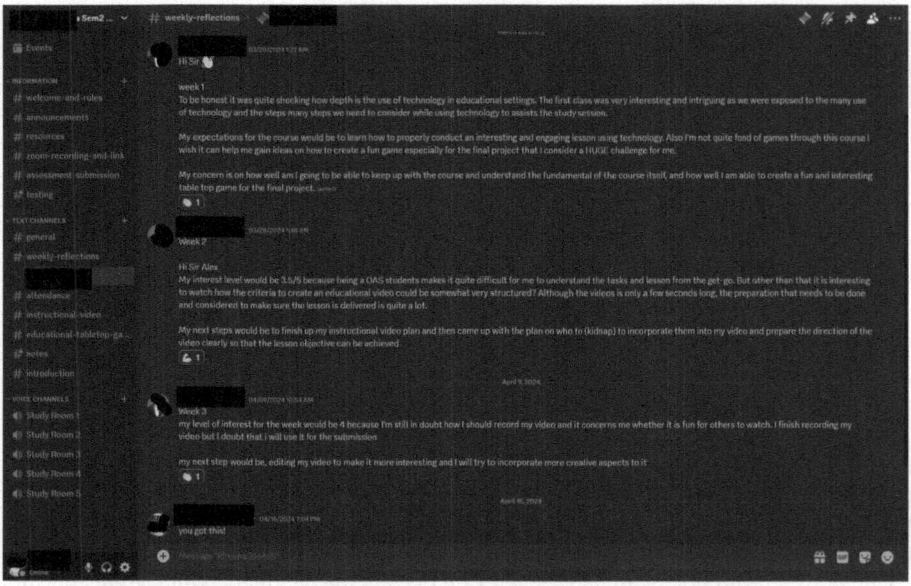

Fig. 2. The various channels of the HyFlex course's Discord server.

Table 3. Description of each channel's purpose in the HyFlex course's Discord server.

Channel	Purpose
Welcome-and-rules	The landing channel presents the server's rules and regulations and a short video on using Discord
Announcements	For students to stay updated with the critical information related to the HyFlex course
Resources	For students to access all course materials
Zoom-recording-and-link	Students can access the Zoom meeting link and the cloud folder link containing all recorded sessions
Assessment-submission	Students can access the assessment submission link
General	This is for formal or casual discussion and conversation between all members
Weekly-reflection	For students to write their weekly reflection under their thread
Attendance	Students can access the attendance link and report their reasons and official documents for their absence
Instructional-video	This is for discussing any information related to the instructional video project
Educational-tabletop-games	This is for discussing any information related to the educational tabletop game project
Introduction	For students to do a self-introduction during the first session
Study rooms	For students to have online synchronous meetings

As shown in Fig. 2, the server is segregated into three channels: information, text, and voice. In the information channels, only the instructor had permission to send messages, and students could only react to the messages with emojis. All server members had more permissions in the text and voice channels, whereby they could send messages and have voice or video calls with each other.

The purposes of the "testing" and "notes" channels in Fig. 2 are not explained in Table 2 because these are the private channels for the instructor to test the functionalities of Discord bots and applications and take notes. The authors acknowledge that Discord could not be utilised as a standalone LMS as it has yet to have the functionality of evaluating students' performance and recording online meetings. Thus, Discord was used alongside other online platforms in this study.

4 Findings

4.1 Demography

Table 4 presents the demographics of the participants. All the participants majored in the Bachelor of Education. Most of the undergraduate students involved in this study were female, with 70.2% female and 29.8% male participants; many were 21 (44.7%) and 22

(40.4%) years old. Almost three-quarters of the participants were unfamiliar with the Discord application, as 38.3% of them had never used Discord before, and 34.0% only used Discord once monthly. Among those who were exposed to Discord before, gaming (38.3%) was their main reason for Discord usage, followed by community activities (17.0%), streaming (4.2%), and, finally, work-related factors (2.1%). This indicates that using Discord as an LMS is novel to almost all the participants.

Table 4. Demographics of the participants.

	Frequency	Percentage (%)
Gender	14	29.8
Male	33	70.2
Female	47	100.0
Total		
Age	1	2.1
20	21	44.7
21	19	40.4
22	4	8.5
23	1	2.1
24	1	2.1
25	47	100.0
Total		
Frequency of previous use of Discord	18	38.3
Never	16	34.0
Once a month	6	12.8
Once a week	4	8.5
Once a day	3	6.4
More than once a day	47	100.0
Total		
Reasons for previous use of Discord	18	38.3
Never	18	38.3
Gaming	8	17.0
Community activities (sports, religion, music, etc.)	2	4.3
Streaming / Content creation	1	2.1
Work / Study / Business	47	100
Total		

4.2 What is the Extent of Perceived Usefulness and Ease of Use of Discord in a HyFlex Course Among Undergraduate Students?

Tables 5 and 6 show the PU and PEOU of Discord among undergraduate students. The overall mean scores of PU and PEOU were 3.98 (SD = 0.63) and 4.04 (SD = 0.61), respectively, suggesting that the students perceived Discord as valuable and easy to use in a HyFlex learning setting.

Table 5. PU of Discord in a HyFlex course among undergraduate students.

Items	Strongly disagree		Disagree		Neutral		Agree		Strongly agree	
	f	%	f	%	f	%	f	%	f	%
1. Using Discord would make it easier for me to understand the learning content	0	0	0	0	14	29.8	18	38.3	15	31.9
2. Using Discord would make me feel more comfortable to connect with my instructor	0	0	0	0	8	17.0	23	48.9	16	34.0
3. Using Discord would make it easier for me to exchange knowledge with my peers	0	0	1	2.1	7	14.9	29	61.7	10	21.3
4. Using Discord in my learning would enable me to complete my assignments more quickly	1	2.1	2	4.3	15	31.9	18	38.3	11	23.4
5. Using Discord would enhance my learning performance	0	0	1	2.1	13	27.7	22	46.8	11	23.4

Based on Table 5, item 2 was most agreed upon by the students because Discord made them feel more comfortable communicating with the instructor. The following PU components that the participants resonated with were both **item-**1 and item-3, whereas item-4 is the least agreed statement among the students among all items. The students' PEOU towards Discord is presented in Table 6. Students reacted most positively to item 8 on the ease of sharing information, followed by item 6 and item 10. On the other hand, almost half of the students felt neutral towards the statement of item 7.

4.3 How Do Undergraduate Students Perceive the Use of Discord in a HyFlex Course?

They organized Academic Community. Among the themes which emerged through analysing the students' comments, the most prominent theme was related to how students

Table 6. PEOU of Discord in a HyFlex course among undergraduate students.

Items	Strongly disagree		Disagree		Neutral		Agree		Strongly agree	
	f	%	f	%	f	%	f	%	f	%
6. Learning to use the Discord server would be easy for me	0	0	0	0	8	17.0	25	53.2	14	29.8
7. I would find it easy to get Discord to do what I want	0	0	0	0	20	42.6	19	40.4	8	17.0
8. Participating in discussions and sharing information on Discord would be easy	0	0	0	0	6	12.8	24	51.1	17	36.2
9. I would feel comfortable using the different features of Discord	0	0	1	2.1	13	27.7	19	40.4	14	29.8
10. It would be easy for me to become skilful at using Discord	0	0	1	2.1	8	17.0	23	48.9	15	31.9

felt that the specific channels within the Discord server were organised according to their academic needs, as expressed by one of the students, "*I love how there are channels in Discord that have been divided according to our academic needs*". Another student commented that it contributed to the ease of obtaining important course information and content, "*...we can have different channels for different purposes. It is easier to find information without having to find everything under one channel, which can be hard, especially after sharing so much information*".

Some students even preferred Discord over other social platforms such as WhatsApp and PutraBlast (the official LMS of the university); a participant stated "*Though I do understand that some lecturers prefer a WhatsApp group with students to deliver information. However, I deeply feel that WhatsApp is not as structured compared to Discord*", and another shared "*We don't have to go to Putrablast to submit/wait for it. [I can] just straight away [go] to Discord and ask/post anything*". One of the students also preferred using Discord over Google Meet and Zoom, "*I also think that the voice channels are easier for me to use compared to Google Meet or Zoom. [I don't] need to create links; I can just come into the channel*".

An Inclusive Space for Students' Input. The students cherished the equal opportunities they had to share their thoughts and feelings, which they attributed to the thread function of Discord, *"having the thread option gives a chance for you to be heard,"* as one of the students shared. Another student also shared the same sentiment: "It's more open and easier because we don't 'fight' using our voices, *which can be too loud and distracting"*. This also created a safe space for students with diverse preferences to share their opinions, *"...being able to voice my opinions without feeling intimidated in the public,"* says the student.

Another student also mentioned that she *"likes to use threads to express my reflections because it's like my diary for each week's session"*. As a result, it enabled students to learn from their peers. As a student mentioned, *"We can clearer understand all our friends' opinions. Thus, we can learn more from their sharing"*. Additionally, another student expressed that *"I can reach my lecturer easier. I'm able to ask for feedback and suggestions, getting faster responses from them which help with my work progress"*, suggesting that students felt more comfortable in approaching their instructor when using Discord, which improved their academic progress.

The students also enjoyed using Discord in learning because *"It's so Gen-Z"*, as stated by a few students. They gave credit to the various features that are relevant to their online socialisation culture; a participant wrote, "You can play songs too, *haha and of course chat with your classmates! using funny [Graphics Interchange Format] (GIF)s and send pictures"*.

Unfamiliarity with Discord. On the contrary, a few students shared that they felt confused when navigating through Discord because they lacked the knowledge of using Discord, as stated by a participant, *"I'm still figuring out where to find the channels at times. The threads as well as I've never used it. I've wrongly replied at the wrong channel at times"*. This issue could be compounded by information overload, *"...too much information and too many channels, which sometimes makes me confused. It's also hard to find my work,"* says one of the students. As Discord is novel to most students, they "rarely check Discord, and it's easy to miss any announcement or messages" stated a student. In contrast with the previous theme, some students preferred other social platforms rather than Discord, *"It's hard for me to be familiar with new functions compared to PutraBlast and WhatsApp"*, a participant expressed.

However, another participant mentioned, *"...this is a learning process, and I am glad that I'm given the opportunity to explore Discord"*. Another participant also shared that using Discord would become easier over time, "At first, it was hard getting used to using Discord because of the features, threads, and all the channels. But after some time, it was getting easier". Regarding this challenge, suggestions such *as "Maybe lessen the number of channels"* and *"Teach students about Discord or make them watch the YouTube video on how to use it rather than ask them to watch it on their own time"* were given by a few students. Some participants also noted that *"...students have to be more responsible to actively participate and frequently check the news and update from the discord server"*.

Technical Difficulties. Furthermore, a few students commented, *"sometimes I missed the information because I don't get any notifications"*. Moreover, a few participants struggled with insufficient storage on their mobile phones for installing Discord; this was reflected in one of the participants' remarks, *"I don't have enough storage in phones,* *so I had to install Discord in my Tab only"*. These technological challenges could hinder their acceptance of Discord as an LMS in a HyFlex course.

5 Discussion

Given the overall high mean scores for both PU and PEOU, the quantitative findings show that undergraduate students perceived Discord as useful and easy to use in a HyFlex course. The results also signify that using Discord as the LMS in a HyFlex course can foster students' interaction with their peers and instructors and enhance their comprehension of the course content. These quantitative results are echoed by the qualitative data, where students cherished the structured design of the Discord server and its facilitation of communication with their peers and instructors. This is in line with the findings by [13, 15–17], who postulated that integrating Discord could heighten connections between learners and educators and increase the ease of obtaining information.

This indicates the importance of designing channels and threaded discussions within a Discord server according to students' academic needs, allowing easier access to course information among the students. It is also important to note that the instructor acknowledged students' sharing by reacting and replying to their messages on Discord, together with displaying and reading out some of their thread's responses during the class session, which the researchers believe allowed students to feel heard and motivated other students to learn from each other. Similar to the findings by [12], which stated that Discord could create a conducive online learning community, this study's findings suggest that that was the case because young students perceived Discord to be closely related to their identity. Additionally, encouraging students to use Discord's various relatable features, such as GIFs, emoji reactions, and music to express their opinions, is pertinent in establishing a more enjoyable and engaging HyFlex learning environment.

Nevertheless, disparities between the statistical findings and qualitative results also emerged. Although the quantitative data indicate high overall scores for PU and PEOU, the qualitative findings signify challenges, such as students' initial confusion with Discord due to its novelty as well as the numerous channels and features of Discord. Qualitative findings also show that the students faced occasional technical difficulties, such as missed notifications and lack of storage space on their mobile devices. This also explains the difference in students' preference for Discord and other online platforms in the qualitative results. This finding aligns with previous studies conducted by [11], who stated the obstacles faced by students without experience using Discord, and [14], who raised concerns about technological issues when using Discord.

While some students express confidence in overcoming these challenges over time, as reflected in the findings by [14], others may continue to struggle with these predicaments, potentially affecting their engagement and participation in the HyFlex course [11]. This highlights the need for a more nuanced understanding of students' experience with Discord beyond aggregate scores to identify specific areas for improvement and address emerging challenges. This finding also implies the significance of providing comprehensive support to learners and educators. To maximise Discord's potential as an LMS, institutions should offer training and resources to help onboard students and instructors, allowing them to get familiar with Discord's interface, features, and notification configurations. Beyond that, ongoing evaluation of students' experiences with Discord is also vital to ensure that the platform remains responsive to the evolving students' needs in a HyFlex course.

6 Conclusion

The quantitative data offer insights into overall students' perceptions of Discord, while the qualitative data provide contextual information and illuminate specific benefits and challenges experienced by the students. In summary, Discord presents the potential for becoming a valuable LMS in HyFlex learning environments. Despite initial confusion with Discord's interface and occasional technical issues, the undergraduate students appreciate the organised channels for easy access to information, the threaded discussions that facilitated inclusive and deeper learning, and the features that could narrow the communication gap between instructor and students.

Like any other educational digital platform, the utilisation of Discord is not a universal remedy, it must be paired with a sound design of instruction and analysis of students' needs. To optimise the potential of Discord as an LMS, institutions, educators, and developers should collaborate in providing support such as user onboarding and training on the integration of Discord into instruction, addressing usability challenges like missed notifications and insufficient storage, as well as conducting continuous evaluation and improvement to create a more engaging and effective use of Discord in a HyFlex course.

As the qualitative findings of this study indicate a gap in the preference for LMS platforms, future studies can compare Discord with other platforms to assess its relative effectiveness in facilitating students' learning. Also, this study focused on student perceptions, including faculty's perspectives and best practices in future research, which would provide a more holistic understanding of incorporating Discord into teaching and learning.

Acknowledgement. The authors of this study are thankful for the support provided by Universiti Putra Malaysia through the grant Geran Putra Inisiatif Putra Siswazah (GP-IPS/2023/9744200).

References

1. Beatty, B.: Hybrid-Flexible Course Design, 1st edn. EdTech Books, London (2019)
2. Eduljee, N.B., Chakravarty, R., Croteau, K., Murphy, L.: Understanding research trends in HyFlex (hybrid flexible) instruction model: a scientometric approach. Int. J. Instr. **15**(4), 935–954 (2022)
3. Lakhal, S., Bateman, D., Bédard, J.: Blended synchronous delivery mode in graduate programs: a literature review and its implementation in the master teacher program. Collect. Essays Learn. Teach. **10**, 47–60 (2017)
4. Wong, B.T., Li, K.C., Chan, H.T., Cheung, S.K.: HyFlex learning research and practice: a longitudinal analysis. Sustainability **15**(12), 9699 (2023)
5. Lafortune, A.M.: Differences in students' perceptions of the community of inquiry in a blended synchronous delivery mode. Doctoral dissertation, Université de Sherbrooke (2018)

6. Raihan, J.P., Putri, Y.R.: Pola komunikasi group Discord PUBG. INDO. FUN melalui aplikasi Discord. eProceedings Manage. **5**(3) (2018)
7. Rakhmawan, A., et al.: Analisis pemanfaatan aplikasi Discord dalam pembelajaran daring di era pandemi Covid-19. Prosiding Seminar Nasional Pendidikan FKIP **3**(1), 55–59 (2020)
8. Efriani, E., Dewantara, J.A., Afandi, A.: Pemanfaatan aplikasi Discord sebagai media pembelajaran online. Jurnal Teknologi Informasi dan Pendidikan **13**(1), 61–65 (2020)
9. Dawley, L.: The tools for successful online teaching. IGI Global (2007)
10. Cao, T.X.L.: Benefits and challenges of using LMS in blended learning: views from EFL teachers and students at a Vietnamese public university. Int. J. TESOL Educ. **3**(3), 78–100 (2023)
11. Arifianto, M.L., Izzudin, I.F.: Students' acceptance of Discord as an alternative online learning media. Int. J. Emerg. Technol. Learn. **16**(20), 179–195 (2021)
12. Dayana, Y.E., Andre, O.M., Andrade-Arenas, L.: Design of the Discord application as an e-learning tool at the University of Sciences and Humanities. In: Larrondo Petrie, M.M., Zapata Rivera, L.F., Aranzazu-Suescun, C. (eds.) Prospective and Trends in Technology and Skills for Sustainable Social Development—Leveraging Emerging Technologies to Construct the Future: Proceedings of the 19th LACCEI International Multi-Conference for Engineering and Technology (Article 9) (2021)
13. Di Marco, L.: User-centred evaluation of Discord in midwifery education during the COVID-19 pandemic: analysis of the adaptation of the tool to student needs. Eur. J. Midwifery **5**(51), 1–6 (2021)
14. Wulanjani, A. N.: Discord application: turning a voice chat application for gamers into a virtual listening class. In Mulyadi, D., et al. (eds.) English Language and Literature International Conference (ELLiC) Proceedings, vol. 2, pp. 115–119 (2018)
15. Jannah, E.I., Hentasmaka, D.: The use of Discord application in virtual English learning: an investigation on students' perceptions. J. Engl. Educ. Appl. Linguist. **10**(2), 183–200 (20210
16. Vladoiu, M., Constantinescu, Z.: Learning during COVID-19 pandemic: Online education community, based on Discord. RoEduNet IEEE Int. Conf. **2020**, 1–6 (2020)
17. Fonseca Cacho, J.: Using Discord to improve student communication, engagement, and performance (Poster presentation). UNLV Best Teaching Practices Expo, Las Vegas, Nevada, USA (2020)
18. Odinokaya, M.A., Krylova, E.A., Rubtsova, A.V., Almazova, N.I.: Using the discord application to facilitate EFL vocabulary acquisition. Educ. Sci. **11**(9), 470 (2021)
19. Creswell, J.W., Plano Clark, V.L.: Designing and Conducting Mixed Methods Research. Sage Publication (2018)
20. Davis, F.D.: Perceived usefulness, perceived ease of use, and user acceptance of information technology. MIS Q. 319–340 (1989)

Organizational Intelligence and Energy Crisis Management: A Case of Jordan

Ata Mousa Tayel Alsheyyab(✉), Fadhilah Mohd Zahari, and Ezanee Mohamed Elias

School of Technology Management & Logistics, Universiti Utara Malaysia, 06010 Sintok, Kedah, Malaysia
sheyyabatah@gmail.com

Abstract. Organizational intelligence has been regarded as a crucial aspect in preparing any organization to embrace the crisis. However, literature on energy crisis management tends to focus on policy and technological issues, thus limiting the attention to organizational factors. This study aims to investigate the role of organizational intelligence in energy crisis management in Jordan. It analyzes the role of organizational intelligence in managing the crisis in developing countries, particularly in nations where electricity resources are scarce, like Jordan. Organizational intelligence is conceptualized as a dimension of strategic planning that includes technology intelligence, competitive intelligence, political environment foresight, and consumer foresight. This quantitative study collects data from organizations involved in the energy sector in Jordan. A total of 185 completed responses were used for analysis using SMART PLS 4. The findings indicate that only two dimensions of organizational intelligence have significant relationships with energy crisis management, specifically political environmental foresight and consumer foresight. While these results are unique to Jordan's context, the study paves the way towards further exploration of the energy management crisis and organizational intelligence issues in the Middle East and North Africa (MENA) region.

Keywords: Organizational Intelligence · Energy Crisis Management · MENA Region

1 Introduction

Emerging economies grapple with energy supply and resources to drive economic growth and modernization. Governments worldwide focus on energy security since it is essential to economic growth and allows vital services to function, such as water distribution throughout a nation, industry, transportation, and other industries. Jordan is no exception; the country strives to ensure energy security to meet the growing demands of diverse industries for economic development and other sectors [1, 2]. The country faces unique challenges in energy due to its limited supply of fossil fuels and unstable regional environment. Fossil fuels make up the majority of Jordan's energy mix. But at the moment,

Z. Mat Aji et al. (Eds.): KMICe 2024, ISEM 49, pp. 29–39, 2025.
https://doi.org/10.1007/978-3-031-91485-0_3

Jordan can only supply 10% of its energy needs. About 94% of the nation's energy comes from imports, and it is primarily reliant on Arab Gulf nations like Egypt and Iraq for the security of its crude oil supply [3]. In addition, this signifies that the Jordanian energy sector is especially vulnerable to regional conflicts. Regional instability has also recently contributed to supply and price volatility, alleviating public concerns about growing electricity prices.

This crisis has pressurized the energy sector in Jordan to look for a strategic approach to manage the issues. Literature on crisis management has well acknowledged the role of organizational intelligence in taking a proactive stance to gather, interpret, and use information and knowledge to assist informed decision-making during crisis. Yet, the evidence mainly contributes to the Western context, which might not reflect the energy crisis in other parts of the globe, especially the MENA region. The energy crisis in Jordan involves multidimensional and dynamic problems that demand substantial attention to understand the issue. Moreover, studies in energy crisis management tend to focus more on technological and policy reform without addressing the organizational strategic development for managing the problems.

This study aims to contribute to the existing literature by investigating the role of organizational intelligence in managing the energy crisis in Jordan. Accordingly, the study offers some insights into energy crisis management at the managerial level, which still needs to be explored in the existing literature. This empirical evidence could also inform the energy sector regarding the vital factors they should consider in their strategic planning to anticipate the crisis and make better decisions. The study is subsequently organized as follows: Sect. 2 reviews the literature, followed by Sect. 3 on methodology used in the study. Section 4 explains the analysis and discussion of results, and finally Sect. 5 presents conclusions and implications.

2 Literature Review

2.1 Crisis Management

Crisis management is a planned, coordinated approach to handling, containing, and resolving emergencies. It involves procedures to protect the organization against unanticipated circumstances or occurrences that might endanger its operations or survival [4]. Thus, to minimise and restrict harm, crisis management supports strategically using organisational resources to prepare for unforeseen situations [5]. As a result, interest in crisis management has escalated the need to ensure the organization's long-term stability and viability.

Literature in crisis management indicates that there exist many frameworks, ranging from three phases [6, 7], four phases [8, 9] and five phases [10]. The modern crisis management perspective favours the five phases more as it includes additional steps that allow for better preparedness, response, and recovery. According to [10], the five phases of crisis management are as follows.

Phase 1: Signal Detection. The crisis always starts with some warning. Signal detection is the phase that drives these warnings. The ability to recognize signals is an attitude and a skill that companies need to embrace.

Phase 2: Preparation/Prevention. The crisis management teams are established, and plans are prepared to deal with potential crises in this phase. Crisis management is approached systematically and continuously. The main objective is to avoid as many crises as possible and effectively manage those that do occur.
Phase 3: Containment of the Damage. It is the phase where the actual management of the crisis takes place. The aim is to contain the crisis as much as possible and limit the damage to the organization and its stakeholders.
Phase 4: Recovery. Efforts are made to resume activities as much as feasible during this phase. Recovery usually takes place in several stages. Short-term recovery aims to get the system back up and running so that a minimum acceptable level of service is achieved. Long-term recovery is achieved by returning operational activities to pre-crisis levels. In some cases, improvements are performed as part of the recovery process to raise operations above their pre-crisis level.
Phase 5: Learning. Learning involves reflection activities in which lessons are learned from the crisis. This phase focuses on improving current operational problems and preventing future ones. Hence, any unnecessary actions that could harm the organization in the future will be avoided as they learnt from the previous crisis.

2.2 Organizational Intelligence

Organizational intelligence results from collective efforts to acquire, manage and interpret information from outside the organization. Essentially, it provides meaningful input for the organization's decision-makers. Organizational intelligence plays a vital role in strategic planning, as it monitors the external environments for emerging trends, shifts in customer behaviour, and market disruptions. This awareness enables the organization to predict potential opportunities and threats, including crises. Forecasting and anticipating uncertainties allow organizations to be agile and adjust their strategies according to current realities.

While there are many definitions of organizational intelligence, this study conceptualized organizational intelligence as a business system that cautions of potential negative surprises and discovers new opportunities, which aims to comprehend and anticipate the future. This concept underscores the use of foresight to form the organization's intelligence in various dimensions. The study adopts the dimensions used by [11] organizational intelligence: technology intelligence, competitive intelligence, political foresight, and consumer foresight.

Technology Intelligence. According to [12], technology intelligence provides decision-makers with an answer to how the ever-changing business environment and technological advances or trends can impact an organisation's competitive position. The successful implementation of technology intelligence, which can reveal significant opportunities for an organisation, requires well-planned efforts. Researchers have therefore advocated using high-tech tools for data collection and analysis, disseminating information to those able and willing to act on it, and hiring individuals with the necessary mix of experience, skills and aptitude for the task at hand. When considering these aspects of data management, it is essential to remember that information is of varying quality and that assessing this quality can help identify problems and errors [13].

Competitive Intelligence. Organizations needed integrated analytical frameworks to identify and evaluate the strategies of current and potential competitors [14]. By systematically retrieving and analyzing data on markets, competitors and technologies, competitive intelligence provides decision-makers with a clear picture of the business landscape. A particular focus is on predicting outcomes and making recommendations based on these predictions [15]. The study by [16] states that it is a systematic and ethical process of collecting, analyzing and managing any combination of data, such as information and knowledge about an organization's business environment. Thus, when this data is strategically utilized, the organization could gain a competitive advantage or informed decision-making.

Political Foresight. Political foresight refers to the discovery, evaluation, and use of data related to laws, the political climate, and changes in the political landscape [11, 17]. According to [18], policy foresight can benefit policymaking in three ways: first, through improved anticipation (the ability to anticipate and prepare for future opportunities and challenges); second, through policy innovation (the promotion of creative new approaches to address these opportunities and challenges); and third, through future-proofing (the process of putting proven policies through their paces). The aim is to provide governments and societies with the tools to constantly explore and prepare for the future to manage better, adapt to and shape future developments through their policies. This applies to national decisions and those made at the level of sectors or policy areas [19]. According to research by [20], politically astute leaders in organizations can better anticipate and adapt to the shocks that inevitably lead to crises.

Consumer Foresight. Customer foresight refers to a method of planning that considers customers' wants and needs alongside the dynamics of change and their anticipations of forthcoming conditions, commodities, and offerings [21]. Putting customer-centric strategy and innovation has rendered foresight progressively challenging. One of its most valuable features is strategic foresight, which offers an intelligent solution for consumers.

2.3 Hypotheses Development and Research Framework

Previous studies have shown a specific correlation between organizational intelligence and crisis management. The study by [22] found that strategic organizational capabilities significantly impact crisis management and that these capabilities influence human resource agility, which in turn affects crisis management. The study by [23] found that conducting competitive intelligence, improving competitive intelligence mechanisms and using a competitive intelligence system are crisis management tools useful for the first stage of crisis alert in risk management. [24] point out a positive relationship between technological intelligence and crisis management, as technological intelligence improves crisis management by enabling rapid response and informed decision-making. The study by [22] and [20] support the significant relationship between the foresight of the political environment and crisis management. The relationship between consumer foresight and crisis management was noted in a study by [20]. Therefore, the following relationships were proposed in this study:

H1: Technology Intelligence significantly affects energy crisis management in Jordan.
H2: Competitive Intelligence significantly affects energy crisis management in Jordan.
H3: Political Foresight significantly affects energy crisis management in Jordan.
H4: Consumer Foresight significantly affects energy crisis management in Jordan.

Accordingly, these hypotheses aid the development of the research framework, as illustrated in Fig. 1.

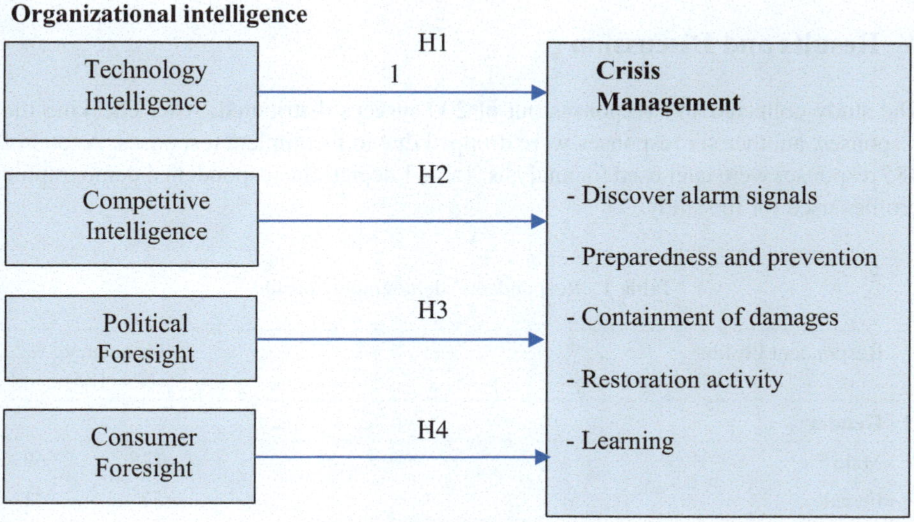

Fig. 1. Research framework.

3 Methodology

The study used a quantitative, cross-sectional approach using a questionnaire for data collection. The study population consists of managers working at energy companies in Jordan. [25] have discussed that managers' opinions are essential in social studies due to their significant role in their organizations, and they are also valuable sources of data, thoughts, views and information. Many researchers believe that managers are the most influential group in organizations in ensuring the role of strategic issues. Eleven power generation companies have been granted licenses to generate electricity from conventional energy sources by the "Energy and Minerals Regulatory Commission" (EMRC), the commission tasked with overseeing the energy industry in Jordan. According to statistics from the Jordanian Ministry of Energy and Mineral Resources (MEMR), there are around 1,622 managers in the eighteen companies. [26] proposed the inverse square root technique in the partial least square structural equation model (PLS-SEM) to determine the sample size of 155 samples required to achieve statistical significance at the 5% level and a route coefficient of 0.2. [27]. To compensate for a potentially low response rate, the researchers added 20% to the minimum sample size suggested by [27], bringing the total number to 200 respondents.

The study uses a ten-point Likert scale ranging from (1) strongly disagree to (10) strongly agree to measure the constructs. Organizational intelligence consists of 16 items that measure the four dimensions (competitive intelligence, technology intelligence, consumer foresight and political foresight) adapted from the studies of [11, 21, 28]. The crisis management construct consists of 19 items that measure the five dimensions (preparedness and prevention, discovering alarm signals, containment of damages, learning and restoration activity), adapted from [29].

4 Results and Discussion

The study collected 191 responses out of 200 surveys distributed. After checking the responses, another six responses were dropped due to incomplete responses. A total of 185 responses were later used for analysis. Table 1 depicts the respondents' demographic profiles used for the study.

Table 1. Respondents' demographic profile.

	Respondent Profile	Frequency (N = 185)	%
1	**Gender**		
	Male	128	69
	Female	57	31
	Total	**185**	**100**
2	**Age**		
	20 – 30 years old	43	23
	31 – 40 years old	78	42
	41 – 50 years old	46	25
	Above 50 years old	19	10
	Total	**185**	**100**
3	**Manager's Qualification/ Education Level**		
	Diploma	4	2
	Bachelors	91	49
	Masters	78	42
	PhD/DBA	13	7
	Total	**185**	**100**
4	**Length of Services**		
	Below 1 year	11	6

(continued)

Table 1. (*continued*)

	Respondent Profile	Frequency (N = 185)	%
	1 – 5 years	63	34
	5 – 10 years	61	33
	Above 10 years	50	27
	Total	**185**	**100**
5	**Status within the Company**		
	Top management (e. g. CEO, MD, or another Board member	65	35
	Middle management (e.g., functional manager or another senior on-board member)	120	65
	Total	**185**	**100**

Overall, the profile suggests that the respondents are among the top and middle managers in the energy service sector, with 60 percent having considerable working experience (five years and above). Therefore, the responses should provide sound insights, as most managers may have a deeper understanding of the company's history, past strategies, successes, and failures in managing the energy crisis in Jordan.

4.1 Data Analysis

In this study, PLS-SEM software was used to analyse the collected data. PLS path modelling is used to create measurement and structural models. The simultaneous regression analyses are used to verify the study's hypotheses and conduct bivariate correlation analysis, while the measurement model is used to assess the construct's validity and reliability.

Measurement Model Assessment. Measurement model assessment involves the examination of individual item reliability, internal consistency of reliability, content validity, discriminant validity, and convergent validity [30]. The outer loadings of each latent variable were examined to evaluate the reliability of individual items [30]. Following the rule of thumb, the items were retained for loadings with values between 0.40 and 0.70 [30]. Moreover, [30] also recommends that the value of the composite reliability coefficient should be at least 0.7 when testing the internal consistency reliability. The results show that all the latent variables in this study had sufficient internal consistency over the 0.7 minimum acceptable level, as presented in Table 2.

Table 3 shows the results of the correlation between variables using the Fornell-Larcker analysis to evaluate the discriminant validity of the measurement model.

According to [31], discriminant validity is achieved when the square root of AVE is greater than the correlation between the components making up each pair. Therefore, the value should be greater than the other off-diagonal elements in the rows and columns. The evaluation shows that the results in Table 3 satisfied the Fornell-Larcker criteria.

Table 2. The constructs' composite reliability and Cronbach's alpha.

Construct	Cronbach's alpha > 0.7	AVE > 0.5
Technology Intelligence	0.879	0.806
Competitive Intelligence	0.837	0.608
Political Foresight	0.752	0.616
Consumer Foresight	0.884	0.743
Crisis Management	0.85	0.626

Table 3. Variables correlation-root square of AVE.

	Organization Intelligence	Crises Management	Political Foresight	Technology Intelligence	Consumer Foresight
Competitive Intelligence	**0.779**				
Crises Management	0.639	**0.791**			
Political Foresight	0.526	0.582	**0.785**		
Technology Intelligence	0.688	0.627	0.629	**0.898**	
Consumer Foresight	0.668	0.574	0.315	0.421	**0.862**

Structural Model Assessment. The structural model assessment was performed to determine the links between latent variables. The analysis used nonparametric evaluation criteria based on a bootstrapping technique using 5000 bootstrap samples to assess the significance of the path coefficients [30]. Table 4 presents the results of hypotheses testing (H1 to H4) using path coefficient measures.

According to the results, the study did not find support for H1, which refers to the effect of technology intelligence on energy crisis management in Jordan. These results are consistent with previous studies, such as the study by [32]. One plausible reason for this is that Jordan's adoption of an intelligence system in the energy sector is still in the early stages. The digital transformation was recently begun in Jordan, and the Internet of Things (IoT) started in companies affiliated with the Ministry of Energy to collect data and establish a comprehensive monitoring centre that can control the network, collect data, and analyze for strategic planning. Therefore, the significant impact of technological intelligence in energy crisis management might not yet be evident.

Similarly, H2 also did not find support for the effect of competitive intelligence on energy crisis management in Jordan. The energy corporations are government-owned, so

Table 4. Structural model assessment.

		Original sample (O)	Standard deviation (STDEV)	t Values	P Values	Decision
H1	Technological Intelligence -> Crises Management	0.014	0.045	0.307	0.760	Not Supported
H2	Competitive Intelligence -> Crises Management	-0.021	0.067	0.307	0.759	Not Supported
H3	Political Foresight -> Crises Management	0.523	0.053	9.868	0.000	Supported
H4	Customer Foresight -> Crises Management	-0.306	0.082	3.746	0.000	Supported

Note: Significant level at * $p < 0.10$, ** = $p < 0.05$, *** $p < 0.01$

their lack of competition could explain the result. Meanwhile, H3 was supported, which suggests that political foresight plays a significant role in energy crisis management. This is consistent with previous studies, such as the study by [20]. Jordan is one of the stable countries in the MENA region that has been in crisis since 2011 and witnessing internal conflicts and tensions between countries. This underscores the importance of political foresight through regional cooperation for effective energy crisis management.

Finally, the study finds support for H4. This indicates that consumers' foresight significantly impacts energy crisis management in Jordan. The consumer is central to the government's planning and strategy because it helps predict the trend and consumption patterns, enabling energy providers and policymakers to plan for sudden surges or reductions in demand and avoid grid overloads or supply shortages. Moreover, communication with consumers in times of crisis is very critical. Therefore, energy service providers can customise their messaging to answer public concerns, elucidate the rationale behind energy-saving efforts, and foster trust by utilising consumer foresight. Effective communication with consumers and stakeholders may encourage people to take the necessary steps to ease the situation, lessen anxiety, and promote cooperation [33].

5 Conclusion

The study investigated the role of organizational intelligence in energy crisis management in Jordan. Accordingly, the empirical evidence shows that only political and consumer foresight have significant influence, while technological and competitive intelligence do not impact crisis management. The study contributes to the empirical evidence in the MENA region, particularly countries with scarce resources and regional conflicts like Jordan. Moreover, it offers insights into organizational factors that are critical for energy crisis management. Jordan's energy sector should focus on and seriously develop organizational intelligence to successfully manage the unexpected and uncertain situations that may lead to a considerable energy crisis that could impact the country's development.

References

1. Hamed, T.A., Bressler, L.: Energy security in Israel and Jordan: the role of renewable energy sources. Renew. Energy **135**, 378–389 (2019)
2. Komendantova, N.: Transferring awareness into action: a meta-analysis of the behavioural drivers of energy transitions in Germany, Austria, Finland, Morocco, Jordan and Iran. Energy Res. Soc. Sci. **71**, 101826 (2021)
3. Shaltout, A.A., et al.: Elemental composition and source apportionment of atmospheric aerosols collected from urban and residential areas of Jordan using multi-secondary targets energy dispersive X-ray fluorescence. Spectrochim. Acta, Part B **170**, 105900 (2020)
4. Sarpong, S.: Crisis management and marketing. In: The Routledge Handbook of Destination Marketing, pp. 100–109. Routledge (2018)
5. Putra, A., Petpichetchian, W., Maneewatt, K.: A survey study of public health nurses' knowledge in disaster management in Indonesia. Open Access Macedonian J. Med. Sci. **9**(G), 328–336 (2021)
6. Smith, D.: Beyond contingency planning: towards a model of crisis management. Ind. Crisis Q. **4**(4), 263–275 (1990)
7. Richardson, B.: Socio-technical disasters: profile and prevalence. Disaster Prev. Manage. **3**(4), 41–69 (1994)
8. Myers, K.: Total Contingency Planning for Disasters: Managing Risk Minimizing Loss Ensuring Business Continuity. John Wiley, New York (1993)
9. Fink, S.: Crisis Management: Planning for the Inevitable. American Management Association, New York (1996)
10. Pearson, C., Mitroff, I.: From crisis prone to crisis prepared: a framework for crisis management. Acad. Manage. Exec. **7**(1), 48–59 (1993)
11. Rohrbeck, R., Battistella, C., Huizingh, E.: Corporate foresight: an emerging field with a rich tradition. Technol. Forecast. Soc. Chang. **101**, 1–9 (2015)
12. Dwivedi, Y.K., et al.: Artificial Intelligence (AI): Multidisciplinary perspectives on emerging challenges, opportunities, and agenda for research, practice and policy. Int. J. Inf. Manage. **57**, 101994 (2021)
13. Sacks, R., Girolami, M., Brilakis, I.: Building information modelling, artificial intelligence and construction tech. Dev. Built Environ. **4**, 100011 (2020)
14. Köseoglu, M.A., Morvillo, A., Altin, M., De Martino, M., Okumus, F.: Competitive intelligence in hospitality and tourism: a perspective article. Tour. Rev. **75**(1), 239–242 (2019)
15. Madureira, L., Popovič, A., Castelli, M.: Competitive intelligence: a unified view and modular definition. Technol. Forecast. Soc. Change **173**, 121086 (2021)
16. McLeana, L., Woodsb, L.A.: Competitive intelligence capabilities. Proc.: Soc. Behav. Sci. **110**, 669–677 (2019)
17. Arokodare, M.A., Asikhia, O.U.: Strategic agility: achieving superior organizational performance through strategic foresight. Glob. J. Manage. Bus. Res. **20**(3), 7–16 (2020)
18. Van Woensel, L.: A Bias Radar for responsible Policy-Making: Foresight-Based Scientific Advice. Springer, Heidelberg (2019). https://doi.org/10.1007/978-3-030-32126-0

19. Dinges, M., Biegelbauer, P., Wilhelmer, D.: The tower of Babylon in the governance of research, technology and innovation: Participatory foresight as a method of policy coordination. Futures: J. Policy Plann. Futures Stud. **100**, 34–44 (2018)
20. Makian, S., Nematpour, M.: Foresight thinking and organizational learning: scenario planning as a DMO crisis management Tool, pp. 107–118 (2021)
21. Schweitzer, N., Hofmann, R., Meinheit, A.: Strategic customer foresight: from research to strategic decision-making using the example of highly automated vehicles. Technol. Forecast. Soc. Change **144**, 49–65 (2019)
22. Rastegari, H., Hosseini, S.H., Ghayoor, H.: Designing and developing the strategic foresight capabilities of medical universities in the accidents and crises of the country with emphasis on human resource agility. Int. J. Logist. Syst. Manage. **36**(1), 124–137 (2020)
23. Yiu, L.D., Yeung, A.C., Cheng, T.E.: The impact of business intelligence systems on profitability and risks of firms. Int. J. Prod. Res. **59**(13), 3951–3974 (2021)
24. Wang, W.T., Wu, S.Y.: Knowledge management based on information technology in response to COVID-19 Crisis. Knowl. Manage. Res. Pract. **19**(4), 1–7 (2021)
25. Sekaran, U., Bougie, R.J.: Research Methods for Business+Learning Space Card a Skill Building Approach. Wiley, Hoboken (2016)
26. Kock, N., Hadaya, P.: Minimum sample size estimation in PLS-SEM: the inverse square root and gamma-exponential methods. Inf. Syst. J. **28**(1), 227–261 (2018)
27. Hair, Joseph F., et al.: Partial Least Squares Structural Equation Modeling (PLS-SEM) Using R: A Workbook. Springer, Cham (2021). https://doi.org/10.1007/978-3-030-80519-7
28. Saayman, A., et al.: Competitive intelligence: construct exploration, validation and equivalence. Aslib Proc. **60**(4), 383–411 (2008)
29. Mishael Obeidat, A., Ben Hamed Al Thani, F.: The impact of strategic leadership on crisis management. Int. J. Asian Soc. Sci. **10**(6), 307–326 (2020)
30. Hair, J., Sarstedt, M., Hopkins, L., Kuppelwieser, V.G.: Partial least squares structural equation modelling (PLS-SEM). Eur. Bus. Rev. **26**(2), 106–121 (2014)
31. Fornell, C., Larcker, D.F.: Structural equation models with unobservable variables and measurement error: algebra and statistics. J. Mark. Res. **18**(3), 382–388 (1981)
32. Alkhaffaf, M., Almomani, H.: Role of intelligent technology in crises management: systematic literature review the case of Corona Virus. In: Alshurideh, M., Hassanien, A.E., Masa'deh, R. (eds.) The Effect of Coronavirus Disease (COVID-19) on Business Intelligence. Studies in Systems, Decision and Control, vol. 334, pp. 3–14. Springer, Cham (2021). https://doi.org/10.1007/978-3-030-67151-8_1
33. Bundy, J., Pfarrer, M.D., Short, C.E., Coombs, W.T.: Crises and crisis management: integration, interpretation, and research development. J. Manage. **43**(6), 1661–1692 (2017)

Knowledge Management as a Support for Cultural Innovation and Promotion: A Study of Knowledge Management Practices – The Case of ADER-Fès

El Hadraoui Douae(✉) and Amrous Naila

Laboratory "Medation – Information – Knowledge – Society" (MIKS) Team "Competitive Intelligence, Analysis and Strategy" (CIAS) School of Information Sciences, Rabat, Morocco
elhadraoui.douae@outlook.com

Abstract. This paper explores the critical importance of knowledge management (KM) in fostering cultural innovation, particularly within the context of ADER-Fès, a key organization in the Fez-Meknes region dedicated to heritage restoration. This research employs a qualitative, exploratory approach, using non-participant direct observation and semi-structured interviews to assess ADER-Fès's KM practices. The study reveals the framework within ADER-Fès for managing historical, technical, and cultural knowledge, which is crucial for the effective conservation of the rich heritage of the Fez-Meknes region. However, the research also highlights critical areas for improvement, such as the enhancement of documentation systems, communication strategies, continuous professional development, technological integration, and organizational change. Focusing on these strategic areas, the paper proposes a forward-looking approach that seeks to address the current gaps and prepare the organization for future challenges in the dynamic field of cultural heritage management. The paper concludes with the expectation that ADER-Fès will significantly improve its operations and contributions to global cultural heritage preservation by implementing these strategies. The organization's commitment to continuous improvement and adaptation is poised to set a benchmark for KM in cultural heritage institutions. The findings and proposals in this paper will serve as a roadmap for ADER-Fès's future strategy, ensuring the preservation and valorization of cultural heritage for generations to come.

Keywords: Knowledge management · Cultural Heritage · ADER-Fès

1 Introduction

Knowledge management (KM) is the process of managing and safeguarding organizational knowledge, including its creation, storage, transfer, and application [1]. However, KM is not extensively applied in cultural organizations, and there is a shortage of academic research on KM strategies tailored to the distinct features of these entities. These organizations, such as ADER-Fès, work in culture and heritage preservation and report to public administrations with their specific processes and policies.

Z. Mat Aji et al. (Eds.): KMICe 2024, ISEM 49, pp. 40–52, 2025.
https://doi.org/10.1007/978-3-031-91485-0_4

In the World Heritage Convention, the United Nations Educational, Scientific, and Cultural Organization (UNESCO) outlines the responsibilities of states to implement proper legal, technical, administrative, and financial strategies to ensure the recognition, protection, preservation, presentation, and restoration of heritage. This instruction emphasizes the need for coordinated efforts in knowledge exchange concerning Cultural Heritage Projects, especially in the context of the Sustainable Development Goals.

Within this framework, local organizations like ADER-Fès contribute significantly to cultural preservation. ADER-Fès is a profoundly active institution in conserving and restoring heritage in the Fez-Meknes region. The organization's operations aim to preserve the architectural integrity and historical authenticity of the area while also promoting sustainable tourism and economic growth. The application of KM practices in cultural heritage organizations remains understudied, particularly in regions like Fez-Meknes, where heritage preservation is crucial. This paper aims to explore whether ADER-Fès employs KM methods. Are these methods effective? Do they foster and sustain cultural innovation?

This study illuminates the region's existing understanding of heritage management and paves the way for exploring how these practices can be refined to encourage cultural innovation and advancement. It ensures that the region's rich historical and cultural fabric is preserved and adapted to modern-day significance.

The primary goal is to thoroughly examine current KM procedures at ADER-Fès, evaluating how knowledge is systematically acquired, structured, exchanged, and disseminated. This investigation is essential for evaluating the efficiency of existing KM systems and pinpointing potential areas for improvement, especially regarding cultural promotion. The second aim is to assess the needs and expectations of the organization's collaborators. This involves thoroughly exploring experts' views on KM methods, their active participation in these processes, and their aspirations for the future of heritage protection and cultural promotion. By tackling these elements, the research provides a holistic view of ADER-Fès' internal knowledge practices, aligning its KM methods with the goals of cultural promotion and national heritage preservation.

This paper is divided into five sections: The first section examines KM ideas, heritage restoration, and the role of KM in such organizations. The second section describes the materials and procedures utilized to conduct this investigation. The third section presents the research findings and outcomes. The fourth section reviews the challenges KM faces at ADER-Fès and proposes improvement solutions. Finally, the fifth section concludes with recommendations and future directions. This study finishes with a review of the issues KM faces in Ader-Fès and solutions for improving it.

2 Background

2.1 Data, Information, and Knowledge

Data, information, and knowledge are three concepts that are usually linked in literature; in fact, there is a distinction between data and information. Raw facts which have not been treated or arranged are referred to as data, yet information is the outcome of processing and arranging data for a specific reason.

According to various researchers [18], knowledge can be classified into two main categories:

- Tacit knowledge: Extracted from subjective experiences and not typically formalized. It remains difficult to store or share as it is an individual's comprehensive knowledge.
- Comprehensive knowledge: Easily captured, stored, and shared.

2.2 Knowledge Management

KM practices involve capturing the collective expertise within an organization to meet the specific needs of different entities [2]. Collective knowledge expertise can encompass various resources such as knowledge bases, intellectual capital, or documents.KM is made up of four important components. These components include the organization's knowledge values and beliefs, the methods for acquiring data and knowledge, and the guidelines that facilitate knowledge transfer [3, 4].

For the past two decades, mainly since the initiation of the Lisbon Process in 2000, which intended to make the European Union economy's knowledge-based the most competitive [5], KM has witnessed global expansion and established itself as a pivotal reference for economic, cultural, and social advancement.

Researchers have extensively studied the field of KM in public organizations [6, 7]. They investigated how the public sector's performance relates to KM. Other authors have comprehensively reviewed twenty articles on this topic, uncovering valuable insights [8]. Management practices in the public sector have undergone significant improvements, resulting in enhanced organizational performance, increased efficiency, and greater productivity.

Integrating KM strategies inside firms has significant advantages [8, 9]. The researchers used structural comparison models to emphasize the importance of KM, including the performance of these firms, especially in public and semi-public Moroccan organizations. These models allow for the examination of several linkages among dependent and independent latent variables simultaneously [10].

However, Organizational development has increasingly prioritized KM as a strategic approach to foster innovation and gain a competitive edge [11]. This enables organizations to leverage their collective knowledge and expertise efficiently. Organizations that are also capable of successfully overseeing and transmitting their knowledge are associated with higher levels of innovation and performance [12].

KM encompasses the concepts, methodologies, and tools that facilitate the capitalization, sharing, and evolution of a knowledge base within a specific domain, including skills and expertise. Typically, this domain is in the economic field and is characterized by a knowledge community comprising recognized specialists and experts, accumulated experience over time, and significant technicality [13].

Strategy is the foundational element underpinning all KM initiatives. Indeed, before embarking on a KM project or strategy, it is prudent for an organization to reflect on the most effective way to leverage it to achieve its objectives. A KM strategy must align with the organization's overall strategy and strategic goals. Moreover, a KM project only adds value when it addresses a broader strategic need. The stakes of a KM approach go beyond merely creating a collective or documentary memory; instead, it aligns with a logic of sustainable strategic development.

Traditional professional domains, like heritage professions, have often been overlooked by the KM movement for specific reasons. Yet, these fields hold substantial economic, social, and cultural importance. Hence, applying and adapting KM approaches to these areas is useful. Several explanations for KM can be found in scientific literature, some outlined in Table 1.

Table 1. Knowledge management definitions.

Author	Definition of KM
[14]	*KM is a managerial task that involves creating and controlling knowledge and using it effectively and efficiently in accordance with the organization's long-term goals*
[15]	*KM involves a structured approach to uncovering, choosing, arranging, condensing, and delivering information to enhance a collaborator's understanding of a specific subject*
[16]	*This approach involves a methodical and intentional process of developing, sharing, applying, refreshing, and enhancing knowledge to achieve organizational objectives*
[17]	*KM is a powerful approach that facilitates the utilization and expansion of organizational knowledge to generate value and enhance competitive advantage*
[1]	*KM is the process of managing and safeguarding organizational knowledge, including its creation, storage, transfer, and application*
[18]	*KM is an essential practice in today's knowledge-based society and globalized world. It involves permanently transferring individual expertise into organizational expertise, allowing individuals to build upon the organization's collective knowledge*

In this context, KM is critical in preserving and restoring national heritage. It involves the continuous transfer of individual expertise into collective knowledge, ensuring the preservation and transmission of the cultural identity to future generations.

2.3 KM's Position in Heritage Restoration

The notion of heritage, legally formalized by the 1972 UNESCO Convention, means, "The totality of the assets inherited from the most distant past to the nearest future in the cultural and natural order ".

Knowledge bases remain essential in this field. Understanding the history and original context of the historical site is crucial. It is essential to have a thorough understanding of the building's construction, its initial purpose, its historical importance, and extensive knowledge of historic architectural styles, construction techniques, and materials used in the past. Furthermore, it allows an extensive understanding of contemporary and conventional conservation methods and materials. Moreover, understanding the legal frameworks and ethical considerations surrounding heritage sites remains crucial.

To date, the impact of KM has been observed in the domain of conventional economics and industry. A pinnacle of this evolution was the publication of an international standard (ISO 30401) in 2018/2019 [19], which outlines the KM requirements for businesses and organizations. The social and societal sphere has also seen developments, notably by establishing novel approaches to human development. This led UNESCO, for example, to establish a "Knowledge Societies" division to address these topics [20].

Knowledge Management (KM) has often overlooked traditional professional domains, like heritage trades and the arts. However, these fields play a crucial role economically, socially, and culturally in many areas. Applying KM strategies to these professions could be both interesting and valuable.

3 Methods

A qualitative exploratory study complemented the theoretical aspect of our work focused on the practices of KM within ADER-Fès, applied in heritage restoration. This approach appears to be a relevant means to answer our research questions, particularly in a context where 'how' and 'why' questions arise, when the investigator has minimal influence over occurrences and when the researcher's attention is focused on a current phenomenon in a genuine social environment [21]. The exploration allows for an understanding of why and how variables are interconnected in the realm of heritage restoration. Throughout our engagement with ADER-Fès, considered one of the essential elements in the preservation of the cultural heritage, we combined various methodological techniques: non-participant direct observation, primarily focused on observing actors in action and the management control process to identify elements of the intangible KL strategy and practices, as well as the various steering tools used. Semi-structured interviews were conducted with managers, based on a pre-established interview guide. Our guide was structured along the following axes: General Information, Practices, Strategic Objectives, Inventory of knowledge assets, and Knowledge management and cultural promotion.

Our data collection comprised one primary technique: in-depth, semi-structured interviews. Interviews, averaging one hour in length, were conducted in a conversational yet structured manner, allowing for predetermined and emergent lines of inquiry. We opted for a non-probabilistic sampling method for respondent selection, interviewing accessible managers within the ADER-Fès departments. The interviews (4 interviews in total) were conducted in December 2023. These individuals were selected based on their leadership roles and direct involvement in knowledge management and cultural heritage projects. Due to their work's specialized nature and deep understanding of organizational processes, they represent the most relevant and insightful respondents for this exploratory study.

A triangulation approach was employed to ensure the reliability of the data. Data collected from semi-structured interviews was cross-referenced with internal documentation such as project reports, meeting notes, and knowledge databases. Additionally, member checking was conducted post-interviews, allowing participants to review and validate the accuracy of the collected data. Member checking was conducted post-interviews, where participants were allowed to review and comment on the findings, thereby enhancing the accuracy and credibility of the data.

In our investigation, we prioritized ethical considerations. Before data collection, all participants provided informed consent, ensuring they were fully aware of the study's goal and rights. Confidentiality and anonymity were scrupulously preserved throughout the research, with sensitive material being securely managed and anonymized in the reporting. Internal documentary analysis (analysis of directives, reports, etc.) was also conducted.

We used manual content analysis for data processing and analysis, ideal for exploratory studies. Qualitative data collected were rich and complex, requiring categorization before meaningful analysis [22]. The analyst's expertise allows deeper insights beyond raw results [23]. We began with essential preliminary steps like transcribing interviews. Instead of focusing on individual responses, we combined relevant information across sources, selecting critical data for each axis [23]. This method helped us descriptively explain the role of intangibles in ADER-Fès's managerial and organizational processes and its impact on managing intangible performance.

4 Results

The outcomes of this investigation shed light on how ADER-Fès has implemented KM practices in the technical and historical knowledge areas. The organization's commitment to preventing knowledge loss through the systematic creation of documents related to projects and procedure standardization demonstrates a proactive approach to protecting and capitalizing on cultural heritage knowledge. Furthermore, the anticipation of integrating current technologies and promoting professional growth demonstrates a forward-thinking strategy.

4.1 Data, Information, and Knowledge

This research reveals that the strategy within Ader-Fès focuses on enhancing innovation, problem-solving, organizational learning, decision-making, collaboration, and knowledge preservation. Specifically, it involves adopting practices for managing knowledge, creating a knowledge database, encouraging collaboration, and facilitating knowledge sharing. The strategy provides detailed guidelines for developing the KM database, including identifying critical knowledge areas, organizing knowledge, and establishing access permissions. Furthermore, it highlights the importance of technology in KM, the implementation of training and awareness programs, and the need for regular maintenance and updating of the knowledge database.

ADER-Fès has developed a robust KM infrastructure that covers data collection, processing, storage, and sharing. This comprehensive system combines operational and support methods for efficient knowledge management using advanced technologies like GIS.

Ader-Fès is dedicated to acquiring and preserving knowledge through project reports and reviews, focusing on protecting intellectual assets with regular updates and validations. Their KM strategies align with their goals in heritage promotion but face challenges in knowledge sharing, quality maintenance, methodology preservation, and legal issues. They utilize infrastructure such as SIG for digital documentation and propose using 3D scanning techniques for monuments, which are essential for preserving the region's history despite facing some challenges.

Ader-Fès uses 3D scanning to gain detailed insights into monument construction, revealing preservation needs and enhancing historical appreciation. Challenges include maintaining high-quality data within technological limits. Sometimes, specific components were used instead of advanced laser scanning. Accurate records are crucial due to potential human-caused changes, requiring continuous monitoring and updates. Risks involve physical damage during collection and digital data loss. Additionally, 3D scanning helps preserve cultural heritage digitally, identifies building materials and issues, and assists in planning restoration efforts.

4.2 Challenges in Knowledge Sharing, Knowledge Access, and Risks Related to Knowledge Loss

This study reveals that collaborators identify their knowledge needs via skills assessments and feedback. However, barriers like communication issues in documentation and resistance to change hinder knowledge exchange and use. This highlights the necessity for better methods to improve knowledge sharing within the company.

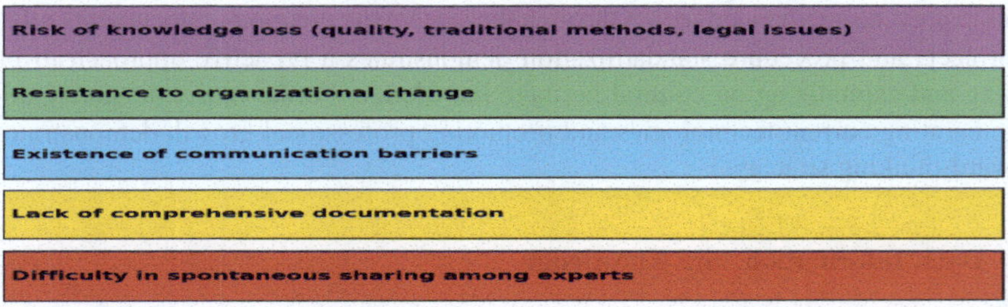

Fig. 1. Barriers and challenges within the KM process.

Survey findings at ADER-Fès reveal that collaborators know their knowledge needs, stemming from talent assessments and feedback from field experience. They understand the crucial knowledge for heritage preservation, including technical details, historical importance, and environmental concerns. This awareness supports the organization's goals of preserving and promoting Medina's heritage and restoring historic sites. However, the study identifies obstacles in transmitting this critical knowledge, mainly due to inadequate documentation processes. Poor recordkeeping makes it hard for staff to access and use essential information, potentially losing valuable insights from past conservation efforts. Figure 1 shows that paperwork issues hinder internal communication, causing delays and inefficiencies. Additionally, Ader-Fès culture shows some resistance to change, whether adopting new conservation methods or using technology for knowledge sharing.

4.3 Current Knowledge Spectrum and Emerging Knowledge Needs in ADER-Fès

Figure 2 showcases critical areas of technical proficiency essential for restoration and conservation, such as brickwork and masonry. The company's expertise combines traditional craftsmanship with modern building methodologies, merging ancient practices with contemporary techniques. This technical know-how is vital for restoring and conserving sites while preserving their authenticity and integrity. Additionally, the company's proficiency is enhanced by a deep understanding of technical aspects, including material analysis and diagnostic technologies, which guide the choice of preservation materials and methods, ensuring durable interventions that protect heritage. Figure 2 also highlights an approach grounded in the expertise and knowledge of ADER-Fès collaborators, emphasizing the importance of multidisciplinary collaboration and experience sharing. Every project benefits from this collaborative environment, fostering free knowledge exchange that brings fresh insights to the team. The figure outlines the fundamental principles and techniques for effectively managing restoration projects, underscoring the necessity of management skills. This involves understanding crucial laws and regulations that ensure compliance and ethically maintain cultural landmarks. Furthermore, it underscores the importance of architectural history and significance, illustrating ADER-Fès's

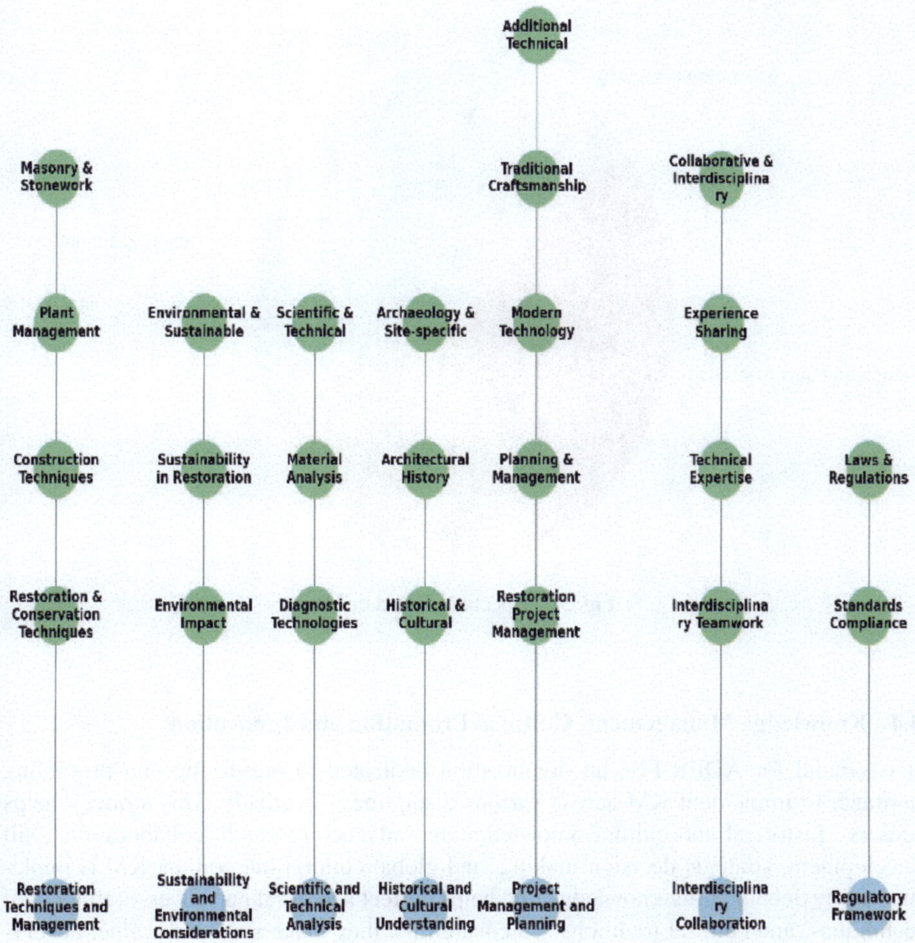

Fig. 2. Knowledge categorization in architectural restoration.

dedication to preserving the physical monuments and their intangible values. Respect for the history of art and architecture shapes ADER-Fès conservation methods, ensuring that restoration efforts are sensitive to each site's uniqueness and contextually appropriate. This map depicts a knowledge ecosystem that guides ADER-Fès's daily mission to protect and promote the Fez-Meknes heritage.

The graph outlines critical areas of ADER-Fès's strategic focus based on input from 40 collaborators, divided into five segments by importance. Improved documentation is the top priority at 30%, indicating a need for better record-keeping for cultural heritage preservation. Improved communication comes at 25%, highlighting the importance of enhancing internal and external interaction for effective workflow and knowledge management. Technology integration stands at 20%, aiming to blend advanced technologies with traditional preservation methods. Constant training allocated 15%, underscores the value of ongoing professional development. Lastly, organizational change receives 10%, signifying structural adjustments are essential but not the focus.

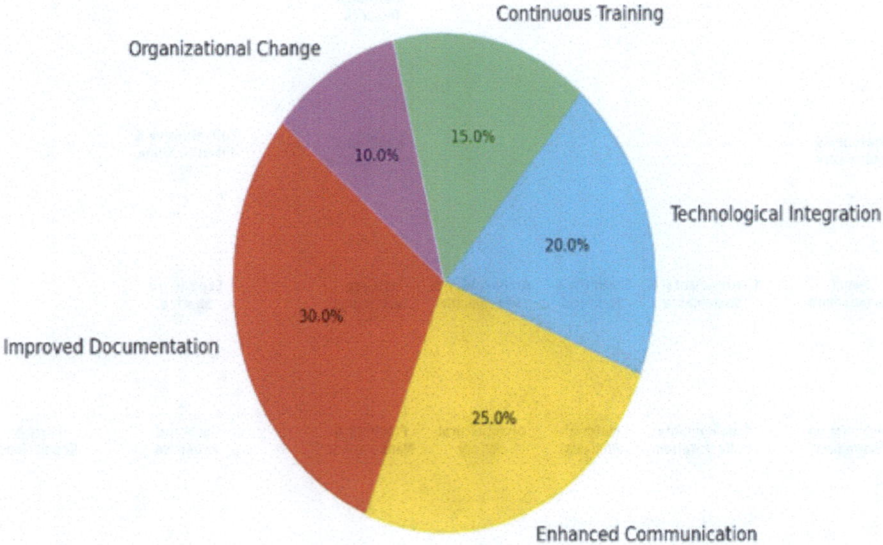

Fig. 3. Expectations and needs.

4.4 Knowledge Management, Cultural Promotion and Innovation

It is crucial for ADER-Fès, an organization dedicated to preserving and promoting heritage, to implement KM across various disciplines effectively. This strategy helps preserve historical and cultural knowledge and advance research, collaboration, skill development, strategic decision-making, and global cultural integration. KM is implemented by documenting knowledge that helps protect historical narratives, architectural techniques, and cultural traditions. To ensure this, they make sure to document everything thoroughly and to digitize it to last a long time and not deteriorate. Therefore, it ensures that knowledge is protected for future generations.

The organization's KM system is crucial for advancing research and development. It is a valuable resource for academics, offering a collection of historical practices and cultural heritage material to support their work. In addition, it also motivates ADER-Fès to produce new and improved techniques for preserving sites and artefacts, helping them constantly improve their methods. The ADER-Fès KM method promotes collaboration among professionals in various subjects such as history, architecture, and archaeology. Moreover, when the community participates in sharing knowledge, it helps to increase awareness and foster respect for the region's rich cultural history. A critical aspect of KM is training and developing the skills of collaborators. This ensures that they have the necessary abilities to conduct preservation work effectively. Additionally, it is beneficial for passing on the skills and trades required to maintain and restore heritage monuments.

At ADER-Fès, strategic decisions are grounded in thorough knowledge. This is crucial for heritage protection, aiding in risk identification and mitigation related to cultural preservation and loss of traditional knowledge. By broadly disseminating information, ADER-Fès raises awareness of the Fez-Meknes region's artistic value and engages in global collaborations to promote its history.

4.5 Future Strategy Expectations

According to the current study, ADER-Fès's future strategy aspirations include a diversified approach to improving KM within the business. A fundamental assumption is that documentation and sharing mechanisms would significantly improve. This advancement is critical for ensuring that knowledge, particularly in the fields of heritage preservation and cultural traditions, is not only efficiently conserved but also easily accessible and shareable among team members and collaborators. The emphasis on a more robust and user-friendly KM system demonstrates a proactive approach to preserving the region's rich cultural heritage.

Besides this, there is a strong need inside the organization for ongoing training and professional development opportunities. This part highlights the necessity of giving collaborators the most recent knowledge and abilities to keep them at the forefront of heritage protection methods. It is anticipated that frequent training sessions and skill-building seminars will be crucial in keeping the crew knowledgeable about the latest developments in historic preservation techniques.

Boosting internal and external collaboration is crucial for ADER-Fès's future strategy. ADER-Fès aims to advance conservation efforts and expand its knowledge base by improving interdepartmental cooperation and strengthening relationships with other organizations. Merging diverse experiences and perspectives will result in more innovative and comprehensive preservation solutions.

Embracing cutting-edge technologies is also anticipated to elevate conservation and restoration work. Incorporating technological advancements is essential for enhancing preservation efficiency and providing new ways to present and understand cultural artefacts. This approach is expected to transform how historical sites are preserved, studied, and showcased, enriching the cultural heritage experience for future generations.

5 Discussion

The following analysis presents a comprehensive overview of the KM practices implemented by ADER-Fès. The organization places significant emphasis on leveraging its capabilities in its KM approach and the robustness of its technology infrastructure. The mentioned approach involves identifying specific areas that necessitate attention to enhance spontaneous knowledge exchange, address obstacles highlighted by collaborators and establish strategies to mitigate knowledge loss.

Integrating collaborators' viewpoints is necessary in establishing a strong framework and aligning ADER-Fès KM procedures with its overarching objectives of global heritage preservation and cultural development. Establishing congruence within an organization is of utmost importance, particularly in the dynamic field of cultural heritage management, to create collective knowledge.

Based on the findings above, it is imperative for ADER-Fès to formulate and implement comprehensive strategies to facilitate effective knowledge exchange. To enhance documentation procedures, it is essential to establish easily accessible digital knowledge bases that capture and store both explicit and implicit organizational knowledge. Enhancing internal communication channels is of utmost importance in KM. This can be effectively achieved by implementing various strategies such as organizing knowledge-sharing meetings, conducting informative seminars, and leveraging digital platforms that facilitate continuous and open exchange of ideas and experiences. In addition, it is imperative to effectively address change resistance, as it significantly influences the evolution of the organization's KM strategy. To effectively accomplish this objective, it is imperative to create a conducive atmosphere that places a high emphasis on the cultivation of knowledge acquisition and the ability to adapt to changing circumstances. It is also essential to encourage data consumption and establish reciprocal trust, enabling data to flow within Ader-Fès to create collective intelligence [12, 24–26]. Promoting a culture that encourages team members to explore innovative ideas and methodologies while fostering creative knowledge and academic research collaborations is highly recommended.

6 Conclusion and Future Work Perspectives

These results from the ADER-Fès use case demonstrate how crucial and effective KM is in protecting and assessing cultural resources. The institution has a wealth of technological, cultural, and historical knowledge that is essential to its goals. However, the data also points out areas that need improvement.

The organization's success and longevity rely on better documentation, enhanced communication, ongoing training, technological integration, and organizational change. Improving documentation methods is crucial for preserving the cultural heritage ADER-Fès aims to protect. Findings indicate that knowledge is often fragmented, highlighting the need for platforms that enhance access and exchange. Better communication fosters a culture of cooperation and knowledge sharing. Continuous training ensures ADER-Fès staff stay informed about the latest heritage management techniques.

The incorporation of technology is recognized as a crucial driver of organizational innovation. Adopting modern technologies and processes will help expedite operations and bring up new opportunities for heritage presentation and interpretation. Finally, while less highlighted, organizational transformation is recognized as essential to support the other key areas and offer a receptive climate for the suggested enhancements.

ADER-Fès aims to enhance its KM system using advanced digital tools, comprehensive staff training, and more robust collaboration networks. Research will evaluate these efforts, while periodic reviews will keep ADER-Fès adaptable in cultural heritage preservation. The goal is to strengthen its role in protecting cultural assets and preserving the Fez-Meknes region's history for future generations.

References

1. Schultze, U., Leidner, D.E.: Studying knowledge management in information systems research: discourses and theoretical assumptions. MIS Q. **26**(3), 213–242 (2002)
2. Hibbard, J.: Knowing what we know. Inf. Week **653**(20), 46–64 (1997)
3. Siadat, S., Aryan, S., Mohammadi, F.: Establishment and implementation of knowledge management case study: Saipa sales and after-sales service organization. Int. J. Educ. Res. **2**(7), 250–262 (2014)
4. Davenport, T.H., Prusak, L.: Working Knowledge: Managing What Your Organization Knows. Harvard Business School Press, Boston (1998)
5. Bouchet, M. H. :La société de la connaissance dans le cadre de la stratégie de Lisbonne, Direction des Journaux officiels, 2005
6. Mc Evoy, P.J., Ragab, M.A., Arisha, A.: The effectiveness of knowledge management in the public sector. Knowl. Manage. Res. Pract. **17**(1), 39–51 (2019)
7. Laihonen, H., Kokko, P.: Knowledge management and hybridity of institutional logics in public sector. Knowl. Manage. Res. Pract. **21**(1), 14–28 (2023)
8. Basso, L.M., Teixeira, A.A., Freitas, W.R.D.S., Oliveira, J.H.C.D.: Green knowledge management in the public sector: synthesis and research agenda. Conhecimento em Ação **5**(2), 50–75 (2020)
9. Bagnoli, C., Vedovato, M.: The impact of knowledge management and strategy configuration coherence on SME performance. J. Manage. Govern. **18**, 615–647 (2014)
10. Payal, R., Ahmed, S., Debnath, R.M.: Impact of knowledge management on organizational performance: an application of structural equation modelling. VINE J. Inf. Knowl. Manage. Syst. **49**(4), 510–530 (2019)
11. Roussel, P.: Methods of developing scales for survey questionnaires. Hum. Res. Manage.: Res. Methods Hum. Soc. Sci. Methods Res. 245–276 (2005)
12. Bastidas, V., Oti-Sarpong, K., Nochta, T., Wan, L., Tang, J., Schooling, J.: Leadership for responsible digital innovation in the built environment: a socio-technical review for re-establishing competencies. J. Urban Manage. **12**(1), 57–73 (2023)
13. Husain, S., Ermine, J. L.: Knowledge Management Systems: Concepts, Technologies and Practices. Emerald Publishing Limited (2021)
14. Xin, X., Miao, X., Chen, Q., Shang, T.: User participation, knowledge management capability and service innovation: e-commerce enterprises in the digital age. Asia Pac. J. Mark. Logist. **34**(4), 629–646 (2022)
15. Hameed, I.: Knowledge management and business intelligence: what is the difference?" (2004). http:/onlinebusiness.about.com/
16. Pillania, R.K.: Strategic issues in knowledge management in small and medium enterprises. Knowl. Manage. Res. Pract. **6**(4), 334–338 (2008)
17. Zhao, J., de Pablos, P.O., Qi, Z.: Enterprise knowledge management model based on China's practice and case study. Comput. Hum. Behav. **28**(2), 324–330 (2012)

18. Liyanage, T., Kilbourne, J., Gerlich, A.P., North, T.H.: Joint formation in dissimilar Al alloy/steel and Mg alloy/steel friction stir spot welds. Sci. Technol. Weld. Join. **14**(6), 500–508 (2009)
19. Mulgan, G.: Big Mind: How Collective Intelligence can Change our World. Princeton University Press (2019)
20. Bindé, J.: Towards Knowledge Societies: UNESCO World Report. UNESCO reference works series. UNESCO, Paris (2005)
21. Aberdeen, T., Yin, R.K.: Case study research: design and methods. Can. J. Act. Res. **14**(1), 69–71 (2013) (2009)
22. Hogan, M., et al.: A group dynamics framework for 21st-century collective intelligence facilitators. Syst. Res. Behav. Sci. **38**(4), 572–576 (2021)
23. Hogan, M., Hall, T., Harney, O.: Collective intelligence design and a new politics of system change. Civitas Educationis **6**(1), 51–78 (2017)
24. Fathullah, M.N., Ulfiah, U., Mulyanto, A., Gaffar, M.A., Khori, A.: Management of digital literacy-based work practice training in the boarding school environment. Munaddhomah: Jurnal Manajemen Pendidikan Islam **4**(1), 1–11 (2023)
25. Reggi, L., Dawes, S.S.: Creating open government data ecosystems: network relations among governments, user communities, NGOs and the media. Gov. Inf. Q. **39**(2), 101675 (2022)
26. Baltzersen, R. K.: Cultural-Historical Perspectives on Collective Intelligence. Cambridge University Press (2022)

Knowledge Transfer Practices: An Insight from Reviewers and Software Engineers Involved in Modern Code Review Projects

Ismail Bello(✉), Mazida Ahmad, and Maslinda Mohd Nadzir

Institute for Advanced and Smart Digital Opportunities (IASDO), School of Computing, Universiti Utara Malaysia, Sintok, Malaysia
esmailbello@gmail.com

Abstract. Effective knowledge transfer between reviewers and developers is essential for the success of Modern Code Review (MCR) process order to reduce waiting time in software development, cutting costs, and enhancing learning experiences. Reviewers play a critical role in assessing code defects, quality efficiency, and alignment with project objectives by providing valuable feedback and suggestions. However, motivation to facilitate knowledge transfer often remains a challenge for reviewers during the MCR process. This study addresses the limited research on knowledge transfer in MCR by conducting a systematic literature review (SLR) to identify and review 13 key factors influencing the process. These factors are categorized into individual (e.g., absorptive capability, motivation, trust), organizational (e.g., communication, feedback mechanisms, leadership), and technological (e.g., infrastructure, collaboration tools) aspects. Experts within the domain reviewed the identified factors. The findings provide practical guidelines for improving knowledge transfer practices during MCR, contributing to enhanced project outcomes, enriched learning experiences, and advancements in software development research.

Keywords: Knowledge Transfer · Software Development Project · Modern Code Review

1 Introduction

Knowledge serves as a fundamental asset for achieving success in sustainable software development engineering. Numerous studies have highlighted the critical importance of effective knowledge transmission for organizational success [1, 2]. However, in the context of Modern Code Review (MCR), various factors, including a lack of motivation, may either encourage or discourage knowledge-sharing among reviewers [3]. MCR is recognized as a crucial practice that ensures software quality and fosters knowledge sharing between experienced and novice developers [4–6]. Conducted using tools like CodeFlow and Gerrit [7], MCR involves reviewers thoroughly evaluating submitted code for correctness, efficiency, adherence to standards, and overall quality, while providing constructive feedback to align the code with project objectives [7].

Z. Mat Aji et al. (Eds.): KMICe 2024, ISEM 49, pp. 53–66, 2025.
https://doi.org/10.1007/978-3-031-91485-0_5

Beyond quality assurance, MCR serves as an effective platform for knowledge transfer, enabling reviewers to share expertise, best practices, and insights that foster continuous learning and improvement within the development team. During the MCR process, reviewers assess newly proposed code changes, identify defects, and provide actionable feedback that authors must address before code integration [5, 8]. Despite its advantages, knowledge transfer during MCR poses challenges for reviewers and authors, such as insufficient documentation, difficulties in reviewer selection, and context-switching [6].

Knowledge transfer in MCR is pivotal for mitigating various inefficiencies, including defects, cognitive load, rework, and delays [3, 9]. However, inadequate knowledge transfer has been identified as the second most significant obstacle faced by software engineers globally, jeopardizing the long-term success of software projects [10]. While previous studies have investigated knowledge transfer in software engineering [11, 12] limited research has specifically addressed its role within MCR [13–15].

To address this gap, this study aims to conduct a Systematic Literature Review (SLR) to identify and validate a set of unique factors influencing knowledge transfer among reviewers and developers in MCR projects. The goal is to minimize waiting time and enhance overall process efficiency. The paper is structured as follows: Sect. 2 provides a summary of relevant background and literature. Section 3 details the methodology adopted for the SLR. Section 4 presents key findings, along with an in-depth discussion of results. Section 5 concludes the study, addressing limitations and suggesting future research directions. Finally, Sect. 6 highlights the study's notable contributions.

2 Background and Related Works

The global economy has transitioned from an industrial to a knowledge-based one, emphasizing knowledge as a critical resource for organizational success [16, 17]. In software development engineering, knowledge transfer is particularly significant due to the knowledge-intensive nature of software development. It involves disseminating expertise within teams, where the willingness of software engineers to share knowledge depends on various factors. Effective knowledge transfer is linked to a software system's ability to meet user requirements [18, 19].

Software engineering projects encompass socio-technical activities such as specification, verification, simulation, development, testing, and MCR [1, 2]. These activities often create opportunities for delays and inefficiencies, termed "project-related waiting waste" [9, 20, 21]. Historically, Fagan introduced a formal code inspection process in 1976 to detect defects systematically, but its intensive and structured nature limited its industrial adoption [22, 23]. In recent years, developers have adopted MCR, an informal yet collaborative approach that not only identifies defects but also enhances code quality, team awareness, and knowledge sharing among team members [5, 6, 20, 23, 24].

MCR is a systematic and collaborative process in which developers submit code contributions for review by peers. The process involves evaluating the code for correctness, efficiency, readability, and quality, addressing defects and identifying areas for improvement [13]. MCR also facilitates communication, feedback, and collaboration through tools such as Gerrit and CodeFlow, making it a valuable platform for knowledge transfer, particularly for junior developers [22, 25].

Despite its benefits, MCR can lead to delays, including waiting time, which negatively affects productivity and project timelines [9, 20, 21]. Addressing these delays requires improved knowledge transfer among reviewers and developers [26]. However, limited research has specifically examined knowledge transfer within the MCR context, highlighting a gap in understanding the factors that influence this process [5, 6]. While prior studies have explored knowledge transfer in broader software engineering activities, little attention has been given to MCR, particularly from the perspective of reviewers. This lack of research has resulted in an absence of clear guidelines for facilitating effective knowledge transfer in MCR settings [20].

This study seeks to address these gaps by identifying and validating factors influencing knowledge transfer among reviewers and developers in MCR. Through a systematic literature review (SLR) and expert review, the research aims to provide a comprehensive framework for understanding and enhancing knowledge transfer in MCR, refining the nomenclature, expressions, and classifications of relevant factors.

3 Methodology

A variety of research activities have been undertaken to systematically identify and organize the factors that influence Modern Code Reviewers to participate in knowledge transfer, specifically targeting the reduction of waiting during the code review process. The methods applied in this study are detailed in the following subsections in accordance with the research question and objectives illustrated in Table 1.

Table 1. Research Questions Goal and Objectives.

Goals	Research Questions	Research Objectives
Systematic Literature Review	What are the key factors for implementing knowledge transfer practices among reviewers and developers involved in MCR projects?	To identify and list factors for implementing knowledge transfer among reviewers and developers involved in MCR project
Expert Validation	How can we validate the factors identified for implementing knowledge transfer practices by software reviewers and engineers involved in MCR projects?	To validate the identified factors for implementing knowledge transfer practices by software reviewers and engineers involved in MCR project

3.1 Systematic Literature Review

A Systematic Literature Review (SLR), following the methodology proposed by Kitchenham and Brereton [27], was conducted to gather and analyze relevant data systematically. The SLR approach ensures a structured and unbiased evaluation of factors influencing knowledge transfer during MCR, providing a detailed examination of current research on the topic. The primary research question guiding this study was:

RQ: What are the key factors for implementing knowledge transfer practices among reviewers and developers involved in MCR projects?

Key terms, such as "knowledge transfer," "modern code review," and "software project engineering," along with their synonyms, were identified and combined using Boolean operators ("OR" and "AND") to refine the search. The search process utilized reputable databases, including AIS, Elsevier, IEEE, Springer, Taylor & Francis, Scopus, and Web of Science, focusing on literature published between 2010 and 2023 to ensure relevance to rapidly evolving technologies [28].

Selection criteria were established based on inclusion and exclusion parameters to ensure only pertinent studies were analyzed. This rigorous process provided a comprehensive foundation for identifying factors influencing knowledge transfer in MCR projects.

i. Key terms and search query: The study's key terms include factors, modern code review, knowledge transfer, and software engineering projects to identify the factors for implementing knowledge transfer practices by reviewers and engineers involved in MCR projects. The search query is formulated using the specified keywords, substitutes, and Boolean operators and is presented in Table 2.

Table 2. Search Strings.

"Factors" OR "Determinants" OR "Elements" AND "Knowledge Transfer" OR "Knowledge Dissemination" OR "Knowledge Sharing" OR "Knowledge Distribution" OR "Knowledge Exchange" AND "Software Engineering Projects" OR "Software Development Project" OR "Information Technology Projects" AND "Modern Code Review" OR "Modern Code Inspection" OR "Lightweight Code Review" OR "Contemporary Code Review" OR "Code Review" OR "Code Inspection"

ii. Database used for data collection: Information is collected from various sources that publish papers related to knowledge transfer in software development projects, as shown in Table 3. These databases were selected based on their extensive coverage of influential peer-reviewed publications.

iii. Selection of Paper: The paper selection process consists of two main steps. The first step is Inclusion and Exclusion, where papers are carefully chosen or eliminated based on predefined criteria. The second step is the Quality Assessment, where the selected papers from the previous step are evaluated for their quality using the checklist provided by Kitchenham and Brereton [27]. The study selection process is illustrated in Table 4.

Table 3. Paper Sources.

Sources	Website
AIS Library	https://aisel.aisnet.org/
Elsevier	https://www.elsevier.com/
Emerald	https://www.emerald.com/
IEEE	http://ieeexplore.ieee.org/
Scopus	http://www.scopus.com/
Springer	http://www.springer.com/
Taylor & Francies	https://www.tandfonline.com/
Wiley	http://onlinelibrary.wiley.com/
Web of Science	http://www.wevofknowledge.com/

Table 4. Selection Criteria.

Inclusion criteria	Research papers written in English
	Research papers published between 2013–2023
	Research papers available in book chapters, conferences, journals, and workshops
Exclusion criteria	Research papers with irrelevant titles
	Research papers without abstracts
	Identical research papers
	Research papers without key defined terms

Despite the comprehensive inclusion and exclusion criteria, assessing the "quality" of primary studies is essential. The checklist outlined by Kitchenham and Brereton [27] has been utilized to evaluate the quality of the studies, as shown in Table 5. Additionally, the studies selected following the initial inclusion and exclusion criteria are further evaluated using the checklist provided by [27].

After identifying the critical studies and evaluating their quality, data is extracted from the selected papers in accordance with the guidelines provided by Kitchenham and Brereton [27]. Following a meticulous data extraction process, the extracted information is synthesised. Techniques such as data coding, continuous data comparison, and grounded theory methodologies [29] are employed to categorize data units and compile a comprehensive list of variables that influence the implementation of knowledge transfer by MCRs. The process of synthesizing data is shown in Fig. 1.

iv. Data Extraction: The data extraction form is structured into two main sections. The first section includes general details about the papers, such as their identification and

Table 5. Assessment of Paper Quality.

Questions	Answers
Is reportage both effective and logical?	Yes/No/Partially
Do various elucidations appear?	-Do-
Are the established points clear?	-Do-
Has research expanded facts?	-Do-
How do the findings advance the literature?	-Do-

methodology. The second section focuses on specific factors that influence knowledge transfer for MCRs. Table 6 shows the form sample.

Table 6. Extraction of Data Form [29].

Features of data	Unique identifier
	Title
	Authors
	Year
	Category (book chapter/conference/journal/report/workshop)
	Publisher Springer, IEEE, Elsevier, etc.)
	Methodology (quantitative/qualitative)
	Selection (inclusion/exclusion) quality
Research question	Factors for implementing knowledge transfer practices among reviewers and developers involved in MCR projects?

v. Data Synthesis: After meticulous data extraction, the obtained data is synthesised by coding, repeated data comparisons, and memoing based on grounded theory [29] to categorise data units and provide a unique list of factor that influence knowledge transfer for MCRs.

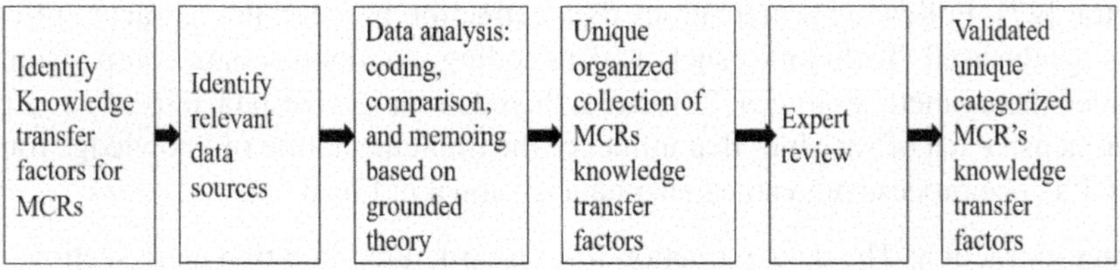

Fig. 1. Process of synthesising data.

3.2 Expert Review

Once the factors influencing knowledge transfer in MCR were compiled, experts were engaged to validate the terminologies, descriptions, and categorizations and suggest additional factors. Experts with 5–10 years of experience in software development, particularly in MCR, were selected through a purposive sampling method. Industry experts were sourced via LinkedIn, email lists, and professional contacts, while academic experts were referred by senior faculty or colleagues (see Table 7. For expert background). Criteria for selection included active involvement in MCR or software engineering projects and substantial academic or professional credentials, such as advanced degrees and relevant publications [30].

A web-based survey was developed to present the factors derived from prior studies. Participants rated the importance of each factor on a 5-point Likert scale and provided feedback for improvement. They could also propose additional factors to enhance the understanding of knowledge transfer in MCR. Verification involved 15 experts, 10 from industry and 5 from academia who reviewed and validated the factors. All participants had at least five years of relevant experience, with academic experts holding Ph.D. degrees in computer science, IT, or software engineering. This expert review ensured the robustness and applicability of the identified factors [30].

Table 7. Background of Expert

Expert Domain	Expert ID	Qualification	Organization	Position	Years of Experience
Academic	Exp1	PhD	A.B.U Zaria	Director I.C. T	9
	Exp2	PhD	UUM Malaysia	Senior Lecturer	8
	Exp3	PhD	AUST Nigeria	H.O. D	11
	Exp4	PhD	Nile University	Associate Prof	16
	Exp5	PhD	Baze University	Lecturer	5
Industry	Exp6	BSc	Ynet Interactive Ltd	Software Tester	5
	Exp7	MSc	Digital Planet X	Code reviewer	5
	Exp8	MSc	Wootlab Innovation	Code reviewer	8

(continued)

Table 7. (*continued*)

Expert Domain	Expert ID	Qualification	Organization	Position	Years of Experience
	Exp9	BSc	Jaraa Software Ltd	Software Tester	7
	Exp10	PhD	Zebitech IT Ltd	Project leader	14
	Exp11	BSc	Media Hooch	Software Tester	12
	Exp12	MSc	The Can Hub	Software Tester	6
	Exp13	BSc	Circle ITC Ltd	Code reviewer	15
	Exp14	MSc	iDegin Technology	Code reviewer	10
	Exp15	BSc	Web Developers Ltd	Developer	8

4 Results

The literature highlights several factors impacting knowledge transfer in software engineering, including trust, motivation, communication skills, and feedback mechanisms [1, 15, 28, 31–33]. Despite these insights, a notable gap exists regarding the role of knowledge transfer in MCR, particularly focusing on the interactions between reviewers and software engineers during the code review process. While general studies address software engineering, there is a critical need for empirical research examining the specific factors that influence knowledge transfer in MCR, as reviewers' expertise is pivotal to project success.

The analysis also reveals potential bias, such as self-selection, where reviewers who declined participation might have differing perspectives on knowledge transfer factors [15, 20]. Nonetheless, the study adheres to international best practices. This section provides an overview of findings, divided into insights from the SLR process and expert recommendations.

4.1 Outcome of SLR and Extracted Results

A systematic review identified 156 papers from databases such as AIS, Elsevier, Emerald, IEEE, Springer, and Taylor & Francis. After excluding 49 non-English papers, abstracts, and general reports lacking relevance, 107 publications remained. Abstract screening excluded 50 more papers, leaving 57 for further analysis. A detailed review of introductions and full texts, guided by inclusion criteria, eliminated 22 papers that lacked focus on MCR or relevant software engineering characteristics. Seven additional articles were excluded based on quality assessment, resulting in 28 selected articles.

These 28 papers underwent detailed evaluation to ensure consistency in terminology, measurement items, and alignment with prior research. A comparative analysis based on research questions, objectives, frameworks, and results yielded a categorized list of 13 factors influencing knowledge transfer in MCR. These factors were grouped into three categories: Individual, Organizational, and Technological.

i. Individual Aspect: From an individual perspective, the most prominent dimension for reviewers and developers engaged in the MCR process is identified [20, 34]. The factors encompassed within this category include absorptive capability [31, 35–39] motivation [13, 20, 28, 31, 35–38, 40–46] reciprocity [44, 47, 48] time Pressure [13, 20, 28, 39, 40, 48] and trust [28, 31, 35–42, 46–51].
ii. Organizational Aspect: The MCR process entails complex interactions among reviewers and authors within the project setting [20, 34]. This aspect encompasses factors such as communication skill [15, 20, 28, 31, 36, 39, 41, 43, 47, 49, 52, 53] culture [20, 28, 31, 33, 35, 36, 39, 41, 42] feedback mechanism [13, 15, 20, 36, 55] reward system [28, 35, 36, 39, 43, 54] team structure [20, 36, 39–41, 45, 50, 52, 55] leadership style [1, 28, 35, 36, 42, 50, 51, 54, 55].
iii. Technological Aspect: The technological aspect facilitates the successful execution of the MCR process [20, 34]. This category encompasses factors such as technological infrastructure [1, 31, 39, 48] and project collaboration tools [1, 28, 33, 36, 39, 41, 45, 53].

4.2 Outcomes of Expert Review

After collecting the data provided by the experts, the next step was to analyse the data using SPSS and test it using a one-sample t-test, as shown in Table 8. This test was employed to compare the population mean (X) with the posited value (X mean) of 3, which was represented by the high significance value on the 5-point Likert scale (1 = very low significance to 5 = very high significance). Hence, the sets of testing values for the variables are as specified below:

a. If the average value of the suggested predictor is greater than 3, this factor is deemed significantly relevant, meaning it impacts knowledge transfer practices [30, 56].
b. Excluded: If the average value of the suggested predictor is less than or equal to 3, the predictor is deemed insignificant, meaning it does not impact the knowledge transfer practices of MCRs and will be omitted from the model [30, 56].

Test value = three significance: 95% level of confidence.

The experts assessed the 13 identified elements and knowledge transfer constructs as crucial and relevant in determining the effectiveness of the MCR process. A component with a mean value greater than 3, a p-value less than 0.05, a t-value greater than 1.96, and a confidence level that does not include zero was considered for inclusion [30, 56]. Based on the data, the communication skill construct had the highest average, while the leadership style construct had the lowest.

Table 8. Results from expert verification

Factors	Mean	Standard Deviation	Decision
Absorptive capability	4.13	0.775	Supported & included
Motivation	4.00	0.892	Supported & included
Reciprocity	3.81	0.885	Supported & included
Time pressure	4.31	0.854	Supported & included
Trust	3.81	0.911	Supported & included
Communication skill	4.51	0.854	Supported & included
Culture	3.81	0.929	Supported & included
Feedback mechanism	4.25	1.263	Supported & included
Reward system	4.19	0.944	Supported & included
Team structure	3.69	0.964	Supported & included
Leadership style	3.52	0.892	Supported & included
Technological infrastructure	3.75	1.109	Supported & included
Project collaboration tool	3.81	0.946	Supported & included

4.3 Comments and Suggestions from Experts

Experts provided valuable insights into factors influencing knowledge transfer in MCR. Expert 6 emphasized MCR's role in enhancing software quality, with reviewer input critical for improving the final product. Expert 15 stressed the importance of timely feedback in meeting code requirements, while Expert 5 noted that financial incentives often motivate reviewers. Conversely, Expert 7 highlighted authors' absorptive capacity as a key determinant of whether feedback is implemented or the code is rewritten, underlining the importance of communication and retention.

Expert 11 observed that many reviewers are motivated more by technical challenges than financial rewards. Experts 8 and 14, however, contended that leadership style and team structure have minimal impact, contrasting with prior literature. Expert 2 flagged collaboration challenges due to diverse project management tools, leading to issues like missing code or incomplete values. Expert 4 emphasized trust as a crucial factor, as reviewers are more likely to share expertise when they believe it will be used constructively. Finally, Expert 1 called for further research into how factors like communication, rewards, infrastructure, trust, and the digital divide interact, to enhance MCR effectiveness and achieve higher-quality software development.

5 Conclusion and Future Directions

This study investigates knowledge transfer practices as a sustainable strategy for reviewers and software engineers in MCR. It identifies key factors influencing knowledge transfer, such as absorptive capabilities, motivation, reciprocity, time pressure, and trust, which are crucial for enhancing code quality and learning experiences. By presenting a categorized inventory of these factors, the study highlights their significance in achieving effective knowledge transfer during the MCR process.

While existing models address aspects of software engineering and knowledge transfer, few focus specifically on reviewers in the MCR context. The identified factors are derived from a SLR and expert validation. Future research should expand this work through empirical studies and incorporate industry-sourced factors to strengthen its applicability. Despite efforts to include all relevant studies, some literature may have been unintentionally omitted.

6 Contribution

This research significantly contributes to the field of software engineering by emphasizing sustainable MCR practices and systematically categorizing factors influencing knowledge transfer among reviewers and software engineers. Academically, it sheds light on the dynamics of knowledge transfer within MCR frameworks, identifying key factors across individual, organizational, and technological aspects. These insights are valuable for reviewers, authors, and developers, enhancing their learning experiences and collaboration within MCR processes.

References

1. Bello, I., Ahmad, M., Mohd Nadzir, M.: Determinants of knowledge transfer for information technology project managers: a systematic literature review. Interdiscip. J. Inf. Knowl. Manag. **18**, 871–891 (2023)
2. Mtsweni, E., Gorejena, K.: Team barriers to tacit knowledge sharing in software development project teams. Electron. J. Knowl. Manage. **21**(1), 59–72 (2023)
3. Zanaty, F.E., Hirao, T., McIntosh, S., Ihara, A., Matsumoto, K.: An empirical study of design discussions in code review. In: Proceedings of the 12th ACM/IEEE International Symposium on Empirical Software Engineering and Measurement - ESEM 2018, pp. 1–10 (2018)
4. Fatima, N., Nazir, S., Chuprat, S.: Knowledge sharing is a key sustainable practice on risk: An insight from modern code review. In: IEEE 6th International Conference on Engineering Technologies and Applied Sciences (ICETAS), pp. 1–6 (2019)
5. Bosu, A., Carver, J.C., Bird, C., Orbeck, J., Chockley, C.: Process aspects and social dynamics of contemporary code review: insights from open source development and industrial practice at Microsoft. IEEE Trans. Softw. Eng. **43**(1), 56–75 (2017)
6. MacLeod, L., Greiler, M., Storey, M.-A., Bird, C., Czerwonka, J.: Code reviewing in the trenches: challenges and best practices. IEEE Softw. **35**(4), 34–42 (2018)
7. Silva, W.A.C.P., Sangkavi, S.V., Wijerathne, L.G.A.T.D, Wijesundara, W.M.K.H, Reezan, S.A.: The effectiveness of code reviews on improving software quality: an empirical study. Int. J. Recent Technol. Eng. **12**(2), 1–10 (2023)
8. Ebert, F., Castor, F., Novielli, N., Serebrenik, A.: Confusion in code reviews: reasons, impacts, and coping strategies. In: IEEE 26th international conference on software analysis, evolution and reengineering (SANER), pp. 49–60. IEEE Xplore (2019)
9. Alahyari, H., Gorschek, T., Berntsson Svensson, R.: An exploratory study of waste in software development organizations using agile or lean approaches: a multiple case study at 14 organizations. Inf. Softw. Technol. **105**, 78–94 (2019)

10. Novikova, L.: State of software development in 2019 (2019). https://medium.com/onebar/poor-knowledge-sharing-is-the-second-biggest-challenge-for-software-development-teams-a4843f9b9aa
11. Dittrich, Y., Vaucouleur, S., Giff, S.: ERP customization as software engineering: knowledge sharing and cooperation. IEEE Softw. **26**(6), 41–47 (2009)
12. Shull, F., et al.: Knowledge-sharing issues in experimental software engineering. Empir. Softw. Eng. **9**(1/2), 111–137 (2004)
13. Fatima, N., Nazir, S., Chuprat, S.: Knowledge sharing is a key sustainable practice on risk: An insight from modern code review. In: IEEE 6th International Conference on Engineering Technologies and Applied Sciences (ICETAS), pp. 1–6 (2019)
14. Pascarella, L., Spadini, D., Palomba, F., Bruntink, M., Bacchelli, A.: Information needs in contemporary code review. In: Proceedings of the ACM on Human-Computer Interaction, 2(CSCW), pp. 1–27, ACM (2018)
15. Fatima, N., Nazir, S., Chuprat, S.: Understanding the impact of feedback on knowledge sharing in modern code review. In: IEEE 6th International Conference on Engineering Technologies and Applied Sciences (ICETAS), pp. 1–5. IEEE (2019)
16. Rehman, M., Mahmood, A.K., Salleh, R., Amin, A.: Work design characteristics and knowledge sharing behaviour among software engineers. In: International Conference on Computer and Information Sciences (ICCOINS), pp. 1–5, IEEE (2014)
17. Ellahi, A.A., Mushtaq, R.: Probing factors affecting knowledge sharing behavior of Pakistani bloggers. Electron. J. Inf. Syst. Dev. Count. **45**(1), 1–14 (2011)
18. Orpipath, N.: Analyzing knowledge sharing behavior for security guards in hospitals. Int. J. Manage. Sci. Bus. Adm. **5**(1), 39–49 (2018)
19. O'Neill, E.: User-Developer Cooperation in Software Development. Springer, Heidelberg (2012). https://doi.org/10.1007/978-1-4471-0355-4
20. Fatima, N., Nazir, S., Chuprat, S.: Knowledge sharing factors for modern code review to minimize software engineering waste. Int. J. Adv. Comput. Sci. Appl. **11**(1) (2020)
21. Sedano, T., Ralph, P., Péraire, C.: Software development waste. In: IEEE/ACM 39th International Conference on Software Engineering (ICSE), pp. 130–140. IEEE (2017)
22. Fagan, M.E.: Design and code inspections to reduce errors in program development. IBM Syst. J. **38**(2.3), 258–287(1999)
23. Kovalenko, V., Tintarev, N., Pasynkov, E., Bird, C., Bacchelli, A.: Does reviewer recommendation help developers? IEEE Trans. Softw. Eng. **46**(7), 710–731 (2020)
24. Greiler, M., Bird, C., Storey, M.C., MacLeod, L., Czerwonka, J.: Code reviewing in the trenches: understanding challenges, best practices and tool needs (2016)
25. Bacchelli, A., Bird, C.: Expectations, outcomes, and challenges of modern code review. In: 35th International Conference on Software Engineering (ICSE), pp. 712–721. IEEE (2013)
26. Winters, T., Manshreck, T., Hyrum, W.: Software Engineering at Google: Lessons Learned from Programming Over Time. O'reilly (2020)
27. Kitchenham, B., Brereton, P.: A systematic review of systematic review process research in software engineering. Inf. Softw. Technol. **55**(12), 2049–2075 (2013)
28. Anwar, R., Rehman, M., Wang, K.S., Hashmani, M.A.: Systematic literature review of knowledge sharing barriers and facilitators in global software development organizations using concept maps. IEEE Access **7**, 24231–24247 (2019)
29. Hoda, R.: Socio-technical grounded theory for software engineering. IEEE Trans. Softw. Eng. (2021)
30. SarkinTudu, S.M., Abd Wahab, A., Ibrahim, H.H.: Predicting key predictors of project desertion in blockchain: Experts' verification using one-sample t-test. Interdiscip. J. Inf. Knowl. Manage. **17**, 497–521 (2022)
31. Srisuksa, N., Wiriyapinit, M., Bhattarakosol, P.: Software project managers' knowledge transfer: an in-depth interview. Electron. J. Knowl. Manage. **20**(2), 78–92 (2022)

32. Habeh, O., Thekrallah, F., Salloum, S. A., Shaalan, K.: Knowledge sharing challenges and solutions within software development team: a systematic review. Stud. Syst. Decis. Control 121–141 (2020)
33. Matthew, A., Dhillon, J.S.: A framework for transferring knowledge between expatriates and local employees in IT-based organisations. In: 8th International Conference on Information Technology and Multimedia (ICIMU), pp. 114–119. IEEE (2020)
34. Nagaria, B., Hall, T.: How software developers mitigate their errors when developing code. IEEE Trans. Softw. Eng. (2020)
35. Akgün, A.E., Keskin, H., Ayar, H., Okunakol, Z.: Knowledge sharing barriers in software development teams: a multiple case study in Turkey. Kybernetes **46**(4), 603–620 (2017)
36. Ghobadi, S.: What drives knowledge sharing in software development teams: a literature review and classification framework. Inf. Manage. **52**(1), 82–97 (2015)
37. Shaqrah, A.A., Al, M.R.: Examining tacit knowledge transfer processes for enterprise system projects success: a conceptual framework. J. Syst. Integr. **9**(4), 29–39 (2018)
38. Nurye, S.A., Molla, A., Desta, T.A.: Factors influencing knowledge transfer in onshore information systems outsourcing in Ethiopia. Af. J. Inf. Syst. **11**(4), 5 (2019)
39. Karagoz, Y., Whiteside, N., Korthaus, A.: Context matters: enablers and barriers to knowledge sharing in Australian public sector ICT projects. J. Knowl. Manage. **24**(8), 1921–1941 (2020)
40. Khoza, L. T.: Measuring knowledge sharing behavior among software development teams. South Af. J. Inf. Manage. **21**(4) (2018)
41. Nidhra, S., Yanamadala, M., Afzal, W., Torkar, R.: Knowledge transfer challenges and mitigation strategies in global software development—a systematic literature review and industrial validation. Int. J. Inf. Manage. **33**(2), 333–355 (2013)
42. Davidavičienė, V., Al Majzoub, K., Meidute-Kavaliauskiene, I.: Factors affecting knowledge sharing in virtual teams. Sustainability **12**(17), 6917 (2020)
43. Khoza, L.T., Pretorius, A.B.: Factors negatively influencing knowledge sharing in software development. SA J. Inf. Manage. **19**(1), 1–9 (2017)
44. Rese, A., Kopplin, C.S., Nielebock, C.: Factors influencing members' knowledge sharing and creative performance in coworking spaces. J. Knowl. Manage. (2020, ahead-of-print)
45. Sani, N.S.B.M., Arshad, N.I.B.: Towards a framework to measure knowledge transfer in organizations. In: International Symposium on Mathematical Sciences and Computing Research (iSMSC), pp. 255–262, IEEE (2015)
46. da Silva, F.P., Mosquera, P., Soares, M.E.: Factors influencing knowledge sharing among IT geographically dispersed teams. Technol. Forecast. Soc. Chang. **174**, 121299 (2022)
47. Alwahdani, A.: The impact of trust and reciprocity on knowledge exchange: a case study in IT outsourcing. J. Inf. Syst. Eng. Manage. **4**(1) (2019)
48. Biloslavo, R., Lombardi, R.: Knowledge transferring and small and medium enterprise's (SME's) effectiveness: emerging insights and future directions. Bus. Process. Manage. J. **27**(6), 1747–1774 (2021)
49. Pietruszka-Ortyl, A., Ćwiek, M., Ziębicki, B., Wójcik-Karpacz, A.: Organizational culture as a prerequisite for knowledge transfer among IT professionals: the case of energy companies. Energies **14**(23), 8139 (2021)
50. Imam, H., Zaheer, M.K.: Shared leadership and project success: the roles of knowledge sharing, cohesion and trust in the team. Int. J. Project Manage. **39**(5), 463–473 (2021)
51. Razzak, M. A., Ahmed, R.: Knowledge sharing in distributed agile projects: techniques, strategies and challenges. In: Federated Conference on Computer Science and Information Systems, pp. 1431–1440. IEEE (2014)
52. Ghobadi, S., Mathiassen, L.: Risks to effective knowledge sharing in agile software teams: a model for assessing and mitigating risks. Inf. Syst. J. **27**(6), 699–731 (2016)
53. Togola, A., Ahmed, S., Jadaan, T.: Barriers of knowledge transfer between globally distributed teams in ICT product development. Scholarspace.manoa.hawaii.edu (2019)

54. Jiang, G., Xu, Y.: Tacit knowledge sharing in IT R&D teams: nonlinear evolutionary theoretical perspective. Inf. Manage. **57**(4), 103211 (2020)
55. Avença, I., Domingues, L., Carvalho, H.: Project managers soft skills influence in knowledge sharing. Proc. Comput. Sci. **219**, 1705–1712 (2023)
56. Yahaya, H., Nadarajah, G.: Determining key factors influencing SMEs' performance: a systematic literature review and experts' verification. Cogent Bus. Manage. **10**(3), 2251195 (2023)

Empowering Knowledge Management Innovation with ChatGPT and Sora

Liu Yonggang, Hapini Awang(✉), and Nur Suhaili Mansor

Institute for Advanced and Smart Digital Opportunities, School of Computing, Universiti Utara Malaysia, 06010 Sintok, Kedah, Malaysia
hapini.awang@uum.edu.my

Abstract. The abstract should summarise the contents of the paper in short terms, i.e. 150–250 words. In November 2022, OpenAI released ChatGPT, which can interact and dialogue with humans and performs excellently in understanding natural language, scalability, writing, programming, responsiveness, etc. In February 2024, Sora was born (though not yet available to the public), which performs very excellently in text-to-video (T2V), ambient rendering, generating complex scenes, maintaining continuity and many other areas. The research objectives of this study are to explore ChatGPT and Sora's potential influences on Knowledge Management, mainly focusing on its possible impact on enterprise data visualisation, enriching knowledge management innovations, industry landscape, effectiveness, staffing structure, economic costs, etc. This study mainly adopts an exploratory research method and a functional analysis method. In the future, GAI, represented by ChatGPT and Sora, will potentially influence some enterprises' KM in various dimensions, such as updating enterprise management philosophies, affecting the enterprise's strategic planning and goal-setting, reducing knowledge management economic costs, enriching knowledge management innovations, reshaping partial industries landscape, improving knowledge management effectiveness, promoting human-computer interaction, reinventing the staffing structure, modernising the enterprise's knowledge management system, affecting companies' traditional organisational structures. Unemployment problems may also be a reality that has to be faced in the aftermath of GAI's onslaught. However, this study is exploratory and still needs further empirical research and data verification for authenticity in the future.

Keywords: Knowledge Management · ChatGPT · Sora · generative artificial intelligence (GAI) · Innovation

1 Introduction

In November 2022, OpenAI released ChatGPT, which can interact and dialogue with humans and perform excellently in understanding natural language, scalability, writing, programming, responsiveness, etc. ChatGPT has garnered significant attention from research institutions, customer service, healthcare, education and other industries within a brief period [1]. In February 2024, Sora was born (though not yet available to the public),

Z. Mat Aji et al. (Eds.): KMICe 2024, ISEM 49, pp. 67–78, 2025.
https://doi.org/10.1007/978-3-031-91485-0_6

which performs very excellently in text-to-video (T2V), ambient rendering, generating complex scenes, maintaining continuity and many other areas. ChatGPT and Sora are outstanding representatives of the most advanced generative artificial intelligence (GAI) in the contemporary world.

Knowledge management (KM) involves using the intellectual capacity of an organisation in a methodical and structured way to enhance efficiency, secure a competitive edge, and stimulate innovation [2]. In today's competitive environment, innovation is widely recognised as a fundamental advantage for the survival and success of organisations [3]. KM is a kind of management philosophy and method emerging in the era of the knowledge economy, which draws on diversified information technologies, various economic theories and different enterprise management ideologies to systematically plan and manage the creation, sharing, preservation, application and updating of the knowledge assets of the whole enterprise, each team and in-dividual employees, and retrieve and use them when needed to create performances and values for the enterprise and safeguard the interests of the enterprise. KM contributes to the enterprise's ability to fully utilise its knowledge assets against market uncertainty and explicit and implicit risks to adapt to market variations and safeguard its benefits.

In an era characterised by the growing interconnectedness of digital globalisation, it is essential to explore how enterprises can embrace digital innovation and utilise knowledge management to enhance their competitive edge in the world-wide marketplace [4]. Knowledge is a distinctive source for companies to attain competitive superiority [5]. KM methods play a crucial role in enhancing AI systems and algorithms [6]. KM is an essential element of enterprise management, and it is precious for enterprises to utilise their knowledge to create a sustainable competitive advantage. Applying the collective wisdom and knowledge wealth of the enterprise will help to improve the enterprise's adaptability and innovativeness to adapt to the fierce market competition. ChatGPT and Sora can play an important role in enterprise data visualisation, industry landscape transformation, process optimi-sation, cost reduction, efficiency improvement, etc.

However, previous studies about ChatGPT and Sora's potential influence on knowledge management innovation are minimal, especially their possible influence in enterprise data visualisation, human-computer interaction, industrial landscape, workflow, efficiency and time, temporal space, and others are still very scarce. The research objectives of this study are to explore ChatGPT and Sora's potential influences on Knowledge Management, mainly focusing on its possible impact on enterprise data visualisation, enriching knowledge management innovations, industry landscape, effectiveness, staffing structure, economic costs, etc.

2 Literature Review

KM has been applied in a wide variety of practices and has made great strides, for example, digital financial innovation [7], healthcare [8], Human Resource Management (HRM) [9–11], digitalisation of innovative education [12, 13], hotel management system [14], supply chain management digitisation [15]. With the help of the hybrid SEM-artificial neural network (SEM-ANN) method, Al-Sharafi et al. (2023) revealed that knowledge application had the most significant impact on the long-term sustainability of chatbots with a normalised relevance of 96.9% [16]. In contrast, perceived usefulness was the second most influential factor, with a normalised importance of 70.7%. In pursuit of the best configuration of collaborative and innovative KM, Bai & Li (2020) realised that the greedy algorithm efficiently assigned resources, and these allocation strategies are locally optimal according to the established optimal metric [17]. For technology-driven organisations, implementing knowledge management elements in all firm departments is necessary for growth in a sustainable landscape [18]. Based on data analysis from 738 respondents, Ozpamuk et al. (2023) found that trust is a critical ingredient in knowledge management, and knowledge is an essential and crucial element in job performance. Drawing inspiration from Heinz von Foersters' observations about the self-regulatory processing of the nervous system, Schaurhofer & Peschl (2005) discovered the necessity to understand the empowerment process as a "knowledge process" and to develop a space for experimentative activities [19]. There was a connection between the success of KM and the congruence between KM and business strategies, whilst the winningest KM programs were motivated by intense business demand [20]. Dneprovskaya et al. (2020) considered that intelligent education research emphasised methodological issues in under-taking training practices, while knowledge management techniques and methodologies could supply the needed speed and quality of instructional materials [12]. After analysing the Wald Test and Odds Ratio (EXP(B)), Ahmad (2023) revealed that the adoption of KM significantly influenced workforce performance [21]. AlMul-him (2023) argued that KM practices served as a crucial intermediary between information technology and administrative management, enhancing the effectiveness of e-government initiatives and mitigating risks [22]. Furthermore, Kordab et al. (2020) confirmed the predictions that learning within organisations had a favourable impact on the procedures of acquiring, storing and sharing, as well as on long-term company performance [23]. Zhao et al. (2022) indicated that by leveraging Big Data, neural networks (NN) technology could examine various organisations' knowledge management techniques and diverse knowledge management strategies [24]. Besides, KM also suffered from a lack of incentives for knowledge sharing, culture variances, unsuitable information structure, and other problems [5]. However, GAI research in promoting KM development remains limited. Although KM has been developed for applications in various fields, there is still a paucity of research on how ChatGPT and Sora combine KM.

As far as theories and models are concerned, there have been some attempts by different researchers. Based on Dalkir's knowledge management model and the UTAUT model, empowered leaders significantly impacted subordinates' knowledge-making, sharing and application in the hotel management system [14]. Borrowing from the Enterprise Knowledge Management (EKM) Capability Model and the Knowledge-Based View (KBV), Iqbal (2021) explored the implications of top management's knowledge value, knowledge-orientated culture and knowledge-based awards for innovation's pace and quality [25]. Based on complexity theory, Faucher et al. (2008) presented a new knowledge cognitive system model and non-linear relationships among variables [26]. Fakhar Manesh et al. (2021) surveyed the conceptual framework and trends in knowledge management in the context of Industry 4.0 [27]. Within the KM theoretical framework, Schniederjans et al. (2020) tried implementing KM into the digitisation of supply chain management. Moreover, Shahzad et al. (2020) investigated the knowledge management process's effect on enterprises' sustainability performance by the resource-based view theory [28]. One investigation indicated that Knowledge generation (KG) and knowledge flow (KF) within a company positively affect productivity. In contrast, the storing of knowledge does not have a straight impact on performance. As a result, KM has been combined with different theories to facilitate the development of relevant theoretical studies and frameworks.

In terms of research methodologies, a questionnaire [7, 21, 29], case studies , content analysis [30], bibliometric analysis [27, 31], interview method [32], focus groups [33], expert evaluation [23] and other methods were utilised in KM-relevant studies. For the first time, Iaia et al. (2024) Commercial communications, AI, and KM viewpoints are linked, and their suggested model is centred on people, processes, and technologies [34]. Tightly integrated with Maslow's Extended Hierarchy of Needs by bringing in innovative PKM constructs and systems, Schmitt (2016) created Personal Knowledge Management for Development (PKM4D) Framework [35]. After processing original data from representatives of 182 high-technology enterprises, Lam et al. (2021) discussed the associations between organisational culture, KM, and innovation ability in open, innovative settings [36]. Another investigation suggested that implementing KM methods had a significant role in fostering innovation as a hierarchy, particularly the utilisation of knowledge, which had the most profound effect on promoting innovation [29]. In this part, some previous re-search approaches and innovative points on KM are reviewed, and these will provide some references and suggestions for this study.

There have been some initial explorations of combining AI with KM, for instance, intelligent knowledge management system [13], data management [30], creativity-oriented human resource management (CHRM) [11], Supply chain digitisation [15], entrepreneurial orientation (EO) [9], knowledge-sharing quality (KSQ) [11], promoting consumer behaviour [37] and so forth. In the context of online communities and the advancement of Artificial Intelligence (AI), Olan et al. (2024) developed a consumer behaviour (CB) meta-framework [37]. In education, Kummar et al. (2021) focused on creating a sophisticated KM system that utilised facial recognition technology to authenticate users, and a chatbot was employed to assist users in effortlessly looking for and retrieving the desired documents [13]. Cope et al. (2021) explored knowledge and its evaluation in AI-enabled learning ecologies [38]. Based on information technology competency theory, Wamba-Taguimdje et al. (2020) examined the impact of AI on the effectiveness of companies, particularly by leveraging the economic worth of projects that involve AI-based transformations. In addition to this, Big Data was emphasised for a variety of applications in the field of knowledge management, including but not limited to healthcare knowledge management systems [8], co-creation strategies [31], considerable data-business strategy interconnection [31]. Thus, the prior various applications of AI and Big Data in the field of KM create favourable conditions for the current study to continue to explore the potential impact of ChatGPT and Sora in the domain of KM.

3 Research Methodology

Exploratory research is a type of study that aims to provide answers to a specific issue or investigate a particular phenomenon [39]. It was suitable when understanding a phenomenon or problem was not yet sufficiently well defined, and it could provide ideas for later quantitative research [40]. Exploratory research offers a more realistic feeling of the world and a flexible approach to collecting, analysing and interpreting information [40]. An exploratory study was applied to assess the adoption of Industry 4.0 technology and identify the main factors that either facilitate or impede its deployment in the context of change management [41]. Another investigation used an exploratory study to determine how higher education institutions (HEIs) might enhance their reputation and visibility by utilising new interactive marketing communication platforms [42]. Apart from that, exploratory research methods were introduced in other studies [43–45]. Since Sora has not been made available to the public until now and re-search on the potential influence of ChatGPT and Sora on KM is scarce, this study primarily used an exploratory research approach and reviewed the literature related to KM. Meanwhile, exploratory research is characterised by flexibility, which is more suitable for this study to explore the potential impact of ChatGPT and Sora on KM. Additionally, based on the techniques and functionalities published by ChatGPT and Sora, the functional analysis method was also taken for this study.

4 Exploration of ChatGPT and Sora's Potential Influence on Knowledge Management Innovation

The content of this section is mainly an exploratory description and analysis based on the historical literature, ChatGPT's functions and Sora's functions, which still need further empirical research and data verification for authenticity in the future. This part mainly explores ChatGPT and Sora's Potential Influence on KM Innovation in different aspects.

4.1 Updating Enterprise Management Philosophies

As a management philosophy, KM directly influences the enterprise's strategic decision-making, long-term planning, human resources system, position settings, work arrangements, and many other aspects of the enterprise. To accommodate the development of GAI represented by ChatGPT and Sora, enterprises should place great importance on their core philosophies and values and give full consideration to the impact and shaping of GAI on the enterprises themselves and their competitors, as well as the implications for long-term development. Positively adapting to the technological development trend represented by GAI and timely updating business leaders' ideas, concepts and management philosophies. For example, some camera film manufacturers, mobile phone brands and charger products that used to have a significant market share in the international market can also fall out of favour quickly within a few years. The GAI represented by ChatGPT and Sora will likely affect the traditional KM ideologies and enterprises' original business management philosophies. Therefore, for some enterprises, it is necessary to adapt to the development of GAI in time, update the philosophy of KM, and improve the art of leadership and management.

4.2 Influencing the Enterprise's Strategic Planning and Goals-Setting

In the future, ChatGPT and Sora may influence some enterprise's strategic planning and goal-setting. The future of KM will probably have intelligent, solid attributes, so it may not be wise for some organisations to ignore the contributions of GAI. Formulation of KM strategy planning and goals is one of the significant works of KM, which needs to consider the GAI developing tendency and be clarified about the enterprise's strategic goal of KM, the current status quo, the principal and secondary tasks, the main advantages and the main disadvantages, and so on. There may be multiple targets for enterprise KM, one of which may be how to deliver the appropriate knowledge to the right people at the proper time, improve an organisation's core competencies, and protect an organisation's core knowledge assets. In the era of high-speed development of GAI, the level and quality of enterprise KM may impact future market share, users' satisfaction, and enterprise profitability.

4.3 Accelerating Enterprise Data Visualization

Sora is equipped to generate high-fidelity, detail-rich and smooth transition videos based on text. GAI, represented by ChatGPT and Sora, is potentially accelerating the future visualisations of training materials, business processes, and management data in enterprises. ChatGPT and Sora potentially enrich business archives and accelerate the visualizations of the various resources within an organization.

4.4 Reducing Knowledge Management Economic Costs

AI techniques will be extensively utilised for the generation of PGC (Professional Generated Content) [46]. One analysis predicted favourable changes in producing material, forecasting that Sora would democratise advertising through videos by increasing the accessibility and cost-effectiveness of video production [47]. By leveraging the power of ChatGPT and Sora, the company's economic costs in terms of pre-advertising, strategical planning, purchasing, operational, promotional, and human resource costs are liable to be reduced.

4.5 Enriching Knowledge Management Innovations

The appearance of Sora and other video-making technologies has broadened opportunities for Imaginative Intelligence [46]. ChatGPT and Sora can generate a lot of innovative product proposals, problem solutions, creative videos and so forth in a relatively short period, providing many options for companies and employees. It has to be admitted that an individual's time, energy, and knowledge are limited. In contrast, ChatGPT and Sora already have a vast amount of knowledge, and with the upgrading of the GAI, they can play a role in many complex ways for ordinary people to reach. An example is a fragment of life generated by Sora about robots in the context of cyberpunk (see Fig. 1). Notably, ChatGPT and Sora have the potential to provide many valuable suggestions or ideas in the development of innovative products. Notably, ChatGPT and Sora have the potential to offer many valuable suggestions or ideas in the research and development of innovative products. Besides, they also carry the potential to play a role in quickly identifying industry gaps, diversifying an organization's knowledge repository, enhancing knowledge maps, and so on.

Fig. 1. A fragment generated by Sora about robots in the context of cyberpunk.

4.6 Reshaping Partial Industries Landscape

The commercialisation of AI-created video facilities will intensify, deeply affecting numerous industries such as advertisements, short-form videos, autonomous driving, education and many others [46, 48]. Sora was a T2V model that had the potential to transform a variety of industries (when published to the public) and was expected to revolutionise how contents are generated [49, 50]. ChatGPT, a recently developed language model known as Generative Pretrained Transformer, is expected to have a significant influence on various industries such as medical services, poetry, songs, training in medicine, biomedical studies and writing about science while being in its early stages [51, 52]. The industry needed to carefully consider how to utilise GAI, which may facilitate the formation of new business models [53]. For some companies, the uncertainty of the environment and the ambiguity of the future may make their survival difficult in many respects. The GAI, represented by ChatGPT and Sora, could reshape the landscape of partial industries. How to actively adapt to the direction of the GAI development trend, enhance the learning abilities of enterprises, enrich the contents of KM and strengthen the core competitiveness is very important to some enterprises.

4.7 Improving Knowledge Management Effectiveness

Through effective KM, it is possible to avoid duplication of efforts and increase teamwork efficiency. Especially with the emergence of GAIs such as ChatGPT and Sora, the original duplication of works and unnecessary procedures may be scaled down, and the efficiency of enterprises may be improved. ChatGPT and Sora can efficiently generate explicit knowledge in the form of text, symbols, graphics, video, computer code and more.

4.8 Promoting Human-Computer Interaction

The impressive advancements achieved by Sora, a sophisticated T2V model, indicate a potential study avenue for high-performance simulation using dynamic visuals [54].

Combined with supervised and reinforcement learning technologies, ChatGPT also delivers topic inspection, naturalistic interaction, and emotion analysis to assist participants in gaining a better understanding of their dialogue partners [55]. Compared to some of the GAIs of the past, ChatGPT and Sora have strong human-computer interaction capabilities. Compared to some of the GAIs of the past, ChatGPT and Sora have strong human-computer interaction capabilities. In creating knowledge, staff can make suitable modifications to unsatisfactory solutions or details through the human-computer interaction function.

4.9 Reinventing the Staffing Structure

ChatGPT had the potential to enhance customer support services by offering clients individualised assistance [56]. Sora may also have a significant impact on some employees' established positions. Some knowledge and cross-technology employees may see the importance of their positions strengthened for a certain period, and some low-skill rate, highly replaceable positions may be relatively hard hit in the long run. For some staff, existing roles may have to change to accommodate GAI. In the long-term, the GAI, represented by ChatGPT and Sora, may bring about a certain degree of unemployment problems and lay-off.

4.10 Modernizing the Enterprise's Knowledge Management System

The modernised KM system needs to be adapted to the trend of GAI represented by ChatGPT and Sora so that the company's knowledge can be promptly transformed into innovations and core competencies based on economic advancement and the consumers' requirements. Under the influence of ChatGPT, Sora and other AI programs, some companies' internal reports, manuals, patents, pictures, video materials and other materials may become more voluminous, and the updating of electronic, informational and digital knowledge may be accelerated. One modern KM system can transform fragmented, dispersed and rough knowledge into a retrievable, reusable, easily transformable, highly efficient, clearly structured and standardised knowledge system.

4.11 Affecting Companies' Traditional Organizational Structures

Many business organisational structures exist, such as linear, functional, divisional, matrix, and flat. In the long term, ChatGPT and Sora also have the potential to impact companies' organisational structure and departmental restructuring. As for the depth and breadth of the implications, they still need to be further explored, taking into account many factors such as the macro environment, the enterprise's growth stage, and the situation's realities.

5 Challenges

Quiroga (2024) thought that technological advances posed great ethical and cultural dilemmas [57]. Baidoo-Anu & Owusu Ansah (2023) argued that ChatGPT had a few limitations, including incorrect information creation and privacy challenges [58]. Data bias and security issues [1] may also occur. Unemployment problems may also be a reality that has to be faced in the aftermath of GAI's onslaught.

6 Conclusion

With the advent and upgrading of ChatGPT and Sora, the future of the KM field may also be affected considerably. This study is an exploratory study that focuses on the potential influence of ChatGPT and Sora on various aspects of KM. In the future, GAI, represented by ChatGPT and Sora, will potentially influence some enterprises' KM in multiple dimensions, such as updating enterprise management philosophies, affecting the enterprise's strategic planning and goal-setting, reducing knowledge management economic costs, enriching knowledge management innovations, reshaping partial industries landscape, improving knowledge management effectiveness, promoting human-computer interaction, reinventing the staffing structure, modernising the enterprise's knowledge management system, affecting companies' traditional organisational structures. Unemployment problems may also be a reality that has to be faced in the aftermath of GAI's onslaught. However, this study is exploratory and still needs further empirical research and data verification for authenticity in the future.

References

1. Ray, P.P.: ChatGPT: a comprehensive review on background, applications, key challenges, bias, ethics, limitations, and future scope. Internet of Things Cyber-Phys. Syst. **3**, 121–154 (2023)
2. Serban, A.M., Luan, J.: Overview of knowledge management. New Dir. Inst. Res. **2002**(113), 5–16 (2002)
3. Khodabakhshi, F., Sajad, N.K., Shiargar, M.: The impact of knowledge management on innovation with the mediating role of empowerment. Life Sci. J. **10**, 1385–1390 (2013)
4. Aripin, Z., Fitrianti, N.G., Fatmasari, R.R.: Digital innovation and knowledge management: the latest approaches in international business. A systematic literature review in the Indonesian context. Kriez Acad. **1**(1), 62–74 (2023)
5. Japhet, I., Olaoluwa, A., Olapeju, I.: A review of knowledge management and its application in the contemporary business environment. Afr. J. Bus. Manag. **15**(10), 274–282 (2021)
6. AlGhanem, H., Shanaa, M., Salloum, S., Shaalan, K.: The role of km in enhancing AI algorithms and systems. Adv. Sci. Technol. Eng. Syst. J. **5**(4), 388–396 (2020)
7. Al-Dmour, A., Al-Dmour, R., Rababeh, N.: The impact of knowledge management practice on digital financial innovation: the role of bank managers. VINE J. Inf. Knowl. Manag. Syst. **51**(3), 492–514 (2021)
8. Phan, A.-C., Phan, T.-C., Trieu, T.-N.: A Systematic approach to healthcare knowledge management systems in the era of big data and artificial intelligence. Appl. Sci. **12**(9) 4455 (2022)
9. Baldegger, R., Caon, M., Sadiku, K.: Correlation between entrepreneurial orientation and implementation of AI in human resources management. Technol. Innov. Manag. Rev. **10**(4), 72–79 (2020)
10. Chowdhury, S., et al.: Unlocking the value of artificial intelligence in human resource management through AI capability framework. Hum. Resour. Manag. Rev. **33**(1), 100899 (2023)

11. Younis, R.A.A., Adel, H.M.: Artificial intelligence strategy, creativity-oriented HRM and knowledge-sharing quality: empirical analysis of individual and organisational performance of AI-powered businesses. In: Proceedings of the Annual International Conference of the British Academy of Management (BAM) 2020: Innovating for a Sustainable Future, London (2020)
12. Dneprovskaya, N.V., Komleva, N.V., Urintsov, A.I.: The knowledge management approach to digitalization of smart education. In: Hu, Z., Petoukhov, S.V., He, M. (eds.) AIMEE2018 2018. Advances in Intelligent Systems and Computing, pp. 641–650. Springer International Publishing, Cham (2020). https://doi.org/10.1007/978-3-030-12082-5_58
13. Kummar, R.G., Shetty, S.J., Vishwas, S.N., Upadhya, P.J.V., Munavalli, J.R.: Edu-bot: an AI-based smart chatbot for knowledge management system. In: 2021 IEEE International Conference on Computation System and Information Technology for Sustainable Solutions (CSITSS), pp. 1–6. (2021)
14. Türkmendağ, Z., Tuna, M.: Empowering leadership and knowledge management: the mediating role of followers' technology use. J. Organ. Chang. Manag. **35**(2), 330–347 (2021)
15. Schniederjans, D.G., Curado, C., Khalajhedayati, M.: Supply chain digitisation trends: an integration of knowledge management. Int. J. Prod. Econ. **220**, 107439 (2020)
16. Al-Sharafi, M.A., Al-Emran, M., Iranmanesh, M., Al-Qaysi, N., Iahad, N.A., Arpaci, I.: Understanding the impact of knowledge management factors on the sustainable use of AI-based chatbots for educational purposes using a hybrid SEM-ANN approach. Interact. Learn. Environ. **31**, 7491–7510 (2024)
17. Bai, X., Li, J.: The best configuration of collaborative knowledge innovation management from the perspective of artificial intelligence. Knowl. Manag. Res. Pract. **22**, 1–13 (2020)
18. Tajpour, M., Hosseini, E., Mohammadi, M., Bahman-Zangi, B.: The effect of knowledge management on the sustainability of technology-driven businesses in emerging markets: the mediating role of social media. Sustainability **14**(14), 8602 (2022)
19. Schaurhofer, M., Peschl, M.F.: Autonomy: starting point and goal of personal and social change: a constructivist perspective on knowledge management in empowerment processes. Kybernetes **34**(1/2), 261–277 (2005)
20. Greiner, M.E., Böhmann, T., Krcmar, H.: A strategy for knowledge management. J. Knowl. Manag. **11**(6), 3–15 (2007)
21. Ahmad, F.: Knowledge management implementation and human resource development on employee performance. Jurnal Manajemen Bisnis **10**(1), 221–234 (2023)
22. AlMulhim, A.F.: The impact of administrative management and information technology on e-government success: the mediating role of knowledge management practices. Cogent Bus. Manag. **10**(1), 2202030 (2023)
23. Kordab, M., Raudeliūnienė, J., Meidutė-Kavaliauskienė, I.: Mediating role of knowledge management in the relationship between organizational learning and sustainable organizational performance. Sustainability **12**(23) 10061 (2020)
24. Zhao, Y., Wen, S., Zhou, T., Liu, W., Yu, H., Xu, H.: Development and innovation of enterprise knowledge management strategies using big data neural networks technology. J. Innov. Knowl. **7**(4), 100273 (2022)
25. Iqbal, A.: Innovation speed and quality in higher education institutions: the role of knowledge management enablers and knowledge sharing process. J. Knowl. Manag. **25**(9), 2334–2360 (2021)
26. Faucher, J.P.L., Everett, A.M., Lawson, R.: Reconstituting knowledge management. J. Knowl. Manag. **12**(3), 3–16 (2008)
27. Manesh, M.F., Pellegrini, M.M., Marzi, G., Dabic, M.: Knowledge management in the fourth industrial revolution: mapping the literature and scoping future avenues. IEEE Trans. Eng. Manag. **68**(1), 289–300 (2021)

28. Shahzad, M., Qu, Y., Zafar, A.U., Rehman, S.U., Islam, T.: Exploring the influence of knowledge management process on corporate sustainable performance through green innovation. J. Knowl. Manag. **12**(3), 3–16 (2008)
29. Obeso, M., Hernández-Linares, R., López-Fernández, M.C., Serrano-Bedia, A.M.: Knowledge management processes and organizational performance: the mediating role of organizational learning. J. Knowl. Manag. **12**(3), 3–16 (2008)
30. Ode, E., Ayavoo, R.: The mediating role of knowledge application in the relationship between knowledge management practices and firm innovation. J. Innov. Knowl. **5**(3), 210–218 (2020)
31. Wamba-Taguimdje, S.-L., Fosso Wamba, S., Kala Kamdjoug, J.R., Tchatch-ouang Wanko, C.E.: Influence of artificial intelligence (AI) on firm performance: the business value of AI-based transformation projects. Bus. Process. Manag. J. **26**(7), 1893–1924 (2020)
32. Serey, J., et al.: Artificial intelligence methodologies for data management. Symmetry **13**(11) 2040 (2021). Art. no. 11
33. Ciampi, F., Marzi, G., Demi, S., Faraoni, M.: The big data-business strategy interconnection: a grand challenge for knowledge management. A Review and future perspectives. J. Knowl. Manag. **24**(5), 1157–1176 (2020)
34. Nurdin, N., Yusuf, K.: Knowledge management lifecycle in Islamic bank: the case of syariah banks in Indonesia. Int. J. Knowl. Manag. Stud. **11**, 59 (2020). Accessed 2024
35. Adhikari, D.R., Shrestha, P.: Knowledge management initiatives for achieving sustainable development goal 4.7: higher education institutions' stakeholder perspectives. J. Knowl. Manag. **27**(4), 1109–1139 (2023)
36. Iaia, L., Nespoli, C., Vicentini, F., Pironti, M., Genovino, C.: Supporting the implementation of AI in business communication: the role of knowledge management. J. Knowl. Manag. **28**(1), 85–95 (2024)
37. Schmitt, U.: Personal Knowledge Management for Development (PKM4D) framework and its application for people empowerment. Procedia Comput. Sci. **99**, 64–78 (2016)
38. Lam, L., Nguyen, P., Le, N., Tran, K.: The relation among organizational culture, knowledge management, and innovation capability: its implication for open innovation. J. Open Innov. Technol. Market Complex. **7**(1), 66 (2021)
39. Olan, F., Suklan, J., Arakpogun, E. O., Robson, A.: Advancing consumer behavior: the role of artificial intelligence technologies and knowledge sharing. IEEE Trans. Eng. Manag. **71**, 1–13 (2024)
40. Cope, B., Kalantzis, M., Searsmith, D.: Artificial intelligence for education: knowledge and its assessment in AI-enabled learning ecologies. Educ. Philos. Theory **53**(12), 1229–1245 (2021)
41. Singh, A.: An Introduction to experimental and exploratory research. Available at SSRN 3789360 (2021)
42. Mbaka, N., Isiramen, D.O.M.: The changing role of an exploratory research in modern organisation. GPH-Int. J. Bus. Manag. **4**(12) (2021). Art. no. 12
43. Herrero, P., Armellini, F., Solar-Pelletier, L.: Change management in the context of the 4th industrial revolution: exploratory research using qualitative methods. J. Mod. Proj. Manag.**7**(4) 155–175 (2020)
44. Royo-Vela, M., Hünermund, U.: Effects of inbound marketing communications on HEIs' brand equity: the mediating role of the student's decision-making process. An exploratory research. J. Mark. High. Educ. **26**(2), 143–167 (2016)
45. Camargo, L.R., Pereira, S.C.F., Scarpin, M.R.S.: Fast and ultra-fast fashion supply chain management: an exploratory research. Int. J. Retail Distrib. Manag. **48**(6), 537–553 (2020)
46. Vegliante, R., Sannicandro, K.: The role of the tutor in the university context and in distance learning: an exploratory research. J. e-Learn. Knowl. Soc. **16**(3), 76–85 (2020)

47. Williams, D., Stephen, L.-A., Causton, P.: Teaching interprofessional competencies using virtual simulation: a descriptive exploratory research study. Nurse Educ. Today **93**, 104535 (2020)
48. Wang, F.-Y., et al.: When does Sora show: the beginning of TAO to imaginative intelligence and scenarios engineering. IEEE/CAA J. Autom. Sinica **11**(4), 809–815 (2024)
49. Mogavi, R. H., et al.: Sora OpenAI's Prelude: Social Media Perspectives on Sora OpenAI and the Future of AI Video Generation (2024)
50. Liu, Y., et al.: Sora: A Review on Background, Technology, Limitations, and Opportunities of Large Vision Models (2024)
51. Kustudic, M., Mvondo, G.F.N.: A Hero or a Killer? Overview of Opportunities, Challenges, and Implications of Text-to-Video Model SORA (2024)
52. Mvondo, G.F.N., Niu, B.: Factors Influencing User Willingness to Use SORA (2024)
53. Alkaissi, H., McFarlane, S.I.: Artificial hallucinations in ChatGPT: implications in scientific writing. Cureus**15**(2):e35179 (2023)
54. Haleem, A., Javaid, M., Singh, R.P.: An era of ChatGPT as a significant futuristic support tool: a study on features, abilities, and challenges. BenchCouncil Trans. Benchmarks Stand. Eval. **2**(4), 100089 (2022)
55. Nah, F.F.-H., Zheng, R., Cai, J., Siau, K., Chen, L.: Generative AI and ChatGPT: applications, challenges, and AI-human collaboration. J. IT Cases Appl. **25**(3), 277–304 (2023)
56. Wang, X., et al.: Sora for intelligent vehicles: a step from constraint-based simulation to artificiofactual experiments through dynamic visualization. IEEE Trans. Intell. Veh. **9**(3), 4249–4253 (2024)
57. George, A.S., George, A.H., Martin, A.S.G.: A Review of ChatGPT AI's Impact on Several Business Sectors (2023)
58. Kalla, D., Smith, N., Samaah, F., Kuraku, S.: Study and Analysis of Chat GPT and its Impact on Different Fields of Study, Rochester (2023). Accessed 2024
59. Quiroga, E.: Toward Solipsism: The Emergence of Sora and Other Video Generation AIs in Audiovisual Creation (2024)
60. Baidoo-Anu, D., Owusu Ansah, L.: Education in the Era of Generative Artificial Intelligence (AI): Understanding the Potential Benefits of ChatGPT in Promoting Teaching and Learning. Social Science Research Network (2023)

Examining User Acceptance of a New Innovative Publication Management Approach for Independent Researchers

Hapini Awang[1(✉)], Nur Suhaili Mansor[1], Nur Anis Shafiqah Mazlan[1], Osman Ghazali[1], Juhaida Abu Bakar[1], İsmail Rakıp Karaş[2], Abdulrazak F. Shahatha Al- Mashhadani[3], and Shakiroh Khamis[4]

[1] Institute for Advanced and Smart Digital Opportunities, School of Computing, University Utara Malaysia, 06010 Sintok, Kedah, Malaysia
hapini@uum.edu.my

[2] Computer Engineering Department, Faculty of Engineering, DemirCelik Campus, Karabuk University, Karabük, Turkey

[3] Faculty of Business, Sohar University, Sohar, Oman

[4] Faculty of Information and Communication Technology, University Tunku Abdul Rahman, Kampar, Malaysia

Abstract. Navigating the journey of research papers, from submission to publication, can be challenging in academia. Researchers frequently encounter difficulties managing numerous documents, comprehending their statuses, and ensuring timely updates. This article addresses the development and assessment of an innovative paper-tracking system known as TrackMinder, which aims to improve the efficiency and transparency of the academic publication process. Following the development, an online survey was conducted to evaluate its effectiveness, gathering responses from 50 researchers. The assessment framework was grounded in the Technology Acceptance Model (TAM) and focused on three key elements: Perceived Usefulness (PU), Perceived Ease of Use (PEOU), and Intention to Adopt (ITA). The quantitative analysis employed descriptive statistics and correlation analysis to investigate these components, while qualitative data were analyzed using thematic and sentiment analysis techniques. The quantitative findings reveal that TrackMinder is both user-friendly and advantageous, with respondents showing a strong willingness to continue using it in the future. Significant relationships exist between PU, PEOU and ITA. Furthermore, sentiment analysis aligns with these results, indicating that most users have a favorable or neutral opinion of TrackMinder. These results suggest that the prototype has been positively received by its intended audience and has significant potential to enhance the academic publishing workflow.

Keywords: Technology Acceptance · Perceived Usefulness · Perceived Ease of Use · Intention to Adopt · Academic Publication

Z. Mat Aji et al. (Eds.): KMICe 2024, ISEM 49, pp. 79–91, 2025.
https://doi.org/10.1007/978-3-031-91485-0_7

1 Introduction

Keeping track of research papers has become crucial in the research process. Hence, various systems have been developed to assist researchers with the complex writing, submission, review, and publication procedures common in academic publishing. While gaining popularity among researchers, these systems increasingly offer features such as simple paper submission, real-time progress updates, and open communication among researchers. However, most of the available systems are at the institutional level, which does not accurately satisfy the needs of independent researchers [1].

Independent researchers are increasingly advocating for an effective and centralised method for monitoring the progress of their research publications. The current landscape presents significant challenges, as there is no practical mechanism available for these researchers to assess the status of their work. At present, they are compelled to manually track their articles' status, whether they are at the writing stage or submitted, rejected, under review, evaluated, or at any other stage of the publishing process. This is particularly crucial for active independent researchers who are actively involved in academic publications. The absence of an efficient tracking system hinders their ability to stay informed about the progression of their academic research papers. As a result, they lost track of their submissions and available drafts in their paper bank. Therefore, there is a pressing need for a user-friendly system that facilitates the tracking of publication phases and provides a comprehensive overview of research status.

Therefore, this study introduces a prototype called TrackMinder, an individual-level paper tracking system designed for researchers to address the common challenges they face in managing their publications. TrackMinder serves as a central tool that simplifies the process of monitoring the status of academic papers as they progress through various stages. With TrackMinder, researchers can easily obtain an overview of their submissions' current condition. This system enhances publication tracking by allowing for straightforward management of works in progress. TrackMinder is a user-friendly system that simplifies the paper-tracking process for researchers. It offers real-time updates, management tools, and data analytics to enhance the efficiency and transparency of academic publishing.

The system features two main functions: "Manage Bank" and "Manage Submission." "Manage Bank" allows researchers to oversee their inventory of papers by status—such as "draft," "rejected," "submitted," and "accepted"—and make updates. "Manage Submission" enables users to submit new papers, update details, and track their progress. This streamlined approach keeps researchers informed and reduces the administrative burden of publishing. This article evaluates the development and acceptance of TrackMinder in improving publication skills among independent researchers. The study uses the Technology Acceptance Model (TAM) proposed by Davis (1993) to assess TrackMinder's effectiveness in terms of perceived ease of use, usefulness, and intention to adopt the system. Additionally, qualitative data analysis is included to identify the system's strengths and areas for improvement.

2 Methodology

This section outlines a systematic approach to developing, testing, and evaluating the TrackMinder prototype. This study utilized the waterfall model, which includes five phases: requirement gathering, design, development, and testing [3]. The first phase, the requirements-gathering phase, was crucial for understanding the specific needs and challenges academic researchers face in tracking their paper submissions. During the design phase, TrackMinder was conceptualized based on the identified requirements. This process involved creating detailed use cases and class diagrams using Unified Modeling Language (UML). These diagrams were essential for visualizing and modeling the system's functionalities and structure.

The TrackMinder system was constructed in the development phase. Its functionalities were developed and integrated using phpMyAdmin for database management and Visual Studio Code for coding. The focus during development was on creating a robust backend to manage paper submissions and a user-friendly interface for researchers. The testing phase ensured the TrackMinder system functioned as intended within its target environment. This phase involved comprehensive integration testing to verify that all system components worked together seamlessly. The system was then evaluated against the requirements defined during the analysis phase to ensure it met the initial goals. Fifty respondents participated in the evaluation to provide feedback on the system's performance and usability.

The evaluation model was developed based on the TAM [2]. This model includes three primary constructs: Perceived Ease of Use (PEOU), Perceived Usefulness (PU), and Intention to Adopt (ITA). For the evaluation process, respondents were asked to complete a detailed questionnaire utilizing an interval continuous scale to measure each of these constructs. Descriptive analysis was conducted to assess the levels of each construct, yielding insights into the system's usability and effectiveness. Additionally, Pearson correlation—a non-parametric method—was used to analyze the relationships between the constructs. This analysis helped in understanding how perceived ease of use and usefulness influenced the intention to adopt TrackMinder.

2.1 Prototype Development

TrackMinder is designed for one type of user: independent researchers. Suitable modeling methodologies and tools are essential to visualize and model the application's requirements effectively. In this study, the UML was used to visualize and model the requirements. Two behavioral diagrams—a use case diagram (shown in Fig. 1) and a class diagram illustrating the system's structural elements—were created for this purpose. The diagrams were developed using Draw.io.

In the context of TrackMinder, a system intended for a single user, it is essential to illustrate the interactions between use cases and actors [4]. The structural components of TrackMinder are represented in a class diagram, which delineates the application's attributes and operations. Furthermore, the diagram explicitly depicts the interactions among the various classes. As shown in Fig. 2, the Manage Bank page allows researchers to manage their entire inventory of papers. It offers features such as viewing lists of papers

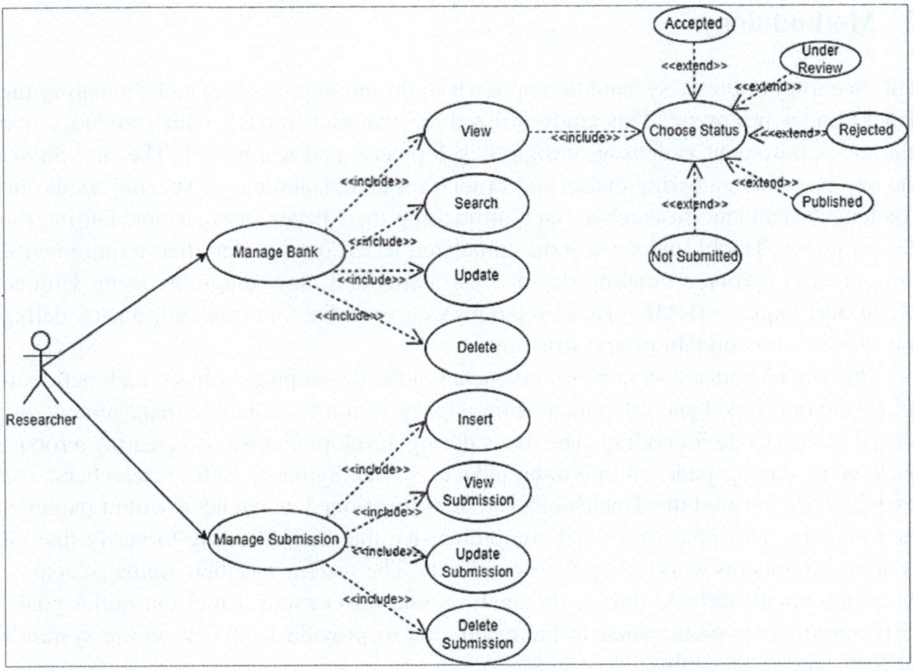

Fig. 1. Use case diagram of TrackMinder

categorised by status (e.g., draft and rejected), searching for documents by title, updating details of papers, and deleting papers from the system.

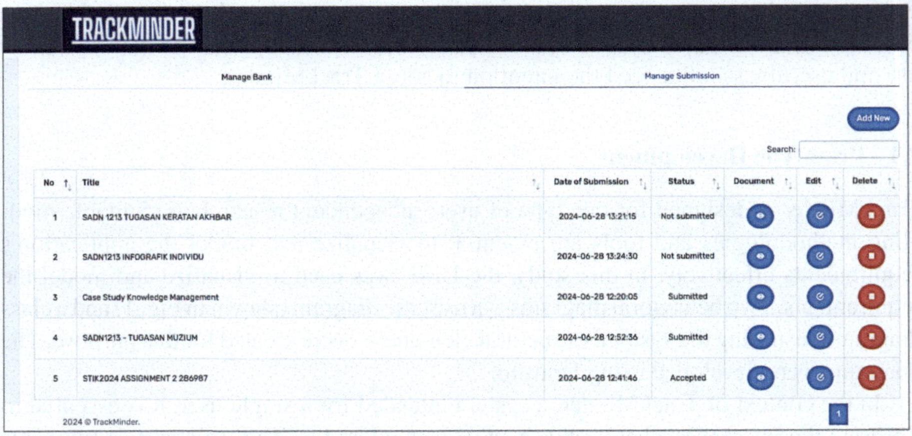

Fig. 2. Manage bank page of TrackMinder

The Manage Submission page, as shown in Fig. 3, is designed to streamline the submission process for researchers. It allows users to submit new papers and enter detailed information about their documents. Additionally, it provides a list of papers organised by submission status, such as not submitted, submitted, and accepted. Users can also update the information for submitted documents and delete paper submissions from the system.

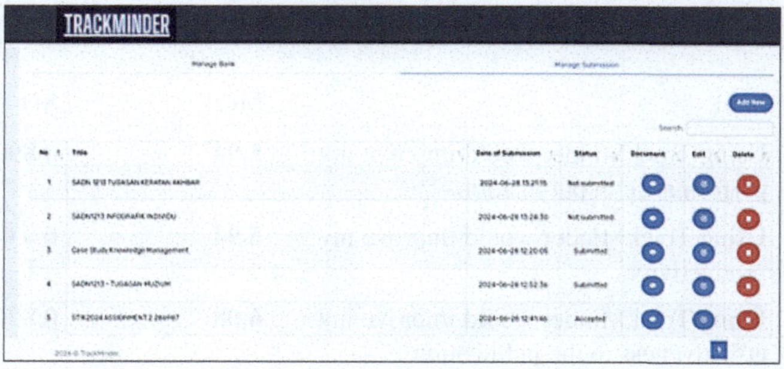

Fig. 3. Manage submission page of TrackMinder

3 Analysis and Findings

The study aimed to investigate users' attitudes and perceptions of TrackMinder, with the intent of informing future development and improvement of the system. Usability was assessed through an evaluation involving potential users, including researchers and lecturers. The evaluation model was based on the Technology Acceptance Model (TAM), as depicted in Fig. 5, and focused specifically on users' intentions to adopt TrackMinder in the future. The analysis concentrated on three key constructs: PU, PEOU, and ITA [5]. Descriptive analysis measured the levels of each construct, while Pearson correlation was used to examine their relationships. Finally, sentiment analysis and thematic analysis were used to examine the qualitative data.

3.1 Descriptive Statistics

All respondents are lecturers from higher education institutions who are actively involved in academic publications. Most of the respondents are female, comprising 54% (27 individuals), while males represent 46% (23 individuals). Most respondents are aged between 21 and 30 years, accounting for 86% (43 individuals). Additionally, 8% (4 individuals) fall within the 31 to 40 age range. Respondents aged 41 to 50 make up 4% (2 individuals), and those aged 51 to 60 represent only 2% (1 individual).

As presented in Tables 1, 2, and 3, PU and PEOU have five measurement items each, while ITA has four. The level of each construct is also calculated using the following formula in Eq. (1).

$$r = \frac{q-1}{x} \tag{1}$$

where r is the range value between each group, q is the original scale point, and x is the total number of groups to be generated. Hence, for a seven-point interval continuous scale, these series have been generated: Mean range: 1.00–3.00 (Low), 3.01–5.00 (Moderate), and 5.01–7.00 (High).

Table 1. Perceived usefulness.

No	Item	Mean	SD
1	Using TrackMinder would improve my performance in publication	**5.94**	0.8901
2	Using TrackMinder would improve my productivity	**5.94**	0.9348
3	Using TrackMinder would improve my effectiveness in the publication	**6.08**	0.8769
4	Using TrackMinder would improve the publishing skills of my paper	**5.9**	1.0351
5	I would find TrackMinder useful	**6.1**	0.7890

Table 1 presents the survey results regarding the perceived usefulness of TrackMinder. The respondents expressed a strong belief in the effectiveness of this publication system. The mean scores for all five items ranged from 5.94 to 6.10, indicating a general consensus that using TrackMinder would enhance their performance, productivity, effectiveness, and publication skills. The overall mean score for perceived usefulness (PU) is 5.992, which suggests a high level of agreement among the respondents. These findings indicate that TrackMinder has the potential to be a valuable tool in the publication process as perceived by its target users.

Table 2. Perceived ease of use

No	Item	Mean	SD
1	Learning to use TrackMinder would be easy for me	6.14	1.0104
2	I would find TrackMinder easy to facilitate my publication	5.92	0.9223
3	It would be easy for me to become skillful in the use of TrackMinder	5.84	0.9553
4	I would find TrackMinder easy to use	6.14	0.8809

For TrackMinder's perceived ease of use (as shown in Table 2), the four measurements indicated positive perceptions, with mean scores ranging from 5.84 to 6.14, resulting in a high average score of 6.01 for perceived ease of use (PEOU). This suggests that respondents generally believe learning and using TrackMinder effectively for publication would be easy. Therefore, it can be inferred that the prototype is widely accepted as an easy-to-use tool for publication.

Table 3. Intention to adopt

No	Item	Mean	SD
1	I think that using TrackMinder is a good idea	6.06	0.9127
2	I have a good perception of using TrackMinder	6.08	0.7783
3	I intend to use TrackMinder	5.8	1.0302
4	I intend to use TrackMinder instead of the other publishing methods	5.96	0.9026

Lastly, Table 3 presents the survey results regarding the intention to adopt TrackMinder. Respondents exhibited a positive attitude towards this prototype, with mean scores ranging from 5.8 to 6.08 across the four items. The overall mean score for intention to adopt (ITA) is 5.975, indicating that respondents generally agree that using TrackMinder is a good idea and intend to use it as an alternative. The findings also suggest that respondents are inclined to incorporate TrackMinder into their publishing practices. Figure 4 illustrates the overall level of each construct, with all constructs achieving a high level of acceptance.

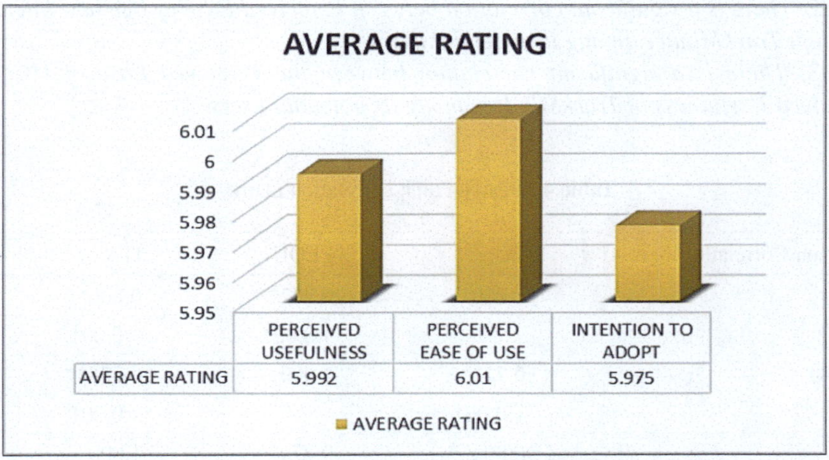

Fig. 4. Level of PU, PEOU and ITA of TrackMinder

3.2 Correlation Analysis

TrackMinder has previously demonstrated high levels of PU, PEOU, and ITA, which is a positive indication of its effectiveness. This has prompted further investigation into how PU and PEOU levels influence TrackMinder's ITA, which is in line with the Technology Acceptance Model (TAM) proposed by Davis (1993). Considering the nature of the gathered data, the most suitable statistical method for this study is the Pearson Correlation. The analysis was based on the TrackMinder evaluation Model, as illustrated in Fig. 5, leading to the formulation of the following three hypotheses.

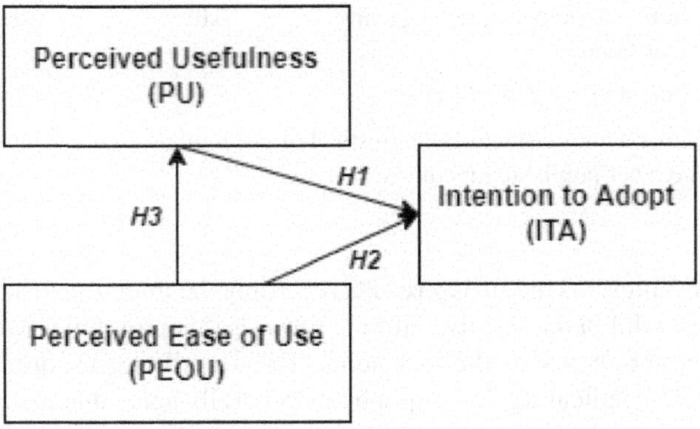

Fig. 5. Evaluation Model of TrackMinder Prototype

H1: *There is a significant correlation between Perceived Usefulness and Intention to Adopt TrackMinder among its potential users.*

H2: *There is a significant correlation between Perceived Ease of Use and Intention to Adopt TrackMinder among its potential users.*

H3: *There is a significant correlation between the Perceived Ease of Use and Perceived Usefulness of TrackMinder among its potential users.*

Table 4. Pearson rank correlation matrix

Pearson Correlation		PU	PEOU	ITA
PU	r	1.000	0.722**	0.806**
	p		<0.001	<0.001
PEOU	r	0.722**	1.000	0.778**
	p	<0.001		<0.001
ITA	r	0.806**	0.778**	1.000
	p	<0.001	<0.001	

N = 50, $p < 0.01$**

Table 4 presents the Spearman Rank Correlation Matrix generated during the analysis. Additionally, the scatterplots in Figs. 6, 7, and 8 illustrate the relationship between PU, PEOU, and ITA. The correlation coefficient is denoted by "r," while "p" indicates the statistical significance of the correlation. Hypothesis H1 proposed that as the perceived usefulness of TrackMinder increases, so too does the intention to adopt it. The results showed a strong positive correlation between PU and ITA ($r = 0.806$, $p < 0.01$, $N = 50$); thus, this hypothesis is supported.

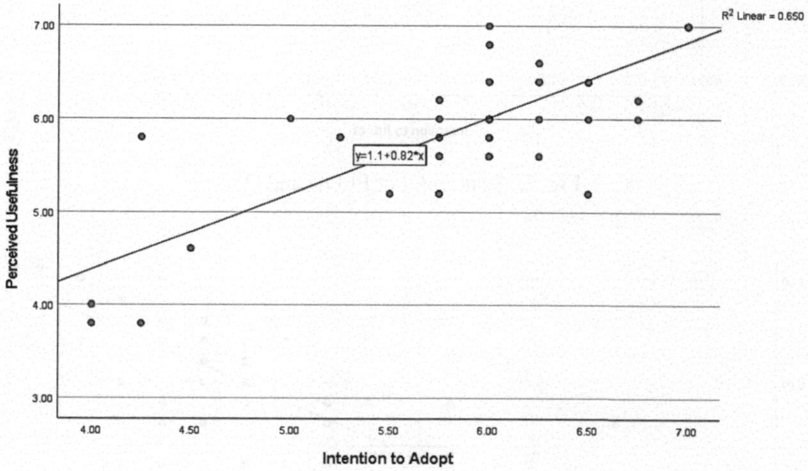

Fig. 6. Scatterplot of PU and ITA

Next, the results indicated a significant positive correlation between PEOU and ITA ($r = 0.778$, $p < 0.01$, $N = 50$), supporting H2. This evidenced that users who perceived TrackMinder as easy to use were likelier to express an intention to adopt it.

Finally, H3 hypothesised that users who perceived TrackMinder as easy to use also tended to perceive it as more practical. The results indicated a significant positive correlation between PEOU and PU ($r = 0.722$, $p < 0.01$, $N = 50$), supporting this hypothesis.

The three hypotheses investigated the relationship between users' perceptions of the usefulness and ease of use of TrackMinder and their intention to adopt the application. The Pearson rank correlation analysis results provide support for all three hypotheses, indicating that the correlation coefficients (r values) are predominantly greater than 0.7, with p values below 0.01. This suggests a significant correlation between these perceptions and the adoption of TrackMinder.

3.3 Sentiment and Thematic Analyses

This study employed open-ended questions to collect qualitative data in conjunction with quantitative data. Both thematic and sentiment analyses were utilised to analyse the qualitative responses. The integration of these quantitative and qualitative methods aims

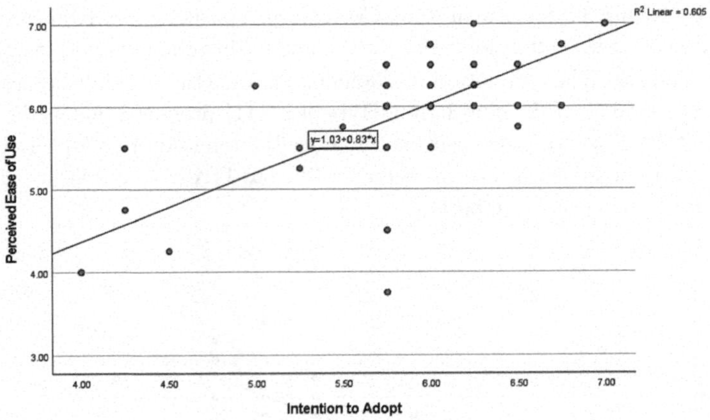

Fig. 7. Scatterplot of PEOU and ITA

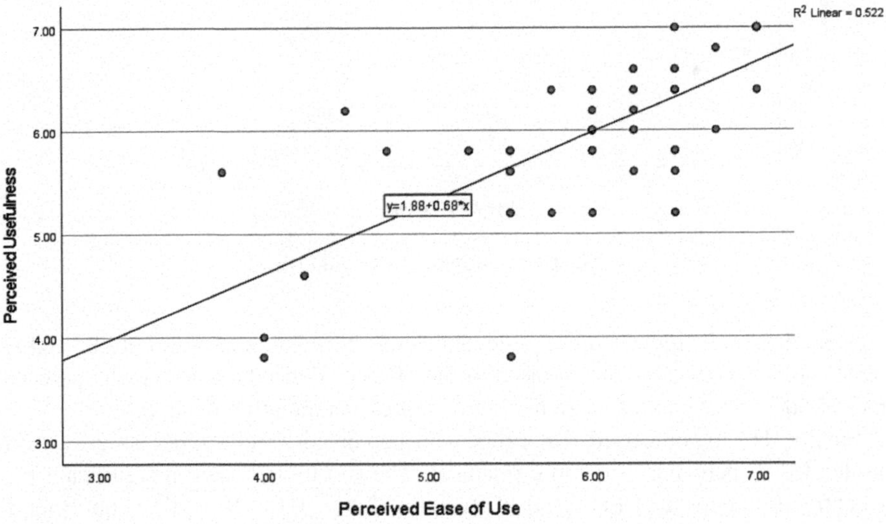

Fig. 8. Scatterplot of PEOU and PU

to provide a more comprehensive understanding of the acceptance of TrackMinder and to identify potential areas for improvement in the future [6]. Thematic analysis facilitates the extraction of underlying themes and patterns within the data, while sentiment analysis offers insights into respondents' overall attitudes toward the prototype. The first open-ended question solicited respondents' general views of TrackMinder, and the findings from its sentiment analysis are presented in Table 5.

Indicators have been employed to evaluate sentiments, specifically focusing on magnitude and sentiment score. The majority of responses were neutral (n = 4) or positive (n = 12). The average sentiment score is +0.563, indicating an overall slightly positive

Table 5. Sentiment analysis

Response	Sentiment	Magnitude	Sentiment Score
Great project!	Positive	0.945	+0.721
The flow is apparent 👍	Positive	0.762	+0.607
Add images	Neutral	0.033	+0.242
Nice work	Positive	0.999	+0.749
Track minder is a good initiative for self-productivity	Positive	0.814	+0.644
It looks simple and easy to understand	Positive	0.866	+0.682
It would be better to explain how the system works in that video	Neutral	0.414	+0.146
I think Track Minder is an excellent initiative to ease the work	Positive	0.538	+0.515
The interface is user-friendly 👍	Positive	0.881	+0.690
A perfect system to manage all the PDFs	Positive	0.814	+0.644
Great job!	Positive	0.964	+0.990
I would like to use TrackMinder in the future	Neutral	0.362	+0.160
Thank you for exposing me to TrackMinder	Positive	0.819	+0.647
Overall look fine	Neutral	0.098	+0.226
The system will be beneficial, especially for lecturers like us	Positive	0.821	+0.648
So far so good	Positive	0.895	+0.697

sentiment. Analysing the individual responses reveals that many potential users found TrackMinder to be both exciting and helpful, while some remained neutral.

The feedback is predominantly positive, with comments such as "Great project!", "The flow is clear 👍," and "Perfect system to manage all the PDFs," demonstrating high levels of satisfaction with TrackMinder. A few neutral comments, like "dd images" and "Overall looks fine," suggest areas for improvement, although they lack strong emotional emphasis. Suggestions such as, "If you include a written explanation in the video about how the system works, it would be better," and "The video should have a narrative," provide constructive feedback for enhancing usability and presentation.

Overall, the sentiment analysis indicates a highly positive reception of the TrackMinder system, with users valuing its functionality, ease of use, and potential to enhance productivity. Nevertheless, it is important to consider the constructive feedback offered in some responses to ensure continuous improvement of the system to better meet user needs. This study also gathered and analysed suggestions for improving TrackMinder using thematic analysis, as presented in Table 6.

Responses were collected as anticipated, and all were deemed valid for analysis. The analysis of these responses identified three primary themes. The first theme emphasized

Table 6. Thematic analysis

Theme	n	%
Positive Impression	6	37.5%
Suggestions for Improvement	2	12.5%
Usefulness and Adoption	8	50.0%
Total	16	100%

the overall positive feedback and appreciation for the TrackMinder system, suggesting that it effectively meets the needs of its intended audience. The second theme concentrated on constructive feedback and recommendations for enhancing the system. Implementing these suggestions has the potential to further improve user experience and increase accessibility.

Finally, the third theme addressed TrackMinder's perceived usefulness and users' intentions to adopt the system. Comments such as "TrackMinder is a good initiative to ease the work" and "This system will be beneficial, especially for lecturers" illustrate its potential to enhance productivity and streamline the academic publishing process. Furthermore, respondents expressed a strong willingness to adopt the system, with statements like, "I would like to use TrackMinder in the future."

4 Discussion

In today's technology-driven world, bridging the digital divide is more important than ever, especially for marginalized communities [7]. TrackMinder has been positively received by its users, who find it valuable and easy to use. Its user-friendly design can significantly benefit researchers, particularly those from under-resourced institutions. By simplifying tracking and managing publications, TrackMinder allows researchers to focus more on their core mission—conducting and sharing their research.

This study investigates the design and development of TrackMinder, a paper tracking system intended to improve publishing practices. The study also examined the acceptance and perceptions of potential users. Notably, participants regarded TrackMinder as a valuable and easy-to-use tool that could enhance their performance, productivity, and effectiveness in publishing. The results indicated a strong connection between TrackMinder's perceived usefulness and ease of use and its acceptance by users. Many participants considered TrackMinder to be an excellent alternative to conventional publishing methods. By incorporating user feedback and providing consistent updates, TrackMinder has the potential to evolve and support a more inclusive and efficient academic publishing process. Ultimately, the goal is to ensure that digital tools like TrackMinder empower all researchers to share their valuable insights with the global academic community.

5 Conclusion

In conclusion, this study presents TrackMinder as a promising tool for enhancing the academic publishing process, particularly for researchers. Users have responded positively to TrackMinder's ability to simplify and improve the management of scholarly work, indicating its potential to democratize access to advanced research management tools. However, there are areas that require further investigation, such as the system's long-term usage and its impact on diverse user groups. Future research should aim to incorporate more customizable features and ensure seamless integration with other research tools. Furthermore, longitudinal studies could offer deeper insights into TrackMinder's sustained effectiveness while examining factors that influence its adoption, including attitudes toward technology, accessibility, and training. Finally, ongoing updates and user-focused improvements will help ensure that TrackMinder remains a valuable asset in the ever-changing landscape of academic research, promoting inclusivity and efficiency within the scholarly community.

References

1. Nicholas, D., et al.: Does the scholarly communication system satisfy the beliefs and aspirations of new researchers? Summarizing the harbingers research. Learn. Publ. **33**, 132–141 (2020)
2. Davis, F.D.: User acceptance of information technology: system characteristics, user perceptions and behavioral impacts. Int. J. Man-Mach. Stud. **38**(3), 475–487 (1993)
3. Prabowo, H., Gaol, F.L., Hidayanto, A.N.: Comparison of the system development life cycle and prototype model for software engineering. Int. J. Emerg. Technol. Adv. Eng. **12**(4), 155–162 (2022)
4. Hernández, U.I., Rodríguez, F.J.Á., Martin, M.V.: Use Processes - modeling requirements based on elements of BPMN and UML use case diagrams. In: ICSTE 2010 - Proceedings of the 2010 2nd International Conference on Software Technology and Engineering (2010)
5. Awang, H., Zahurin, M.A., Wan Rozaini, S.O.: Measuring virtual learning environment success from the teacher's perspective: scale development and validation. In: Proceedings of the 3rd International Conference on Applied Science and Technology (ICAST 2018). American Institute of Physics (AIP), Penang (2018)
6. Mehraliyev, F., Chan, I.C.C., Kirilenko, A.P.: Sentiment analysis in hospitality and tourism: a thematic and methodological review. Int. J. Contemp. Hosp. Manag. **34**, 46–77 (2022)
7. Hushairi, Z., et al.: Bridging the digital divide – the E-Bario and E-Bedian telecommunication framework. In: Khalid, H.M., Helander, M.G., Yeo, A.W. (eds.) Work with Computer System, pp. 277–281, Kuala Lumpur (2004)

Visualising E-Commerce Customer Segmentation Through Clustering Methods

Shakiroh Khamis[1(✉)] and Mazida Ahmad[2]

[1] Department of Digital Economy and Technology, Faculty of Information and Communication Technology,
Universiti Tunku Abdul Rahman, Petaling Jaya, Malaysia
shakiroh@utar.edu.my

[2] Institute for Advanced and Smart Digital Opportunities, School of Computing, Universiti Utara Malaysia, Changlun, Kedah, Malaysia

Abstract. Today's customer understanding has become essential for businesses, especially online once, as the influx of new consumers continues to grow rapidly each day. The challenges for companies are to cater to the requirements of different behaviour groups of consumers and attract them, calling for customer segmentation. This study focuses on analysing customer segmentation in the e-commerce industry. The study addresses the need for effective customer segmentation due to the exponential growth of online consumers, emphasising the benefits for businesses in understanding customer behaviours and tailoring marketing strategies. It involves a detailed examination of customer data types, including income, purchasing behaviour, and demographic information, using advanced data analysis and machine learning techniques. The research employs K-means clustering for customer segmentation, evaluating its effectiveness through the Silhouette Score and Davies-Bouldin Index. The study also highlights the significance of advanced data visualisation tools like Power BI and Python for extracting insights from complex datasets. The project's contributions include improved e-commerce strategies through effective customer segmentation, enhanced decision-making using advanced visualisation, and in-depth analysis of customer behaviours. The research aims to enable businesses to customise marketing strategies, improving customer satisfaction and loyalty. Future work suggests expanding the dataset and exploring alternative clustering algorithms for more comprehensive customer segmentation in e-commerce.

Keywords: Visualization · Clustering · Customer Segmentation

1 Introduction

In today's continuously growing market, customer understanding has become mandatory for businesses, primarily online, to gain a competitive advantage by tailoring their marketing strategies to fulfil customer needs. The number of new consumers is increasing significantly every day, and businesses are being challenged to cater to the requirements

Z. Mat Aji et al. (Eds.): KMICe 2024, ISEM 49, pp. 92–103, 2025.
https://doi.org/10.1007/978-3-031-91485-0_8

of different behaviour groups of consumers and attract them. Hence, customer segmentation is needed to help companies as it becomes the efficient solution for this existing problem [1]. Customer segmentation is crucial to understand consumer behaviour better and foster stronger customer interactions [2].

Customer segmentation is a part of e-commerce strategy, which involves classifying customers based on appropriate criteria to customise marketing tactics and enhance customer experience. Insightful segmentation suggests that companies might not have a clear and organised framework for managing their varied customers. Without a well-defined segmentation strategy, businesses could struggle to identify unique consumer profiles and segments. Due to this unpredictability, general marketing methods may not resonate with specific target populations. Customer behaviour will be impacted by customer segmentation. Customers make various decisions and take actions throughout the buying procedure. E-commerce offers many options that enable businesses to gain comprehensive insight into their customers' behaviour. It helps companies identify which product types are preferred by specific customer segments. Thus, companies might suggest relevant products to customers based on their preferences.

In an e-commerce environment, businesses must deal with data's colossal volume and complexity. Most e-commerce businesses tiny companies have been hesitant to adopt and implement advanced data visualisation tools like Power BI. These technologies provide dynamic and interactive features that can overcome the limitations of traditional reporting methods [3]. The visualisation capabilities required to uncover significant patterns and trends from the vast datasets created by e-commerce operations are limited by the hesitation or delay in switching to advanced tools [4]. Due to traditional reporting methods, it will result in missed opportunities to uncover hidden insights, patterns, and correlations within the data that could inform strategic decision-making.

The study primarily focuses on B2C (Business-to-Consumer) e-commerce platforms due to their rich data availability and significant impact on the retail market. This focus allows an in-depth examination of B2C-specific consumer behaviour patterns, preferences, and segmentation strategies. The study will not include B2B and C2C platforms due to differences in business models and customer interactions. Geographically, the research will expand its scope to include e-commerce activities across multiple international urban areas. Focusing on numerous international locations also helps understand the nuances of customer behaviour in different markets and control external variables that might affect data consistency.

The study aims to show how effective customer segmentation may boost marketing strategies, allowing for personalising marketing approaches to improve the customer experience. Businesses can customise their marketing strategies by identifying distinctive consumer profiles and segments to appeal to specific target audiences. This helps ensure that marketing tactics are not generic but speak to the preferences and actions of distinct customer segments. As a result, overall customer experience improves, leading to higher customer satisfaction and loyalty.

2 Literature Review: Customer Segmentation and Behaviour

Customer segmentation has been an important field of research in marketing and business analytics since the rise of e-commerce, as it allows for the identification of various groups of consumers with similar patterns and behaviours and the development of different organisations' marketing strategies to meet demand [5]. Commonly, customer segmentation is used to determine customer relationship management (CRM). Segmenting customers allows companies to make better decisions by recognising and targeting specific client categories. By focusing on the market segments, businesses can tailor their strategies to match customers' unique needs, making their decision-making more effective and efficient in resource allocation on the right spot for profitability [6, 7]. Machine Learning (ML) clustering methods, which compare the performance of different clustering algorithms, are one of the most utilised consumer segmentation strategies to complete the customer segmentation part. Clustering is putting together data into similar groups based on familiar patterns, known as clusters. On an e-commerce consumer dataset, popular clustering algorithms such as K-means, Hierarchical, and DBScan are used because they are mainly used for clustering method studies [8].

Many studies have been conducted for customer segmentation by using clustering techniques in various areas such as in the banking sector [9, 11], e-commerce [10], financial [12, 13], insurance [14, 15], web content [16] and healthcare sectors [17, 18]. The most popular techniques that many researchers have used are K-means, Hierarchical, and DBScan clustering to segment customers based on their online shopping behavior from the available customer dataset by capturing the patterns and similarities of different groups [19]. Table 1 is the comparison based on the three clustering techniques that most researchers have conducted.

Table 1. Comparison between clustering techniques

Clustering Algorithm	Strengths	Weaknesses
K-Means	Linear computational complexity, suitable for large datasets Simplicity and interpretability Applicability to various data formats	Tendency to get stuck in local minima
DBSCAN	Excellent handling of noisy data Accommodates irregular cluster shapes	Challenges in neighbouring clusters Fixed-radius neighbourhood limitations
Hierarchical Clustering	Versatile and adaptive clustering Provides insights through dendrograms	Sensitivity to outliers Computational complexity in large datasets Tendency to form spherical clusters

In summary, K-means emerges as the topper prime for this specific case. Aside from its power to palm large datasets effectively, K-means produces easily interpretable clusters, which, in turn, are exceedingly beneficial. Despite its limitations in optimising centroid sites worldwide, it remains the most captured quality in this context. Ultimately, the quality of a clustering includes unique characteristics of the information at hand, including the desired figure and configuration of clusters, computational resources, and interpretation.

3 Research Methodology

This study employs the Data Science Life Cycle (DSLC) from Cleveland (2001) [20], which consists of six phases: business understanding, data understanding, data preparation, data modelling, evaluation, and deployment. This research is designed to utilise the methods of Big Data processing and analysis and build a referenced customer segmentation framework for e-commerce.

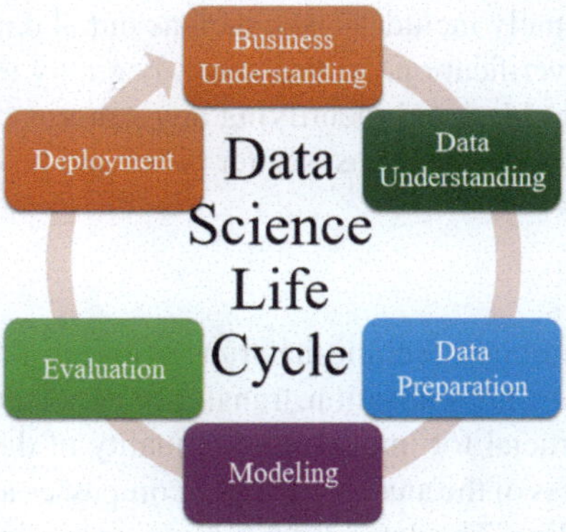

Fig. 1. Flowchart of research methodology

The flowchart depicts a process with six main stages, a data analytics or machine learning project lifecycle. It starts with "Business Understanding", which comprehends the business context and requirements. Following this, the process moves to "Data Understanding," indicating a phase where the data relevant to the business problem is collected and analysed to gain insights. Next, the flow moves to "Data Preparation," where data is cleaned, transformed, and prepared for modelling. From "Data Preparation," the process goes directly to "Modeling," suggesting an iterative relationship between preparing data and developing models. After "Modeling," the process proceeds to "Evaluation," which likely refers to assessing the model's performance against some criteria or testing it with a dataset. The final stage is "Deployment," where the model is put into production, or the findings are implemented into the business operations. The flowchart from Fig. 1 also shows a cyclical nature between "Deployment" and "Evaluation," suggesting that once a model is deployed, its performance is continuously evaluated, and improvements may be made, indicating an iterative process of refinement and deployment.

3.1 Business Understanding

The business understanding for this project is centred around the challenges and opportunities in the e-commerce industry, focusing on customer segmentation, data visualisation, and advanced analytics. With the rise of e-commerce, companies need to understand their customer base better to develop effective marketing plans. The fundamental problems include a lack of customer segmentation insight, limited use of advanced data visualisation tools, and an inability to fully utilise analytics tools for exploring customer behaviour. These issues highlight the need for a strategic approach to manage diverse customer profiles and extract meaningful insights from complex data, enhancing customer interactions and improving the overall customer experience in the e-commerce sector.

3.2 Data Understanding

This stage involves getting familiar with the data, identifying data quality issues, discovering first insights into the data, or detecting exciting subsets to form hypotheses for hidden information. It usually includes tasks such as initial data collection, description, exploration, and quality verification. At this stage, begin by exploring the dataset, getting familiar with the variables, and identifying any patterns or trends. This study uses visualisation techniques like histograms, scatter plots, and box plots.

3.3 Data Preparation

Data preprocessing is a mining technique that transforms raw data into an understandable format. It includes cleaning, normalisation, transformation, feature extraction, and selection. This step is often crucial for improving the quality of data and, consequently, the efficiency and effectiveness of the analysis. This encompasses a wide range of techniques to prepare data for modelling. The data preprocessing steps might include encoding categorical variables, normalising or scaling features, handling missing values, and creating new features through transformations or combinations of existing variables.

3.4 Modelling

Model Selection. As mentioned in the previous literature review, k-mean clustering has been selected as an appropriate algorithm for customer segmentation. Therefore, we used k-mean clustering to perform customer segmentation.

KMeans Clustering. K-means clustering is an unsupervised machine learning method often used to identify underlying patterns by grouping similar data points. The first step of the K-means method is to randomly choose the starting centroids of each cluster. It then uses computations to optimise these centroids iteratively. The process stops when the centroid stabilises or the predetermined number of iterations is reached.

The output of k-means in this study is shown in Fig. 2.

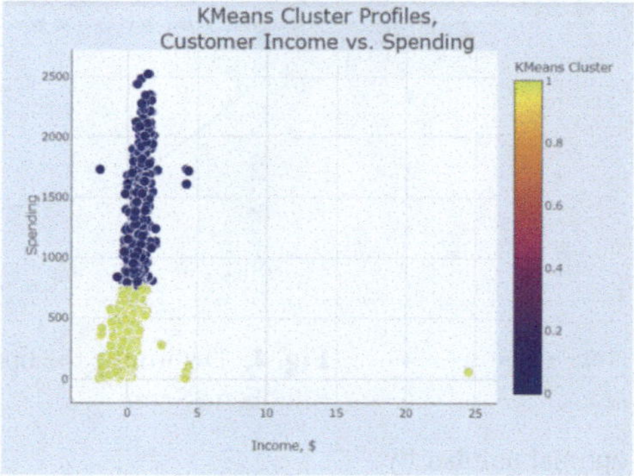

Fig. 2. K-means cluster profiles, customer income vs. spending

Determine the Optimal K Value. At the fundamental step of unsupervised learning, we must determine the optimal number of clusters used in the model. A range of k values was examined and assessed using the silhouette and elbow methods to determine the optimal number of clusters. The Silhouette Score measures each cluster's average distance from points in other clusters. The silhouette score ranges from −1 to 1. The most excellent average silhouette score across all samples indicates the optimal number of clusters. The elbow analysis method plots the explained variance (WCSS) against the total number of clusters. The WSS score (within-cluster sum of squares) is the sum of the squared distances between each data point and the chosen cluster centroid. Typically, the curve has an elbow-like shape. The optimal number of clusters is thought to be when the curve begins to flatten.

For a range of values of k from 2 to 10, the elbow and the silhouette methods were used to identify the optimal number of clusters. Every metric's data were plotted and examined. The analysis revealed that two clusters were the optimal number used in the k-mean clustering algorithm. After thoroughly analysing several methods to produce an effective and trusted solution, the optimal number of clusters was chosen. The output of the optimal k value using the Elbow method and Silhouette Score are shown in Figs. 3 and 4.

Model Training. K-means clustering may be used for various tasks, including market segmentation, anomaly detection, and image processing. It is a helpful tool for exploratory data analysis and pattern identification. In k-mean clustering, a dataset is divided into a predetermined number of groups. The algorithm aims to minimise the squared distance between every data point and its chosen cluster centre, also known as centroids.

To guarantee that every variable had the same scale before using the algorithm, the data was first normalised using z-score normalisation. Using the scikit-learn toolkit, k-means clustering was carried out using random initialisations and varying values of k. Several methods, including the silhouette and elbow methods, were assessed to identify the optimal number of clusters. The optimal number of clusters was determined to be two as it had the highest silhouette score. The elbow-like shape is also at the k = 2 in the elbow method. The identified clusters were shown to be distinct and well-separated

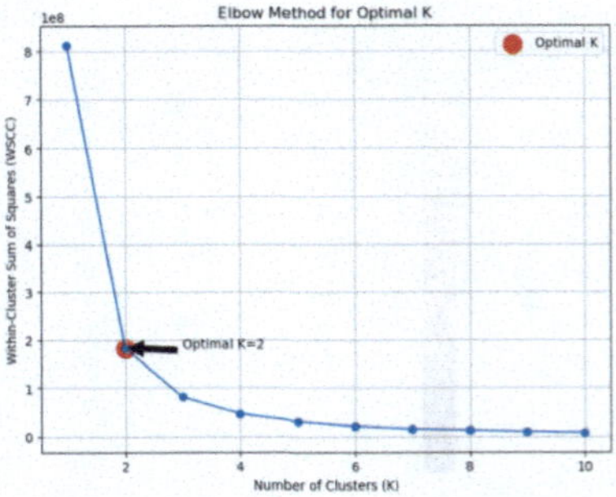

Fig. 3. Determine the optimal number by using the Elbow method.

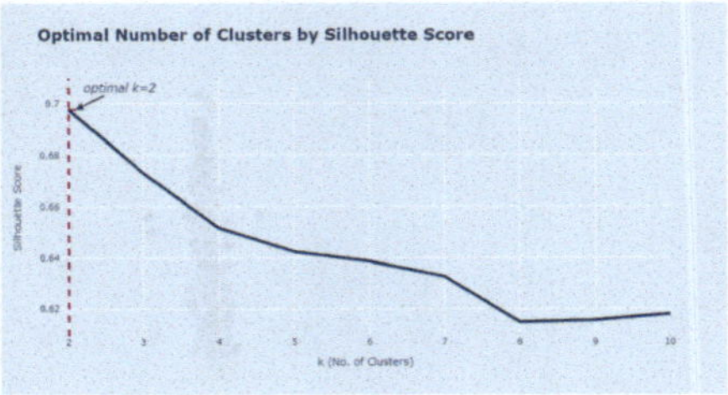

Fig. 4. Determine the optimal number by using Silhouette Score.

by the results. To reduce the algorithm's sensitivity to the original centroid position, the k-means algorithm was performed many times with different initialisations. The number of clusters is significant because it will affect the quality of the final cluster. Therefore, the value of k must be carefully considered in many cases. Model Evaluation.

In our study, the primary focus is not on assessing the model's accuracy. Instead, our focus is determining the optimal number of clusters (k) for customer segmentation. Traditional evaluation metrics such as accuracy and ROC curve were inappropriate for this exploratory investigation since there was no ground truth for comparison. Unlike supervised learning, where specific values are predicted, clustering aims to group customers based on inherent behavioural similarities. Therefore, evaluation metrics such as silhouette score and Davies-Bouldin Index were used to assess the performance of the segmentation algorithms [21].

The Silhouette Score is a quantitative method for assessing the quality of clustering outcomes. By evaluating how well each data point fits inside its allocated cluster and how unique it is from other clusters, it measures the degree of well-defined and uniqueness of clusters. The score provides information about the general homogeneity and separation of the clustering solution by quantifying both cohesiveness within and separation between clusters. When the silhouette score is close to +1, it indicates that it is far away from neighbouring clusters, and when it is close to 0, it suggests that it is on or very near the decision border between two neighbouring clusters. The negative Silhouette values may have been mislabelled to the incorrect cluster. According to Gunay [22], a clustering with an average Silhouette Score of over 0.7 is considered a strong structure, a value between 0.51 and 0.70 is regarded as a reasonable structure, a value between 0.26 and 0.50 is considered a weak structure, and a value less than 0.25 is regarded as no substantial structure.

The validity of a clustering solution may be assessed using the Davies-Bouldin Index (DBI). It calculates the average similarity between every cluster and the cluster that is closest to it. A lower DBI value indicates a better clustering solution. Poorer clustering solutions are correlated with higher values of the DB index. This is because a more excellent DBI value suggests that the clusters are neither compact nor well-separated. On the other hand, a lower DB index value is preferred. It shows that the clusters are compact and well-separated, frequently a sign of a successful clustering solution.

According to the researchers [23], a DBI less than 0.3 indicates a good model because the clusters are well-separated and compact; a DBI between 0.3 and 0.5 is considered a moderate model where separation is acceptable but not optimal, and a DBI exceeding 0.5 is regarded as a weaker model.

Deployment. After training the model, the next step is importing the model into Power BI. The preprocessed dataset was also imported into Power BI for visualisation purposes. The dashboard created in PowerBI includes interactive elements like filters and slicers. Users may dynamically explore and analyse the data by using these features. In addition, it may offer an interactive and customised view of the insights obtained from the trained model, thus improving the user experience.

After we perform the dashboard in Power BI, we also need to evaluate the dashboard. We need to ensure the data in visualisations accurately reflects the insights generated by the clustering model. Besides that, we must test all filters and slicers to confirm that they appropriately adjust the visualisation and ensure the dashboard responds correctly when users interact with these elements. We also need to verify that the dashboard operates as intended when users engage with the dashboard by choosing data points or altering filters.

4 Results and Discussion

The silhouette score results from our models is 0.7, indicating a solid structure. The silhouette score ranges from −1 to 1. The higher values of the silhouette score suggest that the clustering structure is more distinct and well-defined. A silhouette score of 0.7 is considered a strong separation between clusters and a high level of cohesion within clusters. In summary, a silhouette score of 0.7 is considered to indicate that our clustering is a strong structure with well-defined and distinct clusters.

The Davies-Bouldin Score result from our models, which is 0.45, indicates an acceptable model for customer segmentation. The clustering results are reasonable for segmenting customers.

Based on these evaluation metrics, we can conclude that our model is acceptable and reasonable for customer segmentation. The combination of a high silhouette score and a low Davies Bouldin score indicates the reliability and suitability of our model in meeting the purpose of customer segmentation. Besides that, we can suggest that the number of optimal clusters we used in the model performed very well in the clustering model. The low Davies-Bouldin Score and the high silhouette score indicate that the optimal number of clusters was chosen with consideration.

From Fig. 5, we can also observe the optimal number of clusters is 2 using the Davies-Bouldin index because the lowest Davies-Bouldin index is optimal $k = 2$.

Cluster Analysis

Figure 6 shows that in cluster 0, there are 802 customers, and in cluster 1, there are 1438 customers. Therefore, the number of customers in Cluster 1 is more than Cluster 0. The results of every cluster containing cluster 0 and cluster 1 are shown in Figs. 7, 8, 9, 10, 11 and 12.

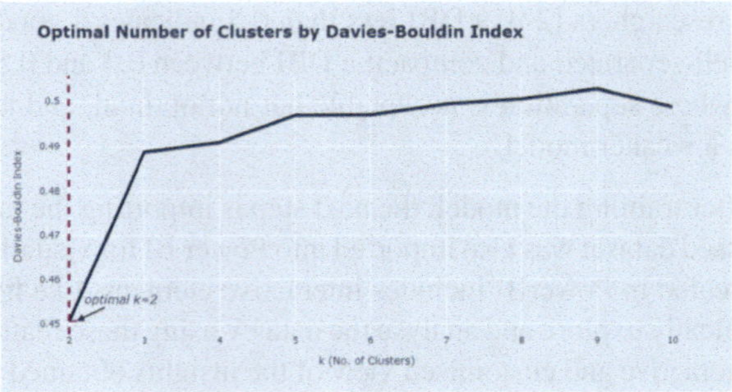

Fig. 5. Optimal number of clusters by Davies-Bouldin index

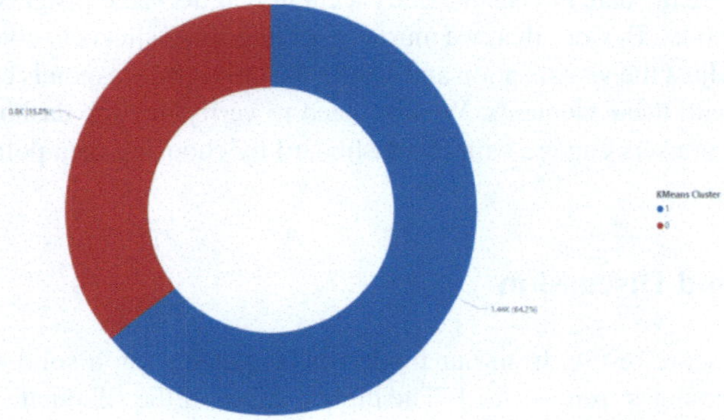

Fig. 6. K-Means cluster

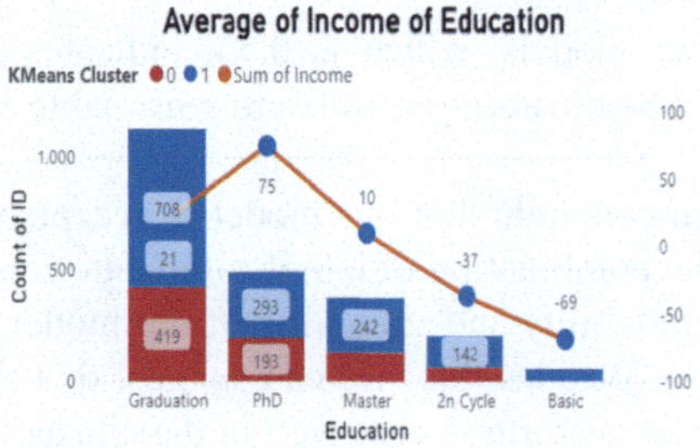

Fig. 7. Average of Income of Education

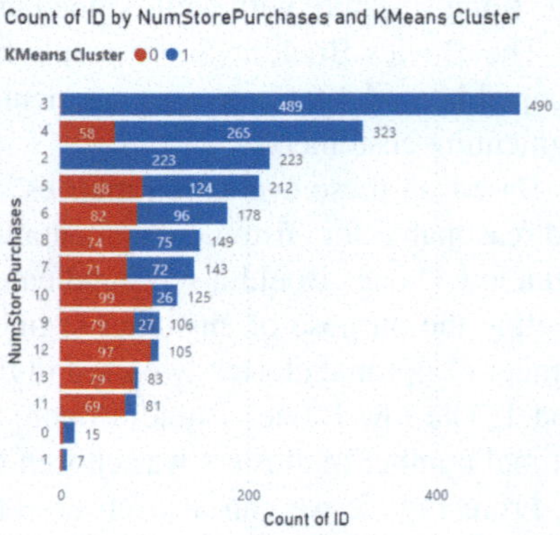

Fig. 8. Count of ID by NumCatalogPurchases and K-Means Cluster

Cluster 0: Highly Active and Moderately Active Customers

Figures 7, 8, 9, 10 and 11 indicate that the customers in this cluster have higher income and spending levels than those in Cluster 1. This cluster consists of most individuals with

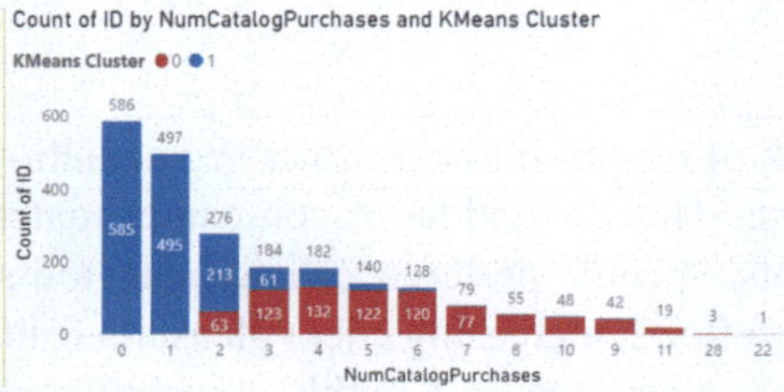

Fig. 9. Count of ID by NumStorePuchases and K-Means Cluster

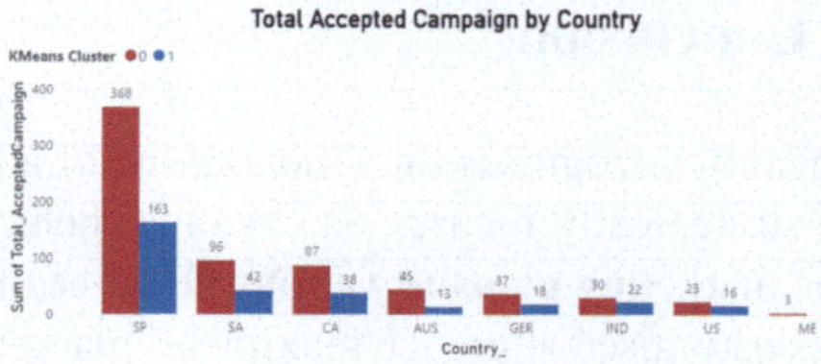

Fig. 10. Total Accepted Campaign by Country

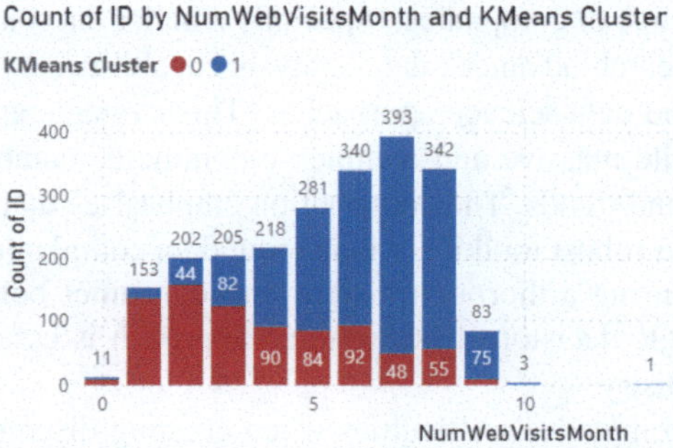

Fig. 11. Count of ID by NumWebVisitsMonth and K-Means Cluster

a professional or postgraduate background, and fewer customers have a fundamental educational background. The customers in this cluster tend to make more purchases through catalogues, which occurs between 3 and 11 times. Their in-store purchasing frequency is higher, occurring between 9 and 13 times. The fewer website visits than the least active customers suggest a preference for other purchasing channels like in-store or catalogues. The number of people accepted in the campaign exceeds the least active customers.

Cluster 1: Least Active Customers

The customers in Cluster 1 have low income and spending. Although individuals of professional or postgraduate backgrounds are in this cluster, their income is comparatively lower. Additionally, this cluster has a higher representation of customers with basic educational backgrounds than cluster 0. The number of in-store purchases occurs between 2 and 5 times, less frequent than in cluster 0. The number of purchases through the catalogue occurs 0 times, followed by 1 and 2 times. The number of website visits is also higher than that of highly active and moderate customers, possibly indicating a trend of browsing without making purchases. The total accepted for the campaign is also lower in this cluster.

5 Conclusion

The study recognises the crucial issue of a lack of insight into consumer segmentation and strategically focuses on customer behaviour. This method helps companies match their marketing plans to various client segments' specific preferences and behaviours. The customised approach maximises marketing efforts, increasing client pleasure, fruitful interaction, and a more distinctive customer experience. In a highly competitive e-commerce environment, the contribution highlights how crucial it is to comprehend all aspects of consumer behaviour to make wise judgments and cultivate enduring customer connections.

The study tackles the challenge of companies, especially smaller ones, not adopting innovative technology. The research advances data analysis capabilities by promoting tools like Power BI, Python, and data science approaches. This strategic approach not only allows companies to handle massive and complex e-commerce databases, but it also uncovers hidden patterns and trends. The contribution emphasises that combining advanced techniques results in a robust toolkit for managing data complexity, avoiding missed opportunities, and acquiring a thorough insight into consumer behaviour and business trends. Making strategic decisions based on this approach is essential in the ever-changing e-commerce industry.

The report acknowledges companies' difficulties when creating discrete customer classifications without a clear segmentation strategy. The study highlights the importance of a well-planned approach that considers consumer behaviour, purchasing patterns, and demographics, which helps create a more precise representation of the customer. Companies can use this method to group customers according to their behaviour, facilitating detailed analysis and carefully customised marketing campaigns. The contribution emphasises that an organised segmentation approach, including customer behaviour, in a competitive market landscape is a critical foundation for creating deeper interactions and a better knowledge of the diverse consumer base. It encourages personalised marketing and experiences, which leads to higher consumer engagement and pleasure.

References

1. Abdulhafedh, A.: Incorporating k-means, hierarchical clustering and PCA in customer segmentation. J. City Dev. **3**(1), 12–30 (2021)
2. Sarkar, M., Puja, A.R., Chowdhury, F.R.: Optimizing marketing strategies with RFM method and k-means clustering-based AI customer segmentation analysis. J. Bus. Manag. Stud. **6**(2), 54–60 (2024)
3. Khamis, S., et al.: Knowledge visualization of internet usage pattern to improve students' academic performance using prescriptive analytic. J. Adv. Res. Appl. Sci. Eng. Technol. **48**(1), 283–298 (2024)
4. Sabri, N., Khamis, S., Zainudin, Z.: Android malware detection using machine learning technique. In: Saeed, F., Mohammed, F., Fazea, Y. (eds.) IRICT 2023. Lecture Notes on Data Engineering and Communications Technologies, vol. 211, pp. 153–164. Springer, Cham (2024). https://doi.org/10.1007/978-3-031-59707-7_14
5. Tabianan, K., Velu, S., Ravi, V.: K-means clustering approach for intelligent customer segmentation using customer purchase behavior data. Sustainability **14**(12), 7243 (2022)

6. Yoseph, F., Ahamed Hassain Malim, N.H., Heikkilä, M., Brezulianu, A., Geman, O., Paskhal Rostam, N.A.: The impact of big data market segmentation using data mining and clustering techniques. J. Intell. Fuzzy Syst. **38**(5), 6159–6173 (2020)
7. Hung, P.D., Lien, N.T.T., Ngoc, N.D.: Customer segmentation using hierarchical agglomerative clustering. In: Proceedings of the 2nd International Conference on Information Science and Systems, pp. 33–37 (2019)
8. Monil, P., Darshan, P., Jecky, R., Vimarsh, C., Bhatt, B.R.: Customer segmentation using machine learning. Int. J. Res. Appl. Sci. Eng. Technol. (IJRASET) **8**(6), 2104–2108 (2020)
9. Raiter, O.: Segmentation of bank consumers for artificial intelligence marketing. Int. J. Contemp. Financ. Issues **1**(1), 39–54 (2021)
10. Alves Gomes, M., Meisen, T.: A review on customer segmentation methods for personalised customer targeting in e-commerce use cases. IseB **21**(3), 527–570 (2023)
11. Chitra, J., Heikal, J.: Customer segmentation using the k-means clustering algorithm in Foreign Banks in Indonesia. Indones. Account. Res. J. **11**(4), 230–241 (2024)
12. Akar, E.: Customers' online purchase intentions and customer segmentation during the period of COVID-19 pandemic. J. Internet Commer. **20**(3), 371–401 (2021)
13. de Lima Lemos, R.A., Silva, T.C., Tabak, B.M.: Propension to customer churn in a financial institution: a machine learning approach. Neural Comput. Appl. **34**(14), 11751–11768 (2022)
14. Abdul-Rahman, S., Arifin, N.F.K., Hanafiah, M., Mutalib, S.: Customer segmentation and profiling for life insurance using k-modes clustering and decision tree classifier. Int. J. Adv. Comput. Sci. Appl. **12**(9), 434–444 (2021)
15. Eckert, C., Neunsinger, C., Osterrieder, K.: Managing customer satisfaction: digital applications for insurance companies. Geneva Pap. Risk Insur.-Issues Pract. **47**(3), 569–602 (2022)
16. Zhou, J., Wei, J., Xu, B.: Customer segmentation by web content mining. J. Retail. Consum. Serv. **61**, 102588 (2021)
17. Ernawati, E., Baharin, S.S.K., Kasmin, F.: A review of data mining methods in RFM-based customer segmentation. J. Phys. Conf. Ser. **1869**, 012085 (2021)
18. Agarwal, R., Dugas, M., Gao, G., Kannan, P.K.: Emerging technologies and analytics for a new era of value-centered marketing in healthcare. J. Acad. Mark. Sci. **48**, 9–23 (2020)
19. Deng, S., Cai, Q., Zhang, Z., Wu, X.: User behavior analysis based on stacked autoencoder and clustering in complex power grid environment. IEEE Trans. Intell. Transp. Syst. **23**(12), 25521–25535 (2021)
20. Cleveland, W.S.: Data science: an action plan for expanding the technical areas of the field of statistics. Int. Stat. Rev. **69**(1), 21–26 (2001)
21. Gupta, R., Subedi, S., Singh, A., Singh, S.K.: Comparative study of unsupervised learning algorithms for customer segmentation. In: 2024 11th International Conference on Computing for Sustainable Global Development (INDIACom), pp. 1054–1059. IEEE (2024)
22. Gunay, D.: Clustering. Medium. https://medium.com/@denizgunay/clustering-49f2894c1c95. Accessed 21 Dec 2023
23. Liu, D., Xuan, T., He, Z., Yao, M., Payri, R.: A study on application of machine-learning on DBI soot diagnostics. Fuel **346**, 128292 (2023)

Trends Analysis in Knowledge Management Practice: Theoretical and Index Keywords Insights (2000–2024)

Mohd Adan Omar[1](✉) and Ammber Nosheen[2]

[1] School of Computing,
Universiti Utara Malaysia, 06010 Sintok, Kedah, Malaysia
adan@uum.edu.my

[2] Institute of Advanced and Smart Digital Opportunities, School of Computing,
Universiti Utara Malaysia, 06010 Sintok, Kedah, Malaysia

Abstract. Knowledge Management (KM) practices have experienced significant evolution over recent decades, driven by changing organizational needs and research developments. This paper offers a comprehensive analysis of the current trends in KM practices, focusing on theoretical frameworks and practical applications within organizations. Effective KM practice is essential because it enhances organizational learning, modernization, and strategic decision-making. Utilizing a keyword trend analysis from the Scopus database, the study identifies enduring research areas, evolving topics, and emerging themes in KM practice from 2000 to 2024 and enriching with quantitative insights of Average Growth Rate (AGR), Average Documents per Year (ADY), and H-Index. The analysis reveals a diverse array of keywords related to KM, yet KM practices will remain central and significant in 2024, highlighting their enduring relevance in contemporary organizational contexts. The findings provide valuable insights into the challenges and opportunities of implementing effective KM strategies, aiming to support organizations leveraging knowledge as a strategic asset. By offering a detailed overview of these developments, this study contributes to the ongoing discourse on KM practice in contemporary organizations, highlighting critical areas of interest and future research directions.

Keywords: Knowledge Management Practice · Trend Analysis · Index Keywords

1 Introduction

Knowledge management (KM) is a process that involves several methods, including the creation of knowledge, its organization, review, and dissemination before its use by other teams. An organization's KM processes should be created accurately for information identification, management, and sharing—especially when doing so securely [1]. KM practice methods in modern enterprises are paramount, as they are critical to maintaining competitive advantage, increasing productivity, and fostering innovation

Z. Mat Aji et al. (Eds.): KMICe 2024, ISEM 49, pp. 104–115, 2025.
https://doi.org/10.1007/978-3-031-91485-0_9

[2]. Effective KM practice has become a vital strategic asset as businesses operate in more complex and dynamic environments [3]. KM strategies help organizations make the most of their intellectual capital in the knowledge-based economy of today, where knowledge and expertise are vital [4]. This intellectual capital includes explicit knowledge, documented and easily communicated information, and tacit knowledge, which is people's unwritten, experiential knowledge [5]. By implementing robust KM systems, organisations can enhance decision-making processes, foster a culture of continuous learning, and ultimately achieve better business outcomes [6].

Designing and developing new services and products and managing the operational process in the modern business world have been shown to benefit significantly from KM practices [7]. Because of this, businesses want to implement fresh approaches to KM practice to accomplish long-term goals [2]. According to Tseng [8], creative ideas derived from human intellectual capital are used to construct modern economies, which promote profitability and sustainability, according to Manesh et al. [9] KM practice helps firms use KM practice with knowledge sharing, knowledge access, knowledge applications, and knowledge storage. The process of retrieving and transferring information and knowledge from a particular system" is known as knowledge access [10]. According to Schiuma et al. [11] Knowledge sharing is the process of disseminating various resources among the individuals taking part in specific activities. According to Islam and Widen [2], a crucial component of technological innovation acceptance and adoption is the desire to share information and knowledge. According to Al-Shahrani [12], knowledge application refers to the process of applying or using information in practical situations. It enables individuals to access information smoothly via efficient storage and retrieval techniques. However, there are software tools that facilitate the application of knowledge by organizing, making knowledge, and sharing it using KM systems for problem-solving and decision-making processes [13].

This study addressed the problem of a lack of comprehensive analysis of evolving trends and vital theoretical insights in KM practices, particularly by examining the inspection of index keywords from 2000 to 2024 and enriching with quantitative insights of AGR, ADY, and H-Index, which considered an understanding of its development and future directions. An overview of KM practices is given in the Introduction section, specifying the investigation's goal, parameters, and format. The section on Theoretical Frameworks in KM Practice that follows next looks at historical viewpoints and summarizes critical theories, including more recent theoretical advancements that have shaped the subject. The Current Trend Index Analysis on KM Practice examines current patterns, highlighting recurring themes and new subjects in the field. The study's main conclusions are finally summarized, and their ramifications for organizations are discussed in the "Conclusion and Future Directions and Implications" section. Additionally, it identifies possible directions for further study and offers workable solutions for efficiently utilizing KM practice in a globalized and technologically advanced world. The primary goal of this study is to investigate new theoretical advancements and analyze current trends in the area of KM practice. It seeks to offer a thorough examination of theoretical frameworks by highlighting the most recent advancements influencing the direction of

KM practice. The study hopes to provide insightful information for academics, professionals, and decision-makers interested in comprehending and utilizing KM practice to improve organizational effectiveness.

2 Methodology

Denyer and Tranfield [14] Proposed three processes that the general research structure applied: 1. database; 2. bibliometric analysis; 3. content Analysis and keywords trend analysis. This research used 460 article journals from a Scopus database work as a unit of analysis and was represented by index keywords. Because of its stellar reputation, Scopus was used in the first search stage for research about KM practice. The terminology chosen for the study was a combination of concepts related to KM and KM practice. Titles, abstracts, and keywords (2000–2024) were examined in the search. The bibliometrics analysis of the trend index keywords is used for the second portion of our investigation, with a focus solely on the analysis of the index keywords. The analysis stage was constructed using a sample of 460 Scopus index journal articles utilized for index keywords analysis from 2000 to 2024. These actions are to find pertinent keyword data and gather processed data useful for additional analysis. The following Figs. 1 and 2 display the findings of index keywords trend analysis results.

3 Theoretical Frameworks in KM Practice

From its inception, the discipline of KM practice has experienced substantial evolution. Early in the 1990s, KM became a formal field due to the increasing understanding of knowledge as a vital organizational resource [13]. Explicit knowledge—information that could be readily documented and communicated inside an organization—was the main focus of early KM practice attempts [15]. The management of tacit knowledge—people's unwritten, experiential learning—became a part of KM's expanding scope over time. This change signalled a significant advancement in KM practice by highlighting the significance of social interaction and human cognition in knowledge processes [16].

3.1 Key Theories and Recent Theoretical Developments

Nonaka and Takeuchi's SECI model, which explains the dynamic process of information conversion between implicit and explicit forms, is one of the foundational theories in KM practice [17]. The SECI model describes four methods of converting information: Combination (explicit to explicit), Externalization (tacit to explicit), Internalization (explicit to tacit), and Socialization (tacit to tacit). This paradigm highlights knowledge creation's cyclical nature and social dynamics' significance in KM practice [18]. Nonaka also introduced the Knowledge Spiral, which illustrates the ongoing and dynamic relationship between implicit and explicit knowledge. The Knowledge Spiral highlights how knowledge is created iteratively through organizational learning cycles, which allow individual expertise to be shared and extended [19].

The complexity of organizational contexts and technological advancements have been included in KM practice theories in recent years. The idea of "Big Data KM," which combines big data analytics with conventional KM practice techniques, is one developing hypothesis. This method improves decision-making processes using large-scale data analysis to produce new information and extract valuable insights [20]. The Knowledge Ecosystem Theory is another recent innovation that places KM practice in a dynamic, linked ecosystem [21]. This theory highlights the value of outside knowledge sources, including partnerships and collaborations, and the contribution of digital platforms to innovation and information exchange [22]. Furthermore, the development of intelligent KM systems that can automate the processes of knowledge capture, distribution, and application has resulted from integrating artificial intelligence (AI) with KM practice [23]. AI-driven KM practice solutions improve the efficacy and efficiency of KM practice procedures by using machine learning algorithms to recognize patterns, forecast trends, and offer personalized knowledge recommendations [24]. The continuous development of KM practices is seen in these recent theoretical developments, fueled by technical breakthroughs and the need to adjust to ever-more complicated organizational environments [13]. These new ideas provide valuable frameworks for comprehending and enhancing KM practice in modern enterprises as KM practice continues to develop [23].

4 Current Trend Index Analysis on Knowledge Management Practice

A thorough explanation of Trends Analysis in KM Practice from 2000–2024 is provided in the paper. This study examined index keywords to give a comprehensive overview of the changing environment of KM practices. This study aims to determine the top 10 to 11 keywords that have consistently influenced the field over the past 20 years, using the Scopus database as the primary source for scholarly articles. The trends' graphical depiction aims to draw attention to the dynamic shifts in keyword rankings, showcasing their significance and influence within KM practice. A thorough analysis of the frequency and importance of several terms associated with KM practice from 2000 to 2024 is given in Fig. 1. The graph shows the total number of documents that contained these terms, emphasizing their historical and current relevance. The percentages show how common these keywords were in the published literature at this period, indicating the keywords' continued importance in KM practice research and practice. Figure 2 does an excellent job of illustrating how KM practice research has evolved. Every section denotes a particular time frame, and the corresponding keyword rankings illustrate changing industry patterns and the changes in keyword emphasis across various five-year intervals from 2000 to 2024. Because of their strong influence on knowledge management techniques and close ranking closeness, this analysis focuses exclusively on the top 10 to 11 terms.

In Fig. 1, the graph illustrates the ranking of various keywords, highlighting "Knowledge Management" and "Knowledge Management Practices" as the first and second most prominent terms, respectively. The keyword "Survey" is ranked third overall, while keywords such as "Developing Countries," "Innovation," "Knowledge Management Systems," and "Education" appear at the lower end of the ranking spectrum. This overall ranking spans the period from 2000 to 2024. However, a more detailed examination of specific five-year intervals, as depicted in Fig. 2, reveals distinct keyword focus patterns. While "Knowledge Management" and "Knowledge Management Practices" consistently maintain their high rankings across all intervals, the prominence of "Survey" is particularly notable during the 2010 to 2014 period, coinciding with the emergence of "Knowledge Management Systems" during the same timeframe. This contrasts their lower general ranking as depicted in Fig. 1. Similarly, terms such as "Innovation" and "Developing Countries" are positioned lower overall but gain significance during specific periods. Keywords like "Competition," "Knowledge-Based Systems," and "Management Science," although ranked highly, are predominantly focused within the first decade, particularly from 2005 to 2009. Conversely, other keywords such as "Knowledge Sharing," "Societies and Institutions," and "Decision Making" hold an overall ranking of 7th, 8th, and 9th, respectively. Yet, the graph in Fig. 2 demonstrates their specific prevalence in distinct periods, such as 2005 to 2009 for "Societies and Institutions" and "Decision Making" and 2010 to 2014 for "Knowledge-Based Systems." This investigation underscores the disparity between the comprehensive overview of keyword rankings and the detailed analysis of focused periods, highlighting how specific intervals reflect concentrated attention on certain keywords.

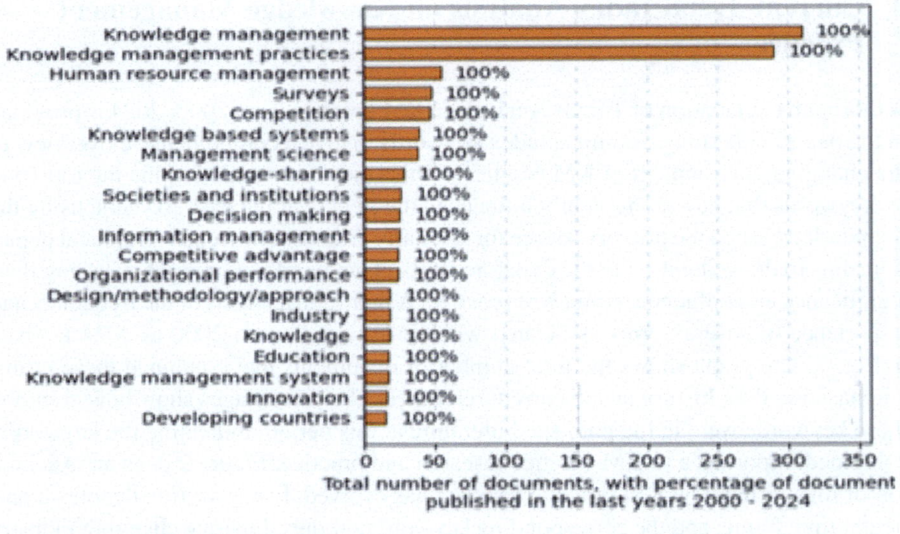

Fig. 1. Trend index analysis comprehensive indication (2000–2024).

The analysis of Fig. 2 illustrates the evolving landscape of KM practices, highlighting how various themes have maintained or shifted in importance over the years. This dynamic trend demonstrates the adaptability and complexity of KM practices across different periods, emphasizing their ongoing relevance in organizational settings. From 2000 to 2024, KM and KM Practices consistently rank highly, reflecting their foundational role in managerial strategy and practice. Information Dissemination, Information Technology, and Management Science have also maintained strong positions throughout this period, underscoring their critical influence on the structure and efficiency of

KM systems. These elements are the backbone of effective KM strategies, supporting organizations to efficiently manage, share, and utilise knowledge.

During the 2000–2004 period, there was a notable emphasis on Risk Management, Industrial Economics, Industrial Management, Computer Systems and Cooperative Work, and Engineering Research. These topics were crucial during this early phase, highlighting the need to integrate technical and economic considerations into KM practices.

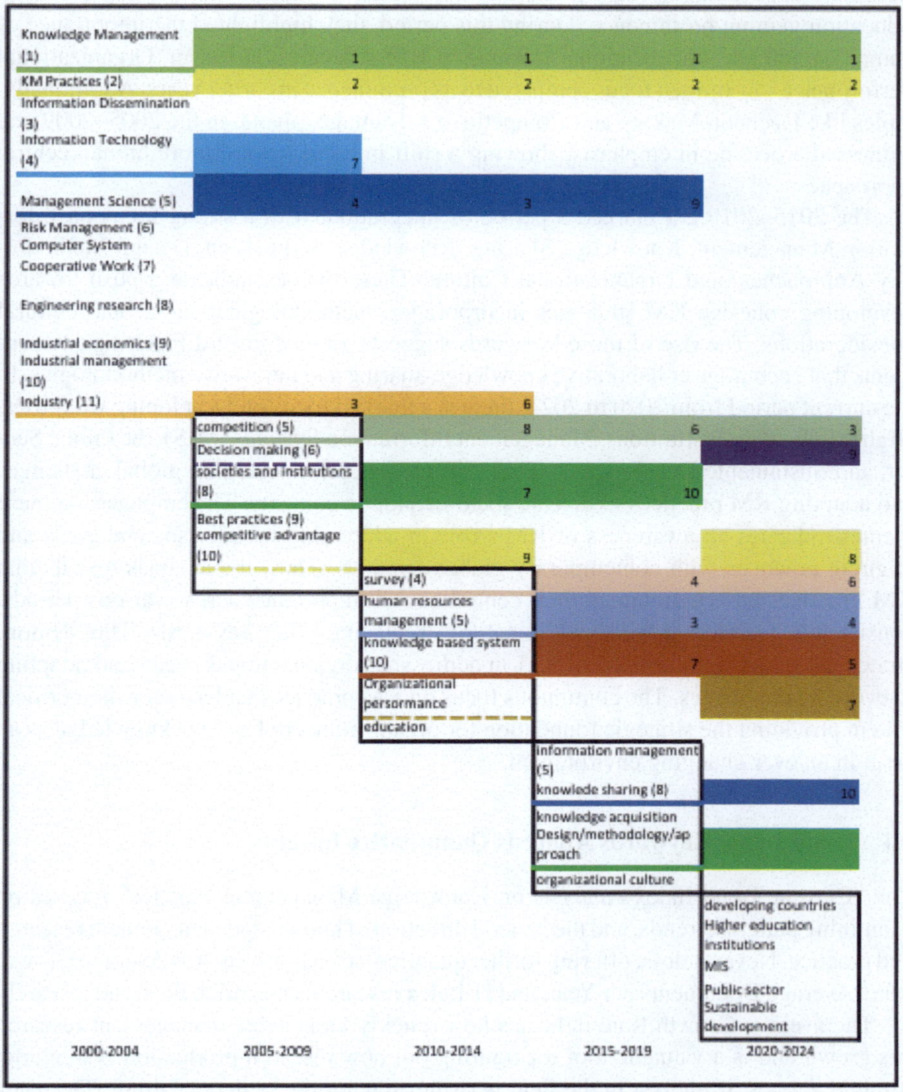

Fig. 2. Key word trends index analysis (2000–2024).

However, this focus diminished later as KM practices evolved to meet new challenges and opportunities. In 2005–2009, there was a significant shift toward themes such as Competition, Decision Making, Societies and Institutions, Best Practices, and Competitive Advantage. These topics began to overlap and replace some of the previous focus areas, indicating a strategic shift in KM practices toward leveraging knowledge for competitive gain and informed decision-making. The rise of these themes reflects an increased emphasis on using KM to drive organizational success and market positioning. 2010–2014 saw further refinement in KM practices, with keywords like Surveys, Human Resources Management, Knowledge-Based Systems, Organizational Performance, and Education gaining prominence. During this period, they highlighted the importance of human capital and organizational learning in KM strategies. However, Organizational Performance saw limited focus compared to its prominence in earlier years. Additionally, topics like Decision Making and Competitive Advantage, pivotal in the 2005–2009 era, witnessed a decline in emphasis, showing a shift in focus toward more human-centric approaches.

The 2015–2019 era marked a period of integration, with a strong focus on Information Management, Knowledge Sharing, Knowledge Acquisition, Design Methodology Approaches, and Organizational Culture. These themes indicate a push towards developing cohesive KM strategies incorporating methodological rigour and cultural considerations. The rise of these keywords suggests a move toward fostering environments that encourage collaborative knowledge-sharing and innovative methodologies. In the current period from 2020 to 2024, there is a marked focus on Developing Countries, Higher Education Institutions, Management Information Systems (MIS), the Public Sector, and Sustainable Development. This shift reflects acknowledging global challenges and adapting KM practices to diverse socio-economic contexts. The emphasis on these themes indicates an awareness of KM's role in addressing broader societal goals and aligning practices with contemporary global demands. Thus, the analysis reveals that KM Practices have maintained their centrality and importance across various periods, consistently interacting with and sometimes replacing other keywords. This demonstrates the enduring relevance of KM in addressing organizational needs and adapting to evolving challenges. The continuous focus on KM practices underscores their critical role in providing the strategic foundation for organizations to leverage knowledge as an asset in an ever-changing environment.

4.1 Trend Index Keywords Analysis Quantitative Insights

The "Current Trend Index Analysis on Knowledge Management Practice" focuses on examining patterns, trends, and the general direction of knowledge management research and practice. Nevertheless, offering further quantitative insights, such as Average Growth Rate, Average Documents per Year, and H-Index research, can enrich the trend research.

The average Growth Rate indicates how quickly knowledge management research has grown and is a valuable tool for figuring out how research production is trending. It might suggest if interest in the field is rising, staying the same, or falling. The average Documents per Year indicator measures the annual constancy of research output, which complements the growth rate. It demonstrates how aggressively the field is being researched throughout time, which further improves the study of trends. H-Index Analysis, which focuses on the quality and effect of knowledge management research, is reflected in the H-index. Including it in the trend analysis exposed the contributions in this field that have changed in quantity and significance over time. Table 1 presents the Average Growth Rate, Average Document Per Year and H-Index Analysis with quantifiable perceptions.

Table 1. Average growth rate, average document per year and h-index analysis.

Index Keywords	AGR	ADY	H-Index
Knowledge management	0.4	12.4	24
Knowledge management practices	0.3	11.6	22
Human resource management	0	2.2	9
Surveys	0	1.9	9
Competition	0.1	1.9	7
Knowledge-based systems	0	1.6	8
Management Science	0	1.5	7
Knowledge-sharing	0	1.1	6
Societies and institutions	0	1	7
Decision making	0	1	6
Information management	0	1	8
Competitive advantage	0	1	3
Organisational performance	0.1	1	6
Design/methodology/approach	0	0.7	11
Industry	0	0.7	6
Knowledge	0	0.7	4
Education	0	0.7	6
Knowledge management system	0	0.7	4
Innovation	0	0.6	6
Developing countries	0	0.6	5

AGR = Average Growth Rate, ADY = Average Document Per Year.

The above table thoroughly examines the main trends and metrics associated with KM practice. Essential terms in the field are highlighted along with the metrics that go along with them: Average Growth Rate (AGR), Average Documents per Year (ADY), and H-Index. The average growth rate (AGR), average number of documents annually (ADY), and h-index for important phrases associated with KM practice are shown in this table. With the greatest AGR of 0.4, ADY of 12.4, and a remarkable h-index of 24, the keyword “knowledge management” dominates the field. “Knowledge management Practice” comes next, showing its noteworthy influence despite slower development with a modest AGR of 0.3, a good h-index of 22, and a respectable ADY of 11.6 With h-indexes of 22, respectively, "Competition" and "organizational performance" both exhibit comparable AGRs of 0.1 with different ADYs of 1.9 and h-index 7 and 6 respectively and indicating their consistent contributions to KM practice research. The keywords "Human resource management" and "Surveys" have a constant output in terms of ADY and H-Index despite a lower growth rate (AGR of 0), suggesting a continuous level of research activity in these domains. Newer subjects with comparatively lower impact scores, such as "Competitive advantage," "knowledge-sharing," and "organizational performance," show moderate research production. The table provides an in-depth assessment of the significant issues influencing KM practice, focusing on existing fields and new trends between 2000 and 2024. These measures shed light on these issues' expansion, significance, and continued applicability in the larger field of research.

5 Discussion

An extensive investigation of index keywords research and trend analysis in knowledge management practice with theoretical underpinnings was conducted. The main goal of this work was to find a bibliometric analysis of the approach over 24 years with an emphasis on the keywords trend index. The primary focus of these keywords was on applying a primary unit of analysis in theoretically grounded knowledge management practice areas. Data was extracted from 460 journal article documents obtained from the Scopus database. The main goal was to identify the top ten keywords that consistently impact "knowledge management" and "knowledge management practice". In computer science domains, "Knowledge Management Practices" and "Knowledge Management" were centred on corporate operations and permanent organizational levels. Based on an examination of the research conducted between 2010 and 2014, the terms "survey" and "knowledge management systems" were the primary research focus with an empirical foundation. Furthermore, the "developing countries" and sustainable development are reflected globally from 2020 to 2024 because of the more significant impact of knowledge management practices. In general, the results showed that sustainable social trends, growing technological advancements, and improved organizational performance all contributed to the success of KM practice. Each trend emphasises how crucial KM practices are to ensuring that knowledge is shared, used, and deployed in a highly integrated and linked environment. This study suggests that discernment in knowledge management practice should be based on an integrative approach to knowledge processes that considers organizational culture from several angles. This study's sole constraint is focusing solely on computer science and selecting specific databases and keywords related to knowledge management and knowledge management practices. Future research can concentrate on various domain perspectives and investigate how the alignment of national and organizational cultures can predict employees' knowledge contribution to their organizations.

6 Future Directions and Implications Practice

Integrating insights from various disciplines can enrich KM practices, leading to more holistic and innovative approaches. For instance, incorporating principles from cognitive psychology can enhance our understanding of how individuals and groups learn and share knowledge. Insights from organizational behaviour can inform strategies to foster a culture of knowledge-sharing and collaboration. Additionally, applying concepts from information science can improve the design and functionality of KM systems, making them more user-friendly and effective. Cross-disciplinary approaches encourage the synthesis of diverse perspectives, fostering innovation and adaptability in KM practices. Globalization has a profound impact on KM practices, necessitating the development of strategies that transcend geographical and cultural boundaries. As organizations increasingly operate globally, they must manage knowledge across diverse and dispersed teams. This requires robust KM systems that support real-time collaboration and knowledge sharing across different time zones and cultural contexts. Additionally, globalization drives the need for multilingual KM tools and systems that can accommodate diverse linguistic needs. Effective KM practices in a global context also involve understanding and addressing the cultural nuances that influence how knowledge is perceived, valued and shared. Additionally, industry-specific regulations can impact KM practices by dictating the types of knowledge that must be retained and the protocols for sharing sensitive information. Organizations must navigate these regulatory landscapes carefully, ensuring that their KM practices comply with legal requirements and enhance trust and transparency with stakeholders.

7 Conclusion

From 2000 to 2024, this study has offered a thorough analysis of KM practices' changing patterns and ongoing importance. The results highlight the crucial influence of modern development theories on KM practices, highlighting their contribution to modernization, organizational learning, and strategic decision-making. By applying a keyword trend analysis using the Scopus database, the study pinpoints important areas of long-term significance and develops subjects and new themes in KM practices. The study emphasizes the importance of KM practices, which will remain at the centre of the field in 2024 despite the wide range of terms associated with KM practice. This prominence exemplifies the KM Practice's ongoing significance and flexibility in response to shifting organizational demands and circumstances. It restates the fundamental role KM Practices play in enabling the effective use and sharing of knowledge as a strategic corporate asset. This study also highlights how important it is to have robust KM procedures that can change with today's dynamic and complicated organizational environments. By understanding these patterns, firms can improve their operational efficiency and competitive advantage by better-utilising knowledge. This study adds to the current conversation about KM practices by thoroughly summarising theoretical advancements and essential topics. It draws attention to how vital it is to keep investigating new issues and potential lines of inquiry since doing so will help to advance KM techniques. Organizations are urged to continue being watchful and aggressive in honing their KM strategies so that they can be in a position to fully utilize information as a strategic advantage in the quickly evolving world of today.

References

1. Orieno, O.H., Udeh, C.A., Oriekhoe, O.I., Odonkor, B., Ndubuisi, L.N.: Innovative management strategies in contemporary organizations: a review: analyzing the evolution and impact of modern management practices, with an emphasis on leadership, organizational culture, and change management. Int. J. Manag. Entrep. Res. **6**(1), 167–190 (2024)
2. Islam, M.A., Widen, G.: Bibliometric analysis of the VINE journal of information and knowledge management systems: 2000–2020. VINE J. Inf. Knowl. Manag. Syst. **53**(3), 467–490 (2023)
3. Durst, S.: A plea for responsible and inclusive knowledge management at the world level. VINE J. Inf. Knowl. Manag. Syst. **54**(1), 211–219 (2024)
4. Yusoff, M. Z.: Knowledge work productivity measurement for software development process in SMEs. In: Knowledge Management International Conference (KMICe) 2021, pp. 1–5 (2014)
5. Gupta, B., Iyer, L.S., Aronson, J.E.: Knowledge management: practices and challenges. Ind. Manag. Data Syst. **100**(1), 17–21 (2000)
6. Saeed, I., Khan, J., Zada, M., Zada, S.: Employee sensemaking in organizational change via knowledge management: leadership role as a moderator. Curr. Psychol. **43**, 6657–6671 (2024)
7. Shea, T., Usman, S.A., Arivalagan, S., Parayitam, S.: Knowledge management practices as moderator in the relationship between organizational culture and performance in information technology companies in India. VINE J. Inf. Knowl. Manag. Syst. **53**(4), 719–747 (2023)
8. Tseng, S.M.: The impact of knowledge management capabilities and supplier relationship management on corporate performance. Int. J. Prod. Econ. **154**, 39–47 (2014)
9. Manesh, M.F., Pellegrini, M.M., Marzi, G., Dabic, M.: Knowledge management in the fourth industrial revolution: mapping the literature and scoping future avenues. IEEE Trans. Eng. Manage. **68**(1), 289–300 (2021)
10. Bresciani, S., Rehman, S.U., Giovando, G., Alam, G.M.: The role of environmental management accounting and environmental knowledge management practices influence on environmental performance: mediated-moderated model. J. Knowl. Manag. **27**(4), 896–918 (2023)
11. Schiuma, G., Kumar, S., Sureka, R., Joshi, R.: Research constituents and authorship patterns in the knowledge management research and practice: a bibliometric analysis. Knowl. Manag. Res. Pract. **21**(1), 129–145 (2023)
12. Mohammed Al-Shahrani, M.: Trends in knowledge management processes and practices. J. Res. Bus. Soc. Sci. **2**(12), 1–8 (2019)
13. Serenko, A.: A structured literature review of scientometric research of the knowledge management discipline: a 2021 update. J. Knowl. Manag. **25**(8), 1889–1925 (2021)
14. Tranfield, D., Denyer, D., Smart, P.: Towards a methodology for developing evidence-informed management knowledge by means of systematic review. Br. J. Manag. **14**(3), 207–222 (2003)
15. Marques, J.E., Gobbo, J.A., Fukunaga, F., Cerchione, R., Centobelli, P.: Use of knowledge management systems: analysis of the strategies of Brazilian small and medium enterprises. J. Knowl. Manag. **24**(2), 369–394 (2020)
16. Mitchell, V.W., Harvey, W.S., Wood, G.: Where does all the 'know-how' go? The role of tacit knowledge in research impact. High. Educ. Res. Dev. **41**(5), 1664–1678 (2022)

17. Mubuyaeta, M.M., Ponera, J.M.: An understanding of the concept of knowledge management for knowledge management system implementation in two mobile telecommunication companies in Namibia. Univ. Dar es Salaam Libr. J. **18**(2), 3–18 (2024)
18. Hu, Y.: The SECI model and external sources of knowledge: a field study on the distribution of search routines. University of New South Wales (2020)
19. Zatsman, I.: Building digital spiral models of knowledge generation. In: 19th Forum (International) on Knowledge Asset Dynamics Proceedings, pp. 2185–2196 (2024)
20. Rosari, R., Ardianto, H., Danarilia, V.O., Mada, U.G.: Exploring the ambidexterity landscape: a comprehensive bibliometric analysis of organizational change research. Jurnal Minds: Manajemen Ide dan Inspirasi **11**(1), 85–110 (2024)
21. Seufert, S., Guggemos, J., Sailer, M.: Technology-related knowledge, skills, and attitudes of pre-and in-service teachers: the current situation and emerging trends. Comput. Hum. Behav. **115**, 106552 (2021)
22. Di Vaio, A., Latif, B., Gunarathne, N., Gupta, M., D'Adamo, I.: Digitalization and artificial knowledge for accountability in SCM: a systematic literature review. J. Enterp. Inf. Manag. **37**(2), 606–672 (2024)
23. Iaia, L., Nespoli, C., Vicentini, F., Pironti, M., Genovino, C.: Supporting the implementation of AI in business communication: the role of knowledge management. J. Knowl. Manag. **28**(1), 85–95 (2024)
24. Abbas, J.: Impact of total quality management on corporate sustainability through the mediating effect of knowledge management. J. Clean. Prod. **244**, 118806 (2020)

Enhancing Student Mathematical Knowledge and Understanding Through Multimedia Videos

Juliana Wahid(✉), Ali Yusny Daud, and Noor Izzah Yahya

School of Computing, University Utara Malaysia, 06010 Sintok, Kedah, Malaysia
w.juliana@uum.edu.my

Abstract. This study investigates using multimedia videos to enhance student understanding and knowledge of mathematics. Traditional teaching methods often rely on text and static diagrams, which may not meet the diverse learning needs of all students. Multimedia, including videos, animations, interactive simulations, and digital presentations, offers dynamic and engaging ways to present abstract mathematical concepts. The study aims to develop multimedia videos, integrate them into the mathematics teaching and learning process, and evaluate their impact on student learning. These videos were incorporated into a classroom setting. Data were collected through pre-tests, post-tests, and observations to measure changes in student performance. Analysis showed significant improvements in student understanding and retention of mathematical concepts when using multimedia videos. This study proves that multimedia videos can effectively improve mathematical comprehension and offer opportunities for integration into the curriculum. The findings support using multimedia to make mathematics more accessible and engaging, suggesting that such approaches can foster more profound understanding and better student learning outcomes.

Keywords: Mathematics Education · Student Understanding · Multimedia Video

1 Introduction

In recent years, there has been a growing interest in leveraging multimedia-based instructional methods to enhance teaching and learning experiences in various subjects, including mathematics. Integrating multimedia elements such as videos, animations, interactive simulations, and digital presentations has shown promise in making abstract mathematical concepts more accessible and engaging for students [12]. These methods can cater to different learning styles, provide visual and interactive representations of mathematical principles, and offer dynamic, real-time feedback that traditional teaching methods often lack [3].

In the context of the STQM1203 Mathematics for Information Technology, which covers two domains of mathematics, i.e. linear algebra and calculus, that are crucial for students' academic and professional development, effective teaching strategies are essential. Traditional teaching methods, which often rely heavily on lectures and static

Z. Mat Aji et al. (Eds.): KMICe 2024, ISEM 49, pp. 116–130, 2025.
https://doi.org/10.1007/978-3-031-91485-0_10

textbook examples, may not sufficiently address the diverse learning needs of all students. Consequently, there is a need to explore innovative approaches to enhance student understanding and retention of mathematical knowledge.

Despite the critical importance of mathematics in the STQM1203 course, many students struggle to grasp and apply key concepts effectively. Traditional instructional methods may not adequately engage students or address their individual learning needs, leading to gaps in understanding and lower academic performance. Specifically, the current approach may fail to make the subject matter relatable and engaging, crucial for fostering a more profound comprehension and appreciation of mathematics [8].

In addition, based on teaching experience and grading the final examination of STQM1203 for more than 5 semesters, it was observed that 50% of students left blank spaces on the final exam paper in the integration topics. This perhaps contributes to 70% of students not achieving more than 50% of the marks in the proficient and advanced category for the second Course Learning Objective (CLO) (Solving appropriate problems using linear algebra and calculus techniques), as shown in Fig. 1.

Semester A221 - (2022/2023 1 - First)
Course STQM1203 - MATHEMATICS FOR INFORMATION TECHNOLOGY
Group

QAT : 70%
70% of the students achieve more than 50% of the marks "proficient" and above for each CLO.

No.	CLO No.	Count of Student	Below Basic [0,25]	Basic (25,50]	Proficient (50-75]	Advance (75,100]	Achieved Target %	Remark
1.	CLO1	18	0 0.00%	3 16.66%	6 33.33%	9 50.00%	15 83.33%	Achieved
2.	CLO2	18	1 5.55%	9 50.00%	4 22.22%	4 22.22%	8 44.44%	Not Achieved
3.	CLO3	18	2 11.11%	1 5.55%	7 38.88%	8 44.44%	15 83.33%	Achieved

LEGEND :

CLO	Domain
CLO1	Describe linear algebra and calculus techniques discussed in this course(C2,A1,P1)
CLO2	Solve appropriate problems using linear algebra and calculus techniques(A5,C3)
CLO3	Use appropriate modern technology to explore linear algebra and calculus concepts(A1,C3)

Fig. 1. Course learning objectives statistic of STQM1203 for first semester 2022/2023.

This study aims to address these challenges by developing and implementing multimedia videos to enhance student understanding and knowledge on the substitution method of integration topic in STQM1203 class. The study will investigate the impact of these multimedia videos on student understanding, confidence, engagement, and overall performance in the course. By integrating multimedia elements into the curriculum, this study seeks to provide a more effective and engaging learning experience that can improve students' mathematical knowledge and ability to apply it in real-world scenarios.

The following section describes previous studies on the impact of multimedia in education, specifically in mathematics, followed by a detailed explanation of the methodology used in this study. Afterwards, the findings and their discussion are presented. Finally, conclusions and future recommendations are summarised.

2 Literature Review

Richard Mayer's Cognitive Theory of Multimedia Learning (CTML) [9] suggests that people learn more effectively from words and pictures than from words alone. Mayer's theory emphasises the importance of designing multimedia instructional materials that align with human cognitive architecture. According to [9], effective multimedia instruction involves three mental processes: selecting relevant information, organising it into coherent structures, and integrating it with existing knowledge. This theory underscores the potential benefits of using multimedia-based materials to enhance student understanding of mathematics.

Research has consistently shown that multimedia can be a powerful tool in mathematics education. Studies by [11] demonstrate that multimedia elements like animations and interactive simulations can improve students' conceptual understanding and problem-solving skills. These multimedia tools provide dynamic and visual representations of mathematical concepts, making abstract ideas more concrete and understandable. For example, a study by [5] found that interactive multimedia in mathematics classrooms significantly improved student engagement and comprehension. The interactive nature of these tools allows students to manipulate variables and observe outcomes, fostering a deeper understanding of mathematical principles.

In the context of higher education, the study by [7] highlighted the positive impact of multimedia on students' understanding and retention of complex mathematical concepts. Multimedia allows for more personalised learning experiences, catering to different learning styles and paces. While the benefits of multimedia in education are well-documented, there are challenges and considerations to be addressed. One challenge is ensuring that multimedia materials are pedagogically sound and aligned with curriculum objectives. As [10] points out, simply adding multimedia elements does not guarantee improved learning outcomes; the design and integration of these elements must be carefully planned.

Videos in teaching mathematics have gained significant attention due to their potential to enhance student understanding, particularly in complex subjects. Previous research has shown that video-based learning media can improve comprehension, as seen in studies where video interventions led to higher post-test scores than pre-tests [1]. Videos allow students to learn at their own pace and revisit challenging concepts, making them a valuable tool in higher education, especially for subjects requiring visualisation, such as mathematics and engineering [6]. Moreover, integrating storytelling, visual aids like hand-drawn diagrams, and interactive video elements can further support students' learning and engagement [4, 15]. These advancements suggest that videos are effective for procedural learning and enhancing cognitive processes, making them an essential resource in modern education.

This study focuses on a different approach by exploring the effectiveness of multimedia videos tailored explicitly for teaching the STQM1203 substitution method of integration in a higher education setting.

3 Methodology

The methodology for this study, as shown in Fig. 2, starts with developing multimedia videos tailored to the substitution integration method. These videos will be used as an intervention to enhance students' understanding. Next, the lecturer demonstrates the substitution method of integration using traditional notes and lectures. After the explanation, along with examples, students will be given integration questions to test their understanding. Students are given 5 min to answer the questions using the substitution method. After the 5-min period ends, students must submit their answers, whether complete or not. Then, students need to take the pretest. Next, the lecturer introduces two videos on solving integration using the substitution method. Students are required to watch the videos while answering new integration questions. Using the videos, students can play or pause if the explanation in the video is too slow or fast. Students are also given 5 min to answer the questions and must submit their answers, whether completed or not, after the 5 min. Then, students need to take the post-test. The answers submitted by students during the pretest and posttest will be assessed by the lecturer using a problem-solving rubric. Pretests and post-tests were carried out to measure changes in student understanding, engagement and confidence before and after the intervention of the multimedia videos. The lecturer's observations on the answers submitted were also collected to determine whether the student has fully grasped the problem's requirements, to evaluate the student's proposed solution method and to assess the final answer provided by the student.

At the end of the semester, during the final exam, each student's marks on the question related to integration using the substitution method are recorded. The aim is to assess the effectiveness of the multimedia videos developed for the substitution integration method. Finally, all the collected data are analysed using statistical methods to determine if there is a significant improvement in students' understanding, engagement, confidence and performance of the substitution integration method. The detailed phases of the methodology are described in the following sub-section.

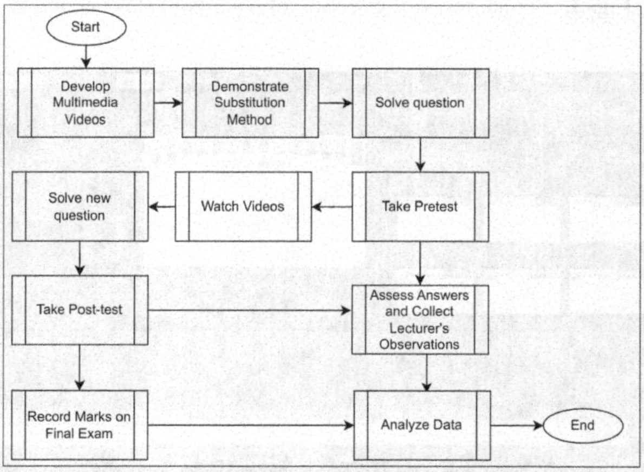

Fig. 2. Methodology flowchart.

3.1 Design and Development Multimedia Video

The process of developing the multimedia videos for the substitution method of integration begins with identifying and selecting a range of integral questions that effectively demonstrate the concept and applications of the substitution method. Once the questions are chosen, a detailed script or outline is prepared, explaining each step of the solution process clearly and concisely. The script includes explanations of the reasoning behind selecting specific substitutions, the algebraic manipulations involved, and the final integration steps. After the script is ready, the video recording process is executed by recording a live presentation with the Microsoft PowerPoint Screen Recording feature (Fig. 3) and a digital pen to write the solutions. This digital pen allows for clear, precise handwriting and the ability to highlight essential steps or concepts, making the tutorial more engaging and more accessible to follow. Using a digital pen also facilitates the creation of diagrams, annotations, and other visual aids that enhance understanding. Post-production involves editing the video using CapCut (Fig. 4) and Audacity (Fig. 5) to ensure smooth transitions, adding voice-over explanations, and including any necessary corrections or clarifications. The video is produced in MP4 format, a widely compatible and high-quality format for online sharing and playback on various devices. Two videos consisted of different questions, as shown in Fig. 6.

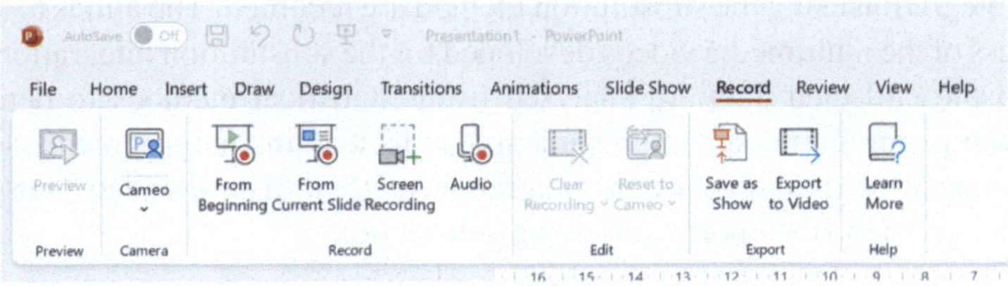

Fig. 3. Screen recording feature of Microsoft PowerPoint.

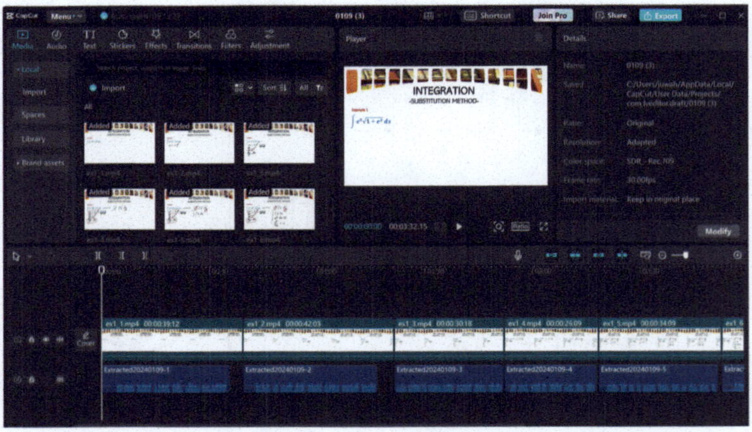

Fig. 4. CapCut application.

3.2 Pre-test and Post-test Design

The pre-test is administered after introducing the substitution method of integration using traditional notes and lectures. It includes four questions (Likert scale type) designed to

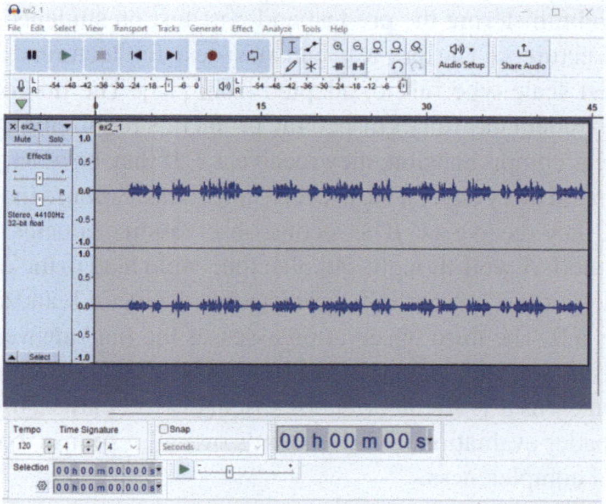

Fig. 5. Audacity application.

Fig. 6. Screen capture of example 1 and example 2 videos.

assess students' understanding, confidence, and engagement and how the conventional instructional approaches have supported the students' conceptual knowledge of the substitution integration method. After introducing the substitution method of integrating multimedia videos, the post-test evaluates how well students have grasped the subject through the videos. The post-test mirrors the pre-test in structure and includes questions regarding the use and experience related to the video examples. The Likert scale responses both in pre-test and post-test include "[Very Low, Low, Moderate, High, Very High]", "[Not confident at all, Somewhat confident, Moderately confident, Very confident, Extremely confident]", "[Strongly disagree, Disagree, Neither agree nor disagree, Agree, Strongly agree]" and "[Very negative, Negative, Neutral, Positive, Very positive]".

In addition to administering the pre-test and post-test on students, the test was also implemented on Lecturers. The pre-test and post-test for lecturers also include four observations (Likert scale type rubric) adapted from [13]. The first observation determines whether the student has fully grasped the problem's requirements. If the student's response shows clear comprehension, they receive a 2. If they have only partially understood or misinterpreted a key aspect, they receive a 1. If the response indicates a complete misunderstanding, they receive a 0. The second observation evaluates the student's proposed solution method. A well-thought-out plan that could lead to the correct solution, if correctly executed, scores a 2. A partially correct plan receives a 1, and a wholly incorrect or absent plan gets a 0. The third observation assesses the final answer provided by the student. A correct answer with proper labelling earns a 2. An answer with minor errors or a partial solution earns a 1. An incorrect or absent answer earns a 0. Finally, lecturers must provide a broader evaluation of the student's overall response, considering clarity, understanding, and completeness.

3.3 Sampling, Data Collection and Analysis

This study employs the substitution method of integrating multimedia videos within one semester for the 2023/2024 first semester (A231) on 144 undergraduate students enrolled in STQM1203 Mathematic for Information Technology subject in Universiti Utara Malaysia. Quantitative data is collected through pre-tests and post-tests both on students and lecturers. In addition, final marks for each student related to integration questions using the substitution method are also gathered. The Likert scale responses of students from the pre-test and post-test were first converted to numerical values based on different response categories, allowing for quantitative data analysis. The conversion was carried out in increasing order, with the lowest value representing the most negative response and the highest value representing the most positive response, e.g. Very Low = 1, Low = 2, Moderate = 3, High = 4, Very High = 5.

Descriptive statistics are employed on all the collected data to analyse survey responses, measuring changes in students' understanding, engagement, confidence, and perceptions of the multimedia videos. Additionally, all the data were analysed using statistical methods of paired t-tests by p-value to compare the mean scores before and after the substitution method of integration multimedia videos. The p-value quantifies the probability of obtaining a pre-test result at least as extreme as the post-test, assuming that the null hypothesis (no effect or no difference between pre-test and post-test responses) is correct. It helps assess the strength of the evidence against the null hypothesis. The threshold of p-value less than 0.05 is used in this study as this value is used in many scientific fields [2, 14]. Finally, to effectively communicate the findings, data visualisation of the analysed data through graphical formats is also presented.

4 Findings and Discussion

This section provides a detailed overview of the participant characteristics, descriptive statistics of the collected data, and data analysis.

4.1 Demographic of Student

Out of 144 students, 108 of them answered the pre-test and only 98 for the post-test. In addition, only 105 students submitted the answers for the integration using the substitution method for the pre-test and post-test. The program and gender related to students who completed both tests and submitted answers were selected to be analysed (Table 1). In total, 94 students completed the pre-test and post-test and submitted the answers.

The respondents had studied different programs, including Bachelor of Arts in Linguistics and Information Technology (B.A (Hons) in LINGUISTICS AND IT), Bachelor of Finance (B.Fin (Hons)), Bachelor of Science (Information Technology) (B.Sc.IT. (Hons)), Bachelor of Business Administration (BBA (Hons)), Bachelor of Operations Management (B.POM (Hons)), Bachelor of Law (LLB (Hons)), and Bachelor of Science Economics (B.Sc.Ec (Hons)).

Table 1. Students' demographic distribution.

	B.A (Hons) in LINGUISTICS AND IT	B.Fin (Hons)	B.Sc.IT. (Hons)	BBA (Hons)	B.POM (Hons)	LLB (Hons)	B.Sc.Ec. (Hons)
Female	13	1	60	1	1	1	1
Male	0	0	15	0	0	0	1

4.2 Data Set

Tables 2 and 3 summarise descriptive statistics of the data collected on students' pre-test and post-test responses, respectively. In contrast, Table 4 summarises descriptive statistics on lecturers' pre-test and post-test observations. The descriptive statistics for the final examination marks obtained by the students are presented in Table 5. Descriptive statistics include measures of central tendency (such as the mean and median) and variability or dispersion (such as standard deviation, variance, and range).

Table 2 summarises the students' self-assessment of their understanding, confidence, engagement, and perceived support from traditional instructional approaches for the substitution method of integration before the introduction of multimedia videos. The mean score for understanding was 3.07, with a confidence level of 2.52. The engagement level with traditional approaches was rated at 3.39, and perceived support for conceptual understanding scored 3.5. The scores showed moderate variability, as indicated by the standard deviations, with confidence showing the highest variability (1.01) among the measured factors. Post-test responses (Table 3) suggested improvements across various aspects of students' experiences with the substitution integration method after exposure to multimedia videos. The mean score for understanding increased to 3.45, and confidence rose to 2.87. Engagement and interest in the multimedia videos were rated at 3.57, with perceived support for conceptual understanding at 3.73. Students felt that these videos helped them better understand the topic than traditional methods, as reflected in a mean score of 3.71. Overall, their learning experience with multimedia videos received a mean rating of 3.84, with relatively low response variability.

Table 4 outlines the lecturers' observations of students' abilities before and after the intervention with multimedia videos. The measured aspects include understanding the problem, planning a solution, getting an answer, and an overall holistic scale. Pre-test scores indicated a mean of 1.52 for understanding the problem and 1.53 for planning a solution. Post-test scores showed slight improvements, with means increasing to 1.73 for understanding and planning and 1.61 for getting an answer. The holistic scale also increased from 4.17 to 5.04, indicating an overall enhancement in students' problem-solving abilities as observed by lecturers. The variability in these observations was generally lower than in the student self-assessments.

As shown in Table 5, the final examination marks had a mean of 1.89 and a median of 2. The standard deviation was 1.64, indicating the extent of variability in the marks, while the variance was 2.71. The range of the marks was 4, showing the difference between the highest and lowest scores.

Table 2. Descriptive statistics of students pre-test responses.

Pre-test Questions	Mean	Median	Standard Deviation	Variance	Range
1. Please rate your current understanding of the substitution method of integration covered in this course	3.07	3	0.69	0.47	4
2. How confident are you in applying the substitution method of integration covered in this course?	2.52	3	1.01	1.02	4
3. How would you rate your engagement and interest in the traditional instructional approaches used in this course?	3.39	3	0.64	0.41	3
4. To what extent do you feel the traditional instructional approaches have supported your conceptual understanding of the substitution integration method?	3.5	3	0.85	0.72	3

Table 3. Descriptive statistics of student post-test responses.

Post-test Questions	Mean	Median	Standard Deviation	Variance	Range
1. Please rate your current understanding of the substitution method of integration covered in this course	3.45	3	0.69	0.48	4
2. How confident are you in applying the substitution method of integration covered in this course?	2.87	3	0.84	0.71	4
3. How would you rate your engagement and interest in the multimedia-based instructional substitution method of integration used in this course?	3.57	4	0.63	0.39	3
4. To what extent have multimedia-based instructional materials supported your conceptual understanding of the substitution integration method?	3.73	4	0.85	0.73	3
5. Compared to the traditional instructional approaches, do you feel that using multimedia-based instructional materials has helped you better understand the substitution method of integration?	3.71	4	0.69	0.48	3
6. How would you rate your learning experience with the multimedia-based instructional substitution integration method?	3.84	4	0.64	0.41	3

4.3 Data Analysis

Table 6 contains p-values from comparisons between different assessments (Pre-Test vs Post-Test, Pre-Test vs Final Marks and Post-Test vs Final Marks) for students' understanding, confidence, engagement, perceived support, video effectiveness, and overall experience.

Table 4. Descriptive statistics of lecturer pre-test and post-test observations.

		Understanding the Problem	Planning a Solution	Getting an Answer	Holistic Scale
Pre-Test	Mean	1.52	1.53	1.26	4.17
	Median	2	2	2	5
	Standard Deviation	0.67	0.63	0.95	1.55
	Variance	0.44	0.40	0.90	2.42
	Range	2	2	2	5
Post-Test	Mean	1.73	1.73	1.61	5.04
	Median	2	2	2	6
	Standard Deviation	0.59	0.58	0.76	1.50
	Variance	0.34	0.34	0.58	2.27
	Range	2	2	2	4

Table 5. Descriptive statistics of student final examination marks.

	Mean	Median	Standard Deviation	Variance	Range
Final Exam Marks	1.89	2	1.64	2.71	4

The p-values for students' understanding across the comparisons are very low, indicating significant differences. Specifically, Pre-Test vs Post-Test: $p = 0.000003652$, Pre-Test vs Final Marks: $p = 0.000000003106$, and Post-Test vs Final Marks: $p = 0.00000000000004843$. These values suggest that the substitution method of integration multimedia videos significantly impacted students' understanding, as measured from pre-test to post-test and final marks. The p-values for students' confidence also show significant differences with p-values where Pre-Test vs Post-Test: $p = 0.006958$, Pre-Test vs Final Marks: $p = 0.001721$ and Post-Test vs Final Marks: $p = 0.\ 0000009431$. The consistent reduction in p-values indicates increased student confidence, which became more pronounced when comparing post-test scores to final marks. The p-values for students' engagement are similarly low, where Pre-Test vs Post-Test: $p = 0.02873$, Pre-Test vs Final Marks: $p = 0.0000000000000501$ and Post-Test vs Final Marks: $p = 0.00000000000000129$. These findings highlight a significant increase in student engagement throughout the course, particularly from the pre-test to the final marks. The p-values for students' perceived support indicate substantial differences across all comparisons, with the lowest being between post-test and final marks ($p = 0.00000000000000022$). The students' responses on video effectiveness and overall experience are only compared between post-test and final marks, with extremely low p-values ($p = 0.000000000000000536$ and $p = 0.00000000000000022$), respectively, suggesting that video resources were highly effective and positively contributed to the overall experience.

In addition, a bar chart to visualise the p-values for each comparison in Table 6, shown in Fig. 7, aims to help communicate the results more effectively. In the bar chart, all comparisons of students' understanding, confidence, engagement, and perceived support assessment have the lowest p-value for Post-Test vs Final marks (red bar), showing significant differences observed after the intervention and final marks.

In Table 7, the p-values are divided into different scales of lecturers' responses, including how students understand the problem, plan a solution, get an answer, and a holistic scale. Significant differences are noted in the p-value on how students understand the problem across all comparisons, particularly between pre-test and post-test ($p = 0.000008221$) and between pre-test and final marks ($p = 0.03953$). Meanwhile, significant improvements are seen in the p-value on how students plan a solution, especially from pre-test to post-test ($p = 0.000004865$) and pretest to final marks ($p = 0.04419$). For the scale on how students get an answer, the p-value differences are significant for pre-test vs post-test ($p = 0.00001453$) and pre-test vs final marks ($p = 0.001246$) but less for post-test vs final marks ($p = 0.1287$). The holistic scale shows highly significant differences across all comparisons, with p-values ranging from 0.000000001059 to 0.000000000000000220.

Table 6. P-value comparisons between different assessments on students.

	Pre-Test vs Post-Test	Pre-Test vs Final marks	Post-Test vs Final marks
Understanding	0.000003652	0.000000003106	0.00000000000004843
Confidence	0.006958	0.001721	0.0000009431
Engagement	0.02873	0.0000000000000501	0.00000000000000129
Perceived Support	0.02244	0.00000000000001122	0.00000000000000022
Video Effectiveness	NA	NA	0.000000000000000536
Overall Experience	NA	NA	0.00000000000000022

The bar chart in Fig. 8, which visualises the p-value in Table 7, highlights significant differences across most metrics, particularly between pre-test and post-test scores and pre-test and final marks (blue and green bar). The Post-Test vs Final marks comparisons show higher p-values in some categories, indicating less significant differences or changes. Overall, the interventions seem to have a substantial impact, particularly in understanding the problem, planning solutions, and overall holistic performance.

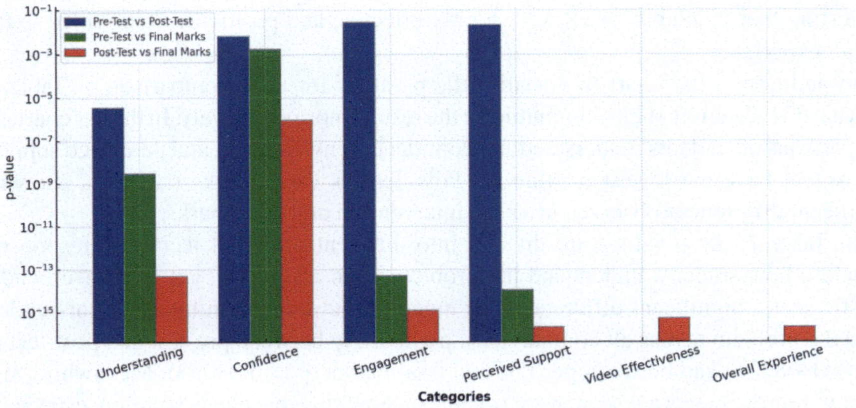

Fig. 7. Comparison of P-value across Different Assessments on Students.

Table 7. P-value comparisons between different assessments on lecturers' observations

	Pre-Test vs Post-Test	Pre-Test vs Final marks	Post-Test vs Final marks
Understanding the Problem	0.000008221	0.03953	0.3784
Planning a Solution	0.000004865	0.04419	0.3784
Getting an Answer	0.00001453	0.001246	0.1287
Holistic Scale	0.000000001059	0.00000000000000022	0.00000000000000022

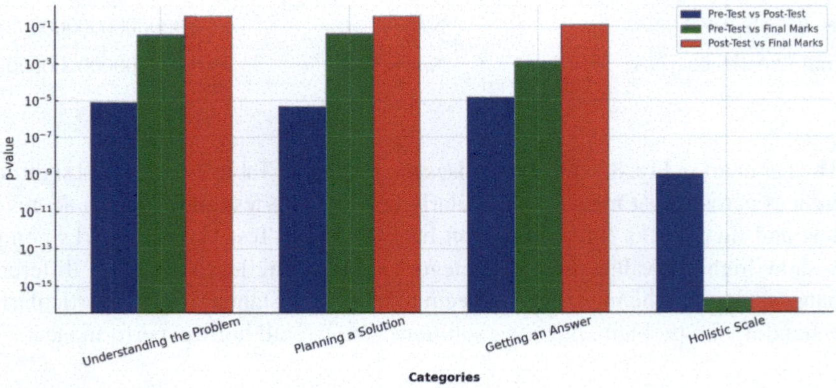

Fig. 8. Comparison of P-value across different assessments on lecturers.

5 Conclusion

In conclusion, this study demonstrates the positive impact of multimedia videos in enhancing student understanding and retention of mathematical concepts. By integrating dynamic visual aids into traditional teaching methods, students could engage more deeply with abstract ideas, resulting in significant performance improvements. The findings suggest that multimedia videos not only make mathematics more accessible but also cater to diverse learning styles, promoting a more inclusive and effective educational environment. This study highlights the potential of multimedia videos as a valuable resource in the mathematics curriculum, encouraging further exploration of their role in fostering deeper comprehension and improved learning outcomes.

Acknowledgements. This work is supported by Universiti Utara Malaysia SoTL Research Grant (S/O code: 21588).

References

1. Afiyah, A.N., et al.: Timestamp-based video integrated by STEAM to improve student's understanding of the composition of function. AIP Conf. Proc. **2926**(1) (2024). Art. No. 020031
2. Benjamin, D.J., et al.: Redefine statistical significance. Nat. Hum. Behav. **2**(1), 6–10 (2018)
3. Boyce, S., O'Halloran, J.: Active learning in computer-based college algebra. PRIMUS **30**(4), 458–474 (2020)
4. Hernandez-Martinez, P., Keane, T.: Learning mathematics and its relevance through a digital storytelling assessment task at university. Int. J. Math. Educ. Sci. Technol. **56**, 1–17 (2024)
5. Kay, R.H., Knaack, L.: Analysing the effectiveness of learning objects for secondary school science classrooms. J. Educ. Multimed. Hypermedia **18**(1), 382–392 (2009)
6. Lu, J.C.C.: Creating effective educational videos on YouTube in higher education. IEEE ICEIB **2023**, 32 (2023)
7. Liao, C.W., Chen, C.H., Shih, S.J.: The interactivity of video and collaboration for learning achievement, intrinsic motivation, cognitive load, and behavior patterns in a digital game-based learning environment. Comput. Educ. **133**, 43–55 (2019)
8. Malhotra, R., Verma, N.: An impact of using multimedia presentations on engineering education. Procedia Comput. Sci. **172**, 71–76 (2020)
9. Mayer, R.E., Moreno, R.: A cognitive theory of multimedia learning: implications for design principles. J. Educ. Psychol. **91**(2), 358–368 (1998)
10. Mayer, R.E.: Cognitive theory of multimedia learning. In: Mayer, R.E. (ed.) The Cambridge Handbook of Multimedia Learning, vol. 2, pp. 31–48. Cambridge University Press, New York (2014)
11. Moreno, R., Mayer, R.: Interactive multimodal learning environments: Special issue on interactive learning environments: contemporary issues and trends. Educ. Psychol. Rev. **19**(3), 309–326 (2007)
12. Nancy, W., Parimala, A., Merlin Livingston, L.M.: Advanced teaching pedagogy as innovative approach in modern education system. Procedia Comput. Sci. **172**, 382–388 (2020)
13. New Jersey Institute of Technology. Math Rubrics. https://web.njit.edu/~ronkowit/presentations/rubrics/samples/math_probsolv_chicago.pdf. Accessed 31 July 2024
14. Wasserstein, R.L., Schirm, A.L., Lazar, N.A.: Moving to a world beyond "$p < 0.05$". Am. Stat. **73**(sup1), 1–19 (2019)

15. Zhang, I., Guo, X.H., Son, J.Y., Blank, I.A., Stigler, J.W.: Watching videos of a drawing hand improves students' understanding of the normal probability distribution. Mem. Cogn. **53**, 262–281 (2024)

Part 2: Blockchain and Cybersecurity with Knowledge Management

Mapping the Landscape: A Bibliometric Perspective of Blockchain Applications and Trust Management

Athirah Rosli[1,2](✉), Suhaidi Hassan[1], Mohd Hasbullah Omar[1], Mohd Samsu Sajat[1], Zainab Senan Attarbashi[3], and Khuzairi Mohd Zaini[1]

[1] Internetworks Research Laboratory, School of Computing, Universiti Utara Malaysia, 06010 Sintok, Kedah, Malaysia
athirah_rosli1@ahsgs.uum.edu.my

[2] Faculty of Computer Science and Mathematics, Universiti Malaysia Terengganu, 21030 Kuala Nerus, Terengganu Darul Iman, Malaysia

[3] Faculty of Information and Communication Technology, International Islamic University Malaysia, Gombak, Malaysia

Abstract. This comprehensive bibliometric study examines the burgeoning research landscape to study the relationship between blockchain applications and trust management. Leveraging the extensive Scopus database, a number of 541 relevant documents spanning 2017 to 2024 were meticulously curated. The investigation unveils critical insights through a multifaceted analysis encompassing publication trends, preeminent authors, influential journals, collaborative networks, citation impact, and emerging research trajectories. The findings illuminate a marked surge in scholarly output, underscoring the escalating academic enthusiasm and industrial adoption of blockchain-powered trust solutions across diverse sectors. Pivotal authors and institutions steering the field are spotlighted, delineating their seminal contributions. Premier publication venues propagating cutting-edge advancements are highlighted, fostering interdisciplinary discourse. This study elucidates the intrinsic research dynamics propelling knowledge dissemination by mapping intricate collaboration patterns and assessing citation resonance. Harnessing powerful analytical tools, including VOSviewer, Microsoft Excel, OpenRefine, Biblioshiny, and Harzing's Publish or Perish, this investigation charts an exhaustive cartography of the existing state-of-the-art while identifying potential research avenues primed for exploration. This bibliometric analysis serves as an essential area for stakeholders navigating the frontiers of trustworthy blockchain integration.

Keywords: Future Internet · Trust Management · Blockchain Management

1 Introduction

Blockchain technology, a decentralized and distributed digital ledger, has emerged as a groundbreaking innovation with the potential to transform various industries. Initially developed to support cryptocurrencies like Bitcoin, blockchain's unique characteristics,

Z. Mat Aji et al. (Eds.): KMICe 2024, ISEM 49, pp. 133–146, 2025.
https://doi.org/10.1007/978-3-031-91485-0_11

such as immutability, transparency and security, have paved the way for its application beyond financial transactions. Blockchain's decentralized nature ensures that no single entity controls the entire data chain, enhancing security and reducing the risk of data tampering. This feature is particularly crucial in trust management, where the reliability and integrity of information are paramount.

Blockchain technology transforms For example, in supply chains, blockchain verifies product authenticity and protects against tampering during transit [1]. In healthcare, it secures patient data, ensuring records are accurate and private [2]. In financial services, it facilitates secure, direct transactions, which reduces costs and improves efficiency [3].

By facilitating direct transactions through its peer-to-peer network, blockchain reduces costs and mitigates fraud risks. It fosters a resilient system less prone to errors and corruption, enhancing the efficiency of processes [4]. For instance, blockchain boosts transparency in government operations, potentially reducing corruption and bolstering public confidence. In digital identity management, it empowers individuals to control their personal data, thereby decreasing identity theft and enhancing privacy [5].

In conclusion, the intersection of blockchain technology and trust management represents a significant area of innovation with far-reaching implications. As blockchain continues to evolve, understanding its impact on trust management through a comprehensive bibliometric analysis can provide valuable insights into the current research landscape, highlight emerging trends, and identify potential areas for future study.

This paper presents a thorough bibliometric analysis of blockchain applications and investigates the literature on trust management in blockchain applications. To accomplish this objective, the following research questions must be addressed:

RQ1:	What are the publication trends in blockchain applications and trust management?
RQ2:	Who are the leading authors, and what are their contributions to the field?
RQ3:	Which journals and conferences are the primary venues for publishing research on blockchain and trust management?
RQ4:	What are the collaboration patterns among researchers and institutions?
RQ5:	What is the citation impact of research in this field?
RQ6:	What are the emerging trends and future directions in blockchain applications and trust management?

The subsequent sections of this document are structured as follows: Sect. 2 will delve into the methodology employed for the bibliometric analysis and the data utilized in the study. In Sect. 3, the findings and analyses of all the research questions are presented and discussed. In conclusion, Sect. 4 elucidates the research's findings.

2 Methodology

2.1 Bibliometric Analysis

The terms "scientometrics," "bibliometrics," "informetrics," and "librametrics" derive from combining "metrics" with "bibliography," "information," "science," and "library," respectively. These terms, often used interchangeably, pertain to the field of library and information science and are fundamentally connected to the measurement of knowledge. They focus on the development of new ideas through effective information communication channels, which aim to facilitate the timely collection and distribution of crucial information for knowledge creators [6].

Bibliometric analysis classifies publications based on various criteria, such as the year of publication, authorship, affiliation, or country of origin. This classification helps identify emergent characteristics in the research field. Metrics like the number of citations, annual citation rate, h-index, and g-index are commonly used to evaluate the impact and performance of publications. Additionally, co-authorship, co-citation, keyword occurrence, and bibliographic coupling indicators can map and visualize the current state of the art in specific research domains [8].

The availability of extensive data from academic databases such as Scopus, Web of Science, and Dimensions, along with tools like VOSviewer, CitNetExplorer, and CiteSpace, has significantly increased the number of studies employing bibliometric analysis. These tools enable researchers to visualize complex networks and identify key trends, contributing to a more nuanced understanding of the research landscape [9].

2.2 Analysis and Results

The data utilized in this study were sourced from the Scopus database on May 13, 2024. Scopus was chosen due to its continuous updates and enhancements, making it one of the largest and most comprehensive abstract and citation databases, comparable to Web of Science and Google Scholar (Falagas et al., 2008; Mongeon & Paul-Hus, 2016). For this study, Scopus datasets spanning from 2017 to 2024 were retrieved and cleaned using OpenRefine software. The query used to extract relevant documents was: ("blockchain" AND "trust management"). This search yielded a total of 541 documents. All extracted data were saved in comma-separated values (.csv) and research information systems (.ris) file formats.

Our document selection was restricted to peer-reviewed sources, including journal articles, conference proceedings, and book series to ensure the highest quality and reliability. The rigorous peer-review process is essential for ensuring the validity and expert endorsement of the content, which is crucial for conducting an objective and comprehensive bibliometric analysis. Specific exclusion criteria were established to refine our dataset further, omitting trade publications, short surveys, notes, reports, and errata during the data searching phase. Our study encompasses a diverse linguistic spectrum, including documents in multiple languages, particularly English and Chinese. The methodology of our search, including the selection and exclusion criteria, is comprehensively illustrated in Fig. 1.

The obtained data were used to create visual representations illustrating the increase in publications and cumulative citations of scientific articles related to trust management in blockchain applications from 2017 to 2024. This research utilized several tools for the visualization of bibliometric networks, including VOSviewer (version 1.6.17), Microsoft Excel, OpenRefine, Biblioshiny, and Harzing's Publish or Perish. These tools were selected for their user-friendly interfaces and extensive visualization features. The following section will present an analysis of the collected data and its alignment with the previously formulated research questions.

3 Results and Discussion

3.1 Current State of the Research

Publication Data. The number of publications per year from 2017 to 2024. The total number of publications has generally increased, starting from just three in 2017 to 142 in 2023. This indicates a growing interest and research activity in trust management in blockchain. In 2022, there was a significant rise in the number of publications, with 131 publications. This suggests that 2022 was a particularly productive year for research in this field.

The cumulative citations show a different trend compared to the number of publications. Citations peaked in 2020 with 2034 citations but have since declined, dropping to 445 citations in 2023. This peak in citations in 2020 indicates that earlier publications were highly influential and frequently cited during that year.

While the number of publications continued to rise, citations decreased after 2020. This could suggest that newer publications are not yet as widely cited or that earlier works were more foundational and thus cited more frequently. The most recent year, 2023, shows the highest number of publications, 142, but a significantly lower number of citations, 445. This disparity might indicate that these recent publications have not yet had enough time to accumulate citations.

The increasing number of publications suggests that trust in NDN is becoming an increasingly important area of study, attracting more researchers and studies. The high citation counts in 2020 imply that specific key papers published earlier significantly impacted the field, likely providing foundational insights or methodologies that subsequent research has built upon. The lower citation counts in recent years may indicate a lag effect, where newer publications have not yet had sufficient time to be cited. It could also suggest that recent research has yet to achieve the same level of influence as earlier work.

The graph in Fig. 2 indicates a robust growth in research output related to trust management in blockchain, peaking in terms of publication in 2023. However, citations peaked earlier in 2020 and have since declined, suggesting a time lag in citation accumulation for newer publications or a shift in the influence of recent research. The increasing trend in publications highlights a growing academic interest and potential for future contributions to the field.

Analysis of the Most Prolific Years. The analysis of prolific years involves identifying the years in which the most significant and impactful research activity occurred within a particular field. Table 1 provides an overview of citation trends and publication patterns from 2017 to 2024. This information is used to analyze the most prolific years of research on trust management in blockchain applications that are actively run.

Based on Table 1, there is a clear upward trend in the number of publications, reflecting a growing interest and research activity in the field. Starting with only three publications in 2017, the number of articles steadily increased, peaking at 142 in 2023 before dropping to 41 in 2024. This growth underscores the expanding scholarly engagement and the rising significance of trust management in blockchain applications. The mean total citation per year further illustrates the citation dynamics over time. The highest mean total citation per year was observed in 2019, at 8.89, indicating that the articles

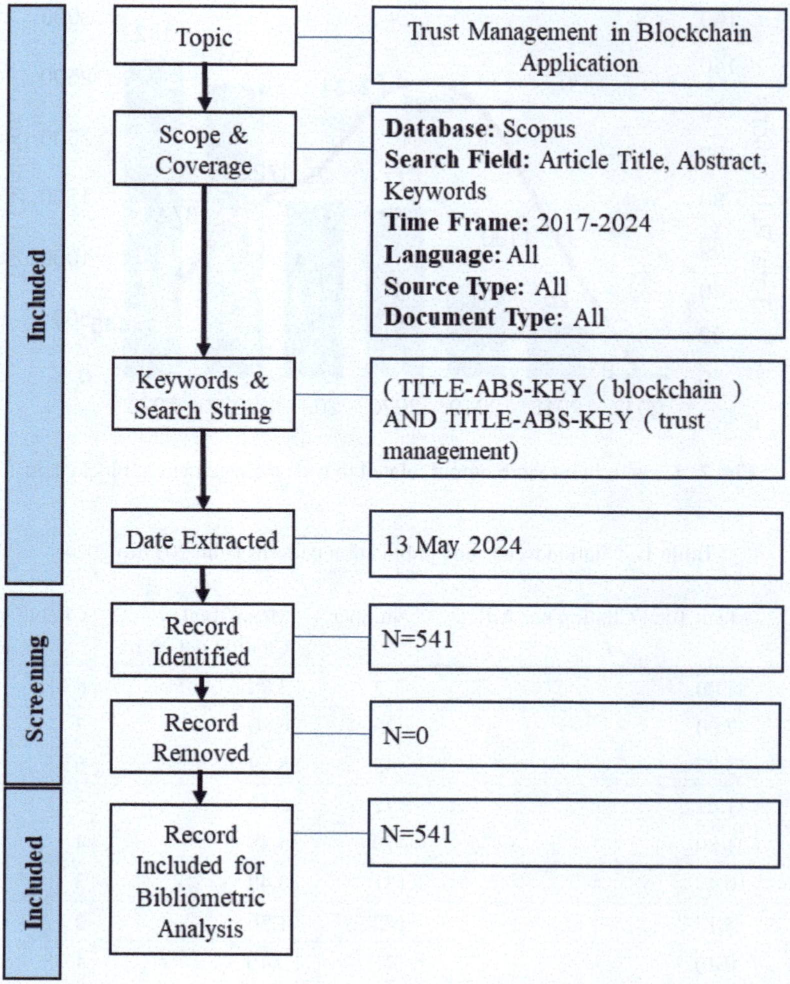

Fig. 1. Flow diagram of the research strategy.

from this year have been consistently cited at a high rate. Subsequent years show a decline in this metric, with 2024 having the lowest mean total citation per year at 0.46. This trend reinforces the observation that newer publications are still being integrated into the academic discourse and are yet to receive significant citations. This pattern underscores the time-lag in citations and highlights a significant rise in research on trust management in blockchain from 2017 to 2023, with earlier works receiving more citations due to their longer presence in scholarly discussions.

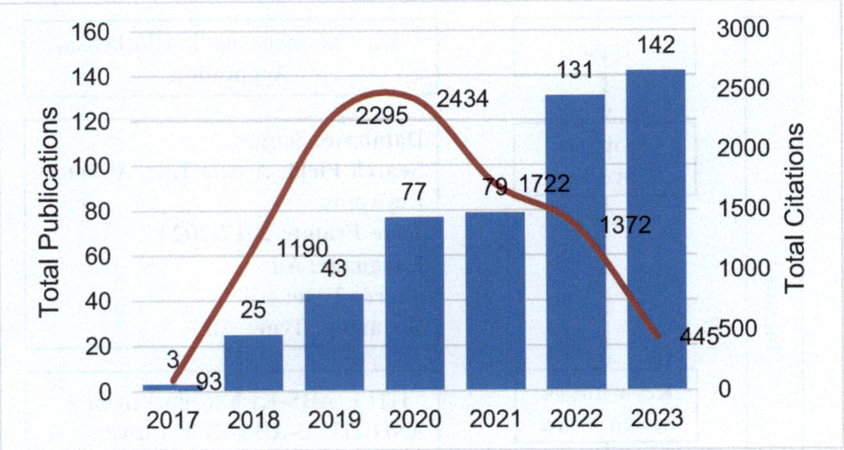

Fig. 2. Growth in research output related to trust management in blockchain.

Table 1. Citation trends and publication patterns from 2017 to 2024.

Year	Mean Total Citation per Article	Number	Mean Total Citation per Year	Citable Years
2017	31.00	3	3.88	8
2018	47.60	25	6.80	7
2019	53.37	43	8.89	6
2020	31.61	77	6.32	5
2021	21.80	79	5.45	4
2022	10.47	131	3.49	3
2023	3.13	142	1.56	2
2024	0.46	41	0.46	1

3.2 Leading Authors and Contributions

Identification of the Most Relevant Authors. Knowing the most relevant authors in the blockchain trust management field is crucial for advancing discipline and fostering scholarly development. This also answers RQ2, which is to identify the leading authors and contributions to the field. These leading authors are typically at the forefront of introducing innovative ideas, addressing critical challenges, and developing new methodologies that shape the field. Their work provides a foundation for subsequent research, guiding new investigators to build upon proven concepts and avoid redundant efforts. Furthermore, identifying these key contributors facilitates collaboration and networking opportunities, enhancing the potential for impactful research outcomes.

Table 2. Most relevant authors.

Authors	Articles
Li, Wenjuan	12
Kanhere, Salil S	10
Javaid, Nadeem	9
Jurdak, Raja	9
Dedeoglu, Volkan	8
Ayed, Samiha	7
Shiaeles, Stavros	7
Meng, Weizhi	6
Putra, Guntur Dharma	6
Wang, Yu	6

Table 2 shows that several authors have emerged as pivotal contributors to trust management research. Wenjuan Li has authored 12 significant articles leading the field, establishing herself as a foremost expert whose extensive research contributions have profoundly impacted the understanding and advancement of trust mechanisms in blockchain systems. Close behind, Salil S. Kanhere, with ten published articles, has made substantial contributions, offering valuable insights into the complexities and innovations in blockchain trust management. Nadeem Javaid and Raja Jurdak, with nine publications, have made notable strides in the field, providing critical research that shapes current practices and future developments in trust mechanisms.

3.3 Prominent Journals and Conferences

Top Journals and Conferences by Number of Publications. To explore RQ3 on key journals and conferences in blockchain trust management, several notable publications emerge. IEEE Access leads with 38 articles showcasing significant advancements, followed by the IEEE Internet of Things Journal with 32 articles emphasizing IoT and blockchain integration. The Lecture Notes series, including Lecture Notes in Computer Science, Artificial Intelligence, and Bioinformatics, contributes 21 articles, enriching the discourse with AI-driven insights. The Lecture Notes in Networks and Systems and the ACM International Conference Proceeding Series also play crucial roles, with 12 and 11 articles, respectively, underscoring their impact in network and blockchain research.

Recognizing these prominent journals and conferences is crucial for researchers and practitioners. These publications not only provide access to the latest findings and technological advancements but also serve as platforms for disseminating high-quality research. Knowing where influential work is published helps researchers stay abreast of the latest trends, methodologies, and breakthroughs, facilitating informed and impactful research. Furthermore, these sources offer opportunities for collaboration, peer review, and academic discourse, which are essential for advancing the field of trust management in blockchain. By focusing on these key journals and conferences, researchers can ensure their work reaches a relevant and engaged audience, maximizing its impact and contributing to the overall growth of the field.

3.4 Collaboration Patterns

Leading Institutions. In the realm of trust management in blockchain, several institutions have emerged as leaders, contributing significantly to the advancement of the field. RQ4 is addressed in this section in which Table 4 shows the leading institutions in the field of trust management in blockchain. Topping the list is Xidian University, with an impressive 44 articles showcasing its strong research capabilities and focus on blockchain technology. Comsats University Islamabad follows with 33 articles, indicating its active role in exploring and addressing trust issues within blockchain networks. Beijing University of Posts and Telecommunications and Nanjing University of Aeronautics and Astronautics have also made substantial contributions, with 25 and 23 articles, respectively. These institutions are pivotal in driving research that enhances blockchain systems' security, efficiency, and reliability (Table 3).

Table 3. Most prominent journals and conferences.

Sources	Articles
IEEE Access	38
IEEE Internet of Things Journal	32
Lecture Notes in Computer Science (Including Subseries Lecture Notes in Artificial Intelligence and Lecture Notes in Bioinformatics)	21
Lecture Notes in Networks and Systems	12
ACM International Conference Proceeding Series	11
Electronics (Switzerland)	9
IEEE Transactions on Vehicular Technology	9
IEEE Transactions on Intelligent Transportation Systems	8
Sensors	8
Communications in Computer and Information Science	6

The National University of Defense Technology and King Saud University stand out with 22 and 20 publications, respectively, leading in developing trust management frameworks. Similarly, Guangzhou University contributed to 18 articles, and Beijing Jiaotong University, Wuhan University, and Wuhan University of Technology contributed significantly with 14 articles. Recognizing these institutions is essential for fostering collaborations, leveraging existing expertise, and building upon proven research, thus supporting further innovation in blockchain-based trust management. This knowledge also aids funding bodies in directing resources to these centres of excellence, promoting stronger academic and industry partnerships to enhance the global implementation of blockchain technologies.

Table 4. Leading institutions.

Affiliation	Articles
Xidian University	44
Comsats University Islamabad	33
Beijing University of Posts and Telecommunications	25
Nanjing University of Aeronautics and Astronautics	23
National University of Defense Technology	22
King Saud University	20
Guangzhou University	18
Beijing Jiaotong University	14
Wuhan University	14
Wuhan University of Technology	14

3.5 Citation Impact

Most Highly Cited Papers. RQ5 focuses on trust management in blockchain, highlighting several pivotal studies listed in Table 5. The paper "Blockchain-based decentralized trust management in vehicular networks" leads these contributions with 652 citations and an annual rate of 108.67 citations, demonstrating its significant impact on blockchain applications in vehicular networks. Another noteworthy paper is "When Intrusion Detection Meets Blockchain Technology", which has 367 citations and 52.43 citations per year, providing an extensive examination of how blockchain technology can enhance intrusion detection systems and serving as a key reference for security advancements in blockchain applications.

Notable studies such as "A privacy-preserving trust model based on blockchain for VANETs", with 284 citations and "TrustChain: Trust management in blockchain and IoT supported supply chains", with 226 citations, highlight blockchain's impact on vehicular networks and supply chains. Other significant contributions include "DITrust Chain: Towards blockchain-based trust models for sustainable healthcare IoT systems" and "BARS: A blockchain-based anonymous reputation system for trust management in VANETs", cited 224 and 204 times, respectively. These papers are critical, offering foundational research that informs and directs new inquiries in the domain, ensuring that future work builds on validated approaches to accelerate advancements in blockchain-based trust management.

Table 5. Highly cited papers.

Paper	Total Citations	TC Per Year
Blockchain-based decentralized trust management in vehicular networks [10]	652	108.67
When intrusion detection meets blockchain technology: A review [11]	367	52.43
A privacy-preserving trust model based on blockchain for VANETs [12]	284	40.57
TrustChain: Trust management in blockchain and IoT supported supply chains [13]	226	37.67
DITrust Chain: Towards blockchain-based trust models for sustainable healthcare IoT systems [14]	224	44.80
BARS: A blockchain-based anonymous reputation system for trust management in VANETs [15]	204	29.14
Blockchain-based secure and trustworthy Internet of things in SDN-enabled 5G-VANETs [16]	197	32.83
BinDaaS: Blockchain-based deep-learning as-a-service in healthcare 4.0 applications [17]	161	40.25
Secure and provenance enhanced Internet of health things framework: A blockchain-managed federated learning approach [18]	147	29.40
A proof-of-trust consensus protocol for enhancing accountability in crowdsourcing services [19]	139	23.17

3.6 Emerging Trends and Future Directions

Recent Topics and Increasing Areas of Interest. The final RQ6 will be discussed in this section, which aims to discover the emerging trends and future directions in blockchain applications and trust management. The developments and growing focus areas illustrated in Fig. 3 demonstrate the dynamic research and innovation topics in this domain. The term "security" consistently appears as a predominant topic, reflecting its critical importance in blockchain applications. The rising interest in "behavioural science" and "blockchain" underscores the interdisciplinary approach adopted to enhance trust mechanisms. Additionally, "smart contracts" and "blockchain" remain focal points, indicating the ongoing efforts to improve transactional security and integrity through automated and secure contractual agreements.

Other notable trends include "trust management," "network security," and "trusted computing," each illustrating the multifaceted efforts to bolster trust frameworks within blockchain networks. The prominence of terms such as "privacy by design," "Internet of Things (IoT)," and "vehicular networks" reveals the expanding application of blockchain technology across various domains. This diversification broadens the scope of trust management research and highlights the need for tailored solutions to address domain-specific challenges. Understanding these recent topics and increasing areas of interest is crucial for researchers and practitioners, as it provides insight into current priorities and emerging challenges in the field. Staying informed about these trends facilitates targeted research, fosters collaboration, and drives the development of innovative solutions that enhance the security and reliability of blockchain systems.

Potential Research Gaps and Opportunities. Figure 3 illustrates the frequency of key terms over time, providing insights into emerging trends and potential research gaps in trust management in blockchain. The consistent rise in terms such as "security," "blockchain," and "smart contracts" highlights their ongoing relevance and the persistent need for robust solutions in these areas. Despite significant advancements, the increasing mention of these terms suggests that challenges remain, particularly in ensuring the security and reliability of blockchain applications. This persistent focus underscores the necessity for continued research into innovative security mechanisms and trust frameworks that can adapt to evolving threats.

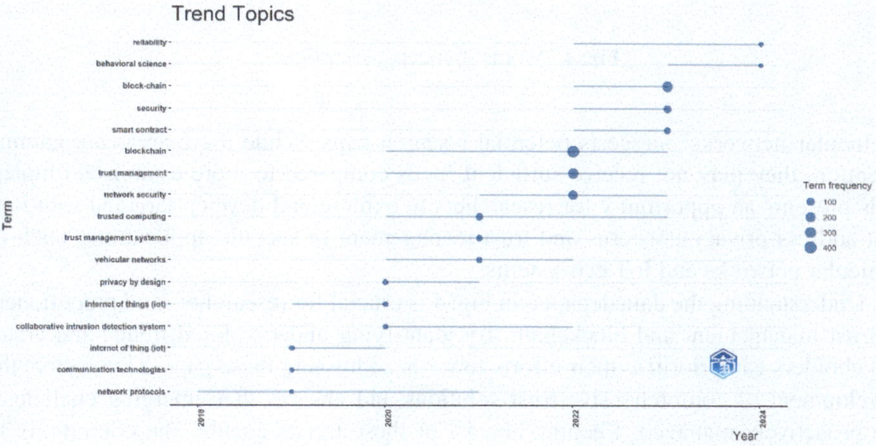

Fig. 3. Trend topics.

Potential Research Gaps and Opportunities. Figure 4 illustrates the frequency of key terms over time, providing insights into emerging trends and potential research gaps in trust management in blockchain. The consistent rise in terms such as "security," "blockchain," and "smart contracts" highlights their ongoing relevance and the persistent need for robust solutions in these areas. Despite significant advancements, the increasing mention of these terms suggests that challenges remain, particularly in ensuring the security and reliability of blockchain applications. This persistent focus underscores the necessity for continued research into innovative security mechanisms and trust frameworks that can adapt to evolving threats.

The figure also reveals a growing interest in interdisciplinary approaches, with terms like "behavioural science" and "trusted computing" showing a notable upward trend. This indicates a shift towards integrating diverse fields to address complex trust issues in blockchain. However, the relatively slower rise in terms such as "privacy by design" and

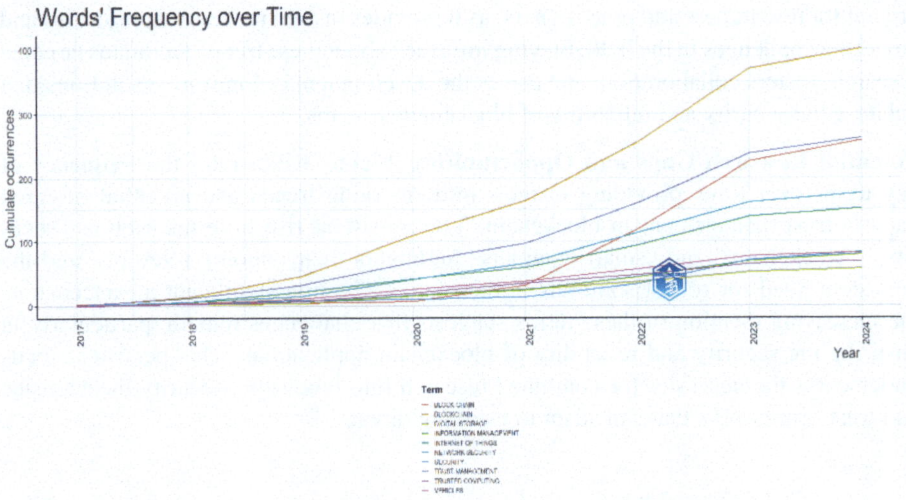

Fig. 4. Words' frequency over time.

"vehicular networks" suggests potential research gaps. While these areas are gaining attention, they may not receive sufficient focus compared to more established topics. This presents an opportunity for researchers to explore and develop targeted solutions that address privacy concerns and trust management in specific applications, such as vehicular networks and IoT ecosystems.

Understanding the data depicted in Fig. 4 is crucial for researchers and practitioners in trust management and blockchain. By identifying underexplored trends and areas, stakeholders can prioritize their efforts towards addressing these gaps. This fosters the development of comprehensive trust solutions and ensures that emerging challenges are proactively managed. Keeping abreast of these trends enables the community to stay ahead of the curve, driving forward the innovation and adoption of blockchain technologies in a secure and trustworthy manner.

4 Conclusion

This study presents a thorough bibliometric analysis of research at the intersection of blockchain applications and trust management, elucidating the current landscape and future directions. Leveraging data from the Scopus database spanning 2017 to 2024, it unveils publication trends, leading authors, prominent venues, collaboration patterns, citation impact, and emerging topics. The findings reveal a significant surge in research output, underscoring growing academic interest and industry adoption of blockchain for trust-based solutions. Influential scholars and institutions driving the field are identified. Key journals and conferences disseminating cutting-edge work are highlighted. Critical insights into research dynamics and knowledge diffusion are uncovered by mapping collaboration networks and assessing citation influence.

The analysis highlights areas that are becoming increasingly important, including the integration of smart contracts, behavioural science, and applications such as the Internet of Things (IoT). Concurrently, potential gaps are recognized, emphasizing the need for targeted investigations into privacy-preserving solutions and domain-specific trust frameworks. This comprehensive bibliometric study is invaluable, providing a holistic view of the state-of-the-art and guiding future endeavours in leveraging blockchain technology for robust, trustworthy systems across sectors like supply chains, healthcare, finance, and government services where trust is paramount.

Acknowledgment. This research was supported by the Ministry of Higher Education (MoHE) through the Fundamental Research Grant Scheme (Ref: FRGS/1/2020/ICT07/UUM/01/1). The content of this article is solely the responsibility of the authors and does not necessarily represent the official views of MoHE Malaysia.

References

1. Kshetri, N.: Can blockchain strengthen the internet of things? IT Prof. **19**(4), 68–72 (2017)
2. Agbo, C.C., Mahmoud, Q.H., Eklund, J.M.: Blockchain technology in healthcare: a systematic review. Healthcare **7**(2), 56 (2019)
3. Gupta, S.S.: Blockchain. IBM Online (2017). http://www.IBM.COM
4. Ølnes, S., Ubacht, J., Janssen, M.: Blockchain in government: benefits and implications of distributed ledger technology for information sharing. Gov. Inf. Q. **34**(3), 355–364 (2017)
5. Zyskind, G., Nathan, O., Pentland, A.: Enigma: decentralized computation platform with guaranteed privacy, p. 14. arXiv preprint arXiv:1506.03471 (2015)
6. Waltman, L., Van Eck, N.J.: A new methodology for constructing a publication-level classification system of science. J. Am. Soc. Inform. Sci. Technol. **63**(12), 2378–2392 (2012)
7. Pritchard, A.: Statistical bibliography or bibliometrics. J. Doc. **25**, 348 (1969)
8. Van Eck, N., Waltman, L.: Software survey: VOSviewer, a computer program for bibliometric mapping. Scientometrics **84**(2), 523–538 (2010)
9. Chen, C.: Science mapping: a systematic review of the literature. J. Data Inf. Sci. **2**(2), 1–40 (2017)
10. Yang, Z., Yang, K., Lei, L., Zheng, K., Leung, V.C.M.: Blockchain-based decentralized trust management in vehicular networks. IEEE Internet Things J. **6**(2), 1495–1505 (2018)
11. Meng, W., Tischhauser, E.W., Wang, Q., Wang, Y., Han, J.: When intrusion detection meets blockchain technology: a review. IEEE Access **6**, 10179–10188 (2018)
12. Lu, Z., Liu, W., Wang, Q., Qu, G., Liu, Z.: A privacy-preserving trust model based on blockchain for VANETs. IEEE Access **6**, 45655–45664 (2018)
13. Malik, S., Dedeoglu, V., Kanhere, S.S., Jurdak, R.: TrustChain: trust management in blockchain and IoT supported supply chains. In: IEEE International Conference on Blockchain (Blockchain), pp. 184–193 (2019)
14. Abou-Nassar, E.M., Iliyasu, A.M., El-Kafrawy, P.M., Song, O.Y., Bashir, A.K., Abd El-Latif, A.A.: DITrust chain: towards blockchain-based trust models for sustainable healthcare IoT systems. IEEE Access **8**, 111223–111238 (2020)
15. Lu, Z., Wang, Q., Qu, G., Liu, Z.: BARS: a blockchain-based anonymous reputation system for trust management in VANETs. In: 17th IEEE International Conference on Trust, Security and Privacy, in Computing and Communications/12th IEEE International Conference on Big Data Science and Engineering (TrustCom/BigDataSE), pp. 98–103 (2018)

16. Xie, L., Ding, Y., Yang, H., Wang, X.: Blockchain-based secure and trustworthy Internet of Things in SDN-enabled 5G-VANETs. IEEE Access **7**, 56656–56666 (2019)
17. Bhattacharya, P., Tanwar, S., Bodkhe, U., Tyagi, S., Kumar, N.: BinDaaS: blockchain-based deep-learning as-a-service in healthcare 4.0 applications. IEEE Trans. Netw. Sci. Eng. **8**(2), 1242–1255 (2019)
18. Rahman, M.A., Hossain, M.S., Islam, M.S., Alrajeh, N.A., Muhammad, G.: Secure and provenance enhanced internet of health things framework: a blockchain managed federated learning approach. IEEE Access **8**, 205071–205087 (2020)
19. Zou, J., Ye, B., Qu, L., Wang, Y., Orgun, M.A., Li, L.: A Proof-of-Trust consensus protocol for enhancing accountability in crowdsourcing services. IEEE Trans. Serv. Comput. **12**(3), 429–445 (2018)

Leveraging Deep Learning for Enhancing Blockchain Smart Contract Security: A Systematic Literature Review

Zhyar Yassin Abdalla[1], Mohammed Nawzad M. Ramzi[2], Mohammed Alsamman[3], and Nur Haryani Zakaria[3(✉)]

[1] Faculty of Engineering and Computer Science, Qaiwan International University, Sulaymaniyah 46001, Kurdistan Regional Government, Iraq
[2] Information Technology Department, Tishk International University, Sulaymaniyah, Iraq
[3] School of Computing, Universiti Utara Malaysia, 06010 Sintok, Kedah, Malaysia
haryani@uum.edu.my

Abstract. Although blockchain smart contracts hold great promise for decentralised applications, the security threats against them are quite prevalent and can be very expensive. Deep learning techniques have started to show prominent prospects in their detection. Despite the increased concern over these weaknesses, the current studies have mainly focused on detection methods based on machine learning without considering the role of deep learning techniques in smart contract vulnerability detection. This paper presents a systematic literature review of 16 recent studies strictly related to deep learning-based smart contract vulnerability detection. This study analysed the various deep learning models applied, covering variants of the LSTM, CNN, GNN, and hybrid models. Their performance is assessed with performance metrics, and their targeted vulnerabilities are considered; their limitations and weaknesses are addressed. The study reveals a promising trend in this review, where custom datasets and models tailored to specific vulnerabilities often yield high accuracy rates. However, there are challenges to generalising these approaches across diverse vulnerabilities and datasets. This review gives the research community an overview of the current landscape, identifies the methods used, and identifies the types of vulnerabilities trending in the current approaches to detection.

Keywords: Smart Contracts · Vulnerabilities · Deep Learning

1 Introduction

Satoshi Naka-moto proposed blockchain, a decentralised ledger system, as the infrastructure for Bitcoin in 2008 [1]. Cost and time savings make the blockchain decentralised network more efficient and innovative than centrally organised networks. Each block contains information about the previous block and provides authentication for transactions without third parties [2]. Vitalik Buterin introduced the Ethereum blockchain in 2014.

The original version of the chapter has been revised. The region name (Kurdistan) in the affiliation of the author (Mohammed Nawzad M. Ramzi) was removed. A correction to this chapter can be found at https://doi.org/10.1007/978-3-031-91485-0_43

Z. Mat Aji et al. (Eds.): KMICe 2024, ISEM 49, pp. 147–160, 2025.
https://doi.org/10.1007/978-3-031-91485-0_12

Buterin's main goal was to create a blockchain that developers could build on and lead to decentralised applications (DApps), which use smart contracts to promise transparent, secure, and immutable transactions without banks, governments, or organisations [3]. Several blockchain platforms focus on decentralised applications. These applications use smart contracts, self-executing code on a blockchain network, to facilitate transaction terms and agreements [4].

Smart contracts are applications written in code using programming languages. Many programming languages, including Solidity, Vyper, Yul, Cairo, Rust, and others, support different blockchain platforms. This technology has vulnerabilities and security issues like any programming language and service [5]. These smart contract vulnerabilities allow attackers to steal cryptocurrencies from digital wallets or assets, as smart contracts govern blockchain or DApp agreements [6]. In 2016, smart contract code vulnerabilities allowed attackers to steal 3.6 million ethers (ETH). Binance Smart Chain, a popular blockchain platform, was vulnerable in 2021, and hackers stole $50 million from their Automated Market Maker (AMM) contract [6]. Many more attacks stole millions of cryptocurrencies and dollars due to smart contract code vulnerabilities.

Developers cannot modify or upgrade blockchain smart contracts [7]. Smart contracts are vulnerable because programming languages and tools are new and underdeveloped [6]. Smart contracts are different from traditional programming because they require security. Since smart contracts store cryptocurrencies and digital assets, many attackers will target them. Increasing smart contracts also increases vulnerabilities. More contracts require more testing tools before deployment. Machine learning's deep learning branch is promising. This algorithm analyses and learns from data using artificial neural networks [8]. Deep learning has many applications in various fields. Attack detection is a popular cyber security area. Deep learning affects cyber security intrusion detection because it can identify unidentified cyberattacks. Deep learning allows intrusion detection systems to analyze large amounts of data and identify complex patterns that may indicate an attack. This reduces breaches and data loss and allows proactive cyberattack defence [9].

Many types of vulnerabilities in smart contract codes can be threatening. Common vulnerabilities include reentrancy attacks, integer underflow/overflow, and timestamp dependence [6]. Reentrancy occurs when the attacker makes a malicious smart contract to attack the victim contract; the attacker finds a loophole to withdraw from it until draining all the currencies from the victim's account. This will happen by accessing the withdrawal function and calling it multiple times without letting the victim contract to change its state [6]. Two more common vulnerabilities are integer underflow and overflow, Solidity can only hold up to 256 bits. The value of an integer variable will be reset to zero when another integer, 1, is added to it if the value reaches its highest point, which is $2^{256} - 1$, by raising or lowering the value of integers until they reach the maximum or minimum value [6]. Another common vulnerability is timestamp dependency; this attack takes advantage of miners' capacity to alter block timestamps, thereby impacting smart contracts that depend on these timestamps for crucial decisions or timing. Such a vulnerability has the potential to result in foreseeable outcomes or financial exploitation arising from contract logic sensitive to timing. This is a sample; this is a sample [6].

Current review and survey papers primarily concentrate on machine learning techniques to analyse the various current vulnerability detection approaches. This is even though there is a growing concern regarding vulnerabilities in blockchain smart contracts. Additionally, the weaknesses and problems that the researchers encountered in their study are not mentioned in any existing reviews on deep learning approaches. This systematic literature review fills the gaps within current research by focusing on using deep learning techniques within smart contract vulnerability detection. The main contributions that are a part of this effort are as follows:

1. Comprehensive overview: The semantic synthesis of state-of-the-art deep learning approaches applied to this problem will enable researchers and practitioners to learn from this helpful resource.
2. Comparative analysis: We compare the efficiency of various deep learning models concerning their strengths and weaknesses across multiple datasets and vulnerability types.
3. Vulnerability focus analysis: We systematically analyze the distribution of research efforts across different types of smart contract vulnerabilities, identifying those that have received the most attention and remain under-explored.

The paper is structured as follows: Sect. 2 reviews the current literature of other literature reviews and surveys. Section 3 explains the study's methodology and how the systematic steps of the study's collection are implemented. Section 4 provides the results and answers to the research questions and discusses the results of the reviewed studies. At the end, the conclusion of the study is presented.

2 Related Literature Reviews/Surveys

This section briefly overviews the secondary studies in blockchain smart contract vulnerability detection approaches using deep learning and summarises the related reviews.

A survey by [6] focused on enhancing smart contract security through machine-learning approaches—the research aimed to identify four machine-learning methods that can improve smart contract security. The study uncovered the benefits and drawbacks of the following methods: Supervised learning. This approach is well known for having high predictive accuracy and is broadly applicable in various fields. A sizable labelled dataset is necessary to train it well, which might be a barrier for some applications. Compared to supervised learning, semi-supervised learning can handle partially labelled data, offering the advantages of high data utilisation and lower labelling costs. However, it has to contend with issues like growing algorithm complexity and reliance on underlying presumptions, which could compromise its functionality. Unsupervised learning: This approach has two main advantages: it can find latent structures in the data and doesn't require labelled data. However, compared to supervised techniques, their predictive performance is typically lower, and assessing the calibre of unsupervised models can be difficult.

A review by [10] focused on the current literature on using deep learning-based approaches as a tool for smart contract vulnerability detection. The study compared current works in this area to traditional smart contract vulnerability detection tools. The study mainly focuses on the technical aspects of these approaches, how the developed models work, and what their stages are. The study categorises these tools based on their open-source status, data input types, and feature extraction methods.

Another study by [11] examined how symbolic execution can identify shortcomings of smart contracts. The study showed how symbolic execution might uncover smart contract vulnerabilities and aid vulnerability identification. The paper also presents and evaluates symbolic execution-based smart contract vulnerability detection approaches. [12] examined how machine learning can identify Ethereum's smart contract challenges. They thoroughly analysed and combined machine learning-based intelligence-detecting methods. They also explored the common traits of machine learning-driven solutions.

The existing literature provides insights into different methods of smart contract vulnerability detection. However, there is a noticeable gap in the existing literature regarding using deep learning mechanisms and approaches to detecting vulnerabilities in blockchain smart contracts. Hence, our work provides an SLR that focuses on using smart contract vulnerabilities and detection with deep learning approaches.

3 Methodology

A systematic literature review (SLR) examines and analyses all relevant research on particular subjects or research questions. SLRs follow a search and review protocol, according to [13]. This protocol guides the search, identification, and evaluation of relevant research works and defines the research questions. A search query was used to search the Scopus digital library. The search was related to current deep learning approaches in blockchain smart contract vulnerability detection. The search query has been defined based on the research questions.

3.1 Research Questions

To accomplish the objective of our systematic literature review (SLR), we have devised the subsequent research questions (RQs):

RQ1: What are the current deep learning approaches and methods for detecting vulnerabilities in blockchain smart contracts and their limitations and weaknesses?

RQ2: How influential are the identified deep learning approaches in detecting vulnerabilities in blockchain smart contracts?

RQ3: What kind of smart contract vulnerabilities are evaluated in the current approach's experiments?

3.2 Inclusions and Exclusions

To ensure that we are retrieving pertinent research, several criteria have been established to help us decide which works should be taken into consideration and which ones fall outside the purview of this review. Finding the studies that use deep learning techniques to find vulnerabilities in blockchain smart contracts is one step in finding the studies. The inclusion criteria chosen to locate the pertinent works are explained in Table 1. Describes all exclusions that have been decided upon to exclude unnecessary works.

Table 1. Inclusions and exclusions criteria.

Inclusion	Exclusion
All years of publication	Conference papers, white papers, and book reviews
Studies related to blockchain smart contracts vulnerability detection with deep learning approach	Studies that are not published in English
Keywords related to Smart Contracts, Vulnerabilities Detection, and Deep Learning	Subject areas that are not Computer Science and Engineering

3.3 Search Strategy

The database used for the collection of studies in this research is Scopus. The steps involved in performing our search query are depicted in Fig. 1. The search string used was "Deep learning" OR "deep learning approach" OR "deep learning approaches" AND"S-mart contract" OR "Smart contract vulnerability detection," and the study contained all years toward the end of 2023. After that, we identified and eliminated duplicates; we were left with just 372 of the 378 documents we had initially recovered. We used additional keyword trimming to focus our results, and 70 papers were remained. To get all the research on using deep learning for vulnerability detection in smart contracts, we have determined some relevant inclusions and excluded irrelevant documents, as shown in Table 1. Consequently, after applying filters, our dataset was reduced to 17 relevant documents. After that, we analysed all the research based on its title and abstract; based on this, four papers were removed. The refinement process produced a final dataset of 16 documents for analysis in this study. Figure 2 on the left side shows the number of studies published after our pruning, which concern vulnerability detection in blockchain smart contracts using deep learning, and on the right side is the publisher's trend for publication in this area of study.

Fig. 1. Search process diagram

4 Results and Discussion

In this section, we present the findings that were retrieved from the analysis of all the relevant work that was done to provide answers to the research questions that were defined in Sect. 3.1.

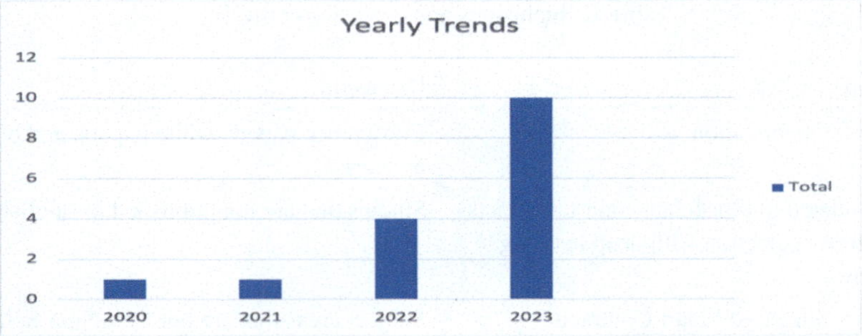

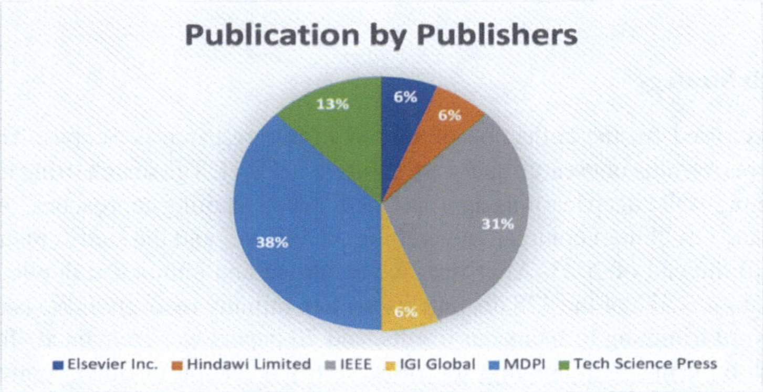

Fig. 2. Yearly and publication trends.

4.1 What are the Current Deep Learning Approaches and Methods for Detecting Vulnerabilities in Blockchain Smart Contracts and Their Limitations and Weaknesses?

Blockchain security researchers have recently investigated deep learning methods for smart contract vulnerability detection. As shown in Table 2, each approach provides each work's proposed model and its limitations and weaknesses. Contract code is analysed using diverse neural network architectures to identify security issues. RNNs and LSTMs can capture sequential code dependencies, including variants with attention mechanisms [14] and bidirectional architectures [18, 24]. For feature extraction, Convolutional Neural Networks (CNN) treat code as text or image data [19, 20, 29]. Smart contract structural information, including control flow and data flow, can be captured by graph neural networks (GNN) [19, 26, 29]. Transformer-based models [20, 27] use attention mechanisms and pre-training on large code datasets in this domain. Hybrid approaches like CNN with RNN [16] or graph features with expert patterns [23] have been developed to capitalise on different architectures.

Despite these advances, these approaches have significant limitations. Many models target specific vulnerabilities, like reentrancy [14, 16], and struggle to detect all vulnerabilities [16, 24, 29]. Poor data quality and imbalance affect model performance and generalizability [15, 22, 23]. Smart contracts become more complex, increasing scalability and efficiency issues [19, 27–29]. Many approaches lack context understanding, especially for multi-contract interactions or dynamic behaviours [18, 21]. Most static code analysis methods may miss runtime or multi-transaction vulnerabilities [20, 24]. Advanced models' lack of interpretability [25] makes their decisions hard to understand and trust. Adversarial attacks may also affect several models [19, 25, 28, 29].

These findings impact smart contract security tools and practice development. The diverse approaches show that the problem is complex and requires multifaceted solutions. Limitations like the focus on specific vulnerability types and scalability issues highlight areas for future research improvement. A secure environment requires reliable detection methods due to the possibility of adversarial attacks. Future research directions are to enhance dataset quality to address imbalance issues, dynamic analysis to capture runtime vulnerabilities, and model interpretability to build trust in automated detection systems. Developing hybrid approaches that combine the strengths of different architectures while mitigating their weaknesses seems promising [16, 24, 27]. Advances in this field could improve blockchain ecosystem security, increasing trust and adoption of smart contract technologies across industries.

4.2 How Practical Are the Identified Deep Learning Approaches in Detecting Vulnerabilities in Blockchain Smart Contracts?

To answer how practical the current approaches are in detecting smart contract vulnerabilities, we had to look at the measurement metrics and see the results of each study. As shown in Table 3, each approach provides measurement results after testing and experimenting with its models based on the datasets used to train the models. These methods use a wide variety of datasets, such as those accessible to the general public, such as SmartBugs and specialised datasets specially designed to address particular vulnerabilities. In addition to reentrancy, timestamp dependency, integer overflow, and other vulnerabilities, the evaluation considers various vulnerabilities. The RSC and RCS approaches, which have accuracies of 73.83% and 73.83%, respectively, for reentrancy detection, are among the findings that constitute noteworthy discoveries. On the other hand, specialised methods such as Opcode sequence analysis (OpSe) and Eth2Vec demonstrate accuracy greater than 95% when detecting multiple vulnerabilities. Custom datasets demonstrate robust performance, achieving accuracy rates that range from 88.55% to 96.76%. In addition to providing valuable insights for the ongoing development of efficient, smart contract security solutions, the findings highlight the nuanced performance variations that affect a variety of vulnerabilities and datasets.

The efficacy of these models depends on the composition of the dataset, the variety of vulnerabilities, and the inherent robustness of the deep learning architectures employing them. It is imperative to acknowledge the need for additional comparative analyses and testing in the real world to validate the applicability and generalizability of these deep learning approaches in practical blockchain security scenarios. Although the results that have been reported are encouraging, it is imperative to acknowledge the need for these tests. Conducting ongoing research and refining these models to ensure they effectively protect blockchain ecosystems will be essential.

Table 2. Limitations of the current models.

Ref	Model	Limitation
[14]	LSTM with an attention mechanism	The study only detects reentrancy vulnerability. In more smart contract vulnerability detection tasks, the authors recommend sequential models. Also, the results are less accurate than those of other methods
[15]	SCVSN (uses Siamese network architecture)	Only smart contract reentrancy vulnerabilities are detected by the model. Creating equal positive and negative samples may not reflect real-world vulnerable and non-vulnerable contract distributions
[16]	SPCBIG-EC (Combination of CNN and RNN for sequence modelling and feature extraction)	The proposed model cannot detect integer overflow vulnerabilities at higher performance levels
[17]	CBGRU (CNN-based feature extractor with bidirectional GRU-based)	The model performs worse on two smart contract shortcomings. The code doesn't show these two vulnerabilities, reducing accuracy. The proposed model's training may lack sufficient data
[18]	CodeNet (CNN-based model)	The limitations of CodeNet include its inability to consider the dynamic behaviour of smart contracts and its difficulty in detecting multi-contract vulnerabilities
[19]	MRN-GCN (Multi-relational nested graph convolutional network)	Insufficient ground truth due to unlabeled data, time complexity with more contracts, smart contract data preprocessing, and potential adversarial attacks on model performance
[20]	GRATDet (Graph representation and Transformer)	The model's limitations include reliance on a small dataset, accurate AST parsing, and difficulty detecting deep contract vulnerabilities due to single transaction analysis
[21]	The work combined source code, operation code, and control flow graph	The model's limitations include data imbalance, scalability challenges, limited generalizability, limited feature extraction, and insufficient coverage of vulnerability types

(*continued*)

Table 2. (*continued*)

Ref	Model	Limitation
[22]	BiLSTM (based on metric learning, triplet loss, and bidirectional long short-term memory)	The model lacks data imbalance, scalability challenges, limited generalizability, incomplete feature extraction, and insufficient coverage of vulnerability types
[23]	CGE (Combining Graph features and Expert patterns)	The model's limitations are that it cannot handle contracts that only have bytecode, cannot detect sharing-variable reentrancy vulnerability, and requires a large dataset of smart contracts to achieve high accuracy
[24]	SCLMF (Solc-based learner-meta-learner framework) uses a convolutional neural network as the base learner and the Model-Agnostic Meta-Learning (MAML)	The limitations of the SCLMF model are that it requires a compiler, is not as robust to adversarial attacks as other methods, and requires a large dataset of smart contracts to achieve high accuracy
[25]	DM (Vulnerability detection mechanism) is a two-stage system using a deep learning model (GNN) and expert knowledge	The model's weaknesses are that it relies heavily on datasets, is not very interpretable, adds time overhead, and has a limited scope
[26]	SolBERT-BiGRU-Attention (A hybrid model that uses a lightweight pre-trained BERT model and a bidirectional gated recurrent unit neural network with a hierarchical attention mechanism)	The model's weaknesses are that it is not as efficient as other methods in terms of detection time, and it can be challenging to evaluate its performance due to the lack of a standardised benchmark dataset
[27]	Con2Vec (Converts opcode sequences into text classification problems using TextCNN)	It is less robust to adversarial attacks and less efficient in detection time than other methods
[28]	TextCNN and Graph Neural Networks to extract syntactic and semantic features from the source code	The model is not robust to adversarial attacks, requires a large labelled dataset, has limited vulnerability coverage, and is less efficient than other methods in terms of detection time
[29]	The Hybrid Attention Mechanism (HAM) model combines single-head attention, Bi-GRU, and multi-head attention to detect vulnerabilities in smart contracts	The model is not robust to adversarial attacks, requires a large labelled dataset, has limited vulnerability coverage, and is less efficient in detection time than other methods

4.3 What Kind of Smart Contract Vulnerabilities are Evaluated in the Current Approaches?

Some current approaches to using deep learning for vulnerability detection focus on making an efficient model that, with high accuracy, detects most vulnerabilities. However, most of the proposed models are still focused on some specific number of vulnerabilities.

Table 3. Performance evaluation of current approaches.

Ref	Datasets	Vulnerabilities Detected	Performance Metrics
[14]	RSC	Reentrancy	Accuracy: 73.83%
	RCS		Accuracy: 73.83%
[15]	SourceCode	Reentrancy	Accuracy: 96.01%
	SmartBugs		
[16]	CESC	Reentrancy	Accuracy: 96.66%
		Timestamp Dependency	Accuracy: 91.11%
		Infinite Loop	Accuracy: 94.87%
	UCESC	Integer overflow	Accuracy: 85.11%
		Callstack depth attack	Accuracy: 90.02%
		Integer underflow	Accuracy: 86.83%
[17]	Smartbugs-wild and dataset by [14]	Various vulnerabilities	Accuracy: 85.50%
[18]	SmartBugs	Timestamp Dependency	Accuracy: 94.35%
		Tx.origin	Accuracy: 99.21%
		Unchecked Calls	Accuracy: 98.79% Accuracy: 98.27%
		Reentrancy	Accuracy: 98.27%
[19]	SmartBugs	Timestamp Dependency	Accuracy: 94.21%
		Reentrancy	Accuracy: 94.43%
		Arithmetic	Accuracy: 88.96%
[20]	Custom Set 1	Various vulnerabilities	Accuracy: 0.9522
	Custom Set 2		Accuracy: 0.876
[21]	ScrawID	Arithmetic	Accuracy: 0.916
		Locked Ether	Accuracy: 0.895
		Reentrancy	Accuracy: 0.909
		Timestamp Dependency	Accuracy: 0.948
[22]	Custom	Arithmetic	Accuracy: 88.31%
		Reentrancy	Accuracy: 93.25%
		Unchecked Calls	Accuracy: 91.85%
		Inconsistent Access Control	Accuracy: 90.59%
[23]	ESC	Reentrancy	Accuracy: 89.15%
		Timestamp Dependency	Accuracy: 89.02%

(*continued*)

Table 3. *(continued)*

Ref	Datasets	Vulnerabilities Detected	Performance Metrics
	VSC	Infinite Loop	Accuracy: 83.21%
[24]	Smartbugs,	Various vulnerabilities	Accuracy: 88.55%
	Smartbugs-wild		Accuracy: 72.36%
	WScrawlD		Accuracy: 98.5%
[25]	Custom	Reentrancy	Accuracy: 89.74%
		Timestamp Dependency	Accuracy: 88.52%
		Code injection	Accuracy: 88.62%
		Call with hardcode gas amount	Accuracy: 90.62%
[26]	Custom	Various vulnerabilities	Accuracy: 93.85%
[27]	Custom	Various vulnerabilities	OpSe Acc. (96.76%)
			OpANSe Acc. (96.44%)
			OpNSe Acc. (96.61%)
[28]	Eth2Vec	Implicit Visibility	Accuracy: 89.00%
		Integer Overflow	Accuracy: 95.85%
		Integer Underflow	Accuracy: 96.42%
		Time Dependency	Accuracy: 98.25%
		Reentrancy	Accuracy: 98.64%
[29]	Custom	Reentrancy	Accuracy: 93.36%
		Arithmetic Vulnerability	Accuracy: 980.85%
		Unchecked Return Value	Accuracy: 82.56%
		Timestamp Dependency	Accuracy: 85.62%
		Tx.origin	Accuracy: 82.19%

Table 3 also shows the kinds of vulnerabilities that the approaches focus on. And What datasets have been used to test the current deep learning models to detect smart contract concepts?

Figure 3 shows the most common vulnerabilities detected in the reviewed approaches. Reentrancy emerges as the most frequently identified vulnerability, with ten instances evaluated across the reviewed approaches. Timestamp Dependency is the second most evaluated one, with 8 times. Some of the studies used datasets with multiple vulnerabilities, and they evaluated their model on various vulnerabilities and provided the average performance, as shown in Fig. 3. Five studies evaluated datasets with Various Vulnerabilities. Arithmetic vulnerabilities were identified 4 times, and Integer Overflow, Underflow, Tx.origin, and Unchecked Calls were assessed twice.

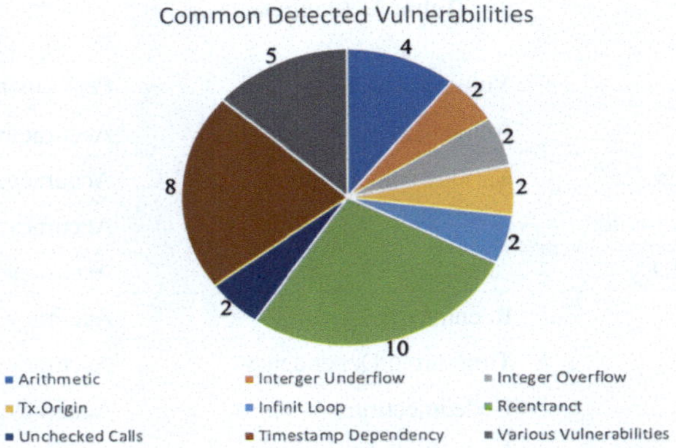

Fig. 3. Number of approaches that detected each vulnerability.

5 Conclusions

This systematic literature review presented and analysed the current deep learning approaches for detecting smart contract vulnerabilities. We examined the selected 16 studies based on three identified research questions:

RQ1: The review revealed a wide variety of deep learning approaches and methods for detecting vulnerabilities in blockchain smart contracts, including LSTMs, RNNs, CNNs, GNNs, and hybrid models. We identified promising trends towards using custom datasets and models tailored to specific vulnerabilities, which often yield high accuracy rates. However, these approaches usually struggle with generalizability across diverse vulnerabilities and datasets.

RQ2: The analysis found that deep learning-based methods often achieve high accuracy when evaluated on specific vulnerabilities and datasets. However, the overall effectiveness of these methods in real-world scenarios remains uncertain due to limited generalizability and the need for further research to validate their applicability.

RQ3: The research examined various vulnerabilities, including reentrancy, timestamp dependency, integer overflow, and other vulnerabilities. The review highlighted a growing trend towards focusing on specific vulnerabilities, with reentrancy and timestamp dependency being the most commonly evaluated vulnerabilities.

The analysis conducted in this SLR indicated a new area of study growing in recent years: using deep learning approaches and models as vulnerability detector tools in blockchain smart contracts. This research also identifies the current datasets used by the researchers to test the current models. This study also combines all of these 16 studies and compares the performance of each study, as well as the different experiments and evaluation approaches. Identifies what kinds of smart contract vulnerabilities have been evaluated by the researchers using the detecting tools they proposed.

Acknowledgement. The authors would also like to extend their gratitude to the Data Management & Software Solution Research Lab, School of Computing, Universiti Utara Malaysia, for their generous support in sponsoring the publication of this article.

References

1. Nakamoto, S.: Bitcoin: a peer-to-peer electronic cash system (2008). https://bitcoin.org/bitcoin.pdf
2. Stephen, R., Alex, A.: A review on blockchain security. IOP Conf. Ser.: Mater. Sci. Eng. **396**, 012030 (2018)
3. Buterin, V.: A next-generation smart contract and decentralized application platform. White Paper 3(37), 2-1 (2014)
4. Huang, Y., Wang, B., Wang, Y.: Research and application of smart contract based on ethereum blockchain. J. Phys.: Conf. Ser. **1748**, 042016 (2021)
5. Hulicki, M.: The legal framework and challenges of smart contract applications. In: Conference System Sciences, pp. 3–4 (2017)
6. Praitheeshan, P., Pan, L., Yu, J., Liu, J.K., Doss, R.: security analysis methods on ethereum smart contract vulnerabilities: a survey. ArXiv Preprint abs/1908.08605 (2019)
7. Zhao, Q., Shang, Z.: Deep learning and its development. J. Phys.: Conf. Ser. **1948** (2021)
8. Ferrag, M.A., Maglaras, L., Moschoyiannis, S.K., Janicke, H.: Deep learning for cybersecurity intrusion detection: approaches, datasets, and comparative study. J. Inf. Secur. Appl. **50** (2020)
9. Jiang, F., et al.: Enhancing smart-contract security through machine learning: a survey of approaches and techniques. Electronics **12**(9) (2023)
10. Wu, H., Peng, Y., He, Y., Fan, J.: A review of deep learning-based vulnerability detection tools for ethernet smart contracts. Comput. Model. Eng. Sci. **140**(1), 77–108 (2024)
11. Wang, Y., Sheng, S., Wang, Y.: A systematic literature review on smart contract vulnerability detection by symbolic execution. In: Chen, J., Wen, B., Chen, T. (eds.) BlockSys 2023. CCIS, vol. 1896, pp. 226–241. Springer, Singapore (2024). https://doi.org/10.1007/978-981-99-8101-4_16
12. Surücü, O., et al.: A survey on ethereum smart contract vulnerability detection using machine learning. In: Disruptive Technology Information Sciences VI, vol. 12117, p. 121170C, SPIE (2022)
13. Kitchenham, B., Charters, S.: Guidelines for performing systematic literature reviews in software engineering. EBSE Technical Report (2007)
14. Qian, P., Liu, Z., He, Q., Zimmermann, R., Wang, X.: Towards automated reentrancy detection for smart contracts based on sequential models. IEEE Access **8**, 19685–19695 (2020)
15. Guo, R., Chen, W., Zhang, L., Wang, G., Chen, H.: Smart contract vulnerability detection model based on Siamese network (SCVSN). Energies **15**(24) (2022)
16. Zhang, L., et al.: SPCBIG-EC: a robust serial hybrid model for smart contract vulnerability detection. Sensors **22**(12) (2022)
17. Zhang, L., et al.: CBGRU: a detection method of smart contract vulnerability based on a hybrid model. Sensors **22**(9) (2022)
18. Hwang, S.J., Choi, S.H., Shin, J., Choi, Y.H.: CodeNet: code-targeted convolutional neural network architecture for smart contract vulnerability detection. IEEE Access **10**, 32595–32607 (2022)
19. Liu, H., Fan, Y., Feng, L., Wei, Z.: Vulnerable smart contract function locating based on multi-relational nested graph convolutional network. J. Syst. Softw. **204**, 111775 (2023)
20. Gong, P., Yang, W., Wang, L., Wei, F., Hailati, K., Liao, Y.: GRATDet: smart contract vulnerability detector based on graph representation and transformer. Comput. Mater. Continua **76**(2), 1439–1462 (2023)
21. Deng, W., Wei, H., Huang, T., Cao, C., Peng, Y., Hu, X.: Smart contract vulnerability detection based on deep learning and multimodal decision fusion. Sensors **23**(16) (2023)
22. Wang, M., Xie, Z., Wen, X., Li, J., Zhao, K.: Ethereum smart contract vulnerability detection model based on triplet loss and BiLSTM. Electronics **12**(10) (2023)

23. Liu, Z., Qian, P., Wang, X., Zhuang, Y., Qiu, L., Wang, X.: Combining graph neural networks with expert knowledge for smart contract vulnerability detection. IEEE Trans. Knowl. Data Eng. **35**, 1296–1310 (2021)
24. Yang, Z., Zhu, W., Yu, M.: Improvement and optimization of vulnerability detection methods for ethernet smart contracts. IEEE Access **11**, 78207–78223 (2023)
25. Liu, Z., Jiang, M., Zhang, S., Zhang, J., Liu, Y.: A smart contract vulnerability detection mechanism based on deep learning and expert rules. IEEE Access **11**, 77990–77999 (2023)
26. Guangxia, X., Liu, L.: Vulnerability detection of ethereum smart contract based on colBERT-big-attention hybrid neural model. Comput. Model. Eng. Sci. **137**(1), 903–922 (2023)
27. Xu, Z., et al.: An efficient code-embedding-based vulnerability detection model for ethereum smart contracts. Int. J. Data Warehous. Min. **19**, 1–23 (2023)
28. Han, D., Li, Q., Zhang, L., Xu, T.: A smart contract vulnerability detection model based on syntactic and semantic fusion learning. Wirel. Commun. Mob. Comput. **2023**, 1–12 (2023)
29. Wu, H., Dong, H., He, Y., Duan, Q.: Smart contract vulnerability detection based on hybrid attention mechanism model. Appl. Sci. **13**(2) (2023)

Framework for Integrating Blockchain into Knowledge Management Systems for Academic Credentials

Omar S. Saleh[1], Osman Ghazali[1](✉), Norbik Bashah Idris[2], and A. V. Senthil Kumar[3]

[1] Universiti Utara Malaysia, 06010 Sintok, Kedah, Malaysia
osman@uum.edu.my

[2] Kulliyyah of Information and Communication Technology, International Islamic University Malaysia, Kuala Lumpur, Malaysia

[3] Department of MCA, Hindusthan College of Arts and Science, Coimbatore, India

Abstract. This paper explores integrating blockchain technology into knowledge management systems for academic credentials to enhance security, efficiency, and trust. Utilising Hyperledger Fabric, the proposed framework incorporates Decentralized Identifiers (DIDs), Verifiable Credentials (VCs), and Self-Sovereign Identity (SSI) principles. The system is designed around five key modules: DID management, schema management, credential definition, claim definition, and role management. Each module plays a critical role in ensuring the integrity and authenticity of academic credentials. The use of smart contracts on Hyperledger Fabric governs the lifecycle of certificates, enabling secure issuance, storage, and verification without relying on intermediaries. This decentralised approach allows universities to issue credentials that students can securely store in their digital wallets, while verifiers can authenticate these credentials by accessing the blockchain. By leveraging these technologies, the proposed system enhances data privacy and security and streamlines the verification process, significantly reducing the time required. Integrating blockchain into academic credential management represents a transformative step towards a more secure and efficient system.

Keywords: Blockchain · Knowledge Management · Academic Certificates · Self-Sovereign Identity · DID · VC

1 Introduction

In the academic domain, knowledge management (KM) constitutes a vital process that involves the structured creation, storage, retrieval, and distribution of information [1]. Robust KM systems are essential for managing valuable academic resources, including research outputs, institutional databases, and student records, with a focus on both security and operational efficiency. Within this scope, academic credentials stand out as a foundational element, providing authenticated proof of a student's academic achievements. These credentials hold substantial value for a range of stakeholders: students,

Z. Mat Aji et al. (Eds.): KMICe 2024, ISEM 49, pp. 161–173, 2025.
https://doi.org/10.1007/978-3-031-91485-0_13

educational institutions, employers, and regulatory agencies, as they support academic advancement, facilitate employment pathways, and maintain alignment with educational standards [2].

Despite the fundamental importance of academic credentials, existing systems for their management and verification encounter considerable challenges. The prevalence of fraudulent practices, such as the falsification of academic certificates, poses a direct threat to the credibility of educational qualifications, eroding confidence in the legitimacy of these records [3]. Additionally, inefficiencies within credential management, particularly the reliance on manual verification, contribute to delays and elevate administrative burdens. The limited transparency in traditional systems further complicates verification efforts, hindering stakeholders from promptly and reliably authenticating academic credentials [4]. Collectively, these issues underscore the need for a more secure, streamlined, and transparent approach to academic credential management.

Blockchain technology, known for its decentralized architecture and immutable data storage, offers a promising alternative for addressing these challenges. Blockchain's security features can safeguard academic records from unauthorized modifications, thus upholding data integrity [5]. Furthermore, the inherent transparency of blockchain facilitates swift and reliable credential verification, which mitigates the risk of fraudulent documentation. Additionally, blockchain can simplify the credential management lifecycle, contributing to greater efficiency and cost-effectiveness. This paper investigates the application of blockchain technology within KM systems for academic credentials, examining both its theoretical framework and practical implications.

2 Literature Review

2.1 Blockchain Technology

Blockchain technology is a decentralized ledger system that enables data storage across a distributed network of computers, ensuring security, immutability, and transparency. Each block in a blockchain contains a series of transactions, chronologically linked to previous blocks, forming a secure, tamper-resistant chain [6]. Key attributes of blockchain technology include decentralization, immutability, transparency, and security [7].

1. **Decentralization:** Unlike traditional centralized databases, blockchain operates on a peer-to-peer network without reliance on a central authority. This decentralization minimizes the risk of data manipulation and eliminates single points of failure.
2. **Immutability:** Data recorded on the blockchain is permanent and cannot be altered or deleted, preserving the integrity and authenticity of stored information.
3. **Transparency:** Transactions on a public blockchain are accessible to all participants, fostering transparency. Even within private or permissioned blockchains, transparency is maintained through controlled visibility and audit capabilities.
4. **Security:** Blockchain secures data through cryptographic methods, where each block is linked to its predecessor using a cryptographic hash. This design makes it computationally impractical to alter a single block without altering all subsequent blocks, enhancing data security.

Blockchain applications span numerous domains, including finance, supply chain management, healthcare, and increasingly, education [8]. Within the educational field, blockchain technology is leveraged to securely store and verify academic credentials, document student achievements, and improve administrative workflows.

2.2 Knowledge Management Systems

Knowledge Management Systems (KMS) facilitate the structured processes of knowledge creation, storage, retrieval, and dissemination within organizations [1]. In academia, KMS are integral to managing extensive data, such as research findings, institutional knowledge, and student records.

1. **Creation and Storage:** KMS support the systematic collection and organization of knowledge, ensuring that essential information is captured and stored methodically.
2. **Retrieval and Dissemination:** KMS offer tools that simplify the retrieval and sharing of knowledge, including search functions, collaborative platforms, and knowledge repositories.

In academic credentialing, KMS manage student records, issue academic certificates, and facilitate credential verification. However, traditional KMS are hindered by issues such as data silos, limited interoperability, and susceptibility to fraud and tampering. The integration of blockchain technology into KMS addresses these challenges by providing a secure, transparent, and efficient solution for managing academic credentials.

2.3 Previous Work

Recent research has increasingly focused on the application of blockchain technology in education and knowledge management, investigating both its benefits and potential challenges for academic credentialing and KMS.

- **Blockchain for Academic Credentialing:** Numerous studies have explored blockchain's utility in the issuance, storage, and verification of academic credentials. For instance, research in [9] analyzed blockchain-based digital certificates in higher education, emphasizing its capacity to improve security and prevent fraud. Similarly, [10] developed a blockchain-powered system for issuing and verifying academic certificates, demonstrating the technology's effectiveness in maintaining data integrity and authenticity.
- **Blockchain for Knowledge Management:** Scholars have also examined blockchain's role in knowledge management. The study in [11] introduced a blockchain-based, decentralized knowledge management system, enabling secure and transparent sharing of information across organizations. Additionally, [12] proposed a blockchain-driven framework for managing academic records, highlighting the advantages of decentralization and immutability in bolstering data security and trust.
- **Challenges and Limitations:** Despite its promise, implementing blockchain in knowledge management systems entails various challenges. Technical issues, such as scalability and interoperability, as well as legal and regulatory concerns regarding

data privacy and compliance, require careful consideration [7]. Studies by [13] and [14] have addressed these challenges, proposing solutions to support scalable and compliant implementations.

In conclusion, existing literature reveals the transformative potential of blockchain technology in academic credentialing and knowledge management. By leveraging blockchain's decentralized, immutable, and transparent characteristics, credentialing processes can achieve greater security, efficiency, and credibility. However, addressing technical, legal, and regulatory challenges is crucial for successful implementation and widespread adoption.

3 System Design and Architecture

3.1 Requirements Analysis

A clear definition of system requirements is crucial for the effective development of a blockchain-based academic credential management system. These requirements are categorized into functional and non-functional specifications, which guide the system's design and operational benchmarks [15].

Functional Requirements

1. Credential Issuance: The system must enable educational institutions to issue academic credentials, such as degrees and certificates, directly on the blockchain, ensuring authenticity from the point of issuance.
2. Credential Storage: Secure storage of academic credentials on the blockchain is essential to maintain immutability and safeguard against tampering, ensuring data integrity over time.
3. Credential Verification: The system should facilitate a verification mechanism for third-party entities, such as employers and other educational institutions, to authenticate the validity of academic credentials efficiently.
4. User Authentication: Robust authentication protocols are required to restrict access, ensuring that only authorized users—such as institutional administrators and students—can issue, view, and manage credentials.
5. User Interface: User-friendly interfaces should be developed for different stakeholders, including a portal for institutions to issue credentials, a platform for students to access and share credentials, and a verification module for third-party verifiers.
6. Revocation and Updates: The system should support the ability to revoke or update credentials as needed, such as in cases of credential fraud, to ensure ongoing accuracy and reliability.

Non-Functional Requirements

1. Security: The system must be fortified against unauthorized access, data breaches, and cyber-attacks to protect sensitive academic data.

2. Scalability: The design should support high transaction volumes and user loads without degradation in performance, accommodating future growth in demand.
3. Interoperability: Ensuring compatibility with existing knowledge management systems and adherence to industry standards is essential for seamless integration.
4. Privacy: The system must adhere to data privacy regulations, such as GDPR, protecting the confidentiality of personal information throughout the credential lifecycle.
5. Performance: High-performance benchmarks are necessary, including optimized transaction processing speed and minimal latency, to provide a responsive and efficient user experience.
6. Reliability: Ensure the system's high availability and reliability with minimal downtime.

3.2 System Architecture

This section elaborates on the proposed solution's system design, which includes key stakeholder and role identification, system architecture, cryptography primitives, and system operations. The model proposed in this work is based on VCs, DIDs, and the Hyperledger fabric Blockchain.

Key Stakeholder and Role Identification. The following are the key stakeholders of the proposed model and their legitimate user roles:

1. Issuers – Universities who will issue certificates (Verifiable Credential) to the students.
2. Verifiers – Could be different players such as Companies (who provide Jobs to the Students) etc., who verify certificates.
3. Holders/Owners – Students who get certificates from Universities and provide these to Companies/other organizations to get the Job/different services.
4. Governance Body – Ministry of Higher Education who authenticates the Issuers and confirms that the Issuer is genuine.

System Architecture Layers. The layered system architecture of the proposed solution is shown in Fig. 1, the system is divided into three layers, which are described below:

1. Application layer: provides the proposed protocols for academic certificate issuance and verification while preserving privacy.
2. The Self-Sovereign Identity layer provides secure and reliable certificate management features and maintains the identities and schemas.
3. The blockchain layer provides a blockchain-based immutable transaction distributed ledger and state database, maintained via a consensus of authorized peer nodes in the network. This layer provides an operating environment for running smart contracts.

In the upcoming sections, we will delve into the specifics of every layer in the proposed system model.

Application Layer. The system's application layer serves as the user interface and functional hub for issuing, storing, verifying, revoking, and updating academic certificates. Access to these functionalities is strictly controlled, ensuring that only authorised parties can modify certificates with the agreement of certificate holders. This approach

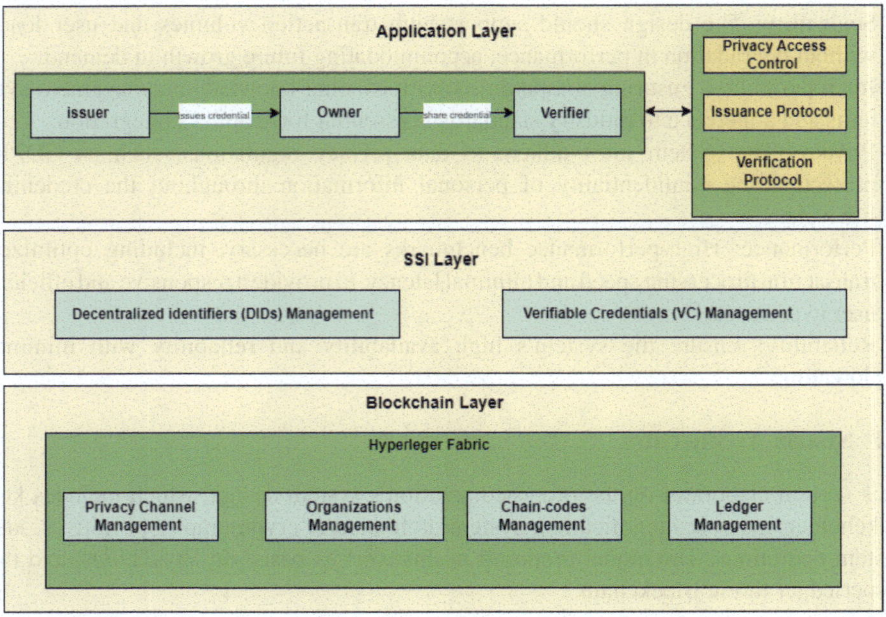

Fig. 1. Layered architecture of the blockchain-based academic credential management framework.

provides a secure and privacy-preserving method for individuals to verify their qualifications while offering employers, educational institutions, and other authorised entities a reliable, tamper-proof means to authenticate these credentials. The Hyperledger Fabric SDK acts as the communication bridge between the blockchain network and the applications within the privacy-preserving system. This architecture caters to the needs of educational institutions (issuers), students (recipients), verifiers, and educational authorities. Institutions can issue and revoke certificates, students can apply for and manage certificates, verifiers can validate diplomas presented by students, and authorities can monitor diploma issuance records across institutions. The foundational structure of the system is a ledger rooted in Self-Sovereign Identity (SSI) principles, made accessible through the Fabric SDK. The blockchain network consists of multiple service nodes, each designated with specific roles such as transaction ordering, fault tolerance assurance, transaction simulation and endorsement, and ledger querying and invocation. Transactions are executed through chaincode programs, which undergo an installation and endorsement process by consortium organizations before being committed to the designated channel. This mechanism enforces consensus among participating organizations regarding the operational scope and functionality of chaincode within the network.

Interactions between the application layer and blockchain infrastructure are facilitated through a suite of APIs, which empower decentralized applications (DApps) to submit transactions, retrieve data, and interact with smart contracts deployed on the blockchain. Middleware components further enhance system capabilities by providing essential services such as authentication, access control, and data processing functions, ensuring a secure and efficient interaction environment across the network layers.

Self-Sovereign Identity Layer. Self-Sovereign Identity (SSI) empowers individuals to control their digital identities, eliminating reliance on centralized systems like certificate authorities [16]. Built on Decentralized Identifiers (DIDs) and Verifiable Credentials (VCs), standardized by W3C, SSI enhances autonomy and privacy [17].

Decentralized Identifiers (DID). A Decentralized Identifier (DID) is a secure, independent identifier that operates without traditional registries or centralized providers [18]. DIDs include:

1. **DID URI Scheme**: Defines the structure of DIDs.
2. **DID Method**: Specifies procedures for creating and managing DIDs.
3. **DID Method-Specific Identifiers**: Ensures unique identities using the DID Method.

Each DID links to a DID document containing metadata like cryptographic keys and authentication methods. Key features include:

1. **Privacy and Security:** Users share only necessary data, maintaining control over their information.
2. **Decentralization:** Reduces identity theft risks by removing centralized authority reliance.
3. **Interoperability:** Enables seamless use across blockchain networks.
4. **Transparency:** Creates tamper-proof academic records.
5. **Cost-Effectiveness:** Offers an efficient alternative to traditional systems.

DID Resolution DID resolution retrieves the associated DID document, enabling authentication, signature verification, and service discovery. Listing 1 illustrates a DID document schema in JSON-LD format, showcasing components such as issuer details, student information, and cryptographic proof.

```
{
 "@context": "https://www.w3.org/ns/did/v1",
 "issuer": {
  "issuerID": "1KoR4pzD59gfD2eUPvFp91KxCFy638EWhS",
  "publicKey": { "type": "Ed25519VerificationKey", "value": "..." },
  "issuanceDate": "2021-03-01 18:37:19"
 },
 "student": {
  "studentID": "1BoBiew5dkyZmAJF5XQApHBrHfrkyCocJw",
  "fullName": "Omar Saad Saleh",
  "email": "omar@malayisa.ac.my",
  "affiliation": { "institutionName": "University Utara Malaysia", "department": "Computer
Science" },
  "personalInfo": { "type": "Ed25519Encryption", "phone": "####", "address": "####" },
  "publicKey": { "type": "Ed25519VerificationKey", "value": "..." }
 },
 "authentication": {
  "type": "EdDSA",
  "signature": "..."
 },
 "proof": {
  "type": "EdDSA",
  "signature": "..."
 }
}
```


Listing 1. Design of a DID document schema in JSON-LD 1 format.

This example demonstrates how DID documents define identity, personal information, and cryptographic proofs using machine-readable and interoperable JSON-LD format.

Verifiable Credentials

Verifiable Credentials (VCs) are digital statements issued by trusted entities, using the W3C VC Data Model and JSON-LD for semantic interoperability. A VC schema includes:

1. @context: Maps document terms to definitions.
2. id: Unique schema identifier.
3. type: Object type (e.g., "VerifiableCredential").
4. issuer: Entity issuing the VC.
5. credentialSubject: Details of the entity receiving the VC.
6. proof: Cryptographic methods verifying authenticity.

Listing 2 illustrates a VC schema for a university-issued certificate:


```
{
  "@context": "https://www.w3.org/ns/did/v1",
  "issuer": {
    "issuerID": "1KUTTG5QSWjXydwyE4w1LP2nET8hvNnMs1",
    "publicKey": { "type": "Ed25519VerificationKey", "value": "..." },
    "issuanceDate": "2022-12-11 18:37:19"
  },
  "subject": {
    "certificateID": "7BCD-8D4C-9G3K-A62N",
    "studentID": "1BoBiew5dkyZmAJF5XQApHBrHfrkyCocJw",
    "fullName": "Omar Saad Saleh",
    "degree": "MSc in Computer Science & Engineering",
    "institutionName": "University Utara Malaysia",
    "department": "Computer Science",
    "score": "4.48/4.50"
  },
  "proof": {
    "type": "EdDSA",
    "signature": "..."
  }
}
```


Listing 2. Example of JSON-LD verifiable credential schema for a certificate issued to a student by a university.

This schema defines key properties like issuer details, credential subject attributes, and cryptographic proofs to ensure secure and verifiable academic records.

4 Hyperledger Fabric Blockchain-Based DID System Design for Certificate Management

Enhancing the security and efficiency of academic certificate issuance and verification is made possible through a blockchain-based framework that integrates Hyperledger Fabric, Self-Sovereign Identity (SSI), Verifiable Credentials (VC), and Decentralized Identifiers (DIDs), as shown in Fig. 2. This approach revolutionises how academic credentials are managed by leveraging decentralised technologies. The foundation begins with establishing a Hyperledger Fabric network, comprising multiple peer nodes managed by universities, accrediting bodies, and other entities involved in the issuance and verification processes. SSI empowers individuals by allowing them direct control over their identity information, while VC provides a secure method for issuing and verifying academic credentials within the network.

Each entity on the network, whether an individual or institution, is uniquely identified using DIDs, ensuring cryptographic security and privacy. Smart contracts deployed on Hyperledger Fabric govern the lifecycle of academic certificates, ensuring authenticity and enforcing access controls. In practical application, a university generates an academic certificate as a Verifiable Credential (VC) on the Hyperledger Fabric network, linking it to a Decentralized Identifier (DID) that uniquely represents both the issuing institution and the credential itself. Certificate holders can securely retrieve their credentials through their DID, which verifiers authenticate by validating relevant smart contracts on the blockchain. This streamlined system allows verifiers to efficiently access and validate credentials without relying on intermediaries or additional verification steps.

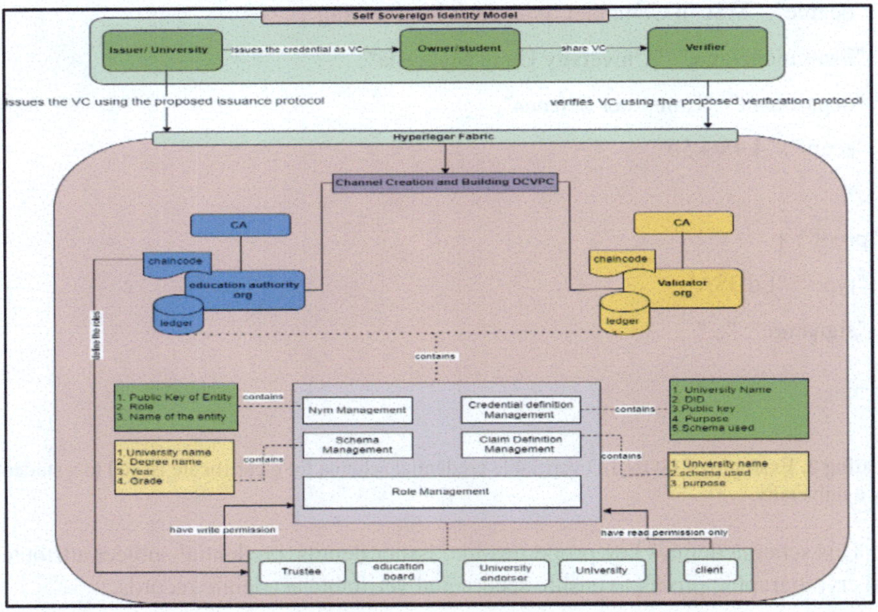

Fig. 2. Hyperledger fabric blockchain-based DID system design for academic credential management in knowledge systems.

To implement these technologies, five core modules are designed and managed: DID Management, Schema Management, Credential Definition, Claim Definition, and Role and Permission Management.

1. **DID Management:** Universities manage their public DIDs with oversight from educational authorities, using Fabric Certificate Authority (CA) and a verifiable data registry. This process includes university registration on the Hyperledger Fabric network,

acquisition of cryptographic credentials, and DID publication to establish institutional control and identity.

2. **Schema Design and Management:** Educational authorities define standardized schemas on the ledger to specify key attributes like UniversityName, DegreeName, Year, and Grade for VCs. Universities apply these pre-approved schemas when generating JSON files for certificate issuance, ensuring uniformity and verifiability across issued credentials.
3. **Claim Definition Creation:** Credential definitions enable secure VC exchanges on the blockchain. Accredited universities create these definitions, detailing the format, structure, and relevant metadata, including the university's name, DID, public key, and schema reference. This approach strengthens trust and facilitates seamless verification for issued credentials.
4. **Roles and Permission Management:** Governance within the blockchain network depends on clearly defined roles, such as Trustee, Education Board, University Endorser, University, and Client. Each role's permissions are administered through smart contracts, ensuring that actions like adding trustees or approving schemas are authorized appropriately.
5. **Workflow of the Proposed Solution:** The solution leverages Hyperledger Fabric's infrastructure, using Fabric CA for identity verification and the ledger for transaction transparency. Issuers (universities) produce VCs that represent academic certificates, embedding cryptographic proofs and essential metadata. Certificate holders (students) securely store their VCs in digital wallets, using DIDs to assert ownership. Verifiers (such as employers or educational institutions) authenticate VCs by referencing issuer details stored on the blockchain, ensuring integrity and eliminating intermediary reliance, as illustrated in Fig. 3.

In this workflow, a trustee oversees the addition of certificate schemas on Hyperledger Fabric, involving three key entities: the education board, Fabric CA, and the ledger.

In this workflow, a trustee manages the addition of certificate schemas on Hyperledger Fabric, involving the education board, Fabric CA, and the ledger (Fig. 3). The process begins with Fabric CA issuing public key certificates to the education board, which then submits a recognition request to the ledger as an education authority. The trustee verifies the board's identity and, upon approval, grants authorization for schema additions. Role and permission checks ensure proper access, enabling authorized boards to create and manage certificate schemas. The ledger records all entities and roles, ensuring transparency and accountability within the credentialing system.

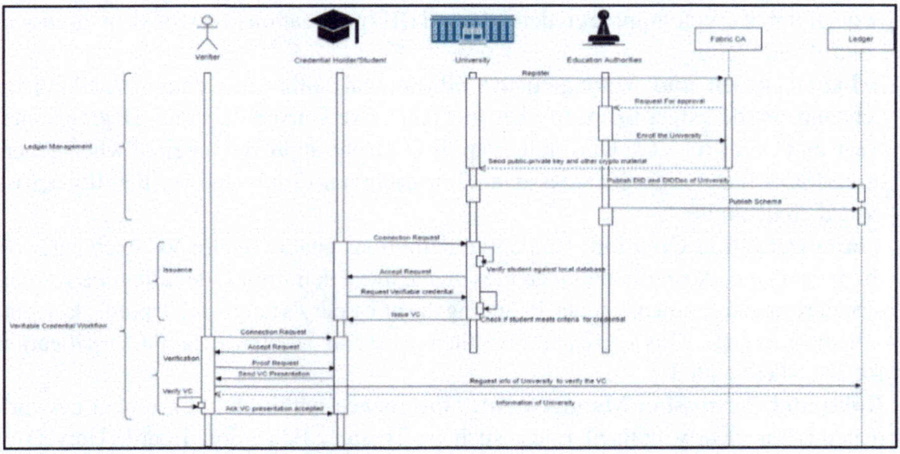

Fig. 3. Workflow of the proposed blockchain-integrated framework for academic credential management within knowledge systems.

5 Conclusion

This paper proposes a comprehensive framework that integrates Hyperledger Fabric, Self-Sovereign Identity (SSI), Verifiable Credentials (VCs), and Decentralized Identifiers (DIDs) to strengthen the security and operational efficiency of academic credential management. By enabling secure collaboration among universities, educational authorities, and credential verifiers within a blockchain-based network, this approach ensures that processes for academic certificate issuance, storage, and verification remain transparent and resistant to tampering. The system's foundation in SSI principles empowers individuals to retain control over their identity data, while VCs provide verifiable assurance of credential authenticity. Future research will involve performance validation through experimental trials, aiming to demonstrate the framework's effectiveness, security, and scalability in practical applications. Such advancements contribute to the wider adoption of blockchain technology in credential management and potentially across various sectors.

References

1. Nair, B.V., Munusami, C.: Knowledge management practices: an exploratory study at the Malaysian higher education institutions. J. Res. Innov. Teach. Learn. **13**(2), 174–190 (2019)
2. Capece, G., Levialdi Ghiron, N., Pasquale, F.: Blockchain technology: redefining trust for digital certificates. Sustainability **12**(21), 8952 (2020)
3. Al Hemairy, M., Abu Talib, M., Khalil, A., Zulfiqar, A., Mohamed, T.: Blockchain-based framework and platform for validation, authentication, and equivalency of academic certification and institution's accreditation: UAE case study and system performance. Educ. Inf. Technol. **1**, 1–30 (2024)
4. Li, J., Kassem, M.: Applications of distributed ledger technology (DLT) and Blockchain-enabled smart contracts in construction. Autom. Constr. **132**, 103955 (2021)

5. Caldarelli, G., Ellul, J.: Trusted academic transcripts on the blockchain: a systematic literature review. Appl. Sci. **11**(4), 1842 (2021)
6. Le, T.V., Hsu, C.L.: A systematic literature review of blockchain technology: security properties, applications, and challenges. J. Internet Technol. **22**(4), 789–802 (2021)
7. Abdelmaboud, A., et al.: Blockchain for IoT applications: taxonomy, platforms, recent advances, challenges, and future research directions. Electronics **11**(4), 630 (2022)
8. Hameed, B., et al.: A review of Blockchain-based educational projects. Int. J. Adv. Comput. Sci. Appl. **10**(10) (2019)
9. Grech, A., Camilleri, A.F.: Blockchain in Education. Publications Office of the European Union, Luxembourg (2017)
10. Turkanović, M., Hölbl, M., Košič, K., Heričko, M., Kamišalić, A.: EduCTX: a blockchain-based higher education credit platform. IEEE Access **6**, 5112–5127 (2018)
11. Sharples, M., Domingue, J.: The blockchain and Kudos: a distributed system for educational record, reputation, and reward. In: Verbert, K., Sharples, M., Klobučar, T. (eds.) EC-TEL 2016. LNISA, vol. 9891, pp. 490–496. Springer, Cham (2016). https://doi.org/10.1007/978-3-319-45153-4_48
12. Chenthara, S., Ahmed, K., Wang, H., Whittaker, F., Chen, Z.: Healthchain: a novel framework on privacy preservation of electronic health records using blockchain technology. PLoS ONE **15**(12), e0243043 (2020)
13. Gandhi, C., Shukla, N., Kaur, G., Yadav, K.: Blockchain technology: concept, applications, challenges, and security threats. In: Choudhury, T., Khanna, A., Toe, T.T., Khurana, M., Gia Nhu, N. (eds.) Blockchain Applications in IoT Ecosystem. EAISICC, pp. 77–104. Springer, Cham (2021). https://doi.org/10.1007/978-3-030-65691-1_6
14. Manoj, T., Makkithaya, K., Narendra, V.: A blockchain-based decentralized identifiers for entity authentication in electronic health records. Cogent Eng. **9**(1), 2035134 (2022)
15. Tang, Q.: Towards using blockchain technology to prevent diploma fraud. IEEE Access **9**, 168678–168688 (2021)
16. Figueroa-Lorenzo, S., Añorga Benito, J., Arrizabalaga, S.: Modbus access control system based on SSI over Hyperledger Fabric blockchain. Sensors **21**(16), 5438 (2021)
17. Abid, A., Cheikhrouhou, S., Kallel, S., Jmaiel, M.: NovidChain: blockchain-based privacy-preserving platform for COVID-19 test/vaccine certificates. Softw.: Pract. Exp. **52**(4), 841–867 (2022)
18. Bouras, M.A., Lu, Q., Zhang, F., Wan, Y., Zhang, T., Ning, H.: Distributed ledger technology for eHealth identity privacy: state of the art and future perspective. Sensors **20**(2), 483 (2020)
19. Vasile, P.C., Dinu, A.: EDIS–Electronic Diploma Integrity Service. Rom. Cyber Secur. J. **3**(2), 59–66 (2021)
20. Merlec, M.M., Islam, M.M., Lee, Y.K., In, H.P.: A consortium blockchain-based secure and trusted electronic portfolio management scheme. Sensors **22**(3), 1271 (2022)

Research Trends in Blockchain and Smart Contracts for Agriculture: A Bibliometric Analysis

Baharudin Osman[1](✉), Hapini Awang[1], Nur Suhaili Mansor[1], Alawiyah Abd Wahab[1], Osman Ghazali[1], Idah Pindai Zengeni[1], Kalthom Husain[1], Hisham Anand Barbhuiya[2], and Bahiah Ab Malek[2]

[1] Institute for Advanced and Smart Digital Opportunities, School of Computing, Universiti Utara Malaysia, 06010 UUM Sintok, Kedah, Malaysia
bahaosman@uum.edu.my

[2] Al-Bukhary International University, Alor Setar, Kedah, Malaysia

Abstract. Blockchain technology and smart contracts in agriculture have attracted significant attention from researchers and industry experts in recent years, leading to a surge in scholarly publications. This study, aimed at tracking research trends and evaluating the evolution of scholarly work in this area, conducts a bibliometric analysis using data from Scopus, one of the top academic databases. Despite being essential for food production and economic activity, the agriculture sector has been lacking in digital transformation, leading to challenges like inefficiencies in the supply chain due to complexities, resource limitations, and increased demand due to population growth and inclusivity of market players. Blockchain and smart contracts present innovative solutions to these issues. Yet, there is limited understanding of the intellectual landscape in this field and how it can resolve the problems in this sector. This analysis explores publication trends, citation connections, and critical research subjects. The findings reveal a substantial increase in publications since 2018, underscoring the growing recognition of blockchain's potential to revolutionise agriculture transparency, traceability, and efficiency. The analysis also identifies the central contributing countries, leading journals, and prolific authors, providing a comprehensive overview of the academic landscape. Notable themes include blockchain application in food safety, supply chain management, and sustainability in agriculture. The study emphasises the transformative potential of blockchain and smart contracts, inspiring the audience with the possibilities for the agricultural sector through technological innovation.

Keywords: Blockchain · Smart Contract · Agriculture · Bibliometric Analysis · Technological Innovation

1 Introduction

The agricultural sector plays a crucial role globally, serving as a primary food source and a critical economic driver. Despite its significance, agriculture has lagged in digitalisation compared to other industries globally [2]. In recent years, the industry has faced growing

Z. Mat Aji et al. (Eds.): KMICe 2024, ISEM 49, pp. 174–186, 2025.
https://doi.org/10.1007/978-3-031-91485-0_14

challenges amplified by population growth and the disruptive effects of climate change. These challenges include increased demands for food production, constraints on natural resources, and inefficiencies within agricultural supply chains [5]. Agricultural supply chains are complex and often riddled with intermediaries, leading to delays, lack of transparency, and increased costs [12]. These inefficiencies impact small-scale farmers and large agribusinesses, food processors, and distributors who face challenges in maintaining supply chain integrity, meeting sustainability goals, and ensuring traceability from farm to consumer.

Blockchain technology offers a promising solution by enhancing transparency, traceability, and operational efficiency throughout the agricultural supply chain [4]. Blockchain's immutable records enable better tracking of agricultural products from farm to consumer, which can significantly improve food safety and traceability. By reducing the number of intermediaries and automating transactions with smart contracts, blockchain can also streamline supply chains and promote inclusivity for small-scale producers [21]. Additionally, blockchain can help promote more sustainable agricultural practices by ensuring that environmental standards are met and verified.

However, despite its potential, there is limited understanding of the intellectual landscape concerning blockchain's application in agriculture. To address this, a bibliometric analysis is performed in this study to track research trends, explore critical subjects, and identify knowledge gaps. This analysis provides insights into how blockchain technology and smart contracts are applied in agriculture, revealing publication trends, key contributing countries, and the most influential research. The study highlights blockchain's transformative potential in addressing the agricultural sector's critical challenges, including improving transparency, sustainability, and efficiency.

Blockchain operates as a decentralised digital ledger, recording transactions across multiple computers within a network [3]. Its versatility has spurred ongoing research and applications across various sectors, including agriculture [4]. Smart contracts are automated agreements triggered by predefined conditions and can complement blockchain technology by streamlining agricultural transactions and agreements [1].

Bibliometric analysis is a systematic review method that facilitates the comprehensive description, evaluation, and tracking of published research. It employs quantitative and visual techniques to analyse the dynamics of scientific publications [13]. Increased availability and accessibility of bibliometric tools and scientific databases have recently gained popularity [6]. This article aims to conduct a bibliometric analysis of blockchain and smart contract research trends within the agricultural sector. By tracing the intellectual terrain of this emerging field, the study seeks to identify key research domains, influential scholarly works, and future research paths to support blockchain technology's integration into agriculture. The paper is organized as follows. Section 2 reviews the literature on blockchain and smart contracts in agriculture, highlighting the key developments and gaps in current research. Section 3 outlines the methods used for the bibliometric analysis, including data sources and analytical techniques. Section 4 presents the analysis results, identifying significant trends, leading contributors, and emerging themes. Finally, Sect. 5 offers the conclusion, discussing the implications of the findings and providing recommendations for future research in this domain.

2 Literature Review

Blockchain technology and smart contracts have surfaced as game-changing tools in numerous industries, including agriculture. However, their adoption and impact within the agricultural sector remain insufficiently examined, especially regarding small-scale farming. Existing literature outlines various potential benefits and challenges, but a detailed analysis exposes notable gaps and unaddressed concerns.

One study points out that agricultural supply chains have become more intricate, resulting in frequent fraud and a widespread lack of transparency [8]. Incorporating blockchain can significantly enhance supply chain visibility and efficiency by providing an immutable ledger that tracks product origins and transactions. The study introduced the Agriculture Supply Chain Management using Blockchain and Internet of Things (Agri-SCM-BIoT) architecture. This architecture combines blockchain and IoT devices to gather real-time data and monitor food throughout the supply chain to optimize storage, scalability, interoperability, and data security and privacy. Nevertheless, the study identified implementation challenges, including stakeholders' hesitance to share information with competitors, scalability concerns, and the lack of uniform standards.

Another study designed a private blockchain-based crop index solution to enable secure and reliable interactions between smallholder farmers and insurers [9]. By utilizing blockchain technology, this solution guarantees that crop data is tamper-proof and transparent, fostering trust between farmers and insurers. Furthermore, it offers a dependable basis for insurance claims and payouts, reducing fraud risks. However, the implementation faces high initial setup costs, technical expertise requirements, and farmers' adaptation to new technologies.

Smart contracts, which automate transactions and enforce contractual agreements without intermediaries, are often praised for improving efficiency in agricultural transactions. It was underscored that smart contracts can mitigate financial risks, reduce administrative and service costs, and improve the efficiency of financial services [14]. A recent study examined the potential and implications of smart contracts in the agro-industrial sector and agriculture. They acknowledge the immense potential in addressing issues such as transaction inefficiencies, transparency issues, and disputes in contractual obligations [11]. Their findings indicated that smart contracts could enhance operational efficiency. However, the novelty of smart contract applications, coupled with the steep learning curve in areas with low digital literacy, underscores the necessity of this bibliometric analysis.

Numerous studies concentrate on theoretical frameworks and potential applications but lack empirical evidence from practical field implementations. For instance, one study proposed a scalable data-sharing system utilising smart contracts and IPFS for smart agriculture to address performance and cost issues [16]. Another study showcased the use of blockchain in soybean traceability, highlighting its ability to enhance efficiency and security without intermediaries [17]. However, a systematic review pointed out that barriers to adoption outweigh facilitators in smart and sustainable agriculture [15]. Although blockchain can enhance trust and collaboration among stakeholders, the absence of global standards and industry best practices impedes widespread adoption.

These studies suggest that while blockchain holds promise for agricultural applications, there is limited empirical evidence of its effectiveness in real-world scenarios, underscoring the need for further practical research and development in this field.

Additionally, most research on blockchain and smart contracts has concentrated on improving traceability and transparency in food supply chains, leaving a substantial gap in understanding how blockchain technology can directly affect and potentially boost farmers' income [18]. This crucial gap highlights the necessity for future research to investigate technological applications and the socio-economic impacts on agricultural stakeholders, particularly emphasising how blockchain innovations can economically empower farmers.

In summary, while the current literature acknowledges the transformative potential of blockchain and smart contracts in agriculture, it frequently fails to offer a complete and realistic view of their applicability and impact. The bibliometric approach delivers a detailed overview of research trends and identifies essential gaps requiring attention. Future research should adopt a more inclusive perspective, considering socio-economic, cultural, and infrastructural challenges. By addressing these shortcomings, scholars can provide more practical insights and aid in developing solutions that benefit all agricultural sector stakeholders.

3 Methods

This study employs a bibliometric analysis to track research patterns in blockchain and smart contracts within the agricultural sector, using data from the Scopus database. Bibliometric analysis is valuable for quantitatively evaluating scientific literature and identifying research trends. Analyzing metadata such as citations, authors, and institutions enables researchers to identify influential works and map research landscapes [19]. Although this method is cost-effective and capable of handling large datasets, it does have limitations, particularly in assessing the qualitative aspects of research [20].

Through quantitative analysis, bibliometric methods examine publication patterns, citation networks, and critical thematic areas [6]. This study analysed data from the Scopus repository to uncover trends in blockchain and smart contract research within agriculture. The methodology employed in this study is outlined in Fig. 1.

The search focused on article titles to identify relevant studies on blockchain and smart contracts in agriculture. The timeframe for the analysis spanned from the inception of blockchain research to the extraction date of May 10, 2024. Only English-language articles were considered to ensure consistency and comparability across studies. Additionally, journal articles were prioritized, but conference papers and other types of publications were not excluded, acknowledging the rigorous peer-review process typically associated with journals.

The search query used "(TITLE-ABS-KEY("blockchain" AND "smart contract" AND "agriculture"))" identified 264 records, which were screened based on specific inclusion criteria, including titles and abstracts. However, none met all eligibility criteria, resulting in no records being included in the final analysis, underscoring a significant gap in the literature at the intersection of blockchain, smart contracts, and agriculture.

This structured bibliometric analysis reveals vital research trends and gaps, highlighting an area ripe for future empirical studies and practical applications of blockchain and smart contracts in agriculture. This approach ensures a systematic exploration of current trends while pinpointing potential research opportunities in this evolving field.

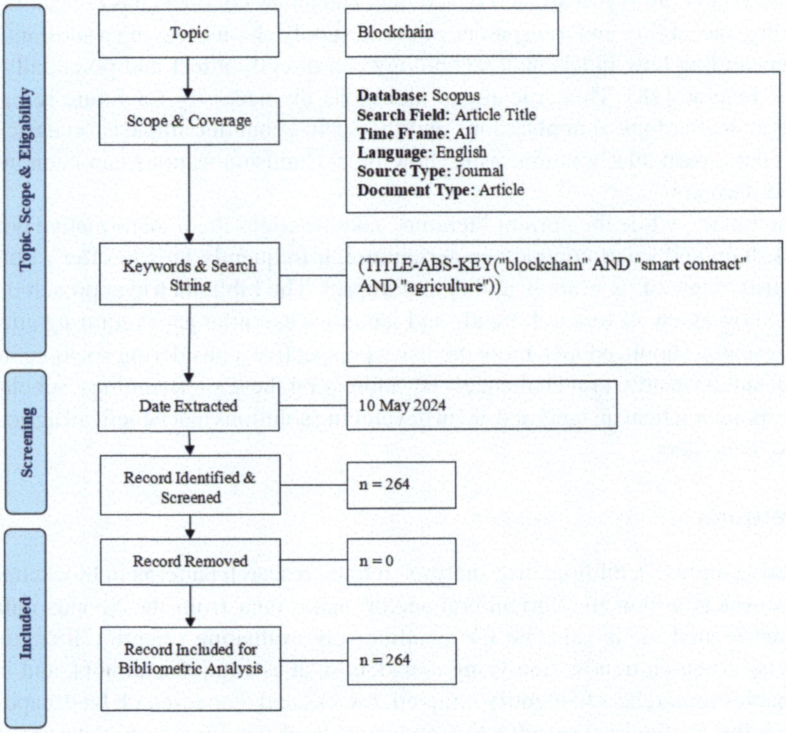

Fig. 1. Flow diagram of the search strategy.

4 Results

This section thoroughly analyses the tables and figures, offering clear insights into the study's findings. It illuminates trends, patterns, and noteworthy observations derived from the data, providing a comprehensive overview of the research results.

Table 1 indicates that out of 264 publications on blockchain, smart contracts, and agriculture, the majority (51.52%) are conference papers. Articles account for 35.23%, book chapters comprise 7.95%, reviews 2.65%, conference reviews 1.52%, and short surveys 0.76%. This indicates a firm reliance on conference proceedings and underlines the necessity for more peer-reviewed articles and empirical studies to advance understanding in these fields.

The data in Table 2 indicates that journals are the primary source for publications on blockchain, smart contracts, and agriculture, accounting for 39.02% of the total 264 publications. Conference proceedings come next at 37.88%, followed by book series at 18.56% and standalone books at 4.55%. This breakdown demonstrates a well-rounded

Table 1. Document type.

Document Type	Total Publications (TP)	Percentage (%)
Conference Paper	136	51.52%
Article	93	35.23%
Book Chapter	21	7.95%
Review	7	2.65%
Conference Review	4	1.52%
Short Survey	2	0.76%
Total	**264**	**100.00**

Table 2. Source type.

Source Type	Total Publications (TP)	Percentage (%)
Journal	103	39.02%
Conference Proceeding	100	37.88%
Book Series	49	18.56%
Book	12	4.55%
Total	**264**	**100.00**

distribution of research across various types of publications, reflecting a diverse scholarly involvement. The significant presence of journal articles and conference proceedings suggests active dissemination of work in peer-reviewed and discussion-based settings.

English is the primary language for blockchain, smart contracts, and agriculture publications, making up 98.11% of the 264 publications. With one bilingual document, Chinese and Spanish account for 1.52% and 0.38%, respectively. This highlights the global reach and accessibility of English research while suggesting the need for more multilingual scholarship to increase inclusivity and impact. Chinese and Spanish publications indicate global interest across different linguistic communities.

Table 3. Top five subject areas.

Subject Area	Total Publications (TP)	Percentage (%)
Computer Science	217	82.20%
Engineering	129	48.86%
Decision Sciences	66	25.00%
Mathematics	47	17.80%
Physics and Astronomy	30	11.36%

When analyzing the subject areas in the publications, some overlap is evident as they intersect across disciplines. Table 3 shows that Computer Science dominates blockchain, smart contracts, and agriculture publications, making up 82.20% of the 264 total. Engineering follows with 48.86%, and Decision Sciences with 25%. Fields like Mathematics (17.80%), Physics and Astronomy (11.36%), and Business, Management, and Accounting (9.47%) have smaller shares. Agricultural and Biological Sciences constitute only 7.95%, highlighting a gap despite blockchain's potential in agriculture. This suggests more collaboration between tech and agricultural experts is needed to address practical challenges and avoid overlooking socioeconomic and environmental aspects.

Table 4. Year of publication.

Year	TP	%	Cumm. TP	Cumm. %	NCP	TC	C/P	C/CP	h-index	g-index	m-index
2018	4	1.66	4	1.66	4	490	122.50	122.50	4	4	0.1429
2019	20	8.30	24	9.96	18	994	49.70	55.22	11	20	0.0917
2020	36	14.94	60	24.90	35	1714	47.61	48.97	16	36	0.0889
2021	40	16.60	100	41.49	35	912	22.80	26.06	15	30	0.0938
2022	61	25.31	161	66.80	52	625	10.25	12.02	12	23	0.0656
2023	80	33.20	241	100.00	47	293	3.66	6.23	10	14	0.0625
Total	**241**	**100**			**191**	**5028**	**20.86**	**26.32**	**68**	**127**	**0.0818**

Table 4 shows a steady increase in publications on blockchain, smart contracts, and agriculture from 2018 to 2023, reaching 80 in 2023. The selection of publications from 2018 to 2023 ensures a comprehensive analysis of the most recent developments in blockchain and smart contracts within the agricultural sector, capturing a complete cycle of research outputs and reflecting the latest trends as these technologies gain traction. Of 241 publications, 191 have been cited 5028 times, averaging 20.86 citations per publication and 26.32 per cited work. The h-index is 68, the g-index is 127, and the m-index is 0.0818. Although not all research has gained significant recognition, the average citations and indices reflect a solid impact and productivity, indicating strong growth and potential for future research in this field.

The Journal of Cleaner Production leads with three publications and high citation metrics, indicating a significant scholarly impact. Other notable contributors include Lecture Notes in Business Information Processing, the International Journal of Scientific and Technology Research, and Scientific Reports, each with varying citation metrics. The diversity in influence among these sources suggests differing levels of recognition.

Over six years, 2,264 papers have accumulated 5,048 citations, averaging 841.33 annually, highlighting active engagement and significant attention, as seen in Table 5. Each paper averages 19.12 citations, reflecting substantial recognition. The average citation per author is 1,459.26, each contributing to 82.33 papers, demonstrating a collaborative research approach. The h-index of 34 and the g-index of 66 indicate a solid scholarly impact, with many papers receiving significant citations and recognition.

Table 5. Citations metrics.

Metrics	Data
Papers	2264
Number of Citations	5048
Years	6
Citations per Year	841.33
Citations per Paper	19.12
Cites_Author	1459.26
Papers_Author	82.33
Authors_Paper	3.77
h_index	34
g_index	66

The top 20 highly cited articles in blockchain, smart contracts, and agriculture primarily focus on innovative applications of blockchain in agriculture, including supply chain management, crop insurance, and sustainability. Most articles are recent and have not yet been widely cited, except for "KRanTi: Blockchain-based farmer's credit scheme for agriculture-food supply chain," which has 18 citations and addresses critical agricultural financing and supply chain management issues. The topics of these articles range from crop disease information systems to IoT-enabled food supply chain transparency, emphasising transparency, efficiency, and sustainability. Despite varying citation counts, these articles are poised to influence future research and practical applications, shaping advancements in blockchain technology within agriculture.

Regarding authorship distribution, it is revealed that 67.80% of publications involve 2 to 4 authors, indicating a trend towards small to moderate-sized research teams that foster interdisciplinary collaboration. Single-author publications are less common, showing individual contributions, while documents with five or more authors represent a smaller percentage, suggesting larger collaborative or interdisciplinary projects. A few documents with zero authors listed highlight cases where traditional authorship is not applicable. Overall, the analysis reflects the collaborative nature of research in this field, emphasising the role of teamwork in advancing knowledge and innovation.

The overlap of keywords is notable, as some authors may use multiple terms. Table 6 highlights key terms in blockchain, smart contracts, and agriculture publications. "Blockchain" is the most common keyword, appearing in 85.98% of publications, emphasising its role in enhancing transparency and security in agriculture. Keywords like "Smart Contract(s)" indicate interest in automating and optimising agricultural agreements. Terms such as "Internet of Things" and "Agricultural Robots" reflect the integration of advanced technologies for efficiency and data-driven decisions. "Supply Chains" and "Supply Chain Management" point to efforts to improve logistics and

Table 6. Top keywords.

Author Keywords	Total Publications (TP)	Percentage (%)
Blockchain	227	85.98%
Smart Contract	125	47.35%
Block-chain	114	43.18%
Agriculture	113	42.80%
Internet Of Things	85	32.20%
Smart Contracts	62	23.48%
Supply Chains	57	21.59%
Agricultural Robots	42	15.91%
Supply Chain Management	41	15.53%
Distributed Ledger	38	14.39%

traceability. "Distributed Ledger" suggests broader uses of decentralised data management. These keywords demonstrate how blockchain and related technologies advance efficiency, transparency, and sustainability in agriculture.

4.1 Visualisation Map

In addition to previous tables and discussions, this study uses visualisation maps to analyse and present complex bibliometric data, providing insights that might be difficult to extract from traditional tables. For example, Fig. 2 presents a network visualisation of global citation relationships in blockchain and smart contracts within agriculture.

This study also includes an analysis of network visualisation maps showing global citation relationships and co-authorship networks in blockchain and smart contracts research within agriculture. The analysis highlights India's significant citation impact, making it the country with the most influence, followed by China and the United States. The maps reveal strong citation relationships and regional research collaborations, particularly between India and China, as well as the UK with Bangladesh and Morocco. This emphasises India's central role and extensive international cooperation in this research field.

The co-authorship networks among countries indicate substantial global collaboration, with significant contributions from critical works and authors. Again, India emerges as a central node, with notable research links to the UK, the US, and Saudi Arabia. These networks reveal the strength and frequency of international research partnerships, highlighting India's pivotal role in advancing blockchain and smart contracts research in agriculture. Clusters formed around prominent countries underscore the collaborative nature of these efforts.

The maps also highlight co-authorships. Among individual authors, distinct clusters of collaborative groups are evident, showcasing focused study areas and cooperation. Notable clusters include those centred on "Mohanta, Bidyadhar Kumar", and

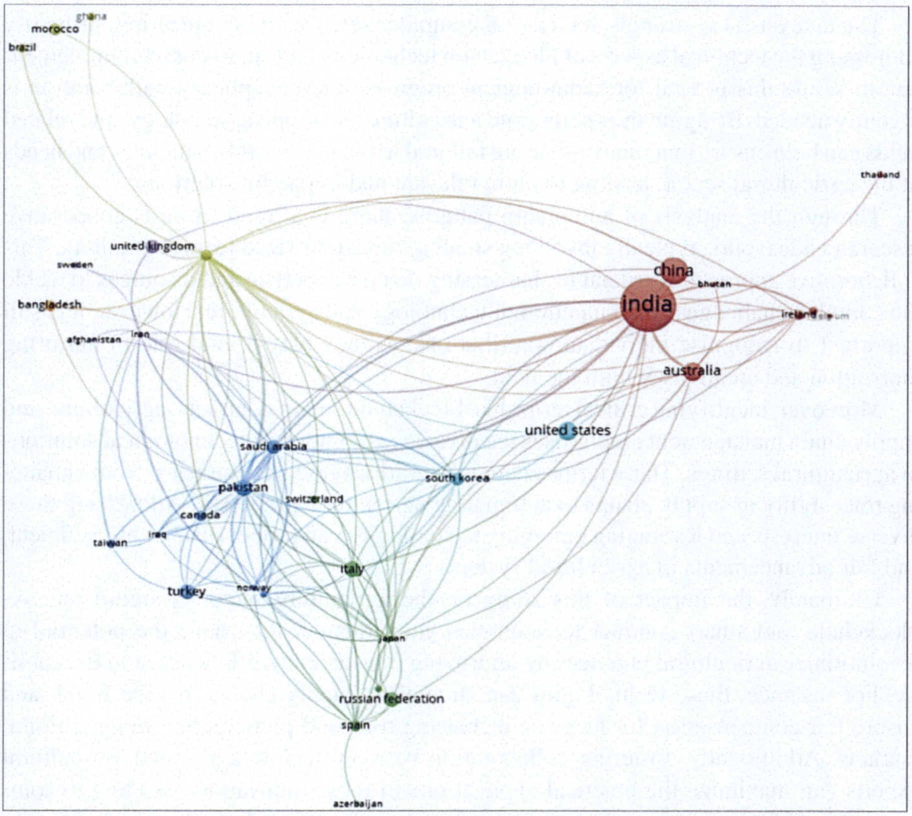

Fig. 2. Network visualisation map of the citation by countries. Note: Minimum number of documents of an author = 1; Minimum number of citations of an author = 5.

"Gadekallu, Thippa Reddy." These clusters illustrate the collaborative nature of research, with specific groups of researchers working together to advance the field. Overall, the findings underscore the interconnected and collaborative efforts driving the research landscape in blockchain and smart contracts within agriculture.

5 Discussion

This bibliometric analysis provides essential insights into the emerging research trends in blockchain technology, smart contracts, and agriculture. The results illuminate significant developments and recurring themes, providing a comprehensive view of academic research's status and future directions in this rapidly evolving field.

A significant observation is the dominant use of conference proceedings to disseminate research findings. While conferences are excellent for sharing preliminary results and fostering collaboration, the high number of conference papers highlights the need for more extensive peer-reviewed articles. This gap offers researchers a chance to conduct more in-depth empirical studies and theoretical analyses, thereby enriching scholarly discourse and advancing knowledge in the field.

The analysis also strongly focuses on computer science and engineering, primarily addressing the technical aspects of blockchain technology and smart contract implementation. While this is vital for technological progress, interdisciplinary collaboration is urgently needed. Bringing in experts from agriculture, economics, sociology, and related fields can help ensure that innovations are tailored to the real-world challenges and needs of the agricultural sector, leading to more relevant and impactful solutions.

Through the analysis of authorship patterns, there is a trend towards cooperative research endeavours, typically involving small to moderate-sized teams of authors. This collaborative approach is crucial for harnessing diverse expertise and resources to tackle the complex challenges at the junction of technology and agriculture. However, it is still important to recognise individual contributions, as they play a vital role in nurturing innovation and creativity within the field.

Moreover, identifying critical terms like blockchain, smart contracts, agriculture, and supply chain management emphasises the diverse exploration of technological solutions in agricultural settings. These terms encompass various research interests, from enhancing traceability in supply chains to automating agricultural processes. Addressing these diverse interests and leveraging emerging technologies can propel sustainable, efficient, and fair advancements in agricultural systems.

Ultimately, the impact of this study on the agriculture sector is significant. As blockchain and smart contract technologies gain traction, they have the potential to revolutionize agricultural practices by improving transparency, efficiency, and traceability. For instance, these technologies can streamline supply chains, reduce fraud, and ensure fair compensation for farmers, increasing trust and participation in agricultural markets. Additionally, fostering collaboration between technologists and agricultural experts can maximize the practical applications of these innovations, leading to solutions that address economic but also social and environmental challenges, ultimately enhancing productivity and sustainability in the sector.

6 Conclusion

In summary, this analysis offers a thorough overview of the developing research field at the intersection of blockchain, smart contracts, and agriculture. It underscores the primary emphasis on technical exploration, especially in the implementation of blockchain and applications of smart contracts and highlights the significant role of conference proceedings in disseminating this research. Despite evident collaborative efforts, the study notes potential limitations due to an over-reliance on quantitative metrics and a lack of qualitative insights. This points to opportunities for future research to adopt diverse methodologies and interdisciplinary approaches.

The field would benefit from deeper stakeholder engagement, wider dissemination of findings through peer-reviewed channels, and a more detailed exploration of blockchain and smart contracts' socio-economic and environmental impacts in agriculture, which is currently lacking. Future research should address these limitations by incorporating qualitative methods and cultivating collaborative research across disciplines. Focusing on the societal impact of technological innovations in agriculture, including blockchain and smart contracts, would enrich the discussion and ensure solutions are technically sound and contextually appropriate. Additionally, efforts to publish findings in peer-reviewed journals and involve industrial stakeholders in co-creative processes would be necessary, as the present trend indicates that higher learning institutions dominate scholarly publications. This deficiency implies that the findings, especially those on the technical application of blockchain technology in agriculture, would be slower without assistance from industrial partners.

Acknowledgement. This research is supported by Universiti Utara Malaysia (UUM) Matching Grant Scheme: S/O Code: 21693.

References

1. Food & Agriculture Org.: E-agriculture in Action: Blockchain for Agriculture: Opportunities and Challenges (2019)
2. Sandeep Kumar, M., Maheshwari, V., Prabhu, J., Prasanna, M., Jothikumar, R.: Applying blockchain in agriculture: a study on blockchain technology, benefits, and challenges. In: Suresh, A., Paiva, S. (eds.) Deep Learning and Edge Computing Solutions for High Performance Computing. EAISICC, pp. 167–181. Springer, Cham (2021). https://doi.org/10.1007/978-3-030-60265-9_11
3. Vaigandla, K.K., Karne, R., Siluveru, M., Kesoju, M.: Review on blockchain technology: architecture, characteristics, benefits, algorithms, challenges and applications. Mesop. J. Cybersecur. 73–85 (2023)
4. Khezr, S., Moniruzzaman, M., Yassine, A., Benlamri, R.: Blockchain technology in healthcare: a comprehensive review and directions for future research. Appl. Sci. **9**(9), 1736 (2019)
5. Kasan, K.T., Kasan, Y.T.H., Fadare, S.A.: Agriculture 4.0: impact and potential challenges of blockchain technology in agriculture and its management. In: Proceedings of International Conference on Law, Patent, and Technology, vol. 11(8s) (2023)
6. Donthu, N., Kumar, S., Mukherjee, D., Pandey, N., Lim, W.M.: How to conduct a bibliometric analysis: an overview and guidelines. J. Bus. Res. **133**, 285–296 (2021)
7. Zupic, I., Čater, T.: Bibliometric methods in management and organisation. Organ. Res. Methods **18**(3), 429–472 (2014)
8. Bhat, S.A., Huang, N., Sofi, I.B., Sultan, M.: Agriculture-food supply chain management based on blockchain and IoT: a narrative on enterprise blockchain interoperability. Agriculture **12**(1), 40 (2021)
9. Omar, I.A., Jayaraman, R., Salah, K., Hasan, H.R., Antony, J., Omar, M.: Blockchain-based approach for crop index insurance in agricultural supply chain. IEEE Access **11**, 118660–118675 (2023)
10. Awan, S., et al.: IoT with Blockchain: a futuristic approach in agriculture and food supply chain. Wirel. Commun. Mob. Comput. **2021**, 1–14 (2021)
11. Nazarov, A., Nazarov, D.: Leveraging smart contracts for enhanced efficiency in the agro-industrial complex and agriculture. BIO Web Conf. **67**, 02028 (2023)
12. Takale, D.G., Mahalle, P.N., Sule, B.: Blockchain-powered direct Farm-to-Consumer supply chains (2024)
13. Alkhayyat, S., Khan, M., Ahmad, T., Haroon, N., Tariq, H., Baig, M.: A bibliometric analysis of the top 100 most cited papers and research trends in breast cancer related BRCA1 and BRCA2 genes. Medicine **101**(38), e30576 (2022)

14. Zheng, Z., et al.: An overview on smart contracts: challenges, advances and platforms. Futur. Gener. Comput. Syst. **105**, 475–491 (2020)
15. Akella, G.K., Wibowo, S., Grandhi, S., Mubarak, S.: A systematic review of blockchain technology adoption barriers and enablers for smart and sustainable agriculture. Big Data Cogn. Comput. **7**(2), 86 (2023)
16. Rahman, M.U., Baiardi, F., Ricci, L.: Blockchain smart contract for scalable data sharing in IoT: a case study of smart agriculture. In: Proceedings of Global Conference on Artificial Intelligence and Internet of Things (2020)
17. Salah, K., Nizamuddin, N., Jayaraman, R., Omar, M.: Blockchain-based soybean traceability in agricultural supply chain. IEEE Access **7**, 73295–73305 (2019)
18. Bartol, T.: Smallholders and small-scale agriculture: mapping and visualization of knowledge domains and research trends. Cogent Soc. Sci. **9**(1) (2022)
19. Lazarides, M.K., Lazaridou, I.-Z., Papanas, N.: Bibliometric analysis: bridging informatics with science. Int. J. Lower Extrem. Wounds 1–3 (2023)
20. Hemmingsen, M.N., Lau, A., Larsen, A., Ørholt, M.: The role of bibliometric analyses in plastic surgery—Advantages and disadvantages. Gland Surg. **12**, 873–874 (2023)
21. Tripoli, M., Schmidhuber, J.: Emerging opportunities for blockchain application in the agri-food industry. Revised Version. Rome and Geneva, FAO and ICTSD (2020)

Exploring the Impact of Online Game Addiction on Cybercognition Among Malaysian Adolescents

Wan Mohd Yusoff Wan Yaacob[1(✉)], Nur Haryani Zakaria[2], and Zahurin Mat Aji[2]

[1] Politeknik Tuanku Syed Sirajuddin, 02600 Arau, Perlis, Malaysia
wanmohdyusoff@ptss.edu.my
[2] Universiti Utara Malaysia, 06010 Sintok, Kedah, Malaysia

Abstract. The study investigates the intricate relationship between online game addiction and cybercognition among adolescents. As online gaming becomes increasingly popular, concerns about its impact on adolescents' cognitive and social development have grown. This study explores how excessive gaming influences cognitive functions such as intention, decision-making, and social cognition, including online communication and social interactions. For the current study, 660 adolescents who play online games living in the four regions in peninsula Malaysia were researched using a quantitative approach. The equipment included a demographic data sheet, an online game addiction scale, and the cybercognition test. The collected data was evaluated using SPSS to determine the relationship between online game addiction and cybercognition using frequency distribution and correlation. The findings revealed that online gamers who remained preoccupied with online game addiction had difficulties in decision-making and exhibited withdrawal symptoms that interfered with cybercognition. Additionally, it shows that online gamers who are involved in online game addiction had a significant deficit in intention and concentration, and this contributed to a lack of security behaviour. A cybercognition enhancement solution was proposed to mitigate the problems.

Keywords: Cybercognition · Security Behavior · Online Game Addiction

1 Introduction

1.1 Online Game Addiction

Online Game Addiction (OGA) has emerged as a significant issue due to the unique characteristics and widespread availability of online gaming platforms. It can be classified as problematic, excessive, and compulsive usage of online gaming that causes a substantial impairment in the ability to function in multiple domains over an extended time [1]. The online gaming environment is designed to be highly engaging and rewarding, often featuring in-game achievements and social interactions. This combination of

Z. Mat Aji et al. (Eds.): KMICe 2024, ISEM 49, pp. 187–199, 2025.
https://doi.org/10.1007/978-3-031-91485-0_15

immediate gratification and continuous reinforcement can make games habit-forming, leading to excessive play. While online gaming can be an enjoyable and immersive pastime, there are specific components that contribute to the development of addiction [2]. In the studies related to the component model of addiction in the biopsychosocial framework and game addiction scale, Lemmens [3] and Griffiths [4] have described the main components of addiction, such as mood modification, salience, tolerance, relapse, withdrawal, conflict and problems. These findings are essential fundamentals to support this study [5].

The recognition and classification of mental illnesses by authoritative agencies such as the American Psychiatric Association's Diagnostic and Statistical Manual of Mental Disorders (DSM) play a crucial role in understanding and treating psychological conditions [6]. While OGA is not officially classified as a mental illness in the DSM, it is essential to acknowledge that the field of mental health is constantly evolving since Berthon, Pitt and Campbell (2019) have revealed that OGA shows similar symptoms to Internet Addiction [8]. The scholars that focused on the factor regarding OGA are limited. For instance, Hadlington [9, 10] stressed the attitudes and factors related to the cybersecurity perspectives and Durak [11] emphasized human values in cybersecurity among adolescents. While these studies investigated the cybersecurity context, their perspectives differed. As such, OGA has garnered increasing attention as a public health concern prompting the need for research, education and intervention to mitigate its impact and ensure responsible gaming habits.

1.2 Online Game Addiction Phases

Khazaal et al. [12] revealed that OGA can be divided into three phases. During the first phase, users experience active enjoyment and are motivated to continue playing for more fun. This is due to the regular release of dopamine in the brain. It is natural to a person as there is a part of the brain which is a natural process regulated by the Homeostasis Hippocampus. The brain's ability to maintain and balance dopamine release plays a crucial role in developing tolerance, enabling the individual to handle increasing levels of stimulation and achieve new effects [13]. In the second phase, once tolerance is developed and the release is achieved, the games will start to be less fun, and the time consumed will need to be increased to maintain euphoria and dopamine. This process decreases the brain's sensitivity, and the user will play the games for no specific reason. Finally, in the last phase, users become trapped in the games and experience a lack of enjoyment and fun. They struggle to disengage from the games and are unable to pursue other activities, which leads to a state of biological addiction. At this time, it will cause a dopamine exhaustion problem as the body excessively relies on dopamine resulting in a significant increase in tolerance towards its release. When the influence of OGA reaches a certain level, the release of dopamine can diminish the effectiveness of fear-based protection motivation. Consequently, fear appeal-based strategies related to the negative emotion when recognizing danger gradually become ineffective, influence the action of security behaviour and fail to safeguard users against addictive behaviour [13–15].

1.3 The Effect of Online Game Addiction Among Adolescents

OGA also has a profound impact on academic performance among adolescents. Considering the results, Durak pointed out that OGA affects adolescent students' time management, which leads to their inability to handle various educational activities. Baturay also emphasized the detrimental effect of OGA on academic performance [16] who mentioned that students were increasingly at risk of developing addictive gaming behaviours that could interfere with their academic responsibilities. According to Griffiths [3] and Lemmens [4], the seven components adapted from general components of addiction: mood modification, salience, tolerance, relapse, withdrawal, conflict, and problems usually influence performance, and these findings also suggest that the issues of OGA are connected to intentional behaviours that can affect academic performance among the students. As online games continue to gain popularity, addiction to online games can lead to a range of negative consequences, including poor time management, reduced study time and increased procrastination. Students addicted to these games often struggle to balance their academic commitments and gaming pursuits. This can result in missed assignments, lower grades or even academic failure. Furthermore, the psychological effects of addiction, such as anxiety and decreased motivation, can further impede a student's ability to excel in their studies [17–19].

OGA also significantly affects physical and mental health. Excessive and prolonged gaming often leads to a sedentary lifestyle, which contributes to physical health problems such as obesity, muscular and skeletal issues and cardiovascular concerns. Moreover, the hours spent in front of screens can lead to eye strain and disrupted sleep patterns, further impacting overall well-being [20]. In the context of health, Montag and Walla [21] have documented the increasing recognition of OGA as a significant issue, necessitating the development of accessible and cost-effective interventions to regain control over one's well-being. According to Kuss et al. [22], the psychological aspect is substantial, with addiction to online games potentially causing anxiety, depression and increased stress levels. OGA can trigger a host of psychological challenges. The phenomenon of Internet dependency is associated with several psychological traits and withdrawal symptoms, such as diminished appetite, sleep disturbances, depressive symptoms and heightened irritability. According to the statistics on online game dependencies, a range of 6% to 15% of the population exhibited indications of high risk for OGA [23]. Wang and Sheng support this claim [24] as well as Peper and Harvey [25] who reports that OGA is associated with users' anxiety, loneliness and depression and can be attributed to the influence of intentional behaviour among adolescents. OGA can also trigger a host of psychological challenges. The constant urge to play and the anxiety that arises when gaming is disrupted can lead to heightened stress levels. Individuals often become so absorbed in virtual worlds that they withdraw from real-life social interactions, which leads to feelings of loneliness, isolation and depression.

1.4 Cybercognition in Online Gaming

Cybercognition refers to studying cognitive processes and behaviours related to using digital technologies and the Internet [26]. It encompasses how individuals perceive, think, learn and make decisions in online environments and how they interact with digital devices and systems. This field explores topics such as digital literacy, online decision-making, and the impact of technology on attention, memory and problem-solving skills. Additionally, cybercognition examines the psychological and social effects of digital engagement, including issues like cyberbullying, digital addiction and the influence of social media. As technology evolves, understanding cybercognition becomes increasingly crucial in navigating the digital world and fostering healthy and responsible use of technology.

The fear appeal was identified as having a lack of effectiveness in mitigating the issues which are influenced by addictive behaviour. Hence, cybercognition was introduced to improve cognitive processes from a behavioural addiction perspective and contribute to security behaviour by integrating cognitive factors and performing better cautious systems individually in online gaming [26, 27]. With the combination of impulsiveness, awareness, and fear appeal, this factor can be considered an essential construct for developing security behaviour in the context of OGA among adolescents. According to Hadlington [9], cybercognition is the factor describing the mental action of acquiring knowledge or cognitive activities of humans conducted in cyberspace (Internet), which influence impulsiveness in decision-making and intentional behaviour. Cybercognition was used to interpret the exploration of mental processes in the online digital environment and deal with psychological aspects. It provides new insight into the relationship between cognitive function and OGA behaviour, which aligns with Yadav [28] who mentioned that gamers who did not stop playing online games showed signs of withdrawal that interfered with their ability to retrieve awareness and had trouble deciding the security issues.

Additionally, this study shows a notable attention and concentration deficit in online gamers who encountered conflict due to their games. The findings from Blackwell et al. [29] regarding extraversion, neuroticism, and fear of missing out (FOMO) have mentioned cybercognition, which is related to the reaction of the user when they are communicating in online games. On the other hand, cybercognition issue also has been observed by McNeese [30] regarding the perspective on the role of cognition in cybersecurity which can be divided into Visuospatial (visual), Episodic Buffer (segmentation) and Phonological Icop (language).

2 Review of Literature

Online gaming has transformed the entertainment and leisure landscape by captivating millions of players worldwide [31]. With the advent of high-speed Internet and advanced gaming technologies, online gaming has become a thriving industry that offers immersive and interactive experiences. It has evolved from simple text-based adventures to visually stunning and socially connected virtual worlds [32]. One of the main attractions of online gaming is the vast availability of games to players. From first-person shooters to massively multiplayer online role-playing games (MMORPGs), including strategy games and sports simulations, there is a game that caters to every interest and preference [33]. Online gaming offers a diverse range of genres, themes and gameplay mechanics, providing the players with endless options for entertainment and exploration. Besides entertainment and exploration, another significant temptation of online gaming among many players is the social aspect in which players can connect and interact with others. Multiplayer games also enable collaborative gameplay, competitive matches and the formation of virtual communities. Hence, online gaming has created a vibrant and interconnected space where players can communicate, cooperate and compete in real-time, fostering social connections [34].

The technological advancements in online gaming have greatly enhanced the overall experience. High-definition graphics, realistic physics engines and immersive sound design bring virtual worlds to life, providing players with visually stunning and immersive gameplay environments [35]. Additionally, the emergence of online gaming communities' platforms promotes the exchange of experiences, strategies and content among players. Eventually, the online gaming platform has created a dynamic and interactive ecosystem where players can engage with others and learn from skilled gamers to showcase their gameplay achievements. Online games typically comprise special features and elements including flow experience, unique simulation, versatility, competition, reciprocal mechanisms and a captivating storyline [36]. Furthermore, online games prioritize social interactions and the formation of teams among players, supported by additional features such as special credit, good team leadership and a mechanism for sharing points (Durak, 2019a). However, several features are considered disadvantages as they can cause users to become addicted gamers. Similarly, Alexiou and Schippers [35] emphasized that plot, aesthetics and game mechanics are essential in influencing adolescents into OGA. The number of scientific studies exploring diverse facets of problems relating to OGA has considerably grown since the early 2000s, in which most scholars focused on the domains that correlate with psychology and physical health and academic performance among adolescents [37–39].

The relationship between intention behavior and OGA is also multifaceted. Intention behavior refers to an individual's planned actions or strategies to engage in certain activities after prior knowledge [40]. In online gaming, intention behavior refers to the actual execution of behavior previously intended to enhance security, which involves setting boundaries, managing time, and prioritizing real-life responsibilities related to online gaming. However, when intention behavior is negatively oriented by addictive behavior, individuals may fail to regulate their gaming habits, leading to excessive gameplay and potential OGA [41]. This OGA, which is characterized by compulsive gaming behavior, can further erode intentional behavior, making it challenging for adolescents to adhere to healthy gaming limits. This cyclical relationship means that as addiction intensifies, the ability to engage in positive intention behavior diminishes, exacerbating the problem. To strengthen the intention behavior, it becomes essential to enhance the integration of protection motivation by incorporating cognitive abilities such as cyber cognition in mitigating behavioral addiction like OGA [42].

2.1 Need for Current Study

OGA and cybercognition are connected through the lens of self-regulation and ethical decision-making in digital environments. In online gaming, individuals exhibiting cybercognition are more likely to set healthy boundaries and avoid excessive gaming. On the other hand, a lack of security behaviour can lead to OGA, where individuals lose control over their gaming habits, potentially neglecting essential aspects of cybersecurity.

Cybercognition also has the potential to combine with information security behaviour, as mentioned by prominent scholars [43–46], which pertains to protecting organizations or individuals from risks and threats. Intention behaviour examines the likelihood of translating intention into actions, encompassing both the level of behaviour and commitment. While many behaviours are habitual and triggered automatically by situational cues, forming intention is crucial for securing long-term cybersecurity goals, as noted by Jenkins, Durcikova and Nunamaker [40]. Thus, performing intention behaviour has become a highly debated issue in the IS research discipline. The inconsistency of intention behaviour, which relates to multiple variables as revealed by previous researchers, contributes to the complicated process of comparing the outcomes of those studies [47]. Considering this, it is urgent to address these issues thoroughly and provide valuable insights into factors such as cybercognition with OGA to form a solid foundation for the research [28].

3 Methodology

The present study examined the relationship between OGA and cybercognition among adolescents.

Objective: To study the relationship between OGA and cybercognition.

Sample: For the present study, 660 online game adolescents were studied in peninsula Malaysia, following a quantitative research design.

Instruments: Two measures were used in this study;

Online Game Addiction Scale (OGAS): To measure respondent's degree of pathological gaming, a seven-item game addiction scale developed by Lemmens et al. [4] was used, which is based on the DSM IV criteria for pathological gambling previously adopted by Griffiths [3]. There are two versions: the extended version has 21 questions, and the short version has only 7. Both have a high level of reliability and in the present study, a 21 21-item scale was used, which had seven criteria (mood modification, salience, tolerance, relapse, withdrawal, conflict and problems) with three items each randomly distributed over the scale. It had items on a 5-point continuum scale ranging from 1 (never), 2 (rarely), 3 (sometimes), 4 (often) and 5 (very usually) [48].

Cybercognition Test (CT): A Cybercognition Test is an assessment tool designed to evaluate an individual's cognitive processes in online game environments. This test typically measures various cognitive functions, including awareness, impulsiveness, fear appeal and decision-making as they relate to activities in online gaming. The test aims to understand how online gaming influences cognitive abilities and behaviors, particularly in adolescents who frequently use digital technologies. By examining how individuals process information, make decisions and navigate social interactions in the online game environment, the CT provides insights into the impact of OGA on cognitive development and functioning. It had items on a 5-point continuum scale ranging from 1 (strongly disagree), 2 (partially disagree), 3 (neutral), 4 (agree) and 5 (strongly agree) [42].

3.1 Procedure

The data was collected using a standardised questionnaire. After the questionnaire was distributed, instructions were given to respondents according to the survey session among adolescents in higher learning institutions in peninsula Malaysia. Responses were scored according to the manual. The collected data was analysed using SPSS to determine the association between online gaming addiction and cybercognition, utilising frequency distribution and correlation.

4 Result

Table 1 shows the characteristics of the data for this study based on frequency distribution and percentage.

Table 1. Characteristics of data.

Demographic	Category	(N) Percentages
Gender	Male	(355) 53.8%
	Female	(305) 46.2%
Age	17	(105) 15.9%
	18	(135) 20.9%
	19	(420) 63.2%
Race	Malay	(522) 79.1%
	Chinese	(43) 7.0%
	India	(36) 5.0%
	Others	(59) 9.0%
Educational Level	Bachelor	(62) 9.0%
	Diploma	(495) 75.0%
	Certificate	(103) 16.0%
HLIs	University	(107) 16.2%
	Polytechnic	(450) 68.2%
	Community College	(103) 15.6%

It shows that most of the sample was male (53.8%) and belonged to learning institutions such as Universities (16.2%), Polytechnics (68.2%) and Community Colleges (15.6%). Almost all respondents (97.0%) have experienced playing online games. From the 660 respondents, 23.0% have spent less than 1 h per day, 33.0% have spent 1–2 h, 24.0% have spent 2–3 h, 10.0% have spent 3–4 h, 5.0% have spent 4–5 h and 5.0% have spent more than 5 h per day. Mobile phone is categorized as the most preferred online gaming platform (74.0%), followed by computers (21.0%), consoles (2.0%), tabs (2.0%) and arcades (1.0%). Most of them prefer games related to strategy (35.6%) as compared to MMO (30.0%), adventures (10.0%), sports (10.0%), simulation (7.4%) and racing genre (7.0%).

Table 2. Characteristics of the sample mean ± SD

Variables	Mean ± SD
Age of onset	17.98 ± 3.28
Time Spent	5.12 ± 6.94
Total Duration (in months)	59.12 ± 30.05

Table 2 shows the sample's clinical characteristics, with mean age of sample onset as 17.98 ± 3.28. The mean time spent per day was 5.12 ± 6.94 a, and the total duration since starting playing the mean was found to be 59.12 ± 30.05.

Table 3. Components of OGA

Component of OGA	Percentages
Mood Modification	72.02%
Tolerance	47.25%
Salience	51.42%
Relapse	42.24%
Withdrawal	36.13%
Conflict	33.06%
Problems	32.09%

Table 3 outlines various components of OGA and their prevalence among affected individuals, measured in percentages. The most common element, Mood Modification at 72.02%, indicates that a significant majority of individuals use gaming to alter their emotional state, either to enhance positive feelings or to escape negative emotions. Salience, present in 51.42% of cases, suggests that for over half of the individuals, gaming becomes the dominant activity in their lives, overshadowing other essential aspects such as work or social interactions. Tolerance, at 47.25%, reveals that many individuals need to play games for increasingly more extended periods to achieve the same emotional or psychological effects, similar to other forms of addiction. Relapse (42.24%) and Withdrawal (36.13%) highlight the challenges in reducing or ceasing gaming, with individuals often returning to previous levels of gaming after attempts to quit and experiencing negative emotional or physical symptoms when unable to play. Conflict (33.06%) and Problems (32.09%) indicate that approximately one-third of individuals face significant interpersonal conflicts and life disruptions, including academic, occupational or social issues, due to gaming habits. These components collectively illustrate the complex nature of OGA, impacting emotional well-being, behaviour and overall life functioning.

Table 4 categorizes individuals based on levels of OGA, with levels 1 through 3 labelled as "non-addicted" and levels 4 through 7 as "addicted." The distribution of

Table 4. Data on the level of OGA (non-addicted and addicted)

Level of OGA	(N)	Percentages
1 (non-addicted)	25	3.79%
2 (non-addicted)	198	30.00%
3 (non-addicted)	199	30.15%
4 (addicted)	145	21.97%
5 (addicted)	70	10.61%
6 (addicted)	18	2.73%
7 (addicted)	5	0.76%

individuals across these levels reveals that the majority fall within the non-addicted category, with 30.00% at level 2 and 30.15% at level 3, suggesting that nearly two-thirds of the sample (63.94%) exhibit low to moderate engagement with gaming without addiction. In contrast, the addicted group constitutes a smaller portion of the population, with 21.97% at level 4, indicating a notable increase in addictive behaviours. The percentages decrease progressively at higher addiction levels, with 10.61% at level 5, 2.73% at level 6 and only 0.76% at level 7, the most severe level of addiction. This distribution suggests that while a significant number of individuals engage in gaming at potentially harmful levels, only a tiny proportion reach the most severe stages of addiction. The data highlights the varying degrees of OGA within the population, indicating that most individuals experience moderate involvement, with the critical segment facing more severe addiction issues.

Table 5. Relationship between OGA and cybercognition

Level of OGA	Cybercognition Percentages
1 (non-addicted)	89.2%
2 (non-addicted)	71.9%
3 (non-addicted)	62.5%
4 (addicted)	58.8%
5 (addicted)	55.5%
6 (addicted)	53.2%
7 (addicted)	58.1%

Table 5 presents a relationship between levels of OGA and cybercognition percentages, indicating how cognitive abilities related to digital literacy and responsible technology use vary across different addiction levels. For individuals classified as non-addicted (levels 1–3), cybercognition percentages decrease as OGA levels increase, from 89.2% at level 1 to 71.9% at level 2 and 62.5% at level 3. This suggests that even among non-addicted individuals, higher engagement in online gaming correlates with a decline in cybercognition.

5 Discussion

OGA can harm cybercognition, which refers to the cognitive abilities and skills required for effective and responsible use of digital technologies. As individuals become more engrossed in online games, they may spend excessive time gaming, often at the expense of other activities. This can lead to reduced engagement in intellectually stimulating activities, limiting the development of critical thinking, problem-solving and information-processing skills. Furthermore, the immersive nature of many online games can blur the line between reality and fantasy, potentially impairing one's ability to discern credible information online. As a result, higher levels of OGA are often associated with lower levels of cybercognition, as individuals may become less capable of navigating and critically engaging with the digital world. OGA among adolescents is increasingly recognized as a significant concern, often linked to changes in cognitive processes. This relationship involved various psychological and social factors. Adolescents may develop addictive behaviours towards online gaming due to the immersive and interactive nature of these games, which can offer immediate rewards and social connections. This addiction can lead to changes in attention, memory and decision-making processes, as well as a potential decline in academic performance and social interactions. Cybercognition, or how individuals think and perceive in the digital world, may be altered as adolescents spend more time in virtual environments, potentially leading to reduced attention span and impaired real-world social skills. Understanding this relationship is crucial for developing strategies to mitigate the negative impacts of OGA on adolescent's cognitive and social development.

At the core of OGA is the concept of reward systems. Many online games are designed to keep players engaged through mechanisms like levelling up, obtaining rare items or achieving high scores. These reward systems can trigger the release of dopamine in the brain, a neurotransmitter associated with pleasure and reinforcement. For adolescents, whose brains are still developing and are particularly sensitive to reward stimuli, this can create a powerful draw, leading to compulsive gaming behaviours. The immediate gratification provided by these online games contrasts sharply with the often slow and effortful nature of academic or social achievements in the real world, making games a more appealing option for many young individuals. Memory and learning are other cognitive areas affected by OGA. While certain games can enhance skills such as problem-solving, strategic thinking and multitasking, they can also lead to a reliance on short-term memory strategies, given the frequent need for quick decision-making in games. This may detract from the development of long-term memory retention and deep learning, which are critical in academic contexts. The rapid processing required in games can foster a preference for immediate information, potentially leading to difficulties in absorbing and retaining more complex or nuanced information that requires time and reflection. Addressing the relationship between OGA and cybercognition among adolescents requires a critical approach. Parents, educators and mental health professionals play crucial roles in helping adolescents develop healthy gaming habits. This includes setting boundaries around gaming time and encouraging a balance between online and offline activities related to the potential risks and benefits of gaming. It is also important to educate adolescents about the design of games, helping them understand how reward systems and in-game mechanics can influence their behaviour and cybercognition.

6 Conclusion

The relationship between OGA and cybercognition among adolescents reveals potential benefits and significant risks. While engaging in online games can enhance specific cognitive skills such as quick decision-making, strategic thinking and hand-eye coordination, OGA can lead to adverse outcomes. These include impaired attention spans, reduced academic performance and altered social cognition. The immersive and often rewarding nature of online games can foster addictive behaviours, particularly in adolescents whose cognitive and emotional regulation systems are still developing. This addiction can impact their ability to focus on non-gaming tasks and may lead to social isolation or altered social interactions. Recognizing and addressing these impacts is crucial for promoting cybercognition and social development in adolescents, ensuring that the benefits of digital engagement do not come at the expense of overall well-being.

Acknowledgement. This research was supported by the Ministry of Higher Education (MoHE) through the Fundamental Research Grant Scheme (Ref: FRGS/1/2020/ICT03/UUM/02/1). The content of this article is solely the responsibility of the authors and does not necessarily represent the official views of MoHE, Malaysia. The authors would also like to extend their gratitude to the Data Management & Software Solution Research Lab, School of Computing, Universiti Utara Malaysia for their generous support in sponsoring the publication of this article (S/O Code:14839).

References

1. Aloi, M., et al.: The potential role of the early maladaptive schema in behavioral addictions among late adolescents and young adults. Front. Psychol. **10**, 1–10 (2020)
2. Jameel, S., Shahnawaz, M.G., Griffiths, M.D.: Smartphone addiction in students: a qualitative examination of the components model of addiction using face-to-face interviews. J. Behav. Addict. 1–14 (2019)
3. Griffiths, M.: A components model of addiction within a biopsychosocial framework. J. Subst. Use **10**, 191–197 (2005)
4. Lemmens, J.S., Valkenburg, P.M., Peter, J.: Development and validation of a game addiction scale for adolescents. Media Psychol. **12**, 77–95 (2009)
5. Denis, C., Fatséas, M., Auriacombe, M.: Analyses related to the development of DSM-5 criteria for substance use related disorders. Drug Alcohol Depend. **122**, 22–27 (2012)
6. Nedelec, J.L.: Individual differences and co-occurring victimization online and offline: the role of impulsivity. J. Personal. Individ. Differ. **133**, 77–84 (2018)
7. Berthon, P., Pitt, L., Campbell, C.: Addictive devices: a public policy analysis of sources and solutions to digital addiction. J. Public Policy Mark. **38**, 451–468 (2019)
8. Hong, Y.R., Do, E.Y.: Factors influencing online game addiction of children **20**, 2173 (2020)
9. Hadlington, L.: Human factors in cybersecurity; examining the link between internet addiction, impulsivity, attitudes towards cybersecurity, and risky cybersecurity behaviors. Heliyon **3**, e00346 (2017)

10. Hadlington, L.: The human factor in cybersecurity. IGI Global **3**, 46–63 (2018)
11. Yildiz Durak, H.: Human factors and cybersecurity in online game addiction: an analysis of the relationship between high school students' online game addiction and the state of providing personal cybersecurity and representing cyber human values in online games. Soc. Sci. Q. **100**, 1984–1998 (2019)
12. Khazaal, Y., et al.: Game addiction scale assessment through a nationally representative sample of young adult men: item response theory graded-response modeling. J. Med. Internet Res. **20**, 1–16 (2018)
13. Stubbendorff, C., Stevenson, C.W.: Dopamine regulation of contextual fear and associated neural circuit function. Eur. J. Neurosci. 1–15 (2020)
14. Wise, R.A., Robble, M.A.: Dopamine and addiction. Annu. Rev. Psychol. **71**, 79–106 (2020)
15. Zul Kamal, N.S., Wok, S.: The impact of online gaming addiction on mental health among IIUM students. Int. J. Herit. Art Multimed. **3**, 01–20 (2020)
16. Baturay, M.H., Toker, S.: Internet addiction among college students: some causes and effects. Educ. Inf. Technol. (2019)
17. Ku, C.H., Kwak, M., Yurov, K.M., Yurova, Y.V.: A study of the influence of gaming behavior on academic performance of it college students. In: 20th Americas Conference on Information Systems, AMCIS 2014, vol. 2, pp. 1–11 (2014)
18. Kakaraki, S., Tselios, N., Katsanos, C.: Internet addiction, academic performance and personality traits: a correlational study among female university students. Int. J. Learn. Technol. **12**, 151–164 (2017)
19. Samaha, M., Hawi, N.S.: Relationships among smartphone addiction, stress, academic performance, and satisfaction with life. Comput. Hum. Behav. **57**, 321–325 (2016)
20. Purwaningsih, E., Nurmala, I.: The impact of online game addiction on adolescent mental health: a systematic review and meta-analysis. Open Access Maced. J. Med. Sci. **9**(F), 260–274 (2021)
21. Montag, C., Walla, P.: Carpe diem instead of losing your social mind: beyond digital addiction and why we all suffer from digital overuse. Cogent Psychol. **3**(1) (2016)
22. Kuss, D.J., Griffiths, M.D.: Online social networking and addiction: a review of the psychological literature. Int. J. Environ. Res. Public Health **8**, 3528–3552 (2011)
23. Young, K.S.: Internet Addiction: A Handbook and Guide to Evaluation and Treatment, pp. 173–189. Wiley (2011)
24. Wang, J.L., Sheng, J.R., Wang, H.Z.: The association between mobile game addiction and depression, social anxiety, and loneliness. Front. Public Health **7**, 5–10 (2019)
25. Peper, E., Harvey, R.: Digital addiction: Increased loneliness, anxiety, and depression. Neuro Regul. **5**, 3–8 (2018)
26. Hadlington, L.: Cybercognition: Brain, Behavior and The Digital World. SAGE Publications (2017)
27. Hadlington, L.: Cybercognition: brain, behavior, and the digital world. Psychol. Learn. Teach. **17**, 323–325 (2018)
28. Yadav, N.: Relationship between cognitive functions and online game addiction. Int. J. Indian Psychol. **9** (2021)
29. Blackwell, D., Leaman, C., Tramposch, R., Osborne, C., Liss, M.: Extraversion, neuroticism, attachment style and fear of missing out as predictors of social media use and addiction. Pers. Individ. Differ. **116**, 69–72 (2017)
30. McNeese, M., et al.: Perspectives on the role of cognition in cyber security. In: Proceedings of the Human Factors and Ergonomics Society, pp. 268–271 (2012)
31. Kaya, A., Türk, N., Batmaz, H., Griffiths, M.D.: Online gaming addiction and basic psychological needs among adolescents: the mediating roles of meaning in life and responsibility. Int. J. Ment. Health Addict. (2023)

32. Rajan, A.V., et al.: Virtual reality gaming addiction. In: 2018 Fifth HCT Information Technology Trends (ITT), pp. 358–363 (2019)
33. Tomicic, I., Grd, P., Schatten, M.: Reverse engineering of the MMORPG client protocol. In: 2019 42nd International Convention on Information and Communication Technology, Electronics and Microelectronics, MIPRO 2019 Proceedings, pp.1099–1104 (2019)
34. Bodi, G., Maintenant, C., Pennequin, V.: The role of maladaptive cognitions in gaming disorder: differences between online and offline gaming types. Addict. Behav. **112**, 106595 (2021)
35. Alexiou, A., Schippers, M.C.: Digital game elements, user experience and learning: a conceptual framework. Educ. Inf. Technol. **23**, 2545–2567 (2018)
36. Balakrishnan, J., Griffiths, M.D.: Loyalty towards online games, gaming addiction, and purchase intention towards online mobile in-game features. Comput. Hum. Behav. **87**, 238–246 (2018)
37. Goncalves, L.L., et al.: Scale to assess leaders' perceptions about their workers' digital addiction. Addict. Health. **10**, 223–230 (2018)
38. Shamsudin, N.N.A., Yatin, S.F.M., Nazim, N.F.M., Talib, A.W., Sopiee, M.A.M., Shaari, F.N.: Information security behaviors among employees. Int. J. Acad. Res. Bus. Soc. Sci. **9**, 560–571 (2019)
39. Mıhcı Türker, P., Kılıç Çakmak, E.: An investigation of cyber wellness awareness: Turkey secondary school students, teachers, and parents. Comput. Sch. **36**, 293–318 (2019)
40. Jenkins, J.L., Durcikova, A., Nunamaker, J.F.: Mitigating the security intention-behavior gap: the moderating role of required effort on the intention-behavior relationship. J. Assoc. Inf. Syst. **22**, 246–272 (2021)
41. Conner, M., Norman, P.: Understanding the intention-behavior gap: the role of intention strength. Front. Psychol. (2022)
42. Hadlington, L.: Exploring the role of work identity and work locus of control in information security awareness. Comput. Secur. (2019)
43. Alohali, M., Clarke, N., Furnell, S., Albakri, S.: Information security behavior: recognizing the influencers. In: Proceedings of Computing Conference 2017, 2018-January, pp. 844–853 (2018)
44. Klein, R.H., Luciano, E.M.: What influences information security behavior? A study with Brazilian users. J. Inf. Syst. Technol. Manag. **13**, 479–496 (2016)
45. Padayachee, K.: Taxonomy of compliant information security behavior. J. Comput. Secur. **31**, 673–680 (2012)
46. Karjalainen, M.: Information security behavior: towards multi-stage models. In: Pacific Asia Conference on Information Systems (2020)
47. Yoon, C., Kim, H.: Understanding computer security behavioral intention in the workplace: an empirical study of Korean firms. Inf. Technol. People **26**, 401–419 (2013)
48. Yaacob, W.M.Y.W., Zakaria, N.H., Aji, Z.M.: Identification of factors contributing to online game addiction among adolescents. J. Inf. Commun. Technol. **4**, 565–597 (2021)

Part 3: Big Data and Software Solutions with Knowledge Management

Design of Aquaculture Data Acquisition Device Based on Internet of Things Technology

Tu Jun[1,2](✉), Azman Yasin[2], and Nur Suhaili Mansor[2]

[1] Hunan Applied Technology University, Hunan, China
tujun792324486@gmail.com
[2] Universiti Utara Malaysia, Kedah, Malaysia

Abstract. This paper describes a design scheme for a data acquisition device for the aquaculture industry. The system aims to realize real-time collection, transmission, processing and display of aquaculture environmental data through the Internet of Things technology. The design includes high-precision sensors for capturing real-time data of water temperature, ammonia, PH value and ammonia in an aquaculture environment, wireless transmission of data to the cloud server using a WiFi module, and connection between upper computer software and cloud server database through JDBC or ORM framework. The upper computer software design uses Java programming language, with a user-friendly interface that can intuitively display the breeding environment data. During the design process, we focus on the stability, security and scalability of the system. Modular design ideas and interface definition, as well as the use of open-source frameworks and libraries, ensure that the system has good scalability and is easy to add and modify subsequent functions. At the same time, we plan to implement a comprehensive exception-handling mechanism and logging to facilitate troubleshooting and fault recovery. The intelligent aquaculture management system design scheme proposed in this paper aims to provide comprehensive, efficient and intelligent management support for the aquaculture industry through the Internet of Things technology, provide farmers with convenient data management and operation experience, and promote the sustainable development of the aquaculture industry.

Keywords: Internet of Things technology · Aquaculture · Data acquisition

1 Introduction

As an important part of agriculture, aquaculture plays an irreplaceable role in ensuring food safety, enriching the food supply and promoting rural economic development [1]. The significance of aquaculture data lies in its ability to precisely guide the breeding process, optimize environmental regulation, develop feeding strategies, and prevent and control diseases [2]. This provides a scientific basis for farmers to enhance breeding efficiency, reduce costs, ensure the quality and safety of aquatic products, and promote sustainable development within the aquaculture industry [3]. Therefore, accurate monitoring of ammonia, pH, water temperature and dissolved oxygen data is the key to ensure

Z. Mat Aji et al. (Eds.): KMICe 2024, ISEM 49, pp. 203–217, 2025.
https://doi.org/10.1007/978-3-031-91485-0_16

the stability of the farming environment and the healthy growth of farming organisms [1]. These parameters directly reflect the state of the pond water quality and are essential for preventing fish poisoning caused by ammonia accumulation [4], maintaining the acid-base balance of the water to prevent fish disease, providing the right water temperature to promote fish growth, and ensuring sufficient dissolved oxygen to support normal fish breathing [5]. Through regular collection and analysis of these data, farmers can timely understand the changes in the pond environment, and accordingly take corresponding measures. However, in the existing aquaculture research, many analyses of farming environment monitoring, disease warning, feed management, etc., often rely on secondary data. Although these studies provide us with many valuable insights, there are some problems in secondary data that cannot be ignored, such as difficulty in guaranteeing the accuracy of data, untimely updates of data, limited access to data and so on. These problems not only limit the depth and breadth of research but also affect the sustainable development of the aquaculture industry.

The Internet of Things (IoT) is an essential revolution in the information technology industry, which connects any object to the network according to a specific protocol through information sensing devices, such as sensors and Radio Frequency Identification (RFID), and realizes information exchange and communication between objects [6]. The main features of the Internet of Things technology include emerging information technology. The Internet of Things technology has the advantages of real-time, accuracy and comprehensiveness, which can provide a new means of acquiring data and managing aquaculture [2].

Therefore, this study aims to design aquaculture equipment based on Internet of Things technology to achieve real-time acquisition and analysis of first-hand data on aquaculture. The equipment is designed to achieve the acquisition of first-hand data. Real-time acquisition and analysis of first-hand data is essential to improve the efficiency and quality of farming. Compared with the secondary data, the primary data has higher accuracy and reliability, can more accurately reflect the actual situation of the breeding environment, and provides a scientific basis for breeding management. It can improve breeding efficiency, reduce resource consumption, reduce disease incidence and provide scientific decision support for farmers to help them better cope with market changes. In addition, the application of the system also helps to promote the sustainable development of the aquaculture industry, reduce environmental pollution, and protect the ecological environment.

2 Overall System Design

2.1 System Architecture Overview

The aquaculture system's IoT architecture is designed to achieve real-time monitoring of the aquaculture environment, reliable data transmission, and intelligent management [7]. The architecture consists of the perception, network, and application layers, as shown in Fig. 1. Each layer is indispensable in ensuring the system's efficient operation and accurate data processing.

Perception Layer. The perception layer is the foundation of the IoT architecture and is responsible for collecting all kinds of data about the farming environment in real-time [8]. By using a variety of sensors, such as temperature sensors, dissolved oxygen sensors, pH sensors, etc., the sensing layer is able to accurately monitor critical parameters such as water temperature, dissolved oxygen content, pH and so on in the aquaculture tank or cage.

Network Layer. The network layer is the bridge between the perception layer and the application layer and is responsible for transmitting the data collected by the perception layer to the application layer [9]. In this architecture, we use WiFi communication technology to ensure the stable transmission of data. At the same time, to optimize the data transmission efficiency and reduce the network load, we choose to use the MQTT (Message Queuing Telemetry Transport) protocol as the data transmission protocol. MQTT is a lightweight publish/subscribe messaging protocol designed for low-bandwidth, high-latency, or unstable networks, making it ideal for use in IoT environments.

Application Layer. The application layer is the top layer of the IoT architecture, mainly responsible for processing and presenting the received data [10]. In this architecture, the main function of the application layer is to realize data visualization. Through data visualization technology, we can intuitively display the data of the farming environment to users in the form of charts, images, etc., to help users quickly understand the real-time status of the farming environment. In addition, users can also screen and analyze the data as needed to provide a scientific basis for breeding management.

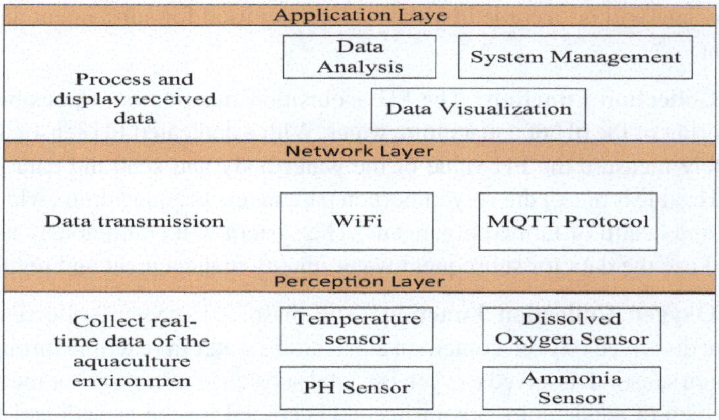

Fig. 1. System architecture.

2.2 System Function Description

Aquaculture data acquisition System is an intelligent monitoring solution that integrates multiple sensors, data transmission and cloud processing technologies. As shown in

Fig. 2, the system mainly collects critical parameters in aquaculture water in real-time, such as water temperature, PH value, dissolved oxygen and ammonia value, and transmits the data to the cloud platform through a WiFi network. Users can view data at any time through the mobile APP to achieve remote data visualization and monitoring, providing a scientific and accurate management basis for aquaculture.

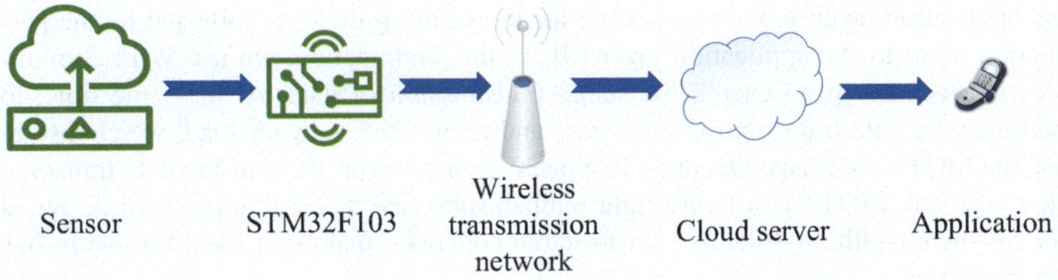

Fig. 2. Functional implementation process.

Water Temperature Collection Function. The water temperature collection function is one of the core parts of the aquaculture data collection system. This function monitors the temperature of the aquaculture water body in real time through the high-precision water temperature sensor and sends the collected data to the data collector. Water temperature sensors are usually installed at different locations in the aquaculture tank to ensure that the collected water temperature data is representative. The collected water temperature data will be recorded by the system and used for subsequent analysis and management.

PH Value Collection Function. The PH acquisition function is responsible for real-time monitoring of the pH of aquaculture water. With a dedicated PH sensor, the system can accurately measure the PH value of the water body and send the data to the data collector. PH value is one of the very important parameters in aquaculture, which impacts the growth and health of farmed organisms. The system will continuously monitor PH changes and use the data for subsequent water quality management and regulation.

Dissolved Oxygen Collection Function. The dissolved oxygen collection function monitors the dissolved oxygen content in aquaculture water in real time through the dissolved oxygen sensor. Dissolved oxygen is a vital substance necessary for the respiration and metabolism of aquaculture organisms and is crucial for the growth and survival of aquaculture organisms. The system accurately measures the dissolved oxygen content in the water through the dissolved oxygen sensor and sends the data to the data collector. Users can view the real-time dissolved oxygen data through the system and conduct corresponding regulation and management according to needs.

Ammonia Collection Function. The ammonia gas collection function monitors the ammonia content in aquaculture water. Ammonia is one of the harmful substances produced in the breeding process. If the ammonia content is too high, it will adversely affect the breeding organisms. The system monitors the ammonia content in the water body in real time through the ammonia sensor and sends the data to the data collector. Users can view real-time ammonia data through the system and conduct corresponding processing and management according to needs to ensure the quality of aquaculture water bodies.

Data Transmission Function. The data transmission function transmits the collected data, such as water temperature, pH value, dissolved oxygen, and ammonia, to the cloud platform through the WiFi network. This function adopts a stable and reliable data transmission protocol to ensure the real-time accuracy of data. Data will be encrypted during transmission to protect user privacy and security. Users can view and manage data anytime through the cloud platform for remote monitoring and management.

Data Visualization Function. The data visualization function visually displays data to users through charts and curves through the cloud platform and mobile APP. Users can view the real-time and historical data of aquaculture water at any time through the mobile APP to understand the changing trends of the aquaculture environment. The data visualization function facilitates users' viewing of the data, helps users better understand the meaning behind the data, and provides a scientific basis for breeding management. At the same time, users can export data as reports or charts according to their needs, facilitating further analysis and management.

3 Module Design

3.1 Microcontroller Selection

In selecting a single-chip microcomputer, the primary consideration is the system's specific needs, cost control budget, development rate and the future expansion potential of the system. Table 1 compares three commonly used microcontrollers of different models—STM32F103C8T6, 89C51, and Arduino. According to the comparison of parameters, STM32F103C8T6 is selected as the single chip microcomputer of this system to develop. The main reasons are as follows:

1) When processing a large amount of sensor data received from the sensor, signal processing, filtering, and algorithm analysis, STM32F103C8T6's information processing speed and computing power are higher than 89C51 and Arduino.
2) The microcontroller needs to store many global variables or even parameters, so the microcontroller is also required to have a larger storage space.

From the results of the comparison of the above parameters, it can be seen that for the sensors required by the system and the requirements of the system, it is necessary to use a larger Flash storage capacity, as well as a higher processing speed and computing power to process the digital signals collected by the sensor. Therefore, STM32 is chosen as the main control chip of this system.

3.2 Sensor Selection and Layout

Water Temperature Sensor. In aquaculture, the choice of water temperature sensor is crucial because it is directly related to the growth and reproduction of fish and the overall efficiency of aquaculture [11]. First of all, considering the specific range of water temperature, the sensor needs to be able to accurately measure the water temperature between 10 °C and 36 °C, especially the temperature change during the optimal growth period of fish between 20 °C and 32 °C [12]. The accuracy of water temperature directly affects the physiological activities and health status of fish [13]. Therefore, selecting high-precision sensors can ensure data accuracy to guide aquaculture management scientifically.

Secondly, because the water may contain corrosive substances, such as dissolved oxygen and chloride ions, the sensor needs to be well-corrosive-resistant to ensure long-term stable operation [14]. At the same time, cost is also an essential factor to consider. You should choose a cost-effective, simple wiring sensor to reduce the farming cost and improve the operation's convenience.

In terms of equipment water resistance, since aquaculture environments are usually humid and may even be in direct contact with water bodies, the sensor needs to have excellent water resistance to prevent equipment damage or data distortion caused by water.

Finally, considering the hassle and cost of equipment replacement, choosing a sensor with high reliability and resistance to damage is crucial. This reduces the frequency of maintenance and replacement and ensures the continuity and stability of the breeding process.

The waterproof DS18B20 water temperature sensor in aquaculture is the ideal water temperature monitoring device due to its excellent performance and advantages. First, the DS18B20 has a wide temperature range, covering a wide range of temperatures from −55 °C to +125 °C, which fully meets the needs of water temperature monitoring in aquaculture. Secondly, its high-precision temperature measurement function, which can be accurate to 0.0625 °C, ensures an accurate grasp of the temperature of the fish growth environment, which helps to optimize breeding management. Moreover, DS18B20 has good corrosion resistance and can resist the erosion of corrosive substances that may exist in water, ensuring the long-term stable operation of the sensor[1]. In addition, DS18B20 adopts a single bus interface mode, simplifying the wiring process, reducing the cost, and improving the system's reliability. More importantly, many sensor modules use the DS18B20 chip and have a waterproof function on the market, which can meet the requirements of waterproof performance in aquaculture. Finally, DS18B20 has a simple structure, high stability, high reliability and long service life, reducing the replacement cost and trouble caused by equipment failure. In summary, the DS18B20 water temperature sensor is the first choice for water temperature monitoring in aquaculture due to its wide temperature measurement range, high precision measurement, corrosion resistance, simple wiring, waterproof function, high reliability and low maintenance costs (Table 2).

pH Sensor. The solution's acidity and alkalinity (PH) is an essential characteristic of aquaculture. The price of PH transmitters on the market is very expensive. The PH test pen is a more mature packaged product; there is no way to carry out the second design and development on the test pen. Therefore, the aquaculture PH acquisition sensor adopts Dfrobot's model SEN0169 Ph sensor, in which a PH composite electrode connects to the PH sensor via a BNC connector. The voltage signal is sampled and processed in

[1] https://wiki.dfrobot.com.cn/_SKU_DFR0024_DS18B20.

the microcontroller and then output. Amplification can be performed by adjusting the 10K blue potentiometer. Note that the clockwise direction is a multiplier of increase, and the counterclockwise direction is a multiplier of decrease. The advantage of this module is that the price is very cheap, easy to buy, very convenient to use, and has high measurement accuracy. The module parameters are as follows Table 3[2].

Table 1. Comparison of microcontroller parameters.

Single-chip microcomputer model	89C51	STM32F103C8T6	Arduino
build	C51	ARM Cortex-M3	AVR
PU	8-bit	32-bit	8-bit
Flash storage capacity	8 KB	64 KB	32 KB
Peripherals	Basic Peripherals	complex peripheral	Basic Peripherals
Clock frequency	12 MHz to 24 MHz	72 MHz	16 MHz
Scalability	Poor scalability	Good scalability	Better scalability

Table 2. DS18B20 parameters.

Item	Specification
Power Supply	3.0–5.5 V
Sensor	DS18B20
Measurement Range	−55 °C–+125 °C
Accuracy	±0.5 °C (from +10 °C–+85 °C)/±2 °C (from −55 °C−+125 °C)
Work Current	1.5 mA
Standby Current	1 μA
Output Pin Sink Current	4 mA (@ VI/0 = 0.4 V)
Response Time	In 750 ms (Converts temperature to 12-bit digital word)
Output	Open-drain 1-wire interface pin
Drift	±0.2 °C on a 1000 h stress test at 125 °C with VDD = 5.5 V
Wire	24AWG/26AWG/Double line wire/Silicon round jacket/PTFE/Fiberglass

Dissolved Oxygen Sensor. When selecting a dissolved oxygen sensor, it is necessary to consider its performance and technical parameters comprehensively. First, ensure

[2] https://wiki.dfrobot.com.cn/search/SEN0169.

Table 3. PH sensor parameters

Item	Specification
Module Power Supply	+5.00 V
Module Size	43 mm × 32 mm
Measurement Range	0–14 PH
Measurement Temperature	0–60 °C
Accuracy	±0.1 pH (25 °C)
Response Time	≤1 min
Industrial Online pH Composite Electrode	BNC connector
PH2.0 Interface	3-pin
Gain Adjustment Potentiometer	Yes
Power Indicator Light	Yes

the sensor has sufficient measuring range to cover the expected change in dissolved oxygen content in a practical application; the sensor may have a measuring range of 20 ppm or wider [15]. Secondly, accuracy is an important indicator to evaluate the sensor's performance, and the selection of high-precision sensors can provide ±0.5%FS (percentage of full scale) measurement results [16]. At the same time, the response time is also a factor that cannot be ignored, especially in scenarios where real-time monitoring of dissolved oxygen changes is required, and the fast response sensor can reflect the water quality promptly. In addition to the above vital elements, other technical parameters should also be considered, such as maximum pressure resistance (10 bar), protection class (IP68) [17], operating temperature range (−5~50 °C), and electromagnetic compatibility.

Based on the above conditions, the device uses Dfrobot's model SEN0237 dissolved oxygen sensor. The device is equipped with measurement electrodes and powered by disposable batteries. Its detection range covers chloride ion concentrations from 0 to 20 mg/L and is suitable for temperature ranges from 0 to 40 °C and pressure environments from 0 to 50 PSI. At 25 °C, the device responds quickly within 90 s, ensuring the accuracy of the measurement. The electrode life is about one year under normal water quality conditions. Still, to maintain the best performance, the membrane cap needs to be replaced every 1–2 months (sewage) or every 4–5 months (clean water) according to the water quality. The electrode-filling solution needs to be changed every month. In addition, the device is equipped with a cable length of 2 m and a connector type of BNC, ensuring compatibility with various systems. The module parameters are as follows: Table 4.

Ammonia Sensor. In aquaculture, selecting an appropriate water ammonia sensor demands thorough consideration. Key factors include a wide measurement range (0.01 to 10 mg/L) to track varying ammonia levels, high accuracy (±5% error range) for reliable data, and fast response time for real-time monitoring [18]. Additionally, the sensor must be waterproof, corrosion-resistant (using materials like 316 stainless steel) with an IP65 or higher protection level, operate within the farming's temperature range (−10 °C

Table 4. Dissolved oxygen sensor parameters.

Item	Specification
Electrode Type	Primary Battery
Detection Range	0–20 mg/L
Temperature Range	0–40 °C
Response Time	90 s to reach 98% response (25 °C)
Pressure Range	0–50PSI
Flow Conditions	0.3 mL/s
Electrode Life	1 year (under normal use conditions)
Maintenance Time	Depending on water quality, replace the membrane cap every 1–2 months for sewage and every 4–5 months for clean water. Also, replace the electrode filling solution every month
Cable Length	2 m
Connector	BNC

to 50 °C), and possess good electromagnetic compatibility (EMC) to avoid interference. These specifications ensure optimal performance and accurate monitoring in aquaculture environments.

Therefore, the Digital RS485 Ammonia Nitrogen Sensor BH-485-NH is selected for the ammonia sensor. The digital ammonia sensor has a wide measurement range, can detect ammonia concentrations from 0.1 to 1000 mg/L and potassium concentrations from 0.5 to 1000 mg/L (optional), and is suitable for water bodies with pH values between 5 and 10. The measurement error of ammonia nitrogen is ±5% or ±0.2 mg/L of the measured value, whichever is greater. The measurement error for potassium ion is also ±5% of the estimated value or ±0.2 mg/L (optional). The accuracy of temperature measurement is ±0.1 °C, and the accuracy of pH value is ±0.1 pH. The sensor is suitable for operating temperature ranges from 0 to 45 °C and communicates over RS485 Modbus RTU. Its resolution is 0.01 mg/L ammonia nitrogen concentration, 0.01 mg/L potassium ion concentration (optional), temperature is 0.1 °C, and pH value is 0.01. The sensor size is 55 mm in diameter and 340 mm in length, and the output signal is 4–20 mA (load <750 Ω, optional). The power supply requirement is DC12 V, the protection level is IP68/NEMA6P, and the cable length of the standard configuration is 10 m, which can be extended to 100 m. The module parameters are as follows in Table 5[3].

3.3 Design of Data Acquisition and Transmission Module

Design of Data Acquisition Unit. In this device, the precise deployment of sensors is the first step to ensure the effectiveness of data acquisition. By surveying the farming area thoroughly, we selected the best locations to install high-precision sensors, including

[3] https://www.boquinstrument.com/digital-rs485-ammonia-nitrogen-sensor.

Table 5. Ammonia sensor parameters.

Parameter	Details
Product	Digital Ammonia Nitrogen Sensor
Measuring range	NH4N: 0.1–1000 mg/L, K+: 0.5–1000 mg/L (Optional), PH: 5–10
Accuracy	NH4N: ±5% of the measured value or ±0.2 mg/L; take the greater one K+: ±5% of the measured value or ±0.2 mg/L (Optional) Temperature: ±0.1 °C pH: ±0.1 pH
Working temperature	0–45 °C
Communication	RS485 Modbus RTU
Resolution	NH4N: 0.01 mg/l, K+: 0.01 mg/l (Optional), Temperature: 0.1 °C, pH: 0.01
Size	55 mm × 340 mm (Diameter * Length)
Output	4–20 mA load <750 Ω (optional)
Power	DC12 V
Protection	IP68/NEMA6P
Length of Cable	Standard 10-m-long cable, which can be extended to 100 m

monitoring equipment for critical indicators such as temperature, pH, dissolved oxygen and ammonia nitrogen. The installation process focuses on stability and connection quality, ensuring data transmission continuity and accuracy. After installation, careful calibration and debugging are carried out to ensure the sensor output data is accurate.

In data acquisition, real-time and periodic modes are integrated to meet the needs of different scenarios. Real-time collection mode ensures immediate response to water quality changes and is suitable for emergency treatment. Periodic collection is carried out according to preset time intervals to meet the needs of daily monitoring and data analysis. The data acquisition module plays a vital role in this process, reading the data from the sensor and transforming the raw data into a format suitable for subsequent processing and transmission through preliminary screening, cleaning and conversion. This process eliminates invalid data and noise interference, improves data quality, and provides reliable data support for subsequent intelligent aquaculture management [19].

Wireless Transmission Technology Selection and Design. The data processed by the data acquisition module must be sent to the cloud server through an efficient and stable data transmission module. We use a single-chip computer as the core controller of data transmission, which is responsible for receiving the data sent by the data acquisition module and sending the data to the cloud server through the WiFi module. The communication between the MCU and the WiFi module is realized through the serial port or other interfaces to ensure the data can be accurately and quickly sent out. At the same time, the MCU is also responsible for monitoring the WiFi module's working status to ensure data transmission's stability and reliability after data is successfully transferred to the cloud server. The database of the cloud server stores and manages the data, providing strong support for subsequent data analysis, processing, and mining.

Through the above design, we realize the complete process from sensor data acquisition to cloud server data storage, providing solid technical support for the intelligent management of aquaculture.

Data Transmission Protocol and Data Format. In aquaculture's data acquisition and transmission process, we use JSON (JavaScript Object Notation) as the data format. JSON is a lightweight data interchange format based on a subset of ECMAScript that stores and represents data in an entirely language-independent text format. JSON is concise, easy to read, write, and parse, making it ideal for use in IoT and cloud computing environments [20]. The format of data collected by each sensor is shown in Table 6, including ID, time stamp, data value, and unit field.

Table 6. Format of data collected.

Field	Meaning of the field	Data type
Sensor id	The number that uniquely identifies the sensor	Int
Timestamp	Time of data collection	datetime
Unit	Units of data values	Varchat
Value	Specific values collected by sensors	double

The single-chip microcomputer combines the data of multiple sensors into an array or list and then sends it to the WiFi module. Once the cloud server receives the data, it stores it in the database. The database tables' design should match the sensor data's structure to efficiently query and analyze the data. The Message Queuing Telemetry Transport (MQTT) is selected as the transmission protocol for data transmission. It is very suitable for data transmission in the Internet of Things environment because of its lightweight, high reliability, and low latency characteristics. We must follow the MQTT packet format specification when using the MQTT protocol for data transmission. MQTT packet consists of a fixed header, variable header and payload. In the fixed header, we need to specify flag bits such as the package type (such as PUBLISH), QoS level, and whether to retain. We can include fields such as package sequence number and properties in the variable header. In the payload, we transmit data in a specific JSON format.

3.4 Upper Computer Design

PC design plays a crucial role in aquaculture data management systems. Java language is used to write the upper computer-mobile app, which can obtain breeding environment data from the cloud server database in real-time and process, display, and store it. Regarding data acquisition, the host computer connects with the cloud server database through JDBC or ORM framework and executes SQL query statements to obtain the required data. At the same time, it adopts a modular design and interface definition to ensure the software is highly scalable and maintainable. In addition, the upper computer also pays attention to security and exception handling to ensure data accuracy and the system's stable operation. The upper computer software written in Java has powerful data processing ability and a user-friendly interface design. It can display the farming environment data obtained from the cloud server database in real-time and visually display it through charts, reports and other forms to help users quickly understand the farming situation. In addition, the host computer also supports data export, parameter setting, alarm notification and other functions to provide users with convenient data management and operation experience. At the same time, its efficient network communication capability and secure data transmission mechanism ensure the real-time security of data (Fig. 3).

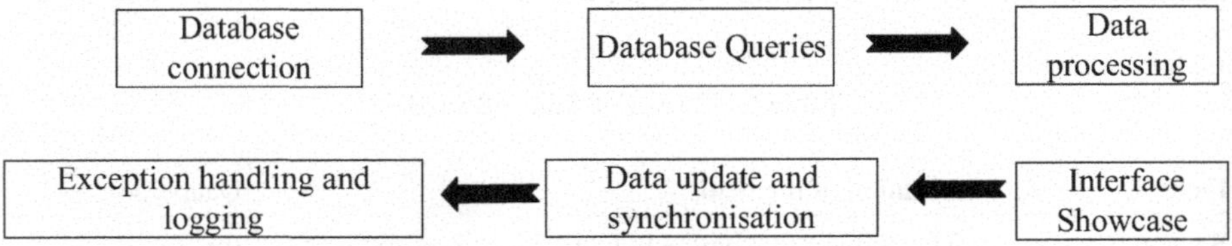

Fig. 3. Upper computer design steps.

This flowchart shows the critical design steps for the data acquisition part of upper computer software written in Java. First, use the JDBC framework to connect to the database on the cloud server by configuring the database URL, port, user name, and password. Next, SQL queries are written and executed based on business requirements to retrieve aquaculture-related data from the database, including real-time and historical environmental data. Then, the data obtained from the database is analyzed, transformed and calculated to meet the needs of computer software display and analysis. Then, Java Swing will be used to design a user-friendly interface that presents the processed data, including real-time data updates and historical data charts. In addition, the latest data on the cloud server is s periodically or in real-time through the scheduled task mechanism to ensure that the data displayed by the upper computer software is the latest. In data acquisition and processing, implement a comprehensive exception handling mechanism and record related logs for troubleshooting and fault recovery. Finally, modular design, interface definition and open-source framework and libraries are adopted to ensure that the upper computer software has good expansibility, making adding and modifying subsequent functions easy.

4 System Test and Evaluation Method

When designing test solutions, we must ensure that software quality is thoroughly and systematically checked. First, test objectives should be defined, typically verifying functional integrity, evaluating performance, checking security, and ensuring system stability and reliability.

To achieve these objectives, we will employ the following specific test methods and tools:

In functional testing, writing detailed test cases and using automated testing tools such as Selenium ensure that all parts of the software work as expected. At the same time, combined with the black box test method, various use scenarios are simulated from the user's point of view to ensure that the software functions meet the requirements.

The JMeter tool simulates the load in a real-world environment for performance testing and evaluates the system's response time, throughput and other key performance metrics. These test tools simulate a large number of users accessing the system at the same time, helping to identify potential performance bottlenecks.

Compatibility testing ensures that the software works well in different environments and configurations. We test our software on various browsers, operating systems, and hardware configurations to ensure users have a good experience in other environments.

In terms of setting up the test environment, ensure that the test environment is as consistent as possible with the production environment. The same network Settings, database connections, and so on are configured as in the production environment to more accurately simulate the real world. At the same time, the version control tool Gi is used to manage the configuration and code of the test environment to ensure the stability and repeatability of the test environment.

During test execution, test results are rigorously documented, and test cases and defects are tracked using the test management tool Jira. These tools help manage the testing process and ensure all testing work is handled correctly. Once the problem is found, the development team communicates promptly, and the defect repair and verification process is tracked.

Finally, after completing the test, the process will be summarized, and a detailed test report will be written. The report comprehensively records the test results, defects found, repair, and test conclusions for the reference of relevant personnel. Through this report, you can clearly understand the software quality status and provide a valuable reference for the subsequent software release and maintenance.

5 Conclusion and Prospect

This research successfully designed an aquaculture management system for aquaculture; the system includes a data acquisition and transmission module and host computer software design. Regarding data acquisition, the critical data of the farming environment is captured in real-time by high-precision sensors, and the data acquisition module is used for preliminary data processing. Regarding data transmission, the single-chip microcomputer is used as the core controller and the WiFi module is used to realize the wireless data transmission to the cloud server. In addition, we have chosen JSON as the data format and MQTT as the transport protocol to ensure the efficient, stable and secure transmission of data.

In the design of upper computer software, we use Java programming language, connect with cloud server database through JDBC framework, and realize real-time data acquisition, processing, display and storage. At the same time, the software has a user-friendly interface design, can intuitively display the breeding environment data, and supports data export, parameter setting, alarm notification and other functions, providing farmers with convenient data management and operation experience. Through system testing and evaluation, we aim to verify the intelligent farming management system's functional integrity, performance, safety, stability and reliability. With the rapid development of the aquaculture industry and the intensification of market competition, intelligent aquaculture management systems will play an increasingly important role. Through real-time and accurate collection and transmission of breeding environment data, the system provides scientific and accurate management decision support for farmers, which helps to reduce breeding costs and improve breeding benefits.

Innovative farming management systems can also contribute to the sustainable development of the farming industry. Through precise control of the breeding environment, the waste of resources and environmental pollution in the breeding process is reduced, and the green and low-carbon development of the breeding industry is realized. In addition, with the continuous development of technologies such as the Internet of Things, big data and artificial intelligence, the smart aquaculture management system will continue to upgrade and improve, providing more robust support for the modern and intelligent management of the aquaculture industry. We believe farming management systems will shortly become integral to the aquaculture industry.

References

1. Ren, Q., Zhang, L., Wei, Y., Li, D.: A method for predicting dissolved oxygen in aquaculture water in an aquaponics system. Comput. Electron. Agric. **151**, 384–391 (2018)
2. Wang, C., Li, Z., Wang, T., Xu, X., Zhang, X., Li, D.: Intelligent fish farm—the future of aquaculture. Aquacult. Int. **29**, 2681–2711 (2021)
3. Boyd, C.E., et al.: Achieving sustainable aquaculture: historical and current perspectives and future needs and challenges. J. World Aquacult. Soc. **51**, 578–633 (2020)
4. Boyd, C.E.: General relationship between water quality and aquaculture performance in ponds. In: Jeney, G. (ed.) Fish Diseases, Chapter 6, pp. 147–166. Academic Press (2017)
5. Franklin, P.: Dissolved oxygen criteria for freshwater fish in New Zealand: a revised approach. NZ J. Mar. Freshwat. Res. **48**, 112–126 (2014)
6. Gokhale, P., Bhat, O., Bhat, S.: Introduction to IOT. Int. Adv. Res. J. Sci. Eng. Technol. **5**, 41–44 (2018)
7. Gao, G., Xiao, K., Chen, M.: An intelligent IoT-based control and traceability system to forecast and maintain water quality in freshwater fish farms. Comput. Electron. Agric. **166**, 105013 (2019)
8. Khattak, H.A., Shah, M.A., Khan, S., Ali, I., Imran, M.: Perception layer security in Internet of Things. Futur. Gener. Comput. Syst. **100**, 144–164 (2019)
9. Mohamed, K.S.: IoT networking and communication layer. In: The Era of Internet of Things, pp. 49–70. Springer, Cham (2019)
10. Ray, P.P.: A survey on Internet of Things architectures. J. King Saud Univ. Comput. Inf. Sci. **30**, 291–319 (2018)
11. Abdelrahman, H.A., Boyd, C.E.: Effects of mechanical aeration on evaporation rate and water temperature in aquaculture ponds. Aquac. Res. **49**, 2184–2192 (2018)
12. Mugwanya, M., Dawood, M.A.O., Kimera, F., Sewilam, H.: Anthropogenic temperature fluctuations and their effect on aquaculture: a comprehensive review. Aquacult. Fisher. **7**, 223–243 (2022)
13. Kwon, I., Kim, T.: Monitoring the effect of water temperature on the heart rate of olive flounder (Paralichthys olivaceus) using a bio-logger. Aquaculture **575**, 739739 (2023)

14. Ubina, N.A., Cheng, S.-C.: A review of unmanned system technologies with its application to aquaculture farm monitoring and management. Drones **6**, 12 (2022)
15. Huynh, T.M., et al.: Electrochemical sensor chips for multiple measurements of dissolved oxygen, pH and oxidation-reduction potential in aquacultural farming. IJNT **15**, 1024 (2018)
16. Lin, J.-Y., Tsai, H.-L., Lyu, W.-H.: An integrated wireless multi-sensor system for monitoring the water quality of aquaculture. Sensors **21**, 8179 (2021)
17. Kamruzzaman, S.M., et al.: Sense-IT: an aquaculture-specific autonomous data acquisition and monitoring system. In: 2022 International Electronics Symposium (IES), pp. 404–409. IEEE, Surabaya (2022)
18. Zhou, L.: Investigations of Ammonia Nitrogen in Aquaculture: The Methodology, Concentrations, Removal, and Pond Fertilization (2015)
19. Krishnamurthi, R., Kumar, A., Gopinathan, D., Nayyar, A., Qureshi, B.: An overview of IoT sensor data processing, fusion, and analysis techniques. Sensors **20**, 6076 (2020)
20. Bourhis, P., Reutter, J.L., Suárez, F., Vrgoč, D.: JSON: data model, query languages and schema specification. In: Proceedings of the 36th ACM SIGMOD-SIGACT-SIGAI Symposium on Principles of Database Systems, pp. 123–135. ACM, Chicago (2017)

Mobile Shopping Apps: Operability Measures in Checklists of Heuristic Evaluation

Nor Laily Hashim[1(✉)], Ihediohamma Raphael Uzoma[2], and Norhanisha Yusof[3]

[1] University Utara Malaysia, 06010 Sintok, Kedah, Malaysia
laily@uum.edu.my
[2] Rafytes Resource Centre, Owerri, Imo State, Nigeria
[3] Department of Information and Communication Technology, Politeknik Balik Pulau, Penang, Malaysia

Abstract. Digital acceleration enhances process efficiency, reduces time and resource wastage, and boosts customer satisfaction and loyalty. Mobile shopping apps, as a critical medium of digital acceleration, serve as knowledge repositories, collecting extensive consumer data to improve user experiences. Despite several heuristic evaluations on mobile apps, more heuristics must be tailored for mobile shopping apps. The existing study applied general mobile application heuristics without tailored guidelines for integrating new and traditional heuristics into checklists, and they did not specifically focus on mobile shopping apps. This study aims to develop a heuristic evaluation checklist for operability measures to identify operability issues in mobile shopping applications. This study addresses this gap through a three-phase approach. Phase 1 involved a comprehensive literature review to identify usability issues and suitable evaluation methods for mobile shopping apps. Phase 2 focused on heuristic identification, categorising heuristics into categories and subcategories, and providing development guidelines. Phase 3 developed and verified a heuristic evaluation checklist, extending Nielsen's heuristics to fit mobile shopping apps. The checklist was refined to include two heuristics, four sub-heuristics, and 14 evaluation questions based on relevance and frequency in the literature. Twelve experts verified the checklist, including usability experts, software/mobile app developers, and mobile shopping app designers. This study presents a validated heuristic evaluation checklist for identifying operability issues in mobile shopping applications, ensuring its applicability and effectiveness in improving user experiences. For future work, additional validation should be made by comparing the proposed heuristic evaluation checklist with other heuristic checklists.

Keywords: Heuristic Evaluation Checklist · Operability Measures · Checklist · Mobile Shopping Apps

1 Introduction

Digital acceleration leads to more efficient processes, reducing time and resource wastage, which leads to higher customer satisfaction and loyalty. Mobile shopping apps are one of the mediums of digital acceleration in online shopping. The data collected

Z. Mat Aji et al. (Eds.): KMICe 2024, ISEM 49, pp. 218–228, 2025.
https://doi.org/10.1007/978-3-031-91485-0_17

from the transactions, products, and customer profiles serve as a knowledge repository, collecting vast amounts of consumer data that can be analysed to improve user experiences. Among the data collected from mobile shopping apps are user profiles and preferences, behavioural data, purchase history, interaction data, feedback and reviews, contextual data, and customer support interactions.

Heuristic evaluation is a usability inspection method in human-computer interaction, where experts assess an interface based on recognised usability principles or heuristics to identify issues. It provides quick, cost-effective design feedback on visibility, control, and error prevention [22]. Heuristic evaluation quickly identifies usability issues in mobile shopping apps by applying established observational standards or principles derived from extensive perception, examination, and experience. These heuristics are widely recognised and validated within their respective fields. However, it is crucial to use specific heuristics tailored to each application. [1] As different applications have their features and requirements. A heuristic evaluation checklist comprises usability-related criteria that UX experts use to determine if a digital or non-digital product is user-friendly.

Mobile shopping applications are unique from other mobile applications and have some significant qualities. In mobile shopping, the objectives are to have the necessary data about a product and for users to have a superior shopping experience. In addition, mobile customers hope to have a comparable or superior experience to a regular shopping centre. Therefore, sufficient thought of necessities is required while designing mobile shopping applications [12].

There are several ongoing works in heuristics evaluation on mobile shopping apps. Proposed new heuristics for evaluating Arabic interfaces, comparing them with Nielsen's set on three Arabic interfaces [2]. However, they are specific checklist items, making it hard for non-experts to assess Arabic interfaces [3]. Expanded on previous research on mobile app heuristic evaluations, focusing on general characteristics, these heuristics must be more suitable for mobile shopping apps [4]. These studies used heuristic evaluation for smartphone launchers for older adults but did not address mobile shopping apps, limiting the generalizability and reliability of their findings. They used general mobile application heuristics without specific guidelines for integrating new and traditional heuristics into application checklists and did not focus on mobile shopping apps.

Many measures will be included in identifying the heuristics checklist. In this study, however, the focus is only on one of the measures: Operability. Operability measures assess how user-friendly a product or system is using metrics from ISO/IEC 25010 [5]. Fundamental dialogue principles for m-shopping apps include software appropriateness, self-descriptiveness, controllability, user conformity, fault tolerance, and customisation as defined in ISO/IEC 25023:2016 [6]. Identifying actionable operability indicators for mobile shopping apps enhances their evaluation and improvement [7].

This paper aims to identify suitable heuristics under the operability measure to create a checklist for conducting heuristic evaluations, supporting the digital acceleration of mobile shopping applications. Section 2 discusses related works. Section 3 details the methodology used in the study. Section 4 introduces the proposed Operability checklist, while Sect. 5 covers the discussion. The final section provides the conclusion, limitations, and future research directions.

2 Related Work

Existing studies focus on developing and using a heuristic checklist in mobile, e-commerce and mobile shopping applications.

A review of domain-specific heuristic sets and checklists used in previous studies reveals that most research has created new lists or updated Nielsen's heuristics to address areas previously overlooked. Recent studies have introduced various heuristics for evaluating mobile application usability [8]. For instance, one study presented a heuristics checklist tailored to mobile application interfaces [9] but did not focus on mobile shopping apps. Therefore, incorporating additional measures at different levels could effectively identify many usability issues in mobile shopping applications.

We have compiled a comprehensive list of guidelines to check mobile application usability [10]. The selected applications were tested utilising the heuristic evaluation that depends on the established usability guidelines. The designers of different mobile applications, which could have helped reveal important heuristic categories and subcategories for mobile applications, should have been consulted in this study. Thus, an expert review must ensure that essential heuristics for mobile applications are included. This will ensure the applicability of the list of heuristics on mobile applications [2]. Focused on proposing a fresh batch of heuristics for evaluating the usability of a website and frameworks that utilise Arabic interfaces. The study did not map the fresh batch of heuristics into a list of comprehensive checklist items, resulting in difficulty for novices in having a preceding evaluation of Arabic interfaces.

[8] Has invented different versions of usability heuristics by modifying Nielsen's heuristics and adding additional heuristics to evaluate specific characteristics and successfully analyse a particular area. Meanwhile, [1] concentrates on defining methodologies that could be used to design a new set of heuristics for specific regions. The method applied in this study has eight phases that were repeatedly applied to refine, improve, and validate the fresh badge of proposed heuristics. Nonetheless, these studies did not justify the new sets of heuristics by applying the methodology to acquire the user experience, which is crucial for more updates regarding specific domains. While these heuristic checklists can assist interface designers in making usability judgments in general, they do not provide a single usability score or allow for a comparison of design options and applications. Categorically, most heuristic checklists in these studies have only been evaluated in a restricted way with a few devices and applications due to the new nature of research in this area. These instruments highlight the necessity to communicate the complicated concept of usefulness concisely and understandably, such as a single score. Furthermore, these checklists demonstrate that evaluating usability for mobile applications, especially mobile shopping applications, presents new issues that necessitate the adaptation of the "conventional" usability evaluation checklist and further study.

Existing usability evaluation checklists for mobile applications need approval to determine their usefulness, accuracy, and applicability in real-world conditions. This lack of validation can undermine evaluators' confidence and the comprehensiveness and validity of the results. Consequently, using these models to evaluate mobile shopping applications may not produce precise and reliable outcomes for decision-making [13]. Thus, it is crucial to acknowledge that current heuristic evaluation checklists and approaches do not adequately describe usability issues and cannot provide adequate solutions [14].

3 Methodology

This study was conducted in three phases. Phase 1 involved identifying problems and objectives through a comprehensive literature review to uncover usability issues in mobile shopping applications. This phase also examined existing evaluation methods to identify suitable heuristic evaluation checklists for mobile shopping apps from existing literature. Activities in this phase resulted in a thorough analysis of existing and related studies, distinguishing evaluation methods for mobile shopping applications, and in-depth research on previous heuristic evaluation checklists.

In Phase 2, heuristic identification processes were carried out, including exploring various heuristic studies, categorising heuristics by categories or subcategories, and providing guidelines for reviewing and developing heuristic categories. This phase produced a classification of heuristic categories and subcategories, highlighting and examining the most suitable heuristic categories.

In Phase 3, heuristic evaluation checklist development was conducted, which involved determining the empirical study approach and instruments, extending Nielsen Heuristics to identify suitable categories and subcategories, and developing a heuristic evaluation checklist. The evaluation instrument was designed and later verified by experts. Outcomes from this phase included the development of a proposed heuristic evaluation checklist for mobile shopping applications, instrument development, and creation a verification form for expert review. The study involved 12 experts: 5 usability experts, four software/mobile app developers, and three mobile shopping app designers. They verified the proposed heuristic evaluation checklist and their feedback was analysed using descriptive analysis.

4 Result

The study involved 12 experts: 5 usability experts, four software/mobile app developers, and three mobile shopping app designers. They verified the proposed heuristic evaluation checklist and their feedback was analysed using descriptive analysis. Table 1 presents the consensus among the experts on the proposed heuristics, sub-heuristics and evaluation questions. Item H4-2 received less favourable agreement among the experts, while the rest of the heuristics, sub-heuristics and evaluation questions received 83% and above.

The following is the feedback from the experts' evaluation related to the evaluation questions, which suggest their suitability, generality, and redundancy.

Table 1. Initial heuristics/sub-heuristics of operational measures and their verification of heuristic evaluation checklist by the experts

Heuristics/Sub-heuristics	Evaluation Questions	Percentage %
Heuristics – Flexibility and efficiency of use	**H4-1:** Are there shortcuts when using the mobile device for commonly used sequences of actions?	92%
	H4-2: Did the library's instructions in mobile shopping applications start to feel lengthy or intrusive while browsing the site on a mobile device?	67%
	H4-3: Is the number of pages a mobile device user must navigate kept to the minimum?	83%
SH1 – Search	**H4SH1-1:** Is the search box easily accessible when using a mobile device?	92%
	H4SH1-2: Is the search box easily recognisable when using a mobile device?	92%
	H4SH1-3: Is there an advanced search option for mobile shopping applications?	92%
SH2 – Navigation	**H4SH2-1:** Can mobile device users cancel out of operations in progress?	92%
	H4SH2-2: Can mobile device users move between fields or options forward and backwards?	92%
	H4SH2-3: Do the mobile shopping applications provide an undo function after initiating the primary action?	92%
Heuristics – Help and documentation:	**H5-1:** Are the help functions visible in mobile shopping applications, such as a key labelled HELP or a special menu?	92%
	H5-2: Are frequently asked questions (FAQ) area available or needed in mobile shopping applications?	92%
	H5-3: Is contacting support in the mobile shopping application through email or a web form easy?	83%

(*continued*)

Table 1. *(continued)*

Heuristics/Sub-heuristics	Evaluation Questions	Percentage %
SH1 – Understandable Categorisation of Information	**H5SH1-1:** Does every interface of mobile shopping applications begin with a title or header that describes page contents?	92%
	H5SH1-2: Are the mobile shopping applications' data entry screens and dialogue boxes supported by navigation and completion instructions?	92%
	H5SH1-3: Do the elements in the mobile shopping applications clearly show which can and cannot be clicked?	92%
SH2 – Multilingual support	**H5SH2-1:** Do the mobile shopping applications support multiple languages?	83%
	H5SH2-2: Are the mobile shopping application's icons, images, and colours adequately designed to understand people of different languages?	92%
	H5SH2-3: Do mobile shopping applications have alternatives/options for people who cannot comprehend the language of use?	83%

5 Verified Operability Measures for Heuristic Evaluation Checklist

Operability measures are utilised to assess the extent to which a product or system has features that make it formal to operate, measured using operability metrics [5]. There is a need to identify trailing and leading indicators of operability measures for m-shopping applications that are holistic and actionable. Hence, such operability measures will encourage a practical evaluation of mobile shopping applications and aid in pinpointing where improvements need to be made [7]. The following are verified heuristics and sub-heuristics produced from Phase Two (refer to Table 3), verified by eleven experts described in Phase Three of the methodology. Items suggested to be removed, as stated in Table 2, were updated, and the verified Operability Heuristic checklist is presented in this section and their description.

Table 2. List of dropped evaluation questions based on experts' feedback

Heuristic Categories	Subcategories	Dropped Evaluation Questions	Heuristic Categories
H4-2 & H4-3 Flexibility and efficiency of use		**H4-2:** Did the library's instructions in mobile shopping applications start to feel lengthy or intrusive while browsing the site on a mobile device?	The question amount needs to be minimised. Below experts' approval (Experts 2, 3, 6, 8, 10. & 12)
		H4-3: Is the number of pages a mobile device user must navigate kept to the essential minimum	Redundant items should be removed (Expert 8)
	SH1-2 Search	**H4SH1-2:** Is the search box easily recognisable when using a mobile device?	The question amount needs to be minimised (Experts 2, 3, 6, 8, 10, & 12). It is less important
	SH2-2 Navigation	**H4SH2-2:** Can mobile device users move between fields or options forward and backwards?	Redundant items should be removed (Expert 8 & 12)
	SH2-3 Navigation	**H4SH2-3:** Do the mobile shopping applications provide an undo function after initiating the primary action?	The question amount needs to be minimised (Experts 2, 3, 6, 8, 10, & 12). This is a general question

5.1 Heuristic 1: Flexibility and Efficiency of Use

This is all about allowing users to approach tasks in various ways. Embracing this heuristic in mobile shopping applications means allowing user customisation, not being prescriptive about core task steps, and adding unobtrusive accelerators that power users can discover and employ efficiently [15]. It is a process that caters to inexperienced and experienced users and allows users to tailor frequent actions [16–18]. Almost every software like Adobe Photoshop Users of Sketch has access to shortcut commands. While several shortcut commands, such as copy, cut, redo, and undo, are practically universal, just a few are software-specific. For instance, if a customer wants to shop on a mobile application, giving them the list of all the items available within their search saves time and mental effort.

Sub-heuristic search is one of the significant ways m-shopping customers interact with websites. A conventional search aid must consider the 'human' side of searching, which involves managing spelling errors and synonyms [19, 20]. The evaluation questions related to this subcategory are: Is the search box accessible using a mobile device?

Is the search box easily recognisable when using a mobile device? Is there an advanced search option for mobile shopping applications?

Sub-heuristic-Navigation: These are heuristics and sub-heuristics directed to understanding an application's structure, such as m-shopping applications and how the user can use that application based on this structure [16]. It is convenient and obvious for expert and novice users to move between related pages and sections [9, 20]. It can be seen as providing alternative interaction and navigating techniques; for example, to duplicate an element, a novice user can edit and click copy [21, 22].

5.2 Heuristic - Help and Documentation

This is a process in which information is simple to find, concentrated on the user's task, has a list of concrete actions to follow, and is manageable. For example, even before looking at these help papers, mobile shopping app users prefer voice calls, so adding a button to contact customer service would be an intelligent step [16, 23, 24].

Sub-heuristic-Understandable categorisation of Information: This involves how well software categorises information in ways that are intuitive and convenient for the intended users and their specific tasks [6, 17, 25].

Sub-heuristic-Multilingual Support: Multilingual support enhances the customer experience for individuals who do not speak English or prefer to communicate in another language. This service allows customers to choose their preferred language when they contact your support line or visit your company website [22, 25, 26].

6 Discussion

The proposed heuristic evaluation checklist comprises heuristics, sub-heuristics, and evaluation questions presented in an orderly and related explicitly to measuring the operability of mobile shopping apps. They have been explained respectively based on their inclusion in the heuristic evaluation checklist and usefulness in evaluating mobile shopping applications according to ISO25023:2016. They are also explained thoroughly, following the essential characteristics incorporated to address the operability issues of mobile shopping applications. Such specifications are uncommon to be discovered in existing heuristic evaluation studies, particularly on mobile shopping applications. This fact attaches greater significance to the proposed heuristic evaluation checklist; thus, domain experts can apply the guideline to mobile shopping applications and effectively conduct heuristic evaluations that can produce reliable results.

This study has overcome the unclear descriptions of each heuristic and sub-heuristic included in the heuristic evaluation checklist that needs to be clearly defined in existing studies based on the findings through content analysis and experts' verification and validation process. Furthermore, all the heuristics and sub-heuristics generated are properly and respectively defined to empower usability experts, software and mobile application developers, and mobile shopping application designers to identify specific usability issues and categorise them accordingly. Hence, improving existing mobile shopping

Table 3. Operational measures in checklists of heuristic evaluation for mobile shopping Applications

Heuristics (H)/Sub-heuristics (SH)	Evaluation Questions
Heuristics – Flexibility and efficiency of use:	**H4-1:** Are there shortcuts when using the mobile device for commonly used sequences of actions?
SH1 – Search	**H4-SH1-1:** Is the search box easily accessible when using a mobile device?
	H4-SH2-1: Can mobile device users cancel out of operations in progress?
SH2 – Navigation	**H4-SH2-1:** From the homepage, is it straightforward where to go and how to browse/search pictures on mobile shopping applications?
	H4-SH2-2: From each page, is it clear where you are and where you can go from the current location using the mobile device?
Heuristics – Help and documentation:	**H5-1:** Are the help functions visible in mobile shopping applications, such as a key labelled HELP or a special menu?
	H5-2: Are frequently asked questions (FAQ) area available or needed in mobile shopping applications?
	H5-3: Is contacting support in the mobile shopping application through email or a web form easy?
SH1 – Understandable Categorisation of Information	**H5SH1-1:** Does every interface of mobile shopping applications begin with a title or header that describes page contents?
	H5SH1-2: Are the mobile shopping applications' data entry screens and dialogue boxes supported by navigation and completion instructions?
	H5SH1-3: Do the elements in the mobile shopping applications clearly show which can and cannot be clicked?
SH2 – Multilingual support	**H5SH2-1:** Do the mobile shopping applications support multiple languages?
	H5SH2-2: Are the mobile shopping application's icons, images, and colours adequately designed to understand people of different languages?

(*continued*)

Table 3. (*continued*)

Heuristics (H)/Sub-heuristics (SH)	Evaluation Questions
	H5SH2-3: Do mobile shopping applications have alternatives/options for people who cannot comprehend the language of use?

applications is guaranteed, providing more opportunities for better development of future mobile shopping applications for the targeted users.

7 Conclusion

This study developed a heuristic evaluation checklist for operability measures to identify operability issues in mobile shopping applications. The checklist was created through content analysis of recent studies, generating and reviewing heuristic categories and sub-categories. An expert review verified its applicability, answering the research question on developing the checklist for mobile shopping apps. Initially, two heuristics, four sub-heuristics, and 18 evaluation questions were identified from research articles. However, this was refined to 2 heuristics, four sub-heuristics, and 14 questions based on frequency and relevance in the literature to ensure efficacy and simplicity. Usability experts, software and mobile app developers, and mobile shopping application designers, all with at least three years of experience, reviewed the checklist to ensure originality and consistency. The experts agreed that the checklist was comprehensive and valuable, achieving the study's objective of developing operability measures heuristic evaluation for mobile shopping apps.

For future work, additional validation should be made by comparing the proposed heuristic evaluation checklist with other heuristic checklists to determine further the relationships between them and their usefulness to the interface/application designers and developers. By applying this checklist conducted by developers and UX practitioners on mobile shopping apps, customer satisfaction with using these apps can be alleviated, indirectly contributing to supporting digital acceleration achievement.

References

1. Quiñones, D., Rusu, C.: How to develop usability heuristics: a systematic literature review. Comput. Stand. Interfaces **53**, 89–122 (2017)
2. Muhanna, M.A., Amro, R.N., Qusef, A.: Using a new set of heuristics in evaluating Arabic interfaces. J. King Saud Univ. Comput. Inf. Sci. **32**(2), 248–253 (2020)
3. Joyce, G.: Adaptation of heuristic evaluation for mobile applications and the impact of context of use (2021)
4. Petrovčič, A., Rogelj, A., Dolničar, V.: Smart but not adapted enough: heuristic evaluation of smartphone launchers with an adapted interface and assistive technologies for older adults. Comput.-Hum. Behav. **79**, 123–136 (2018)
5. ISO/IEC: ISO/IEC 25010:2011 Systems and software engineering – Systems and software quality requirements and evaluation (SQUARE) – System and software quality models (2011)

6. ISO/IEC: ISO/IEC 25023: Systems and software engineering – Systems and software quality requirements and evaluation (SQuaRE) – Measurement of system and software product (2016)
7. Mobile eCommerce Stats in 2019 and the future online shopping trends of mCommerce. https://www.outerboxdesign.com/web-design-articles/mobile-ecommerce-statistics. Accessed 10 May 2022
8. Salah, M.S., Jusah, S., Muhanna, M.A.: The development of usability heuristics for Arabic m-commerce applications. In: IEEE Jordan International Joint Conference on Electrical Engineering and Information Technology (JEEIT 2019), pp. 779–784 (2019)
9. Gómez, R.Y., Caballero, D.C., Sevillano, J.L.: Heuristic evaluation on mobile interfaces: a new checklist. Sci. World J. **2014**, 1–19 (2014)
10. Ballantyne, M., Jha, A., Jacobsen, A., Hawker, J.S., El-Glaly, Y.N.: Study of accessibility guidelines of mobile applications. In: MUM 2018: Proceedings of the 17th International Conference on Mobile and Ubiquitous Multimedia, pp. 305–315 (2018)
11. Abubakar, H.I., Hashim, N.L., Hussain, A.: Usability evaluation model for mobile banking applications interface: model evaluation process using experts' panel. J. Telecommun. Electron. Comput. Eng. (JTEC) **8**(10), 53–57 (2016)
12. Ajibola, A.S., Goosen, L.: Development of heuristics for usability evaluation of M-commerce applications. In: ACM International Conference Proceeding Series, Part F130806, pp. 1–10 (2017)
13. Nielsen, J., Molich, R.: Heuristic evaluation of user interfaces. In: Proceedings of the SIGCHI Conference on Human Factors in Computing Systems, pp. 249–256 (1990)
14. Grigera, J., Garrido, A., Rivero, J.M., Rossi, G.: Automatic detection of usability smells in web application. Int. J. Hum. Comput. Stud. **97**, 129–148 (2017)
15. Usability 101: Introduction to Usability. https://www.nngroup.com/articles/usability-101-introduction-to-usability/. Accessed 24 Mar 2026
16. Da Costa, R.P., Canedo, E.D., De Sousa, R.T., De Oliveira Albuquerque, R., Villalba, L.J.G.: Set of usability heuristics for quality assessment of mobile applications on smartphones. IEEE Access **7**, 116145–116161 (2019)
17. Swaid, S.I., Suid, T.Z.: Usability heuristics for m-commerce apps. Adv. Intell. Syst. Comput. 79–88 (2019)
18. Abulfaraj, A., Steele, A.: A novel approach to heuristic evaluation: mapping usability heuristics to action model and usability components. In: Proceedings of the XXI International Conference on Human-Computer Interaction, pp. 1–5 (2021)
19. Al-Khalifa, H.S., Al-Twaim, B., AlHarbi, B.: A heuristic checklist for usability evaluation of Saudi government mobile applications. In: ACM International Conference Proceeding Series, pp. 375–378 (2016)
20. Usability expert reviews: Beyond heuristic evaluation. https://medium.com/@userfocus/usability-expert-reviews-beyond-heuristic-evaluation-c377cea7b2aa. Accessed 18 Feb 2024
21. Beyk, B.: Enhancing the usability of e-commerce platforms by utilising the usability factors: an investigation into user preferences. Linnaeus University (2015)
22. Panda, S.K., Swain, S.K., Mall, R.: An investigation into usability aspects of e-commerce websites using users' preferences. Adv. Comput. Sci. Int. J. **4**(1), 65–73 (2015)
23. Nielsen, J.: Usability Engineering. Elsevier Science (1993)
24. Dourado, M. A. D., Canedo, E. D.: Usability heuristics for mobile applications: a systematic review. In: ICEIS 2018 - Proceedings of the 20th International Conference on Enterprise Information Systems, pp. 483–494 (2018)
25. Hasan, B.: Exploring gender differences in online shopping attitude. Comput. Hum. Behav. **26**(4), 597–601 (2012)
26. Usability 101: Introduction to usability. https://www.nngroup.com/articles/usability-101-introduction-to-usability/. Accessed 11 Jan 2024

Preschooler's Game-Based Digital Literacy Assessment Application Development

Nor Iadah Yusop(✉), Nur Adlina Kamilia Azhari, and Zahurin Mat Aji

Universiti Utara Malaysia, 06010 UUM Sintok, Kedah, Malaysia
noriadah@uum.edu.my

Abstract. Digital literacy is a crucial skill for preschoolers, equipping them with the cognitive and technical abilities to use technology effectively. However, assessing their digital literacy objectively can be challenging due to age-specific limitations. To address this, this paper presents the design considerations for developing a game-based digital literacy assessment application tailored to preschoolers. The design prioritises enhancing engagement and motivation while simultaneously evaluating learning. Through a design-based research approach, the application was developed and evaluated. Key deliverables include a use case diagram, storyboard, and application. Evaluation results demonstrate the game's ability to assess preschoolers' hardware knowledge and enhance cognitive skills such as attention, memory, and visual discrimination. A game-based digital literacy assessment application is a promising tool for evaluating preschoolers' digital literacy. Consequently, this helps them embrace the digital technology available to them. Future iterations can be enhanced through progressive difficulty levels, visual representations, picture-matching games, and educational components to further optimise preschoolers' learning and assessment experiences.

Keywords: Digital Literacy Assessment · Game-based Assessment · Design-based Research

1 Introduction

Preschool education in Malaysia is a formal education for children aged four to six. Although not mandatory, many parents enrol their children in preschool as preparation for primary school. Preschool education is crucial for children's development as they are the country's future leaders. Institutions should prioritise delivering high-quality education, which can be enhanced using digital technologies. Digital technology offers endless opportunities for expanding knowledge, continuous learning, and unlocking potential.

Globally, the adoption of the Internet and digital technology has increased rapidly, especially during the Covid-19 pandemic. Malaysia's Internet users increased during this period [1]. Due to the movement control order (MCO) and social distancing, many people, including young children, spend more time online conducting various business transactions, educational activities, social interactions, and other daily routines. The

Z. Mat Aji et al. (Eds.): KMICe 2024, ISEM 49, pp. 229–243, 2025.
https://doi.org/10.1007/978-3-031-91485-0_18

reliance on technology has become a phenomenon in which people worldwide, regardless of age, get engaged and connected to technology individually or in groups.

The use of digital technology offers numerous benefits. According to [2], 46.9% of students indicate that technology allows them to organise better and manage the logistical necessities of their studies. ICT also helps to increase students' performance and other educational matters [3, 4]. Similar phenomena also transpire in the younger generation, including preschoolers [5, 6]. Being digitally native, they grow up immersed in a constant flow of digital information. Computers and the Internet are integral aspects of daily life for this generation. As digital natives, they use technologies more often than previous generations of their age.

The rapid growth of digital technology has transformed modern society, requiring a shift in educational approaches. Early childhood education, a vital stage of development, is no exception. As digital natives, young children are increasingly exposed to digital environments, emphasising the need for strong digital literacy skills. While the importance of digital literacy is acknowledged, effective assessment methods for preschoolers remain a challenge. A game-based digital literacy assessment application can contribute to social innovation by addressing a pressing social need, fostering collaboration, and promoting equal access to education. Game-based assessment is a promising method for evaluating preschoolers' digital literacy skills. Understanding the necessary components for developing such an application is crucial to realise this potential. This paper presents the design considerations for developing a game-based digital literacy assessment application for preschoolers. The development follows a design-based research (DBR) methodology, combining scientific research with technological development.

2 Digital Literacy

Digital literacy encompasses the skills required to learn, work, and live in a community where information and communication are facilitated through digital technologies, such as mobile devices, social media, and internet platforms [7]. In line with this, [8] regarded digital literacy as the awareness, attitude, and ability of individuals to appropriately use digital tools and facilities to identify, access, manage, integrate, evaluate, analyse, and synthesise digital resources, construct new knowledge, create media expressions, and communicate with others, in the context of specific life situations, to enable constructive social action, and to reflect upon this process.

The American Library Association (ALA) defines digital literacy as the ability to use technology to search for, evaluate, create, and communicate information using cognitive and technical skills. While these definitions share a common foundation, they each emphasise different aspects. Adobe focuses on business-oriented applications, whereas ALA's definition is more comprehensive and inclusive. The Cambridge Dictionary adds that digital literacy involves understanding basic concepts like binary and reading. The definition of digital literacy by [9] emphasises having a skill set beyond technical proficiency and includes critical thinking, digital citizenship, and safety awareness. As a framework for children's digital literacy, it acknowledges the need to consider children's age, cultural background, and developmental stage when defining and promoting digital literacy. Essentially, digital literacy for children is about empowering them to be active, critical, and responsible participants in the digital world. Despite these variations, all definitions converge on the core idea of using technology to search, evaluate, create, and communicate information.

The digital age has transformed societies worldwide, and Malaysia is no exception. The COVID-19 pandemic sped up this shift, making technology a central part of life for all ages. Preschoolers today are growing up in a digital world, making digital literacy an essential skill for their future [5, 6, 9]. This goes beyond basic technological skills to include critical thinking, problem-solving, communication, and responsible digital behaviour—all vital for full participation in society.

To equip Malaysian preschoolers with these skills, a well-designed assessment framework is essential. Traditional methods may miss the nuances of digital literacy, but game-based assessments offer an effective alternative. They embed learning goals into engaging activities, providing educators valuable insights into children's digital skills while encouraging a positive attitude toward technology. Integrating digital literacy into the preschool curriculum and using targeted assessment practices can help educators support children's development. Investing in early digital literacy is, ultimately, an investment in Malaysia's future, fostering a generation of responsible, tech-savvy citizens.

3 Game-Based Assessment

Assessing digital literacy in preschoolers can be challenging due to their developmental stage. Game-based assessments offer a fun and engaging way to evaluate these skills, fitting well with young children's natural learning style. This paper discusses essential factors in designing these assessments, ensuring they are developmentally appropriate, engaging, and effective in measuring digital literacy. The focus is on core digital literacy skills, with design elements that boost engagement and motivation.

Preschoolers learn through social interaction, imitation, practice, and exploration, building knowledge about sounds, visuals, objects, emotions, language, and stories [10]. Games that incorporate object movement, sounds, and visuals work well for assessing these skills. Game-based assessment combines entertainment with evaluation, using technology or online platforms to make assessment enjoyable and effective.

Examining tablet app usage among preschoolers, it sheds light on how augmented reality can be employed to foster playful and creative learning experiences [11]. The analysis synthesises the essential findings and evaluates the implications of developing game-based learning assessments, and children's tablet behaviour and interactions with various applications were meticulously examined. The study reveals that 65% of children could independently swipe the screen to change photos, suggesting that the swipe feature could be effectively integrated into game-based assessments. Additionally, 60% of children could drag items across the screen without assistance, while 27% required some help, and the remainder were unaware of the activity. These findings highlight the potential of touch-based interactions in educational game design. The study also categorises children's applications, including learning, creativity, gaming, video, and reading apps. A higher percentage of children could independently open learning (49%), creativity (47%), and gaming apps (43%), compared to reading and video apps. The lower usage rates for reading and video apps are likely due to the children's limited reading and video-making skills. This categorisation underscores the importance of designing apps that align with young children's developmental stages and abilities. In line with [11, 12] emphasises the importance of interactive features in educational apps. [12] demonstrates how real-time feedback and embodied learning can be significantly improved by introducing advanced movement technologies.

The analysis also critically examines similar games for preschoolers, highlighting their strengths and weaknesses. "Kids Coding" is praised for its diverse topics, subcategories, numerous levels, and access to instructional videos, all presented in a simple layout suitable for young children. However, the game's limited interaction and extensive levels within a single category may lead to boredom among preschoolers. Similarly, the game "School" is noted for its high levels of interaction and live-action elements, with captions highlighted during student interactions. Despite these strengths, the game has a cluttered interface and lacks clear instructions for activities like colouring or drawing. Lastly, "Khan Academy Kids" offers various learning topics and creative activities, providing a comprehensive and engaging experience. However, the animations could be improved with more written instructions and additional language options.

Undeniably, there was a visible gap in assessing preschoolers' digital literacy. Education in Malaysia also added digital skills as a part of education that needs to be learned. There were gaps in the tools for exposing digital literacy as early as possible to preschoolers, resulting in the unequal access of preschoolers to digital technology, the learning gaps, and the unpreparedness of the preschoolers for future learning. In the preschool context for this project, the content focused on whether preschoolers could access digital literacy, specifically on basic hardware concepts suitable for their age and understanding. Preschoolers' access to digital literacy can be assessed through games, and they answer the questions through visualisations and sounds. This can be supported by games-based learning, a compelling learning technique that can improve the student's experiences in terms of motivation, enjoyment, and emotional engagement [13]. These suggest that educational games should highlight the importance of designing tablet applications that are developmentally appropriate, engaging, and capable of leveraging advanced movement technologies. These elements are crucial for creating compelling and enjoyable learning experiences for preschool children.

Game-based assessment (GBA) offers an opportunity to combine assessment and learning. The convergence of game and assessment design creates opportunities for interactive and engaging evaluation experiences [14]. Through the game Posterlet, [15] explored how GBA can foster self-directed learning and assess students' critical thinking capacities. The study demonstrated a correlation between these metacognitive strategies and academic performance by analysing students' design choices and feedback-seeking behaviours. While successful in middle and college-level contexts, the generalizability of these findings to younger populations warrants further investigation.

In contrast, research by [16] highlights the importance of procedural justice in GBA, particularly in high-stakes assessment contexts. The study emphasises the need for transparent assessment criteria and opportunities for performance demonstration to ensure fair and equitable evaluation. These findings underscore the critical role of assessment design in shaping perceptions of fairness and ultimately influencing participant motivation and engagement.

When considering the application of GBA to preschoolers, it is essential to adapt the assessment to the unique developmental characteristics of this age group. While the principles of engagement and formative assessment identified in studies of older learners are relevant, the specific game mechanics and assessment criteria must be carefully tailored to preschool children's cognitive, social, and emotional needs [17].

Traditional assessment methods for preschoolers often fall short in providing comprehensive and accurate developmental profiles due to their reliance on subjective observations, standardised tests, or self-reported data [17, 18]. In contrast, game-based assessment (GBA) offers a promising alternative by leveraging the inherent engagement of play to gather objective and rich data on children's cognitive, social, and emotional development [18].

GBA aligns with modern assessment approaches by integrating evaluation into the learning process, offering formative feedback and diagnostic insights. Advances in game design now allow "stealth assessments" to unobtrusively track children's behaviours and skills in real-time. Adaptive learning within these games also personalizes the assessment experience. Research shows that GBA fosters independent learning, enhances critical thinking, and provides insights into metacognitive skills, with transparent criteria that promote fairness. Though GBA for preschoolers is still developing, early studies highlight its potential to deliver meaningful data on young children's development while keeping them engaged.

To fully realise the potential of GBA in preschool settings, careful consideration must be given to age-appropriate game design, alignment with early childhood education frameworks, and the development of valid and reliable assessment metrics [9, 19]. By addressing these challenges, GBA can become a cornerstone of practical early childhood evaluation. Hence, the following critical considerations for preschoolers' GBA design are derived: (i) Aligning assessment goals with early childhood education frameworks, (ii) Ensuring age-appropriate challenges and rewards, (iii) Prioritizing playful and exploratory learning experiences, (iv) Incorporating opportunities for social interaction and collaboration; and (v) Collecting data on a variety of developmental domains.

4 Methodology

The game-based assessment application was developed following the methodology proposed by [20] and employed by [21], namely the Design-Based Research (DBR). Following the DBR, which combines scientific research and technological development, the development was divided into three phases: preliminary study, development, and evaluation.

Phase 1 – Preliminary Study

The initial phase involved a preliminary study to develop a game-based assessment application for children aged four to six. A literature review was conducted to establish the context for the application, and existing and relevant applications were analysed to identify appropriate design-related features. The development context was derived from the reviews, and the digital literacy scope suitable for preschoolers was identified. Computer hardware was chosen as the assessment scope due to its physical recognizability, tangibility, and accessibility. The list of requirements for the application was subsequently identified, and the corresponding Use Case Diagram was produced.

Phase 2 – Development

In this phase, the core features and requirements of the project were defined. Workflows and algorithms were designed to match preschoolers' cognitive levels, avoiding overly complex structures. Basic digital literacy skills, such as simple navigation and problem-solving, were integrated into the design. Use case models and specifications were developed, and design elements were selected to suit young children. The user interface featured large buttons, many images, audio, bright colors, and simple backgrounds. A storyboard was created to visualize the interface, incorporating these elements for a familiar feel. Once completed, the application was made installable on Android devices and prepared for evaluation.

Phase 3 – Evaluation

The evaluation phase focused on testing the functionality and usability of the application prototype. Functional testing involved creating test cases to specify input conditions, expected results, and test execution steps. Usability testing was conducted with nine preschoolers aged four to six to validate the application's design. This took place in a kindergarten classroom to provide a familiar environment—comfortable, yet free from distractions. The session was recorded to capture the children's responses, with the observer remaining neutral to avoid influencing actions, while being patient and friendly during interactions. Observing any hesitation or unusual behaviour was a key focus.

Participant observation was used as the data collection technique, which is effective for small groups and brief sessions. This method captures information in context, records nonverbal cues, and works well for young children who may struggle to express their thoughts [22]. To support the observation process, a data collection protocol and checklist were developed, as shown in Tables 1 and 2.

Before conducting usability testing, the kindergarten's administrator's approval was obtained. Participant demographic information was collected, and a comfortable atmosphere was ensured. Children were briefed on the instructions before testing the application. The usability testing session consisted of three stages: registering an account, playing Game 1 (Choose Right Words), and playing Game 2 (Flipping O' Cards). Children's names and ages were required for account registration, with assistance for those who needed it. The "Choose Right Words" game was adapted from a traditional question-type game with a reversed format. Instead of picture and word answers, questions were presented as words with audio, and answers were pictures. This game assessed preschoolers' knowledge of hardware terms by determining their ability to match the correct term with the corresponding picture.

"Flipping O' Cards" was chosen for the game-based assessment due to its extensive use of pictures, making it suitable for beginner preschoolers. This game exposed preschoolers to various hardware components and enhanced their attention span and concentration through memory-based gameplay. The game also helped develop visual recognition of similar hardware structures and language skills. The session concluded

Table 1. Data collection protocol.

Data Collection Protocol	
1	The usability tests were conducted in a regular space that was not too crowded or too private because a crowded place might distract the participants from the testing. If in a private setting, it might make the participants nervous as if they were in a test
2	The testing session should be recorded as the participant's reaction during the testing session needs to be captured
3	The observer must remain neutral to avoid affecting the participant's actions when using the application and playing the game
4	The observer must be patient and friendly with the participants when asking questions or handling the testing session
5	The observer must focus on the participant's actions and take notes when the participant hesitates repeatedly or does anything strange

with closing questions, appreciation, and tokens for the children. Data collection focused on observing the children's use of the prototype. The evaluation aimed to obtain data on the user experience, analyse the data to evaluate the prototype, and make decisions about necessary changes for future versions.

5 Results

The results of each phase are presented separately and sequentially, as each phase influences the next.

Phase 1 – Preliminary Study

The project began with a literature review on children's learning with technology and games, alongside an analysis of existing game-based assessment tools for preschoolers to set the development context. The appropriate digital literacy scope for preschoolers was defined, with a focus on hardware. Two games were chosen: *Flipping O' Cards* and *Choosing Right Words. Flipping O' Cards*, using picture-based interactions, was ideal for beginning readers, helping preschoolers identify hardware components while enhancing attention, concentration, visual recognition, and language skills. *Choosing Right Words*, a question-based game, assessed preschoolers' knowledge of hardware terms and their ability to match words with images.

The following list of requirements (Fig. 1) and Use Case Diagram (Fig. 2) for the game application were produced from the derived context.

Phase 2 – Development

In phase 2, the corresponding use case specifications were developed for each use case in the Use Case Diagram. Figure 3 shows an example of the use case specification. Before the code, the general storyboard for the application was created as in Fig. 4. The storyboard provides a visual representation of the user interface and the user experience.

Table 2. Observation checklist.

Observation	Answer	Remarks
The participant recognises the "/" green button	Yes/No	
The participant can recognise the "start" button on the main menu screen	Yes/No	
Game: Choose Right Words		
The participants know how to play the game	Yes/No	
The participant needs help to initiate playing the game	Yes/No	
How many levels does the participant complete?	1/2/3/4/5/6	
The participant can recognise the timer that indicates the duration of a game. (in level 3,4,5,6)	Yes/No	
How long does the participant focus on the game?		
Any problems or errors that occurred to the participant:		
Game: Flipping O' Cards		
1. The participants know how to play the game	Yes/No	
2. The participant needs help to initiate playing the game	Yes/No	
3. How many levels does the participant pay attention to?	1/2/3/4/5/6	
4. The participant can recognise the timer that indicates the duration of a game. (in level 3,4,5,6)	Yes/No	
5. How long does the participant focus on the game?		
6. Any problems or errors that occurred to the participant:		
Pause and Resume		
The participant knows how to pause the game	Yes/No	
The participant knows how to resume the game	Yes/No	
General		
1. The participant knows how to use the back button	Yes/No	
2. The participant understood a bit about how the framework works. (use button, where to click, when to drag, when to go back)		
3. Any problems or errors that occurred to the participant:		

It maps out each screen or state of the application, detailing the layout, design elements, and transitions between screens. As depicted in the storyboard, each screen's detailed visual and interactive elements were transformed into the interface design, including layout, colours, typography, buttons, icons, and other interactive components. The actual interfaces of the application are illustrated in Fig. 5.

No.	Requirement	Description	Priority
1.	**View Profile**	**This is to let the user view the player's profile.**	
	1.1 Add Player	Adding a new player requirement allows users to have many accounts on one device.	O
	1.2 Change Player	Changing the chosen player allows the users to change to their desired player.	O
	1.3 Delete Player	Delete the chosen player allows users to delete the account they do not want.	O
2.	**View Game Menu**	**This is to let the user view the game menu.**	
	2.1 View Scores Board	View scores boards allow the guardian to view the user's average scores and recommendations.	M
	2.2 Choose Game Pattern	Choose games pattern allows the users to choose what game to play.	M
	2.3 Select Level	Selecting a level allows users to choose any unlocked level.	M
	2.4 Calculate Score	Calculating the score is the system algorithm to calculate users' scores for each level.	M
	2.5 Gives Recommendation	Gives recommendation is the system algorithm that provides recommendations of where the users need to improve.	M
	2.6 Set Time Duration	Set Time Duration is for the system to set the allocated time duration for level 3 to level 6 for both games.	D
3.	**View Settings**	**To let the user view the setting.**	
	3.1 Quit Game	To let the users quit the game.	O
	3.2 Set Sound Volume	To let the user set the sound volume.	O
	3.3 Set Brightness	To let the user set the brightness.	O
	3.4 Quit Level	To let the user quit the level.	O
	3.5 Resume Game	To let the user resume the game.	O
	3.6 Pause Game	To let the user pause the game.	O

Fig. 1. List of requirements.

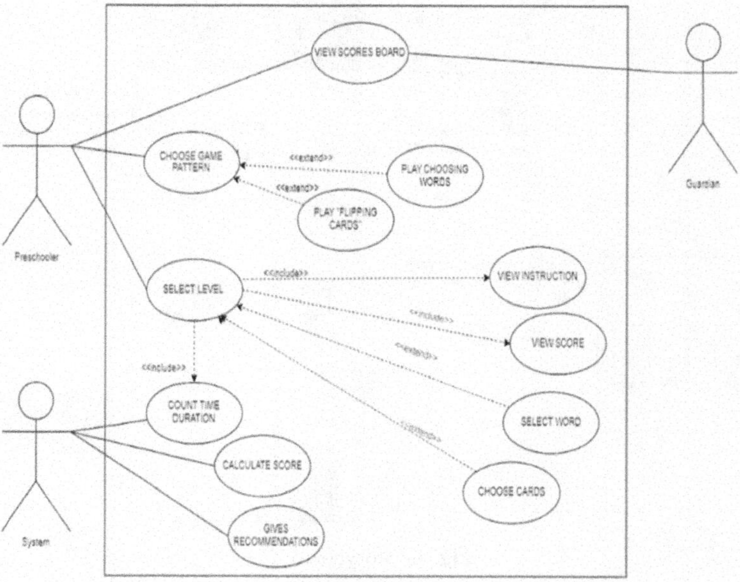

Fig. 2. Use case diagram.

Choose Game Pattern GADLP_02_02	This use case allows users to choose the desired game pattern they want to play. a. Pre-conditions: A player is registered with the device account. b. Characteristics of Activation: Event's Driven. c. Flow of Events: i.Basic Flow: 1. The user clicks the start button on the home screen. 2. The system directs users to the game's main menu page. 3. The user chooses the desired game pattern, such as Choosing Right Words. ii. Alternative Flow: The user has another choice of games, such as Play Flipping Cards. ii. Exceptional Flow: - d. Post-Conditions: The user is directed to the game's level page. d. Rule(s): - d. Constraint(s): -

Fig. 3. A sample of a use case specification.

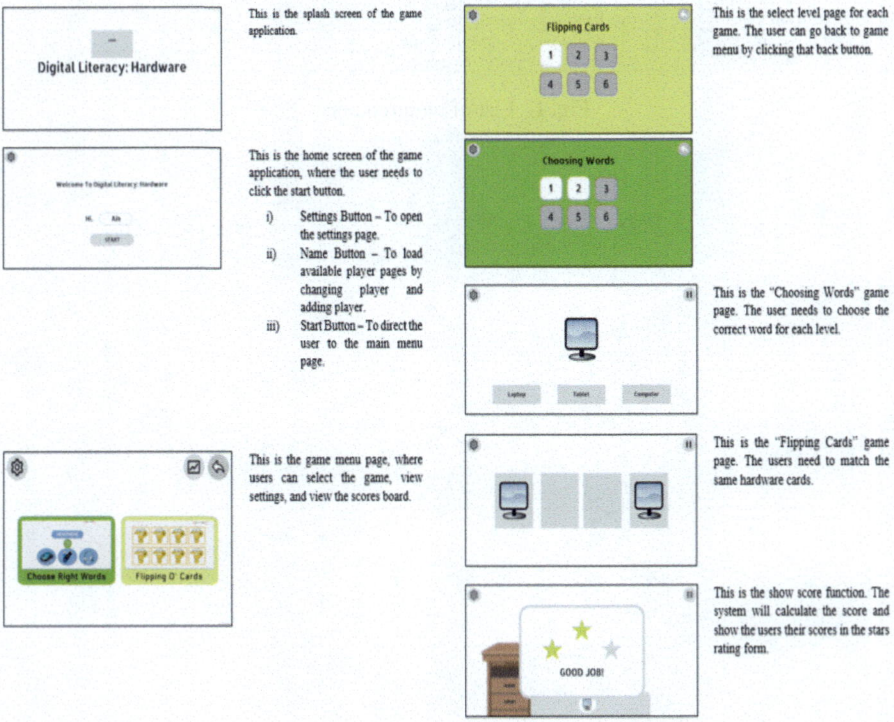

Fig. 4. Storyboard.

Phase 3 – Evaluation

The evaluation phase involves two types of testing: functionality and non-functional. The results of the testing will be elaborated next.

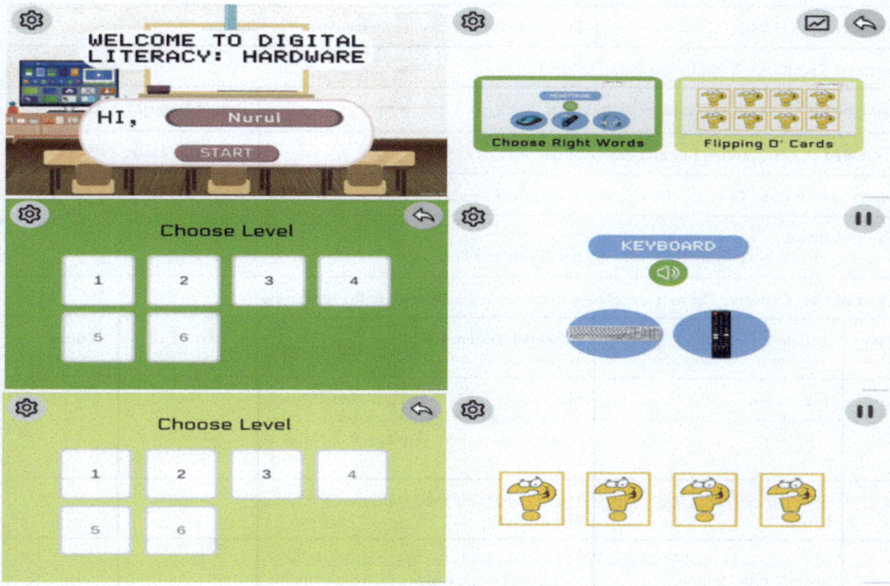

Fig. 5. Interface design.

Functionality Testing

The functionality evaluation of the application precedes the test case design. Twenty-nine test cases were developed to guide the testing process. Figure 6 presents one of the test cases with the testing process results. The testing process was conducted manually for all the test cases. The results indicate that the requirements for the application are fulfilled.

Non-functional Testing

Only the application's usability was tested for the non-functional testing, and nine preschoolers were selected to participate. Four were five years old, two were four years old, and the rest were six years old. Among the four participants with reading ability, two were six years old, and two were five. Given their ages, it is likely that the four-year-olds were at a pre-reading stage, focusing primarily on oral language development. In contrast, the older participants were in the early stages of literacy acquisition and may have possessed basic reading skills. Of the nine participants, five played games and digital devices frequently.

Based on the created observation checklist in Table 2, only two participants did not recognise the function of the buttons even though the design does follow the standard buttons for game applications. This may be due to the lesser frequency of playing games and using digital devices. Six of the participants can play the game without any assistance. Smoothly know the basics of the game, from the main menu to the level of each game, while the others do not. The others need to be instructed to play the game.

Figures 7 and 8 show the focus period, the level of game completion, and the average score for each player for both games: Choosing Right Words and Flipping O' Cards.

Test Case #: GADLP_02_02_02	**Test Case Name:** Choose Game Pattern Requirement
System: TechKids – View Game Menu (GADLP 02)	
Designed By: Nur Adlina Kamilia Binti Azhari	**Design Date:** 15/5/2024
Executed By: Nur Adlina Kamilia Binti Azhari	**Execution Date:** 28/5/2024
Short description: Tests the choose game pattern requirement by choosing the "Choose Right Words" game.	
Pre-conditions: 1. There is a player registered with the device account.	
Overall Pass Criteria: The user can choose the game and is directed to the level scene.	

Steps	Action	Expected System Response	Pass / Fail	Comment
1.	The user clicks on the "TechKids" application.	The system loads the application's "Main Menu" scene with the account that had been chosen before and displays the player's name.	Pass	
2.	The user clicks on the "start" button.	The system loads the application's "Game Menu" scene.	Pass	
3.	The user clicks on the "Choose Right Words" button.	The system loads the application's desired game-level scene and displays the available levels.	Pass	
4.	Check post-condition:		Pass	

Post-conditions: The user is successfully directed to the "Choose Right Words" level scene.

Fig. 6. Sample of a completed test case.

The average score achieved by each participant indicates that the application can assess the preschoolers' digital literacy knowledge. Both games consist of six difficulty levels. Before moving to a higher level in the game, the players must complete the current level within a specified timeframe and achieve a minimum score. The completion of each level reflects an increase in their digital literacy knowledge. For instance, as depicted in Fig. 7, those who completed the game up to levels six and five scored high, between 83 to 77 points, with an average focus time of 3.85 min. Similarly, in Fig. 8, those who completed levels five and six scored high of 100 with an average focus time of 4.51 min.

Player	P1	P2	P3	P4	P5	P6	P7	P8	P9
How long does the player focus on the game? (minutes)	3	2	5	0.5	6	2.10	3	4	3
How many levels does the participant pay attention to?	2	4	6	1	6	5	6	6	4
Average Score	50	59	83	0	83	83	83	77	75

Fig. 7. Player's attention and score on choosing the right words.

Player	P1	P2	P3	P4	P5	P6	P7	P8	P9
How long does the player focus on the game? (minutes)	2.30	4	7	2.30	8	3	4	5	5
How many levels does the participant pay attention to?	3	6	5	4	6	6	6	6	6
Average Score	100	100	100	75	100	100	100	100	100

Fig. 8. Player's attention and score on flipping O' cards.

6 Conclusion and Recommendations

The proposed game-based application offers a promising solution for teaching digital literacy to preschoolers, focusing on basic hardware concepts. Using the design-based research (DBR) methodology, the application combines scientific research and technological development to bridge the gap between research and practice. This approach contributes to the broader body of knowledge.

Based on the analysis of the development context, a list of requirements and a use case diagram were created. Each use case was detailed in a use case specification. A storyboard was developed to outline the application's user interface and visual experience. The game-based assessment employs images and sounds to measure preschoolers' knowledge of basic hardware terms. The application aims to enhance their understanding and contribute to social innovation in knowledge management by exposing them to correct terminology. It addresses a pressing social need by providing an equitable and inclusive way to assess digital literacy skills, thus helping them to embrace digital technology rightly. The game fosters collaboration among educators and researchers, promoting shared knowledge and improved educational practices. Ultimately, it contributes to promoting equitable access to education by providing a valuable tool for assessing and supporting young children's digital literacy development.

Future iterations of the application can be improved by implementing progressive difficulty levels, incorporating visual representations of hardware within game environments, and including a picture-matching game and an educational component focused on hardware terminology. These enhancements have the potential to make the application a valuable tool for promoting digital literacy and fostering a love of technology in young children.

References

1. Malaysian Communications and Multimedia Commission (MCMC), Internet Users Survey. Malaysian Communications and Multimedia Commission 9–25. https://www.mcmc.gov.my/skmmgovmy/media/General/pdf/IUS-2020-Report.pdf. Accessed 28 Sept 2022
2. Henderson, M., Selwyn, N., Aston, R.: What works and why? Student perceptions of 'useful' digital technology in university teaching and learning. Stud. High. Educ. **42**(8), 1567–1579 (2017)

3. Timotheou, S., et al.: Impacts of digital technologies on education and factors influencing schools' digital capacity and transformation: a literature review. Educ. Inf. Technol. **28**(6), 6695–6726 (2023)
4. Qureshi, M.I., Khan, N., Raza, H., Imran, A., Ismail, F.: Digital technologies in education 4.0. Does it enhance the effectiveness of learning? A systematic literature review. Int. J. Interact. Mob. Technol. (IJIM) **15**(04), 31–47 (2021). https://doi.org/10.3991/ijim.v15i04.20291
5. Bay, D.N.: The perspective of preschool teachers on the use of digital technology. Southeast Asia Early Child. J. **11**(2), 87–111 (2022)
6. Ogegbo, A.A., Aina, A.: Early childhood development teachers' perceptions on the use of technology in teaching young children. S. Afr. J. Child. Educ. **10**(1), 1–10 (2020)
7. Western Sydney University: Study smart: build the skills for success. Academic e-book series. https://www.westernsydney.edu.au/__data/assets/pdf_file/0005/2022692/Digital_Literacy.pdf. Accessed 02 July 2024
8. Stergioulas, L., Drenoyianni, H.: The pursuit of digital literacy and e-inclusion in schools: curriculum development and teacher education. In: Sotiriou, S. (ed.) Proceedings of the Designing the School of Tomorrow Symposium 2007, pp. 165–168. Athens (2007)
9. Nascimbeni, F., Vosloo, S.: Digital literacy for children: exploring definitions and frameworks: SCOPING PAPER. UNICEF Office of Global Insight and Policy. UNICEF-Global-Insight-digital-literacy-scoping (2019)
10. Grazzani, I., Ornaghi, V., Conte, E., Pepe, A., Caprin, C.: The relation between emotion understanding and theory of mind in children aged 3 to 8: the key role of language. Front. Psychol. **9**, 724–733 (2018)
11. Marsh, J.: The digital literacy skills and competences of children of pre-school age. Media Educ. **2**, 178–195 (2016)
12. Starcic, A.I., Lipsmeyer, W.M., Lin, L.: Using motion capture technologies to provide advanced feedback and scaffolds for learning. In: Parsons, T.D., Lin, L., Cockerham, D. (eds.) Mind, Brain and Technology. ECTII, pp. 107–121. Springer, Cham (2019). https://doi.org/10.1007/978-3-030-02631-8_7
13. Hartt, M., Hosseini, H., Mostafapour, M.: Game on: exploring the effectiveness of game-based learning. Plan. Pract. Res. **35**(5), 589–604 (2020)
14. Kim, Y.J., Shute, V.J.: The interplay of game elements with psychometric qualities, learning, and enjoyment in game-based assessment. Comput. Educ. **87**, 340–356 (2015)
15. Cutumisu, M., et al.: Eye tracking the feedback assigned to undergraduate students in a digital assessment game. Front. Psychol. **10**, 1931 (2019)
16. Ellison, L.J., McClure Johnson, T., Tomczak, D., Siemsen, A., Gonzalez, M.F.: Game on! Exploring reactions to game-based selection assessments. J. Manag. Psychol. **35**(4), 241–254 (2020)
17. Verma, V., Baron, T., Bansal, A., Amresh, A.: Emerging practices in game-based assessment. In: Ifenthaler, D., Kim, Y.J. (eds.) Game-Based Assessment Revisited. Advances in Game-Based Learning, pp. 327–346. Springer, Cham (2019)
18. Mavridis, A., Apostolidis, I., Tsiatsos, T.: Is game-based assessment more attractive and less stressful for the students than the traditional exams? In: Mikropoulos, T.A., Papachristos, N., Tsiara, A., Chalki, P. (eds.) Proceedings of the 10th Pan-Hellenic and International Conference "ICT in Education", pp. 91–97. HAICTE, Ioannina (2016)
19. Mishra, P., Koehler, M.J.: Technological pedagogical content knowledge: a framework for integrating technology in teacher education. J. Teach. Educ. **57**(1), 60–70 (2006)

20. Plomp, T.: Educational design research: an introduction. In: Plomp, T., Nieveen, N. (eds.) Educational Design Research. Institute for Curriculum Development, pp. 10–51. The Netherlands, Enschede (2013)
21. Laranjeiro, D.: Development of game-based m-learning apps for preschoolers. Educ. Sci. **11**(5), 229 (2021)
22. Creswell, J.: Collecting qualitative data. In: Educational Research: Planning, Conducting and Evaluating Quantitative and Qualitative Research. Pearson Education, London (2012)

GREEN Minds: A Digital Recycling Toolkit to Empower Women and Children

Jhamie Tetz Infante-Mateo[1], Musyrifah Mahmod[2](✉), Suwannit Chareen Chit[2], and Conchita La Madrid – Palencia[3]

[1] ISU Regional GAD Research Learning Center, Isabela State University, 3309 Echague, Isabela, Philippines

[2] School of Computing, Universiti Utara Malaysia, Sintok, 06010 Kedah, Malaysia
musyrifah@uum.edu.my

[3] Isabela Green Ladies Association, Inc., 03 Rosemarie Street Brgy District 1, Cauayan City, Isabela, Philippines

Abstract. To address the never-ending issue of solid waste management, the "Green Minds" initiative proposes a digital recycling toolkit tailored to empower women and children in Malaysia and the Philippines. The toolkit aims to leverage technology and community engagement to enhance environmental awareness, promote sustainable waste management practices, and create economic opportunities for marginalized populations. The project recognizes the critical role of women as social capital within households and local communities and seeks to strengthen their involvement in waste management activities. The digital platform will provide educational resources, interactive tools, and a centralized system for waste collection and recycling, empowering women and children to become active agents of change in recycling activities specifically. This paper presents the development process of the Digital Recycling Toolkit, emphasizing on the process of understanding the problems, followed by requirement gathering and design strategies. The paper continues with a discussion on future work and best practices for solid waste management to be incorporated.

Keywords: Recycling · Women Empowerment · Women and Recycling · Recycling Awareness and Solid Waste Management

1 Introduction

Waste production and management are pressing global issues exacerbated by rapid development, population growth, and evolving consumption patterns. Due to their speedy urbanization, population growth, and shifting consumption habits, Malaysia and the Philippines, in particular, confront significant waste production and management challenges. Regarding waste production and composition, the Philippines is one of the top waste generators in Southeast Asia, with a significant proportion of plastic waste ending up in oceans and waterways [1]. Municipal solid waste (MSW) in the Philippines primarily comprises food waste, plastics, and other recyclables. Similarly, Malaysia

Z. Mat Aji et al. (Eds.): KMICe 2024, ISEM 49, pp. 244–255, 2025.
https://doi.org/10.1007/978-3-031-91485-0_19

grapples with a growing waste problem exacerbated by illegally importing plastic waste from developed countries. The composition of solid waste in Malaysia mirrors that of the Philippines, with a high organic content and a considerable number of recyclable materials [2].

1.1 Best Practices in Solid Waste Management

Efforts in recent years have reflected a range of strategies and approaches tailored to address the issue of solid waste management. In a report by the United States Environmental Protection Agency (USEPA), several strategies for solid waste management – tailored based on specific contexts and challenges - can be adopted [3]. These strategies are briefly discussed and illustrated below. (See Fig. 1.)

The first strategy strongly emphasizes the prevention and reduction of solid waste in households and businesses. Under the Solid Waste Management Hierarchy concept, the public is urged to minimize or reduce domestic and industrial waste [3]. The public is also encouraged to reuse, recycle, recover, and practice proper waste disposal.

A robust stakeholder engagement paves the way to enhance the success of solid waste management. Solid commitment from all parties, particularly the local communities, governmental agencies, private sectors and other non-governmental Organizations (NGOs) in drafting policies, campaigns and programs, presents opportunities for environmental preservation through solid waste management to be realized [4].

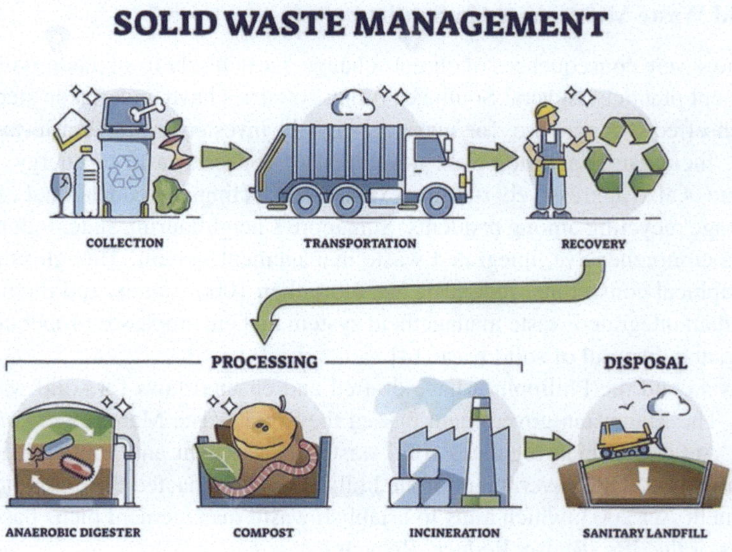

Fig. 1. Solid waste management.

A sound planning system that considers local conditions, waste generation rates, and conservation resources forms the backbone for comprehensive solid waste management planning. This includes sourcing internal revenue resources, Public-Private partnerships (PPP) initiatives, and external financing for conservational efforts.

Policymakers should be firm in drafting and implementing policies and strategies to reduce waste generation. Policies should focus on eliminating or reducing the types and quantity of waste generated while promoting sustainable consumption and adopting waste-reducing technologies.

On the domestic front, household and business owners should actively conduct solid waste recycling and recovery efforts, which include segregation, composting and anaerobic digestion, which reduces the volume of waste transported to landfills while providing us with precious resources such as composts and biogas. The last-mile problem can be addressed by adopting efficient waste collection systems, which include source separation, collection and transportation, and transfer stations to facilitate waste collection [3].

Solid waste management efforts should also consider landfill management. With global issues such as climate change plaguing the world, shifting or transitioning from open dumpsites to controlled dumpsites is timely. It is worth also considering sanitary landfills to mitigate and further cushion the impacts of environmental and health risks. Finally, another approach to solid waste management is energy recovery. All stakeholders should seek to embrace and adopt waste-to-energy technology where energy is generated from waste materials, thereby reducing solid waste volume while providing alternative sources of energy [3].

1.2 Solid Waste Management in Southeast Asia

Noting the severe consequences of climate change, particularly from inadmissible waste management practices, several Southeast Asian countries have undertaken steps to mitigate such effects. Singapore, for instance, heavily invested in its "Waste-to-Energy" initiative. Incinerator facilities were developed to convert waste to energy. The tiny island state is also aggressively running extensive recycling and educational campaigns to encourage recycling among residents. Singapore's neighbouring state, Indonesia, has adopted a comprehensive, integrated waste management system. Though impeded by its geographical constraints, Indonesia has more than 1000 islands, and the nation has developed an integrated waste management system that encompasses proper collection, recycling, and disposal of solid waste [4].

Malaysia and the Philippines have drafted and enacted laws for solid waste management. The Malaysian government passed the Solid Waste Management and Public Cleaning Act 672, which regulates solid waste management and public cleansing for proper sanitation. The Government of the Philippines has enacted the Ecological Waste Management Act 2003, which aims to establish waste management plans based on the principles of the 3Rs (Reuse, Reduce, Recycle).

On the other hand, Thailand has initiated the Zero-Waste initiative, which encourages local authorities to promote and encourage minimal waste generation and intensify organic waste composting. The Thais have also implemented policies to conduct waste separation at the source. Vietnam outlined its Waste Management Masterplan, which extends to 2050, to ensure all types of solid waste are collected and reused based on specific conditions and requirements of each locality [5].

In retrospect, all governments also promote and conduct community-based programs to educate and create awareness of solid waste management among its people. Figure 2 illustrates the efforts undertaken by Governments in Southeast Asia on solid waste management [5].

Fig. 2. Efforts undertaken by governments in Southeast Asia.

1.3 Women and Children in Waste Management

It remains pivotal for public health, environmental quality and sustainable development that waste management is efficient. In this domain, among the neglected but essential dynamics of a global waste solution is attending to how women and children—whom we must acknowledge without stereotyping, as they are critical actors with diverse motivations–perform gender-specific roles when producing and managing our waste. More importantly, the roles of women and children are also gender-specific in terms of waste production and management, which is highly relevant to what occurs not only just in the Philippines but also in Malaysia. For example, among other household waste management tasks, including source separation, recycling, and composting, women often play an important role [6]. Sustainable practices like these will become second nature as children grow up—setting an incredible example for a better environment. Teach your children important lessons in resource conservation, waste reduction and the responsibility of managing one's own waste. Once these behaviour patterns are introduced to a child early, they often become second nature. So, it is more plausible that such behaviours would lead through with them into adulthood, producing positive trickle-down effects upon waste management throughout communities and societies.

Although women and children are important players in waste management, their roles are rarely acknowledged or incorporated into policy- and project-level designs. The government and policymakers in the Philippines and Malaysia also take note of gender-specific dynamics around waste generation and management, thus including women or children, for a few years to come up with and guide design waste control [7]. Including women and children as crucial stakeholders in the waste value chain will provide each country with capable hands that can contribute valuable solutions to address unique challenges and build more robust, more enduring programs.

Differences in the roles of women and children in waste management also stood out between Malaysia and the Philippines. Women are the agents and critical decision-makers in household waste management, informal recycling, and various community-based initiatives in the Philippines. In the Philippines, children engage in waste reduction and recycling efforts at their schools or community-wide. Nevertheless, women and children may find themselves in precarious situations, as the harmful consequences of improper waste management are especially health risks for the youngest. In Malaysia, women's involvement in formal aspects of waste management is minimal, but they are pretty active at the community level regarding recycling and environmental advocacy. However, even in relatively similar neighbouring countries like Malaysia, children have been participating in meetings on environmental education and awareness, implying that they should be educated early about sustainable living.

1.4 Empowering Recycling Awareness Among Women and Children

One of the biggest challenges in waste management is awareness and education, especially about recycling practices, which are essential for contributing to a circular economy. The Philippines and Malaysia struggle to encourage successful waste management processes from industrial and social sitting areas, applying the practices to household and individual settings.

One of the main obstacles to better recycling practices in Malaysia and the Philippines is the lack of environmental educational awareness and understanding among the general public. Most people in both countries have not been adequately exposed to the importance of recycling, its environmental benefits, and the practical steps they can take to recycle effectively in their daily lives [8]. Addressing this awareness gap in environmental education and consciousness is crucial for driving a cultural shift towards more sustainable waste management behaviours at the individual and household levels.

Despite government initiatives like Clean Malaysia and the Action Plan for Beautiful Malaysia being introduced since 1988, there has been little public participation in recycling campaigns [9]). The promotion of recycling awareness and behaviours is aimed at mothers and children. According to [10], children are the future; they represent a new generation that has the potential to establish environmental ethics in their families. According to Utami and Godjali [11], women are frequently the primary caregivers for household waste, which means they have the power to impact the family's recycling habits. There's a possibility that waste management will have a considerably wider socio-economic impact if women and children are taught how to sort their own trash.

Mobile applications, social media platforms, and instructional websites are examples of digital tools or digitally connected learning resources/devices that can be used to increase the reach and efficacy of recycling awareness [12] campaigns in Malaysia and the Philippines. By facilitating information access, encouraging user interaction, and sharing best practices, these digital resources will empower more people to adopt sustainable recycling behaviours.

The Philippines and Malaysia may make significant progress in addressing the issues of recycling awareness and education by emphasizing women and children and utilizing digital resources [13].

2 Digital Recycling Toolkit

Addressing the recycling challenges in the Philippines and Malaysia, a digital solution through the development of a Digital Recycling Toolkit is being proposed to support the Women on Waste Project efforts by the Province of Isabela Green Ladies Association Incorporated (PIGLAI). This solution aligns with and addresses the objectives of Sustainable Development Goals (SDGs) 11 and 12—Sustainable Cities & Communities and Responsible Consumption and Production, respectively—while placing significant emphasis on advocating Gender and Development (GAD) through the involvement of women in waste management, with the potential to inspire and bring hope to many.

The Digital Recycling Toolkit, designed with user-friendly interfaces for children specifically and available in English and Filipino languages, can serve as a centralized platform offering educational resources, best practices, and innovative techniques for waste management. It will provide training modules on waste segregation, recycling methods, and waste-to-energy technologies designed for different communities' specific needs. By leveraging digital technology, this toolkit ensures accessibility and scalability, reaching a broader audience, including women and children who play critical roles in waste management.

The Women on Waste Project can utilize this toolkit to enhance public-private partnerships (PPPs), combining resources and expertise from private companies, local government units (LGUs), and civil society organizations (CSOs). The project can establish efficient waste collection, recycling, and waste-to-energy facilities through this collaboration, integrating advanced technologies such as sensor-based waste sorting systems and innovative solutions like biodegradable waste processing.

Moreover, the toolkit will support community engagement by providing a digital platform for local communities to participate in planning and decision-making. It will empower women and children with knowledge and tools, fostering inclusive and effective waste management strategies. The toolkit will also feature a component on policy advocacy, helping CSOs and communities advocate for stronger waste management regulations and ensuring strict enforcement.

This solution promotes sustainable waste management practices, reduces environmental impact, and improves public health. The Digital Recycling Toolkit, supported by the Women on Waste Project, in collaboration with Universiti Utara Malaysia and Isabela State University, offers a comprehensive and scalable approach to tackling waste management challenges, ensuring customized solutions for diverse communities while empowering women and fostering sustainable development. This focus on reducing environmental impact is intended to instil a sense of responsibility and commitment in the audience.

3 Requirements and Design Strategies of the Digital Recycling Toolkit

The development of the Digital Recycling Toolkit started with requirements gathering and involving all the stakeholders. In-depth interviews and discussions with diverse vital stakeholders, including representatives from local government units, municipal authorities, the Women on Waste NGO, and even school children, were performed. The research team sought to gain a multifaceted understanding of the challenges, needs, and aspirations surrounding recycling efforts in the local community through this comprehensive stakeholder engagement approach.

Discussions with local government units and municipal authorities revealed their need for a platform to streamline recycling efforts, enhance public engagement, and provide data-driven insights to inform policy decisions. The team also engaged with Women on Waste (WoW) representatives, who highlighted the importance of educational initiatives and self-assessment tools to empower individuals and communities to adopt sustainable recycling practices [14]. The research team also went for site visits to existing recycling centres, where the team observed their activities and operations and discussed the challenges and opportunities faced by these facilities see Fig. 3, the WoW recycling centre. The research team also visited the school and participated in recycling activities such as fashion shows from recyclable items (see Fig. 4). These observations and insights gathered from the stakeholder interviews provided a holistic understanding of the requirements the digital recycling toolkit needed to address.

3.1 Functional and Non-Functional Requirements.

Based on the requirement gatherings process, the research team identified **TWO (2)** major components of the Digital Recycling toolkit which includes Digital courseware to educate on sustainable recycling practice and self-assessment to evaluate and test their recycling understanding. The requirements were spelt out and elicited into functional requirements and non-functional requirements. The key functional requirements included the main interaction functions for developing a digital courseware component and implementing self-assessment tools to empower individuals and communities to evaluate their recycling efforts and identify areas for improvement. Table 1 shows some of the functional requirements of Digital courseware for Recycling. The priority for both

Fig. 3. One of WoW's local recycling centre.

Fig. 4. School children recycling activities.

types of requirements must be set as either Mandatory (M), Desirable (D) or Optional (O) to show the level of urgency.

The identified non-functional requirements focused on ensuring the usability, scalability, and accessibility of the digital recycling toolkit. This included incorporating user-friendly interfaces, handling large volumes of data, and providing multilingual support to reach diverse communities.

Table 1. Functional requirements for courseware.

No	Requirement	Priority
1	**Start Courseware**	
	1.1 The user must double-click on the courseware to start	M
	1.2 User shall be able to select language (English, Tagalog, Malay)	M
2	**Choose Topic**	
	2.1 The user shall be able to select topic waste segregation, topic recycling methods, or topic waste-to-energy technologies	M
3	**Perform Assessment**	
	3.1 The user shall be able to choose an answer	M
	3.1 The user shall be able to ask for hints	O
	3.2 The system will calculate a score for the correct answer	M

3.2 Digital Toolkit System Analysis and Design

The system analysis and design of the digital recycling toolkit involved using various Unified Modelling Language (UML) diagrams to model the structure and behaviour of the system. Use case diagrams were employed to capture the critical functionalities of the toolkit, identifying the primary actors (See Fig. 5 and Fig. 6) for use case diagrams of both parts of the Digital Recycling Toolkit and their interactions with the system. Sequence diagrams were then utilized to illustrate the flow of events and interactions between these actors and the toolkit's components, providing a detailed understanding of the system's dynamic behaviour [15]. Furthermore, class diagrams were developed to model the underlying object-oriented structure of the toolkit, defining the various classes, their attributes, and the relationships between them. This laid the foundation for the toolkit's modular and scalable architecture, facilitating future expansion and maintenance. The research team also created storyboards to enhance the tool's user experience and storytelling aspects. These visual representations mapped out the user journeys, interface designs, and interactions, ensuring a seamless and engaging user experience that would resonate with the diverse stakeholders.

During the process of designing the Digital recycling toolkit, the research team had decided to create a persona of MamChit. MamChit is a persona inspired by the characteristics of Madam Conchita La Madrid-Palencia, a prominent advocate for women's empowerment in the Province of Isabela. Known for her passion and dedication, Madam Conchita has actively worked to encourage and uplift women and children, promoting their involvement in various community and development programs. Her unwavering commitment to empowering women has made her a respected figure, driving change and inspiring others in the region. Persona is a fictional character [16] created to represent the typical user of the digital recycling toolkit. By developing a persona like MamChit (See Fig. 7.) to be incorporated into the tool, the design team can better understand the needs, behaviours, and motivations of the stakeholders involved in the local recycling efforts. This helps ensure the digital toolkit is tailored to address the real-world challenges and

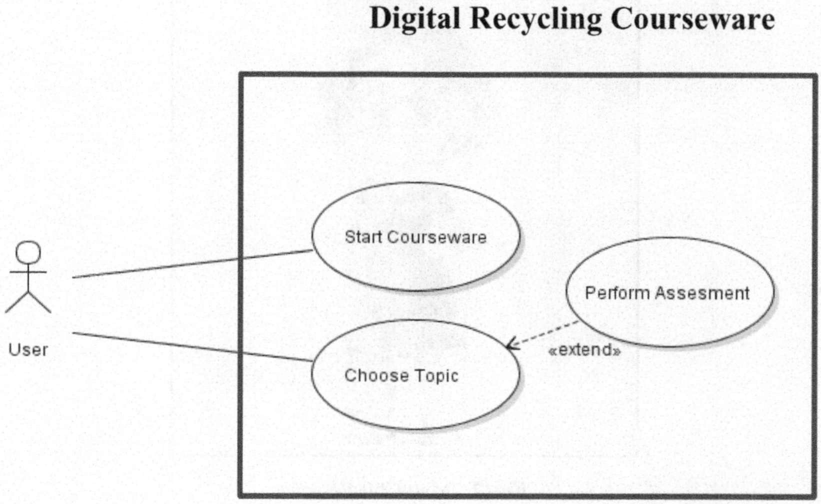

Fig. 5. Use case diagram for Courseware.

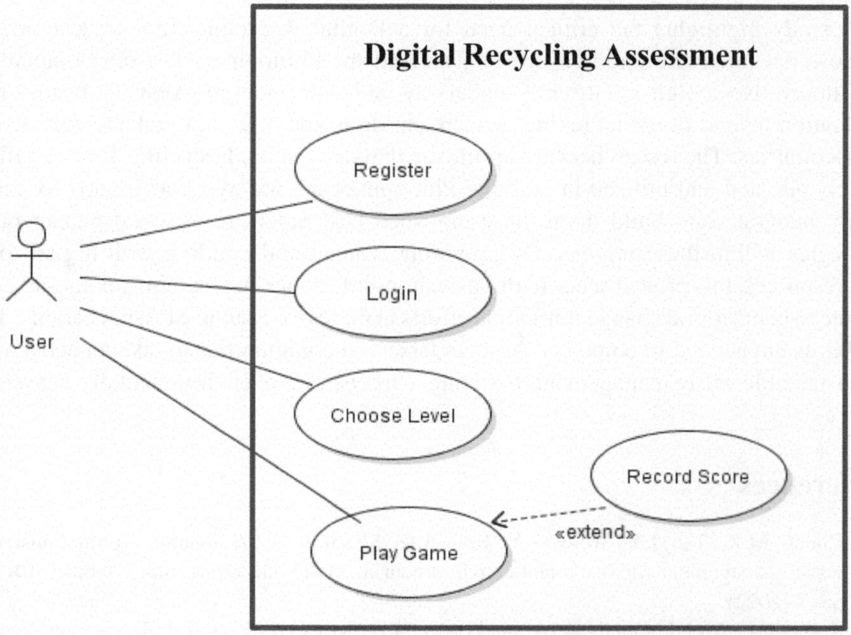

Fig. 6. Use a case diagram for Assessment.

aspirations of the dedicated individuals working to make sustainable waste management a success in the community.

Fig. 7. MamChit Persona

4 Discussion and Future Work

This study highlights the critical need for a Digital Recycling Toolkit designed to empower women and children in Malaysia and the Philippines. The development of this innovative toolkit is currently underway and will undergo extensive testing and evaluation by end-users, including women, children, and other key stakeholders across both countries. The researchers are optimistic that this Digital Recycling Toolkit will be widely adopted and utilized in both the Philippines and Malaysia, as it aims to seamlessly integrate and build upon the established best practices in waste management strategies within these regions. By providing women and children with digital tools and resources, this project seeks to drive meaningful change in recycling habits and contribute to behavioural change mitigation efforts in these two Southeast Asian nations. The toolkit is envisioned to empower these underserved communities to take an active role in sustainable waste management, fostering a greener, more environmentally conscious future.

References

1. Cheng, M,K., Tan, J.Y., Wong, S.Y., Koo, A.C., Sharji, E. A.: A review of future household waste management for sustainable environment in Malaysian cities. Sustainability **14**(11), 6517 (2022)
2. Zulkipli, F., Mohd Nopiah, Z., Ahmad Basri, N.E., Kie, C.K., Khalid, I.: Recycle evaluation using system dynamics for solid waste management in Malaysia. Int. J. Recent Technol. Eng. (IJRTE) **8**(2S11), 657–659 (2019)
3. HI Abdel-Shafy MSM Mansour 2018 Solid waste issue: Sources, composition, disposal, recycling, and valorization Egypt. J. Petrol. **27**(4), 1275–1290
4. Institute for global environmental strategies: Best practices for solid waste management. https://www.iges.or.jp/en/publication_documents/pub/policyreport/en/11066/Best+Practices+for+SWM_Guide+for+Decision+Makers+in+Developing+Countries_Oct+20_0.pdf. Accessed 26 June 2024

5. Somneuk P.: Enabling circular economy in local solid waste management - The case of Muang Kalasin Municipality, Thailand. Master's Thesis, Uppsala University (2020)
6. Chen, L.H., Nath, T.H., Lechner, A.M.: The plastic waste problem in Malaysia: management, recycling and disposal of local and global plastic waste. SN Appl. Sci. **3**(4) (2021). https://doi.org/10.1007/s42452-021-04234-y
7. Jayashree, S., Govindan, M., Chinnasami, M., Kadiresu, I.: Solid waste management in Malaysia – A move towards sustainability. Waste Management- An Integrated Vision. InTech (2012)
8. Ramayah, T., Lee, J.W.C., Lim, S.: Sustaining the environment through recycling: an empirical study J. Environ. Manage. **102**, 141–147 (2012)
9. Soomro, M.A., Ali, M.H., Zailani, S., Tseng, M., Makhbul, Z.K.M.: Understanding barriers and motivations in solid waste management from Malaysian industries: a comparative analysis. Environ. Sci. Pollut. Res. **30**(3), 5717–5729 (2022). https://doi.org/10.1007/s11356-022-22558-z
10. Vicente, P., Reis, E.: Factors influencing households' participation in recycling Waste Manage. Res. **26**(2), 140–146 (2008)
11. Utami, W.K., Godjali, M.R.: The role of women in empowerment and improvement of community literation through waste management J. Gov. **5**(2), 164–178 (2020)
12. Wang, C., Zhu, T., Yao, H., Sun, Q.: The impact of green information on the participation intention of consumers in online recycling: an experimental study Sustainability **12**(6), 2498 (2020)
13. Tiew, K., Basri, N.E.A., Deng, H., Watanabe, K., Zain, M., Wang, S.: Comparative study on recycling behaviours between regular recyclers and non-regular recyclers in Malaysia J. Environ. Manag. **237**, 255–263 (2019)
14. Michalec,O., Fodor, M., Hayes, E., Longhurst, J.: Co-designing food waste services in the catering sector Br. Food J. **120**(12), 2762–27771 (2018)
15. Yildirim,Ü., Campean, F.: Functional modelling of complex multi-disciplinary systems using the enhanced sequence diagram Res. Eng. Des. Sci. Bus. Media **31**(4), 429–448 (2018)https://doi.org/10.1007/s00163-020-00343-8
16. D'Souza, E.M., Lincoln, H.N.: A persona-centric approach to developing complex computer systems: lessons from the field. In: Proceedings of the Human Factors and Ergonomics Society Annual Meeting, vol. 48, no. 5, pp. 917–921. SAGE Publishing (2004)

Designing and Developing a Web-Based Final Year Project Management System Using the Theory of Shame

Ahmad Hanis Mohd Shabli(✉), Muhammad Haikal, and Suwannit Chareen Chit

Universiti Utara Malaysia, Sintok, 06010 Kedah, Malaysia
ahmadhanis@uum.edu.my

Abstract. Poor project management is a plaque that has adverse effects on productivity. In this paper, we address this problem specifically in monitoring final year project implementation by students at Universiti Utara Malaysia. A web application is designed and developed with the aim of improving final year project management by leveraging technology to enhance communication between students and teachers. The system is designed using a Waterfall-Scrum hybrid methodology, where we integrate the benefits of both approaches to ensure a comprehensive and optimum system performance. The web application integrates the Theory of Shame and is designed to address the challenges posed by poor final year project management in public schools in the United States, particularly in the post-pandemic era. This web application provides a framework for managing final year projects and highlights the importance of providing countermeasures and system design experience to create a complete, loosely coupled, and manageable system while placing expectations for project completion.

Keywords: Theory of Shame · Final Year Project · System Design · Web Application · Public Denunciation · Color Psychology · Web Mailing Technologies

1 Introduction

Sophisticated web applications is prevalent and highly used in many sectors. The onset of the COVID-19 pandemic in 2020 however, have led to the development of various applications [3], and in some sectors such as education, it has become more important than ever as sound monitoring practices strongly affects student's academic performance [4]. To date, many project management tools have been developed and used to monitor student progress in the final year projects. To the knowledge of the authors however, most of these tools do not incorporate 'guilt' into effective monitoring [5] thus rendering monitoring practices inefficient.

Procrastination is a significant issue in the context of both office and classroom setting in many institutions. This is driven by the fact that some individuals often believe

Z. Mat Aji et al. (Eds.): KMICe 2024, ISEM 49, pp. 256–270, 2025.
https://doi.org/10.1007/978-3-031-91485-0_20

that tasks can be "put off" until tomorrow or when they feel they are "in the mood" to perform specific tasks resulting in short-term gains over long-term consequences [4–6].

Universitas Jambi has pioneered on addressing this and implemented a system to monitor response time among staff using the Theory of Shame. The implementation of this monitoring system has improved response times among staff by 39.94% [6]. Realizing the importance of project management, Universiti Utara Malaysia adopted the concept implemented by Universitas Jambi to develop a system which monitors final year project implementation particularly in managing and measuring the student-lecturer response time.

In this paper, we describe how a system can be designed to monitor and measure the effectiveness of project monitoring using the Theory of Shame. Specifically, it addresses the following:

RQ1: How can the system be designed incorporating the practical and theoretical elements of the Theory of Shame?

RQ2: What instances can be taken from other Theory of Shame-related literature that can be implemented in the system?

RQ3: What steps should be taken to develop the web application with the prepared system design?

2 Related Works

Over the years, project management tools have been used extensively in various sectors. For context, this paper elaborates some of the project management tools used particularly within the educational realm.

2.1 Universitas Jambi Online Letter Management System

The Universitas Jambi Online Letter Management System was developed to monitor the response time of its staff when instructed to perform specific tasks. Incorporating the Theory of Shame [5], staff are required to complete their tasks within a stipulated time or risking the effects of "shyness and shame" if their names were mentioned in the online system for not completing the tasks given to them. As a result of this implementation, it was reported that staff performance grew by almost 40% [6].

2.2 Jira

Jira is an online project management tool developed by Atlassian tailored for developers using the Agile methodology to track their projects. Jira comes with extensive features such as bug tracking, issue tracking, and project management. Jira is widely used both for open-source and closed source projects [7, 8] Although touted as an effective tool, Jira does create urgency effects to its users as one sources suggests that some open-source project communities that uses Jira still have a varied average fixing time that ranges from 0.7 up to 600 days [7] with a similar distribution of issue type and priority.

2.3 Asana

Asana is a web application that is accessible online to organize and manage the course of a project, featuring collaboration and interaction among parties. Featuring a Kanban board and Gantt chart, it allows users to upload various attachments that is relevant to the task from various cloud services (I.E Google Drive, OneDrive, etcetera) to relevant tasks. The software also features communication tools, team management, and reporting tools [9].

The closest feature Asana can create a "sense of urgency" to users is the team management tool, where milestones and goals are reported to the superiors. This feature, though is useful, does not bring any significant effects to the team members but only the superiors.

3 Methodology

This section consists of two parts, namely defining this project's research methodology, and the software's development methodology.

3.1 Research Methodology

This paper adopts the rapid literature review method [10] as a stepping stone to determine the best approach to develop the web-based application. Often, the rapid literature review is conducted within a narrow timeframe with simplified procedures whilst maintaining a similar level of transparency and minimizing bias.

In this paper, the authors extracted and aggregated the keywords to the extent of using a thesaurus to provide synonyms of the keywords relevant to the study [10]. The digital libraries that were scrounged are ResearchGate, ACM Digital Libraries, IEEE Xplore, and JOIV. In total, 18 articles were retrieved.

3.2 Software Development Methodology

In this paper, the developmental process uses the Software Development Lifecycle (SDLC) principles, where software development typically engages in six core activities [11], namely, requirements analysis, system design, implementation, testing, deployment, and maintenance. The SDLC consists of many methodologies that can be used for development. Usually, each of these methods employs their own use cases. In this project, the hybrid methodology of Waterfall [11] and Scrum were chosen [2, 12].

The Scrum framework was chosen to conduct code reviews during the coding process to ensure the standardization and quality of the code. This approach aims to keep the system lightweight and modular as possible. In contrast, the Waterfall model is employed to maintain a clear scope and to prevent revisions or scope changes, as it needs to stay focused on its original objectives and avoid scope creeping. In this paper, we cover the first two phases of the software development, which consists of requirement analysis and system design.

4 System Architecture

In the realm of software development, the SDLC outlines the steps for the creation of the software, and as stated in the previous section, the requirements planning and system design of this web-based application [12]. In this paper, we cover the first two phases of the software development which consists of requirement analysis and system design.

4.1 Defining the Scope

Carrying out final year projects (FYP) is one of the most quintessential parts of an institution to determine whether a student have proven its worth to graduate. According to Bozdo and Banushi [4], 72% of students in the United States were affected by the Covid-19 pandemic and this is attributed to the behavior of the students particularly when they were tasked to conduct projects. As a result, students were reported to under-perform in many instances of their studies.

Similar trend is also observed in many parts of the world, and Malaysia faces similar circumstances as well. Students were affected by the prolonged lockdown, hindering their learning process. Web access to a myriad of learning application provided a means to facilitate learning. As such, web applications prove to be more beneficial than other platforms due to their ease of access and user-friendliness [13, 14], as most electronic devices in the modern world have a built-in web browser.

4.2 Generating the Use Case

The user requirements of this application must include a method to psychologically influence users. One of the elements is 'shaming' them for late submissions. Adapting the approach of Universitas Jambi, color codes such as red, yellow and green will influence the user to change their shy attitude without feeling pressured or overwhelmed [5, 6]. This can be featured on the implementation.

Most final year projects commonly tend to have three basic administrative roles namely the user, supervisor, and the assessor [15]. The activities that tend to include is supervision times, task managing, project managing, and exhibitions and workshops (general events). These final year project students usually come in batches according to their year of intake and are required to showcase their project at the end of their final year. In such cases where a conflict or disagreement occurs between a student and a supervisor due to incompatibility or other cases [16], a complaint feature can be put into the module to create a medium for a student and the coordinator to communicate if issues arose, these main features can be seen in the following the Fig. 1. Other than that, the essentials of a web-application need to have the essential authentication features which are login, register, and basic profile [14].

4.3 System Architecture Design

System architecture is essential for both the users and developers to produce and use fully cohesive software. This section will cover the system architecture, namely the database and design architecture that is used between the application programming interface (API) and database.

Database Architecture. A database collects, stores, and allows retrievals in a structured and efficient way [17, 18]. A database is essential in this web-based application as it stores various information. To design the database, a use case that has been created meticulously. As a result, the database can be narrowed down to six main tables, which are: User, AcademicStaff, Student, Event, Batch, Project, and Task. These tables are then added with additional tables to create a many-to-many relationship [18] and file storing to indicate various files within the system. Figure 1 shows the diagram of the full-fledged system.

Design Patterns. The most obvious and essential design pattern that can be conducted in this web-based application is a Singleton [21] connection from the API to the database. A Singleton is defined as a well-suited database object for managing shared resources, such as database connections, where having multiple instances could lead to issues like inconsistent data or resource overuse. By ensuring there is only one instance of the database connection manager, the Singleton pattern can help regulate the behavior of the application and prevent conflicts that may arise from multiple instances of the same class.

The Composite design pattern is a structural pattern that allows users to compose objects into tree structures to represent part-whole hierarchies. It lets clients to treat individual objects and compositions of objects uniformly [21] thus the Composite design pattern would be an excellent choice for structuring the main items in the Final Year Project system, such as events, projects, tasks, and batches.

Facades is another form of design pattern that can be utilized to count the shame points and provide the table on the client-side with information about the students who have been submitting late in the program. Façade is a structural design pattern to simplify many libraries, frameworks, complex set of classes and operations into one bundle [21].

Choosing the Optimum Stack. A tech stack, also known as a technology stack, refers to the combination of software tools, programming languages, frameworks, and other technologies used to develop and run an application or a product. Tech stack commonly consists of the client-side, server-side, database, and the web server.

In this paper, we have selected Postgres, Express, React, and NodeJS (PERN) stack as our base stack. Many attributes were chosen for the optimum stack with the consideration of time-constraints, budget, and scope of the project, justifications are in the following sections:

Client-side. React and Angular are both popular JavaScript frameworks used for building user interfaces, but they differ in their approaches and learning curves. React, developed and maintained by Facebook, is a library for building user interfaces. It uses a component-based architecture and a virtual DOM (Document Object Model) to efficiently update the user interface [22, 24]. React has a relatively lower learning curve compared to Angular, as it focuses on a specific aspect of web development and has a simpler syntax. Angular, on the other hand, is a full-featured framework maintained by Google. It provides a complete solution for building web applications, including features like two-way data binding, dependency injection, and a built-in router [23, 24] (Fig. 2).

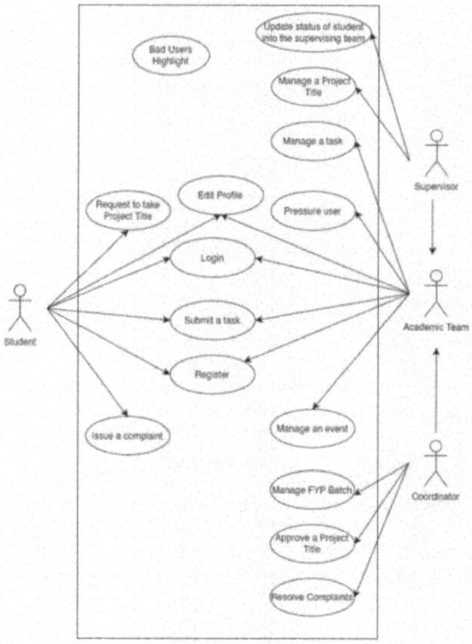

Fig. 1. Use case diagram of the web-based final year project management system.

Server-side. When considering the development of real-time applications that demand seamless communication and instant updates, Node.js emerges as the superior runtime compared to traditional server-side languages such as PHP.

This superiority is attributed to Node.js's event-driven, non-blocking I/O model, as well as its ability to leverage asynchronous programming and real-time communication capabilities. One of the key advantages of Node.js is its real-time environment, which is well-suited for building applications that require instant updates and real-time interactions [25] and in this case, it will be used to build the real-time table of shame. Node.js's event-driven architecture allows it to handle a large number of concurrent connections efficiently [25, 26], making it an ideal choice for real-time applications such as chat applications, real-time dashboards, and multiplayer games.

Web Server. After carefully comparing Nginx and the Apache web server, we have selected Nginx as the optimal choice for our project due to several key factors. Firstly, Nginx offers a more user-friendly and efficient approach when it comes to splitting the server into separate blocks. This simplifies the configuration process and makes it easier to manage different components of the system.

Additionally, our project requires setting up three distinct ports: one for the client-side interface, another for the API, and a third for the WebSocket functionality. Nginx's event-driven architecture and single-threaded event loop make it well-suited for handling this setup efficiently [14], ensuring smooth communication between these three critical elements of our application. In contrast, while Apache is highly configurable and supports a wide range of modules, its traditional process-based or thread-based approach may not be as optimal for managing the expected volume of concurrent connections in our project.

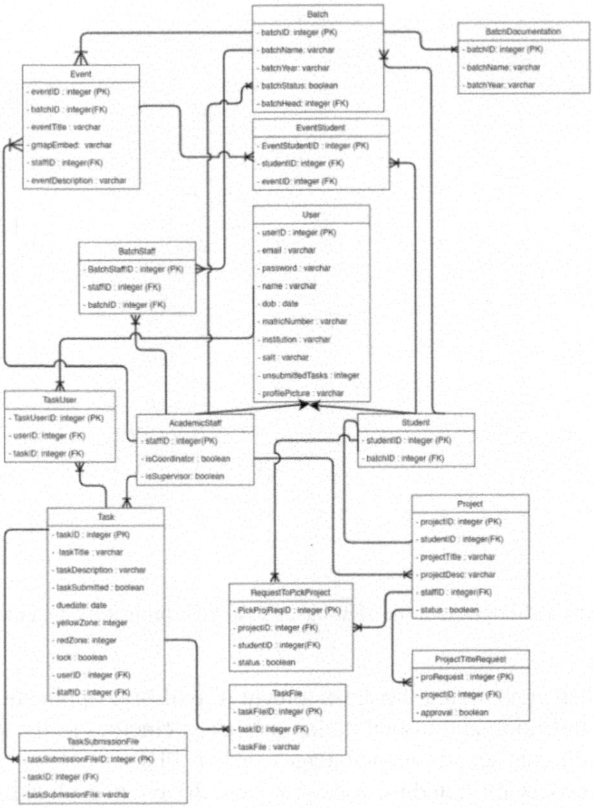

Fig. 2. Database diagram of the web-based final year project management system.

Nginx's superior performance in this area was a key factor in our decision to select it as the webserver for our application.

Database. When selecting a database for a project, the choice between SQL and NoSQL databases is crucial. In this case, Postgres, an SQL database, has been chosen as the optimal solution due to its lightweight nature and cost-effectiveness compared to NoSQL databases such as MongoDB.

One of the primary advantages of SQL databases is its lightweight architecture, which allows for efficient resource utilization and faster performance [17, 18] compared to many NoSQL databases. Moreover, Postgres is an open-source database, which means it is available for free, unlike many NoSQL databases that require significant financial investment. This cost-effectiveness is particularly beneficial for projects with limited budgets or those that need to scale up quickly without incurring high licensing fees.

5 System Development

With the system design done, the development commences with the previous instructions. The scope is set with the stack is set; the first step is the creation of a repository in GitHub [27]. After all of that is done, the method of development must be chosen whether to be top-down or bottom-up, in this section the top-down method was the optimal approach, choosing client-side to server-side.

5.1 Creating Client-Side Web Application

The creation of the client-side application consists of 4 parts, the creation of a landing page, dashboard for each user, and basic authentication screens (login and registration screen).

Landing Page. One of the essentials of an attractive landing page is a clear and compelling headline that communicates the unique value proposition and grabs the visitor's attention, an engaging sub-headline that supports the main message and addresses any objections, and an attention-grabbing hero image or video that aligns with the value proposition and instantly communicates the offer. The landing pages that were developed consist of the Hero (the premise of the application), the description of the application, the Hall of Shame, credits, and the footer (see Fig. 3 and Fig. 4).

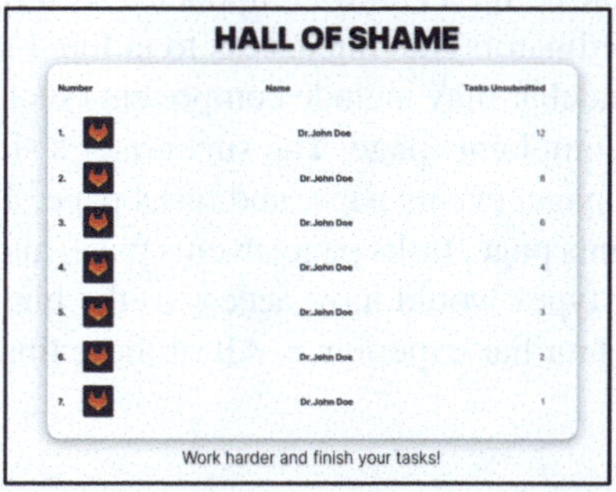

Fig. 3. Initial landing page prototype, bottom-section

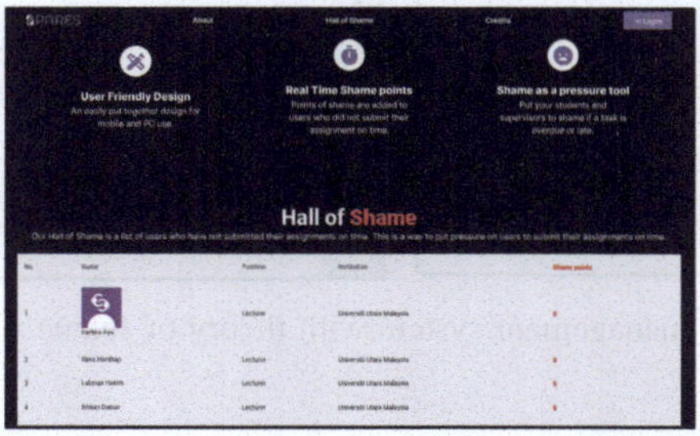

Fig. 4. Final landing page prototype, middle-section

The prototype is then given a facelift to improve its aesthetics as shown in Fig. 5. When designing a web application, the color scheme plays a crucial role in creating a visually appealing and user-friendly interface. Research suggests that dark color schemes can be more eye-friendly and provide a more comfortable viewing experience, especially in low-light environments or for extended periods of use [29, 30] Additionally, a dark theme can help reduce eye strain and improve readability, leading to a more enjoyable user experience.

The use of red, a color often associated with danger or warning, can effectively highlight the severity of the actions being displayed. This color choice, combined with a dark background, can create a striking contrast that draws the user's attention [30] to the shameful behavior. To further enhance the user experience, incorporating real-time functionality into the Hall of Shame table can provide a more engaging and informative experience. By updating the table in real-time, users can stay informed about the latest shameful actions and their consequences. This feature can also serve as a deterrent, as users may be less likely to engage in unethical behavior if they know it will be immediately visible to others, as a study from 2019 in Computers in Human Behavior looked at the relationship between social media use, fear of missing out (FOMO), and anxiety. They found that FOMO mediated the link between internet-use and anxiety symptoms [31]. This additional context can help users understand the gravity of the situation and the potential consequences for the individuals involved.

Creating the Dashboard. A similar approach to navigation and user roles can be employed. The sidebar can be the central component, with different user types (e.g., students, supervisors, coordinators) having access to tailored sets of features and functionalities. The student's sidebar may include components such as a projects page, tasks page, events page, and a complaints page. The supervisor's sidebar, on the other hand, could feature the projects page, events page, and tasks page. The coordinator's sidebar might include the complaints page, tasks page, events page, and a batches page. Despite these differences, all user types would have access to the main page and profile page, ensuring a consistent and familiar experience. All of these final UI design may be seen in the following figures.

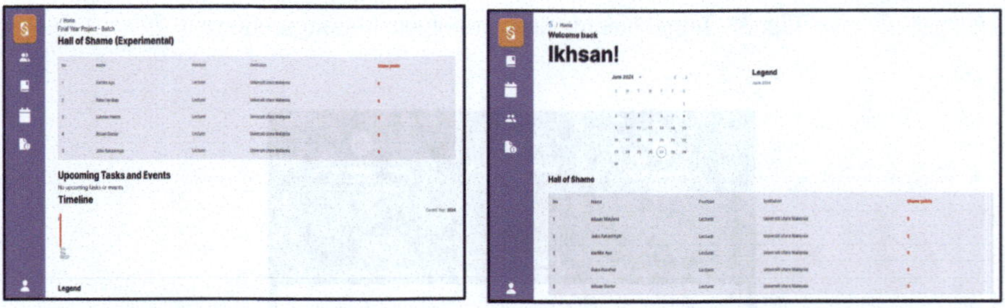

Fig. 5. Final year project management system with theory of shame's UI student & supervisor dashboard design.

The hall of shame still functions all the same, although with a different parameter where the one outside of the dashboard would not be depending on anything, the one that would be inside of each user's dashboard would be with their own institutions.

Each of the create, read, update, and delete (CRUD) functions are now created to be displayed. This paper aims to highlight the features of the task creation, and batch creation. The batch serves as the composite design pattern [21] where the hierarchy goes as Fig. 8.

Figure 6 demonstrates that a batch may include many projects, coordinator assigned tasks, and coordinator assigned events, and projects. The project is another composite [21] of other components which are the supervisor assigned tasks and the supervisor assigned events. The batch has also 3 states, which are preparation, ongoing, and finished. These states function as different points of times and limit what activities may be done or is undone.

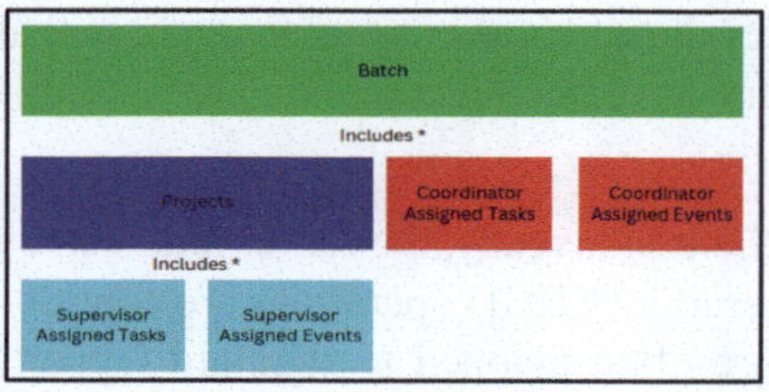

Table 1. Batch states and its permitted activities

	Preparation	Undergoing	Finished
CRUD Tasks	✓	✓	✗
CRUD Events	✓	✓	✗
CRUD Projects	✓	✗	✗

Fig. 6. The hierarchy of the batch component and CRUD.

Figure 6 also shows the full CRUD functions and activities that may be done during the all-batch states. During the preparation phase, supervisors and coordinators are granted the ability to CRUD (Create, Read, Update, Delete) tasks and events. Additionally, supervisors are permitted to CRUD projects, enabling them to create new projects, update existing ones, and delete projects if necessary.

Now, highlighting the task part, a task creation must include the creation of a zone-based warning phases, as Universitas Jambi has proven its ways with the creation of this system to reduce the number of late responses [6]. These zones are segregated into 3 categories, the yellow-zone, the red-zone, and the after-due zone (or the black-zone). The component has been created in a user-friendly way where lecturers may now see a Gantt chart providing the length of each zone (see Fig. 7).

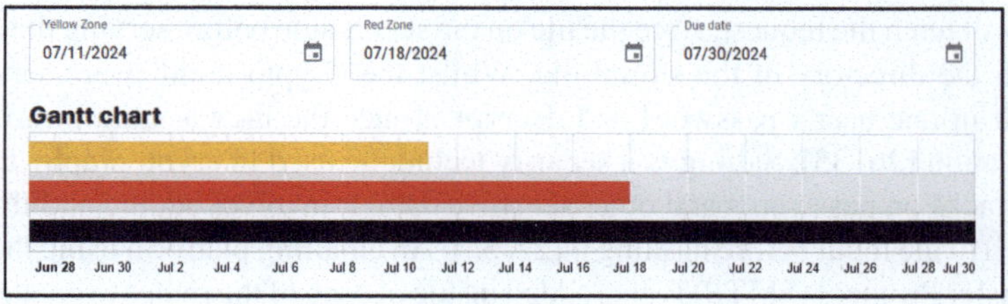

Fig. 7. Zones of a task.

Each zone having its consequences, Table 1 shows the zones and its consequences to the user who has submitted late.

Table 1. Task zones and their shame point consequences.

Zone	Points added
Yellow zone	1
Red zone	3
Black zone	5

Table 2 shows the shame points that will be added on each late tasks, of each task submitted late, the points will be added and in consequence the user's name will go up in the rank of the Hall of Shame, that may trigger the user's anxiety as an email will also pop up to send a notification to the user that they have been incremented depending on their zone [6, 31].

Creating Server-side Application. With the client-side ready, the server-side will consist of all the create, read, update, and delete operation. CRUD operations are commonly used in relational database applications and are often mapped to standard Hypertext Transfer Protocol (HTTP) methods, or Structured Query Language (SQL) statements. This section will cover the recorded steps on the creation of the fully functional server-side application for the Final Year Project Management Using the Theory of Shame web-application.

Database Connection. First, a Singleton [17, 21] design pattern is a standard to establish a connection between the server-side and the database. The Singleton in this instance will be created using a pool to create multiple running instances to not overload the database connection. A thread pool is a collection of pre-created and reusable worker threads that are managed by a pool to efficiently execute tasks. The main purpose of a thread pool is to optimize the use of threads and improve the performance and scalability of an application [25].

API Routes. The routes of the server-side application consist of 6 main routes: user, projects, tasks, events, and complaints. Thus, CRUD operations are simplified for a better developer experience.

The Multer middleware is utilized to handle file uploads sent via HTTP requests, where it will fetch the request, store the file on the server side before serving the database table with the directory of the stored file. Whilst the Crypto middleware serves as to create salt in one user's password and decrypt it once the user is wanting to login to the application [26, 35]. Salting is a security technique used in cryptography to protect against attacks on passwords and other sensitive data. It involves adding additional data (the "salt") to the input before hashing it [22, 35]. An emailing platform using the Simple Mail Transfer Protocol (SMTP) is also added inside of each of the routes to create mailing functions. This SMTP server is located outside of the main server-side application as it was an external tool.

Web Socket and Realtime Functions. To enable real-time functionality in the web application, the server-side API should be designed to support WebSocket connections. By integrating WebSocket capabilities within the `/ws` route, the server can establish persistent, bidirectional communication channels with clients, allowing for immediate data updates and real-time interactions for the hall of shame feature that is planned in the Systems Requirement Specification (SRS) document. An algorithm was also created to make a lightweight query ever 0.1 s to fetch 10 user ids at a time and check to calculate their shame.

This query is done inside of an infinite for loop inside of the server-client to make sure a seamless experience for the user on its shaming, denouncing them on each late submissions they take, where in consequence will influence them to be more active in their educational activities [1, 5, 6].

This cohesion of this code will ensure the effectiveness of the real time updates whilst keeping it low in the cost of memory and processor, suiting the needs of a low-budget server.

Integrating the Server-side with the Client-side and Keeping the Application Secure. At the heart of this integration lies the use of a.env (environment) file, which serves as the central repository for all encryption keys, secret URLs, and other sensitive configuration parameters. By storing these critical details in the.env file, the application can ensure that this sensitive information is kept secure and isolated from the main codebase. Crucially, the.env file should never be committed to a public repository, as it could expose the application's security credentials to potential attackers. This approach allows the application to decouple the database-specific settings from the core application logic, making it easier to manage and update the configuration without modifying the source code.

6 Discussion

6.1 Public Denunciation as a Tool

Public denunciation, within the context of this project management application, is leveraged as a psychological motivator to enhance user accountability. The concept is grounded in the theory of shame, where users are deterred from undesirable behavior due to the fear of public exposure [5, 31].

The Hall of Shame feature exemplifies this by publicly listing users who fail to meet deadlines or exhibit procrastination. This public display serves as a constant reminder of the consequences of their actions, effectively promoting more responsible behavior. The inclusion of real-time updates in the Hall of Shame table ensures that users are immediately aware of their standing, further reinforcing the deterrent effect. Additionally, providing detailed information about the individuals, such as their role and institution, adds a layer of context that magnifies the impact of the public denunciation.

6.2 Color Psychology Implying on a User's Shame

Color psychology plays a significant role in influencing user behavior and evoking feelings of shame. The design of the user interface (UI) incorporates specific color schemes to create a psychological impact on users who engage in undesirable behavior, such as submitting tasks late [30, 31].

The UI design includes a “Hall of Shame” table that is updated in real-time. This table prominently features users who have accumulated shame points due to late task submissions. The shame points are categorized into three zones based on the degree of lateness: yellow zone, red zone, and black zone. Each zone corresponds to increasing levels of severity, with the red zone and black zone indicating more severe delays. The accumulation of shame points results in the user’s name rising higher in the Hall of Shame, which can trigger anxiety and encourage better time management and adherence to deadlines. This approach leverages the psychological impact of color to not only make the UI more visually engaging but also to induce a sense of urgency and accountability [19, 30]. The emotional response elicited by the color red, combined with the public display of the Hall of Shame, serves as a deterrent to procrastination and encourages timely task completion.

6.3 A More Simplified System Architecture

The system architecture of the project management application is designed to be both robust and user-friendly, emphasizing simplicity and efficiency. The architecture includes a well-structured database that supports essential functionalities such as user management, project tracking, and real-time updates.

Key design patterns like the Singleton and Composite patterns are employed to ensure efficient resource management and streamlined user interactions. The decision to use Nginx as the web server is based on its ability to handle multiple ports and concurrent connections effectively, which is crucial for the application’s performance. By focusing on a clean and minimalistic design, the application reduces cognitive load on users, making it easier for them to navigate and utilize its features. This approach not only enhances usability but also ensures that the system can be easily maintained and scaled as needed.

Despite all these features, the web application experiences bloat due to the inclusion of TypeScript, which becomes cumbersome because of its restrictions and rules for static typing. This added complexity turns development into a form of type gymnastics, ultimately providing a less favorable developer experience.

7 Conclusion

In summary, the development and implementation of the web-based Final Year Project Management System using the Theory of Shame has demonstrated significant potential in enhancing student engagement and accountability. By leveraging psychological principles of shame and real-time updates, the system provides a unique approach to encouraging timely task submissions and active participation in educational activities.

Referencing the literature that have been reviewed, the strategic use of color psychology, particularly the use of red to denote severity and urgency, plays a crucial role in influencing a user’s behavior [6, 30]. The Hall of Shame, updated in real-time, publicly lists users who submit tasks late, categorizing them into yellow, red, and black zones based on the degree of lateness. This visual and public display effectively fosters a sense of urgency and responsibility among students, leveraging the psychological impact of color to induce feelings of shame and accountability [5, 6].

The Final Year Project Management System presents a comprehensive solution that not only potentially addresses the administrative needs of managing final year projects but also positively influences student behavior through innovative psychological techniques. Future work can focus on expanding the system's functionalities and exploring additional methods to further enhance student engagement and educational outcomes.

References

1. Lake, R.: Time is running out: COVID-19 set back older students the most, study finds, The 74, pp. 1–12 (2024)
2. Mirza, M.S., Choday, A.V., Datta, S.: Let's do feature driven Scrum. In: Proceedings of the 2019 3rd International Conference on Software and e-Business (ICSEB 19), Association for Computing Machinery, pp. 110–114 (2019)
3. Ratten, V.: The post COVID-19 pandemic era: Changes in teaching and learning methods for management educators. Int. J. Manag. Educ. **21**(2), 100777 (2023)
4. Bozdo, R., Banushi, X.: Post-pandemic students' behavior in HEIs (2022)
5. Martinčeková, L., Enright, R.D.: The effects of self-forgiveness and shame-proneness on procrastination: exploring the mediating role of affect. Curr. Psychol. **39**, 428–437 (2020)
6. Marzal, J., Budiman, W., Adrefiza, A., Hutabarat, B.F., X Kurniawan, V.: Improving the response time of online letter management application users: an application of social representation theory of shame. Int. J. Inf. Vis. **5**(1), 39–45 (2021)
7. Ortu, M., Destefanis, G., Kassab, M., Marchesi, M.: Measuring and understanding the effectiveness of jira developers communities. In: Proceedings of 2015 IEEE/ACM 6th International Workshop on Emerging Trends in Software Metrics (WETSoM), pp. 10–16 (2015)
8. Gori, E.: Full safe V4.5 implementation in Jira (2018)
9. Kamila, J., Marzuq, M.: Asana and Trello: a comparative assessment of project management capabilities. Int. J. Inf. Vis. **8**, 207 (2024)
10. Smela, B., et al.: Rapid literature review: definition and methodology. J. Market Access Health Policy **11**(1), 2241234 (2023)
11. Hossain, M.: Software development life cycle (SDLC) Methodologies for information systems project management. Int. J. Future Media Res. **5**(5) (2023)
12. Adi, P.: Scrum method implementation in a software development project management. Int. J. Adv. Comput. Sci. Appl. **6** (2015)
13. Geeks Insights: The benefits of using web-based applications | Geeks Insights. https://www.geeks.ltd.uk/insights/articles/the-benefits-of-using-web-based-applications. Accessed 26 June 2024
14. Farrukh, S.: Modern and responsive mobile-enabled web applications. Procedia Comput. Sci. **113**, 108–113 (2017)
15. Healey, M., Lannin, L., Stibbe, A., Derounian, J.: Developing and enhancing undergraduate final-year projects and dissertations. York: Higher Education Academy (2013)
16. Sarkar, A.: The learning experience of students in middlesex university business school (MUBS): why do they enjoy some modules/lectures and dislike others?, ResearchGate, (2013). https://www.researchgate.net/publication/237462428_The_Learning_Experience_of_Students_in_Middlesex_University_Busness_School_MUBS_Why_do_they_enjoy_some_moduleslectures_and_dislike_others. Accessed 26 June 2024

17. Date, C.J.: An Introduction to Database Systems, 8th edn. Pearson, Boston (2003)
18. Silberschatz, A., Korth, H.F., Sudarshan, S.: Database System Concepts, 6th edn. McGraw-Hill, New York (2011)
19. Nielsen, J.: Usability Engineering. Morgan Kaufmann (1993)
20. Eeles, P., Cripps, P.: The process of software architecting. Addison-Wesley Professional, 2009. Dive into design patterns, refactoring guru. https://refactoring.guru/files/design-patterns-en-demo.pdf. Accessed 26 June 2024
21. React: React - A JavaScript library for building user interfaces. https://reactjs.org/. Accessed 26 June 2024
22. Angular: Angular - one framework. mobile & desktop. https://angular.io/. Accessed 26 June 2024
23. Ramaswamy, S.: React vs. Angular: which javascript framework is better for your project?, DigitalOcean. https://www.digitalocean.com/community/tutorials/react-vs-angular. Accessed 26 Jue 2024
24. Dahl, R.: Node.js: Asynchronous, event-driven JavaScript runtime. In: Proceedings of the ACM SIGPLAN International Conference on Object-Oriented Programming, Systems, Languages, and Applications (OOPSLA), p. 1 (2012)
25. Lerner, R.M.: At the forge: node.js. Linux J. **2011**(209), 1 (2011)
26. GitHub. (n.d.). https://github.com/
27. Meyer, D.: 10 essential elements of a high-converting landing page, LinkedIn (2021). https://www.linkedin.com/pulse/10-essential-elements-high-converting-landing-page-dustin-meyer. Accessed 27 June 2024
28. Gong, M., Tarasewich, P.: Guidelines for handheld mobile device interface design. In: Proceedings of DSI 2004 Annual Meeting, pp. 3751–3756 (2004)
29. Elliot, A.J., Maier, M.A.: Color psychology: Effects of perceiving color on psychological functioning in humans. Annu. Rev. Psychol. **65**, 95–120 (2014)
30. Wegmann, E., Oberst, U., Stodt, B., Brand, M.: Online-specific fear of missing out and Internet-use expectancies contribute to symptoms of Internet-communication disorder. Addict. Behav. Rep. **5**, 33–42 (2017)
31. Nielsen, J.: Usability 101: Introduction to Usability. Nielsen Norman Group (2012)
32. Norman, D.A.: The Design of Everyday Things: Revised and Expanded Edition. Basic books (2013)

AduDUN@Kedah: A Digital Solution for Handling Complaints at Government Agencies

Fauziah Baharom, Shafinah Farvin Packeer Mohamed, Nur Farhana Shahul Hameed, Nur Mursyidah Shahul Hameed, and Haslina Mohd(✉)

School of Computing, Universiti Utara Malaysia, Sintok, 06010 Kedah, Malaysia
{fauziah,haslina}@uum.edu.my

Abstract. The current complaint handling methods are facing inefficiency issues such as difficulties in complaint submission and complaints status tracking by citizens. Additionally, the State Legislative Assemblies (ADUN) members in Kedah, Malaysia, face difficulties in performing comprehensive analysis regarding the complaints. Thus, the AduDUN@Kedah system, a web-based platform, was developed to optimize the management of citizen complaints directed towards the ADUN. Employing the Rapid Application Development (RAD) methodology, this study started with the analysis and quick design and ended with the deployment. The study engaged a total of 33 participants to evaluate the system, which assessed the system's usability in terms of usefulness (85%), ease of use (85.8%), learnability (84.2%) and overall satisfaction (89.2%) using a structured questionnaire. These findings underscore the system's readiness for broader implementation, highlighting its potential to enhance the efficiency and transparency of complaint management processes within government agencies, specifically State Legislative Assemblies (DUN) in Kedah. The citizens of Kedah could direct their complaints through this system and track them more systematically and efficiently. Theoretically, this study contributes to understanding the system requirements and design models for e-complaint systems. Future work for this system development could focus on enhancing data visualization to improve user engagement and expanding the platform to all Malaysian states to establish a one-stop center for citizens to voice their complaints to their respective ADUN.

Keywords: Digital Solution · E-Complaint · Government

1 Introduction

State Legislative Assemblies in Malaysia, in Malay, known as Dewan Undangan Negeri (DUN), is the legislative branch of the state governments that play an essential role in the country's democracy by making laws, allocating resources and creating rules which impact citizens' lives. DUNs are central hubs for addressing Malaysian citizens' needs and complaints. To maintain a strong democracy, there should be an easy way for citizens

Z. Mat Aji et al. (Eds.): KMICe 2024, ISEM 49, pp. 271–285, 2025.
https://doi.org/10.1007/978-3-031-91485-0_21

to share their problems and seek assistance from their elected representatives. In this digital era, a government agency must be effective and responsive in providing good services to the community. One form of public service is receiving and handling complaints, in which the government must provide adequate facilities and infrastructure to receive and manage complaints promptly and effectively [1]. A robust complaint management system is particularly critical in Malaysia. The Malaysian Public Complaints Bureau (PCB) reports a steady increase in citizen complaints over recent years. Statistics from the PCB show an essential rise in complaints between 2021 and 2022, with 10,455 complaints received in 2021 and a surge to 16,652 complaints in 2022 [2]. These complaints cover various areas, such as government services, public utilities and law enforcement, emphasizing the urgent need for a more efficient, comprehensive and responsive system [2]. The methods of complaint management may be traditional (paper-based) or digital (electronic) [3].

The current methods involve the citizens having to visit government offices or address their problems through inconsistent tools such as Google Forms, which often results in ineffectiveness and inefficiency in receiving and handling complaints. According to Komarudin [1], the authorities still do not appropriately handle or even ignore many complaints. Besides, with the current methods, citizens often have to wait a long time due to a complicated process to get proper complaint handling, and sometimes they do not know the status of their complaints. This lack of a systematic approach delays addressing citizen concerns, causing frustration and inefficiency in government services. Efficient complaint management systems enhance organizational performance by providing a channel for customers to voice their concerns, provide feedback and hold organizations accountable for their actions [4].

Handling complaints is a complex process due to the diverse nature of issues and the involvement of various government departments. A systematic and integrated system is needed to assign complaints efficiently to the relevant departments for resolution. Besides, managing complaints manually is inefficient in overcoming these problems. Aziz et al. [5] designed the ADUN e-Community Portal to improve the citizen-government connection for Pulau Pinang state. However, this system did not include assistance for citizen welfare and feedback features, which are essential. Moreover, "Sistem Pengurusan Aduan Awam Kerajaan Negeri Kedah" (SISPAA KEDAH) was developed to enable citizens to submit various feedback types, including complaints, suggestions, appreciation and inquiries [6]. Nevertheless, the system does not focus on the specific-based DUN and is less user-friendly regarding mobile view.

Furthermore, Alve et al. [7] focused on developing a system that emphasizes reducing paperwork, providing better insights into problems, easy tracking of complaints, locating problem areas in the organization, efficient resource utilization, minimized processing time, effective record management, ease of access and showcasing the organization's concern the users. However, this system lacks a comprehensive analysis of the complaint data. On the other hand, a data-driven approach is crucial for making informed decisions regarding the complaints issued by the citizens as it enables government agencies to use the complaint data to identify recurring issues and long-term trends. An example of a data-driven system by Cai et al. [8] is used in the smart healthcare system to make decisions, analyze a significant amount of health data in various forms and provide users with meaningful information that enables them to facilitate clinical diagnosis and treatment to discover their business values and insights. The DUN also faces the above-mentioned problems in managing the complaints in Kedah.

Consequently, a web-based complaint management system known as AduDUN@Kedah: E-Complaint was developed to bridge the gap by streamlining the complaint management process and providing the necessary tools for DUN representatives to conduct in-depth analyses. This system also addresses the challenges posed by the rising number of citizen complaints, simplifies the process of lodging and tracking complaints, and provides analysis tools to enhance the efficiency and transparency of governance in Malaysia, particularly with the DUN. The system modernizes and improves the complaint management system, creating a user-friendly, efficient and transparent platform that promotes better engagement between the government and citizens. Additionally, system efficiency can be increased by incorporating knowledge management principles into complaint management systems. The system helps DUN representatives make decisions by methodically collecting, storing and analyzing complaint data. This allows the system to offer insightful information for continual improvement. Through proactive problem solving and more efficient resource allocation, knowledge management techniques enable the pattern identification of complaints. More importantly, by developing the system, this study contributes to understanding the system requirements and design models for e-complaint systems, which may be applied to creating a comparable system. This paper discusses the design, development and evaluation of the system.

The outline of the paper is as follows: Sect. 2 describes the literature review, followed by the case study profile of AduDUN@Kedah in Sect. 3. Next, the methodology is explained in Sect. 4. Section 5 presents the findings and discussions, covering the design, development, and usability evaluation results. The final section concludes the study and discusses future work for the system.

2 Literature Review

This section serves as an overview of existing research and literature relevant to the study, establishing a solid foundation for the significance and necessity of the work. The discussion starts with digital transformation and continues with the previous works regarding e-complaints. At the end of this section, the usability evaluation is discussed.

2.1 Digital Transformation

The digital transformation significantly impacts the administration of citizen complaints in the state of Kedah, which gradually changes how complaints are handled. "Digital transformation" refers to the institutional changes made in the government sectors to enhance government agencies' performance by advancing digital technologies and the Internet. Improving the quality of public administration and services is crucial to strengthening efficiency [9]. Complaints made by the citizens are essential to improve the services offered by the government. With technology-enabled services, citizens can create and monitor complaints more quickly and efficiently. According to a study by Alex and his team, traditional complaint systems frequently have longer response times, consume unnecessary manpower and are inefficient [10]. The absence of a database, insufficient security, poor performance and inaccurate data are significant issues. These issues result in missing or manipulated documentation and inconsistent complaint tracking. Therefore, there is a need for a digital transformation in complaint handling to overcome these issues. Other researchers have faced similar situations, as explained in Sect. 2.2. A digital system would make accurate record-keeping, improved safety, better performance, and effective complaint handling possible [10]. This highlights the significance of adopting effective e-complaint systems to enhance consumer satisfaction and service quality. A similar situation is faced by the ADUN in Kedah state. Therefore, the ADUN of Kedah state is gradually utilizing digital technologies by adapting the manual process of handling complaints and other services. This process is part of the digital transformation journey [11].

2.2 Previous Works on E-Complaints

Citizen engagement is crucial for governments, as it ensures that citizens enjoy their fundamental human rights and have the ability to participate in government. As a vital component of engaging with citizens, many countries have implemented complaint management solutions that allow citizens to submit complaints. However, according to Pathmanathan and Poulier [12], government institutions often encounter difficulty handling complaints systematically. The citizens commonly address complaints such as potholes, flooding, faulty street lights and fallen trees via social media. They usually take photos or videos as evidence of the problem and upload them to social media.

This practice introduces problems such as the inability to readily track the history of issues due to a lack of centralized record storage. Additionally, duplication of work arises when multiple citizens submit the same complaints, which are handled separately by different officials because of manual filing systems. Another significant problem is the difficulty in tracking the status of a complaint in real time. Thus, citizens are unclear if the issue has been addressed unless they contact government agencies. These limitations have caused low levels of transparency in government operations [12].

E-complaint is a web application used to gather information from citizens regarding complaints and suggestions, aiming to increase the transparency and accountability of local governments. Once the complaints are captured in a centralized system, government officials can analyze them, forward them to the relevant departments, and monitor their status in real time. Citizens are also given access to this system, enabling them to track the progress of their complaints. E-complaints have been utilized in various sectors of government and private agencies. Among them are e-complaint system for police [13], water supply [14], customer complaints [15], students' complaint [7], garbage management, water supply, electricity management, road repairs or layering of roads and threatening of animals [16]. Two previous works are closely related to this study concerning e-complaints that have been used as the base study: Aziz et al. and Alve et al. [5, 7], as well as the SISPAA KEDAH system [6].

Aziz et al. designed the ADUN e-Community Portal for Pulau Pinang to improve the citizen-government connection [5]. The project's primary objective is to develop an efficient and user-friendly web-based platform that can eliminate the need for citizens to physically visit government offices to file complaints [5]. Like AduDUN@Kedah, the system currently supports three types of users: admin, ADUN, and regular users. However, there are some differences, such as the absence of assistance for citizen welfare, compared to the more comprehensive features offered by AduDUN@Kedah.

Alve et al. developed a web application to improve complaint tracking in organizations [7]. The system sorts the complaints and forwards them to the appropriate authorities for resolution [7]. However, it has certain limitations, with only a primary user interface and limited complaint status options ('pending' and 'complete') [7].

The SISPAA KEDAH system, a complaint management system, has been launched by the Kedah State Government [6]. This system enables citizens to submit various types of feedback, including complaints, suggestions, appreciations, and inquiries, to the relevant authorities [6]. However, the system does not focus more on the specific-based DUN since citizens must select the appropriate agency when filing a complaint. Moreover, the system is less user-friendly in terms of mobile viewing.

These three primary resources have been compared based on the system's main functionalities, as depicted in Table 1. It indicates the functionalities adapted in AduDUN@Kedah system, which are user registration, complaint submission, tracking and complaint resolving. Besides, some limitations have been addressed in the proposed system, which includes a specific-based ADUN system and feedback feature. Moreover, AduDUN@Kedah offers an additional feature, which is the citizen assistance hub. This feature enables citizens to apply for aid from the state regarding health, welfare, life

Table 1. Comparison of related work with AduDUN@Kedah.

Requirement	ADUN e-Community Portal	Web Application for Complaint Tracking and Resolving	SISPAA KEDAH	AduDUN @Kedah
FUNCTIONAL REQUIREMENTS				
Specific-based ADUN	✓	✗	✗	✓
User Registration	✓	✓	✓	✓
Complaint Submission	✓	✓	✓	✓
Complaint Tracking	✓	✓	✓	✓
Complaint Resolving	✓	✓	✓	✓
Citizen Assistance Hub	✗	✗	✗	✓
Data Analytics	✓	✗	✓	✓
Feedback	✗	✓	✓	✓
NON-FUNCTIONAL REQUIREMENT				
User-friendly Interface	✗	✗	✗	✓

sustainability and education. This feature is not offered by any of the other compared systems. On top of that, AduDUN@Kedah also provides a data analytic feature.

2.3 Usability Evaluation

Usability evaluation is a well-defined sequence of tasks that involves collecting information about end-user engagement with an application to discover usability concerns [17]. It is a critical issue that affects willingness to use a system [18]. The user experience with the system will determine the usability evaluation's success or failure. According to ISO 9241-11, usability is "the extent to which specified users can use a product to achieve specific goals with effectiveness, efficiency and satisfaction in a specified context of use" [19]. This study adapted the USE questionnaire proposed by Lund [20]. It is used to evaluate the usability of a system's user interface. The questionnaire contains 30 items divided into four sections: usefulness, ease of use, ease of learning and satisfaction. A seven-point Likert scale was used for these items.

3 Case Study Profile

AduDUN@Kedah is a web-based complaint management system developed to streamline the citizen complaint submission, tracking and analysis process within DUN in Kedah, Malaysia. The system requirements and design were generated using the Unified Modelling Language (UML) and developed using the Rapid Application Development (RAD) model. Figure 1 shows the architecture of AduDUN@Kedah. It encompasses four components: user, module, database and data analytics. The user component comprises the complainant, officer, ADUN, admin and village head council (MKK). The module component includes a user profile, complaint, aid, announcement and feedback. The database module includes user, ADUN, agency, complaint, aid and announcement. Lastly, the data analytic module consists of complaint categories, complaint categories based on district, resolution time and user satisfaction. These components interact with each other to provide an effective and efficient e-complaint management system.

4 Methodology

This application was developed using Rapid Application Development (RAD), a highly adaptive and iterative software development methodology that prioritizes close communication between developers and end-users. It also focuses on rapid prototyping. Suryasari et al. state that the RAD technique is less time-consuming since it focuses on one development requirement at a time [21]. Furthermore, by involving users early in the development process, RAD reduces risks [22]. This methodology has four phases: analysis and quick design, prototype cycles, testing and deployment. Figure 2 shows the RAD phases. These phases are explained next.

In the analysis and quick design phase, requirements were gathered through a survey with 35 potential users in Kedah, where they were asked close and open-ended questions regarding their opinions of the current complaint handling methods, the types of

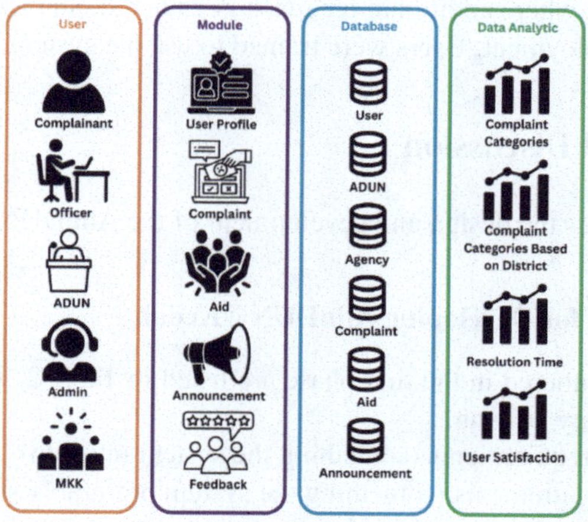

Fig. 1. AduDUN@Kedah Architecture.

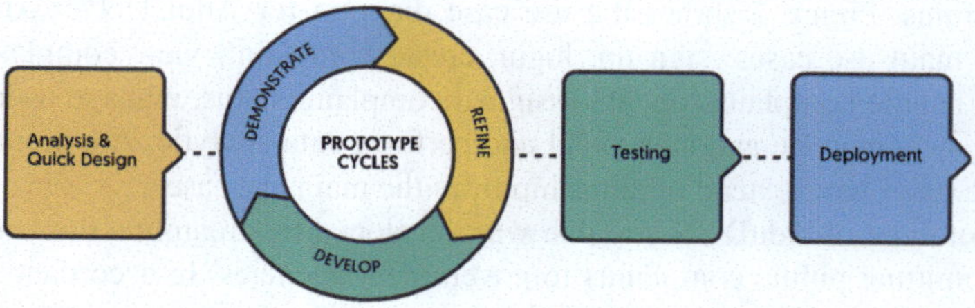

Fig. 2. Rapid Application Development Model (RAD).

complaints they were faced with, the limitations of the systems as they currently stand such as manual processes and the features they would want to see in a new system. Next, a comprehensive review of existing e-complaint management research and web applications was performed to support the survey data. Research papers were searched from online databases. Furthermore, a critical analysis of current systems, including SISPAA KEDAH [6], the ADUN e-Community Portal [5] and the Web Application for Complaint Tracking and Resolving [7], was conducted. This study revealed the existing systems' capabilities and constraints and contributed to creating the requirements list for AduDUN@Kedah. Next, the design was created using UML diagrams.

During the prototype cycles phase, low-fidelity prototypes were developed, demonstrated to stakeholders and refined based on their feedback. This ensures that the system is aligned with users' expectations and requirements. These prototypes were then converted into a working system. The system's front end consists of HTML, CSS and JavaScript, while Python and the Django framework power the back end. phpMyAdmin is used to administer the database. The testing phase involved field testing conducted with users to ensure usability. Three ADUNs and 30 citizens from Kedah evaluated the system through a structured online questionnaire. Finally, during the deployment phase, the system was hosted on the Internet to allow users to access it. The deployment phase is ongoing in a staging environment where additional feedback is gathered, and adjustments are made before full-scale deployment. Users were trained to use the system.

5 Findings and Discussion

This section describes the design and development of the AduDUN@Kedah system.

5.1 Requirements for Developing AduDUN@Kedah

The requirements gathered in the first phase are listed in Table 2. The following terms are used in the priority column.

M – mandatory requirements (something the system must do)

D – desirable requirements (something the system preferably should do)

O – optional requirements (something the system may do)

The next step involves converting the system requirements into diagrams to visualize the system design using UML. The design tool used was Visual Paradigm to create UML diagrams. Figure 3 shows the use case diagram for AduDUN@Kedah. There are eleven main use cases: sign up, login, create complaint, view complaint, verify complaint, handle complaint, update assigned complaint status, manage user account, manage agency, manage announcement and perform data analytic. Additionally, there are five use cases have extend relationship from the main use cases.

The prototype of AduDUN@Kedah was developed to streamline the lengthy process of submitting public complaints to government agencies. In accordance with the requirements gathered, the system includes functionalities intended to improve public-government communication, utilizing the Internet as a communication tool, as stated by Merang et al. [14]. Figures 4 and 5 show the two interfaces of AduDUN@Kedah.

5.2 Usability Evaluation Result

The developed prototype underwent an evaluation to test its usability and adherence to requirements. This evaluation involved 33 participants, of which three were ADUN, and 30 were citizens of Kedah. Thirty participants completed a structured online questionnaire of seven Likert points, ranging from strongly disagree to strongly agree. Although the system has other user roles, such as MKK and officers, the usability evaluation focused on ADUN and citizens to prioritize understanding the interaction and feedback between these two key groups, reflecting the core objective of strengthening the connection between ADUN and the citizens. The evaluation focused on four aspects: usefulness, ease of use, learnability, and overall user satisfaction. Feedback from the real world was used to refine and optimize AduDUN@Kedah. Ahmad et al. stated that usability testing is necessary for software development companies to determine whether their products are usable [23].

According to Fig. 6, the majority of respondents (70%) strongly agreed that implementing AduDUN@Kedah saves time in complaint management. The findings indicate a significant level of agreement regarding the usefulness of AduDUN@Kedah in several aspects of complaint handling.

Table 2. List of requirements for developing AduDUN@Kedah.

No	Requirement ID	Requirement Description	Priority
	ADK_01	**SIGN UP**	
1	ADK_01_01	Users must sign up before entering the system by inserting their IC number, name, address, phone number, email address, security keyword and password	M
2	ADK_01_02	System can verify all the information users enter, whether it is valid or not	D
3	ADK_01_03	System will be able to display an error message for missing or incomplete information	D
4	ADK_01_04	System will be able to display a message if the user has already signed up	D
5	ADK_01_05	Users can choose the "Login" menu to cancel sign up	O
	ADK_02	**LOGIN**	
6	ADK_02_01	Users must log in using a valid IC number and password	M
7	ADK_02_02	System can verify for valid or invalid IC number and password	D
8	ADK_02_03	System will be able to display an error message for an incorrect IC number or password	D
9	ADK_02_04	System will be able to display a message if the user has not signed up	D
10	ADK_02_05	Users can recover the password if they forgot	O
	ADK_03	**CREATE COMPLAINT**	
11	ADK_03_01	Complainants shall create complaints by filling out the complaint's details such as title, selecting the category of the issue, description and attaching an evidence file	M
12	ADK_03_02	System must store the complaint details including user ID, complaint ID, date and time in the complaint database	M
13	ADK_03_03	System will be able to display an error message if complainants enter incomplete information or an empty field while creating a new complaint	D
14	ADK_03_04	Complainants can cancel to create complaints	O
15	ADK_03_05	Complainants may also choose to apply for aid if eligible, such as aid for senior citizens, single parents etc	O
16	ADK_03_06	Complainants may provide feedback on complaint resolution by choosing the rating scales such as fully satisfied, partially satisfied and not satisfied together with their comments	O
	ADK_04	**VIEW COMPLAINT**	

(*continued*)

Table 2. (*continued*)

No	Requirement ID	Requirement Description	Priority
17	ADK_04_01	Complainants shall view the list of their submitted complaints	M
18	ADK_04_02	System must give notification to the complainants every time the complaint status is updated via the system	M
19	ADK_04_03	Complainants may click the "Lihat" option to view the complaint's details and status of their complaints	O
	ADK_05	**VERIFY COMPLAINT**	
20	ADK_05_01	Officers shall view submitted complaints details including name, IC number, complaint category, title, description, supporting attachment, date and time	M
21	ADK_05_02	Officers can accept complaints	O
22	ADK_05_03	Officers can eliminate complaints	O
23	ADK_05_04	Officers can validate aid for aid applications submitted	O
	ADK_06	**HANDLE COMPLAINT**	
24	ADK_06_01	ADUN shall handle the complaints by clicking the "Pilih Agensi" or "Pilih MKK" option to assign the complaints	M
25	ADK_06_02	System must display agency information after ADUN clicks on the "Pilih Agensi"	M
26	ADK_06_03	System must display MKK information after ADUN clicks on the "Pilih MKK" option	M
27	ADK_06_04	ADUN shall assign the relevant agency or MKK to resolve the complaint for further action	M
28	ADK_06_05	ADUN must update the complaint status and enter the justification by clicking the "Kemaskini Status" option	M
29	ADK_06_06	ADUN can view the progress of the complaint updated by the relevant agency and MKK	O
30	ADK_06_07	System will be able to display an error message if ADUN enters an incomplete or empty justification field	D
	ADK_07	**UPDATE ASSIGNED COMPLAINT STATUS**	
31	ADK_07_01	Agency shall view assigned complaints	M
32	ADK_07_02	Agency shall update the status of the assigned complaint	M
33	ADK_07_03	System will be able to pop up a message on updating the assigned complaint status	D
	ADK_08	**MANAGE AGENCY**	
34	ADK_08_01	Admin must add a new agency by filling out details such as agency name, address, contact number and email address	M

(*continued*)

Table 2. (*continued*)

No	Requirement ID	Requirement Description	Priority
35	ADK_08_02	Admin can edit existing agencies	O
36	ADK_08_03	Admin can delete agencies	O
37	ADK_08_04	Admin can cancel adding, editing and deleting agency processes	O
38	ADK_08_05	System shall display an error message for incomplete or an empty field during the adding and editing process	D
	ADK_09	**PERFORM DATA ANALYTIC**	
39	ADK_09_01	ADUN must analyze complaint categories, user satisfaction and time taken to resolve each complaint	M
40	ADK_09_02	ADUN can generate reports on complaint categories, user satisfaction, and resolution time by clicking on the "Jana Laporan" button	O

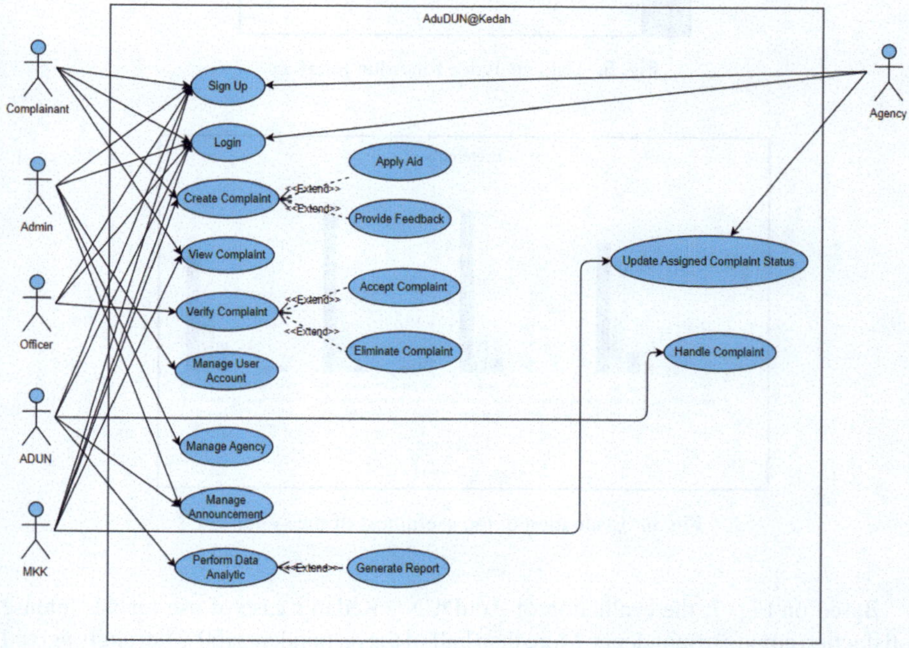

Fig. 3. Use case diagram of AduDUN@Kedah.

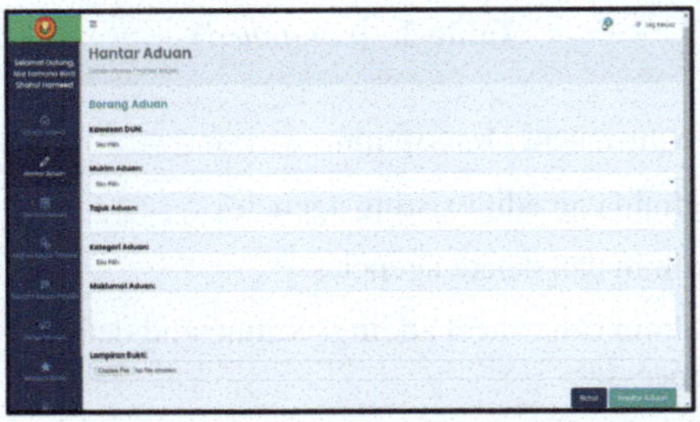

Fig. 4. Complaint form interface

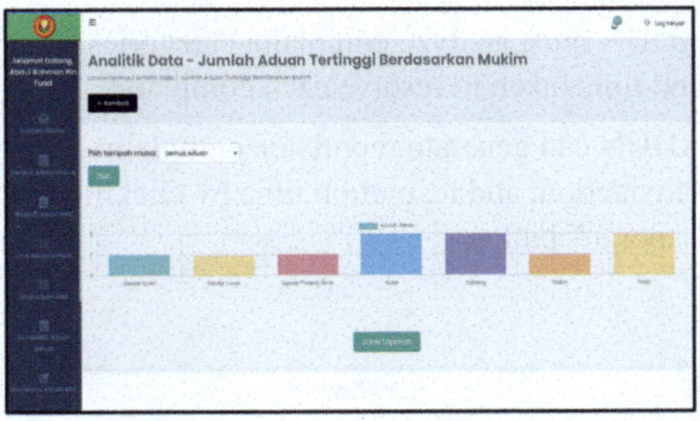

Fig. 5. Data analytics for Adun interface

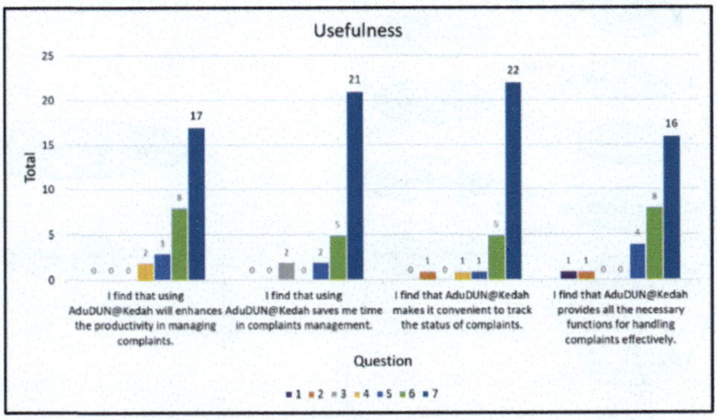

Fig. 6. Evaluation of the usefulness of the system.

Based on Fig. 7, the evaluation of AduDUN@Kedah's ease of use reveals intense satisfaction among respondents. More than half of the respondents (60%) strongly agreed that AduDUN@Kedah is easy to use. Despite some disagreements, these data highlight the system's achievement in providing a highly rated and user-friendly system.

Referring to Fig. 8, respondents have provided positive feedback on the learnability of AduDUN@Kedah. More than half of the respondents (56.7%) strongly agreed that they can learn the system's functions fast on their first try. These findings underscore AduDUN@Kedah's success in being easy to start, learn and contribute to a positive user

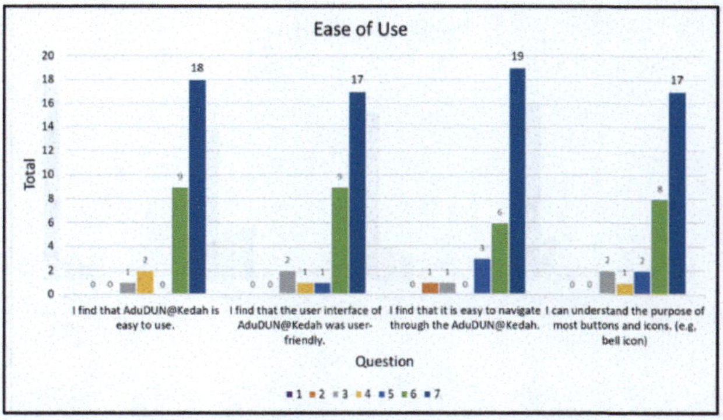

Fig. 7. Evaluation of the system's ease of use.

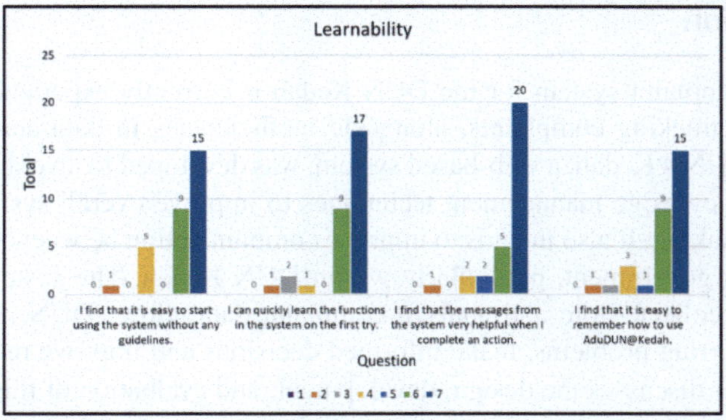

Fig. 8. Evaluation of the system's learnability.

experience. More importantly, Fig. 9 illustrates the evaluation of overall user satisfaction of the AduDUN@Kedah, revealing positive feedback. The majority of respondents (73.3%) strongly agreed that AduDUN@Kedah is the right platform for complaints management. These results demonstrate that AduDUN@Kedah is well-regarded for its overall user experience, useful features, design and suitability as a complaint management platform.

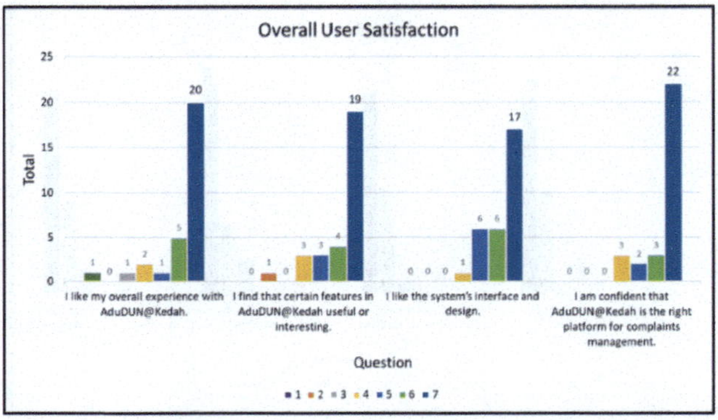

Fig. 9. Evaluation of overall user satisfaction with AduDUN@Kedah.

6 Conclusion

The current complaint system for the DUN Kedah is currently experiencing issues in managing and tracking complaints, alongside inefficiencies in data analysis. Consequently, AduDUN@Kedah, a web-based system, was developed to leverage data-driven insights and knowledge management techniques to improve overall system efficiency and decision-making. It also intends to improve communication between Malaysian citizens and their government, particularly within DUN Kedah. The system's ability to systematically collect, store and analyze complaint data allows DUN representatives to identify recurring problems, make informed decisions and improve resource allocation. This paper discusses the design, development, and evaluation of this system. The outcome of the evaluation reveals that the participants agree that the system is encouraging and promising. As a future work, data visualization for AduDUN@Kedah could be enhanced to facilitate better engagement with system users. Furthermore, it would be beneficial to expand this e-complaint system to all states in Malaysia, thereby establishing a one-stop centre and centralized platform where citizens can voice their complaints and grievances to their elected representatives (ADUN).

References

1. Komarudin, K.: Information system development for receiving and handling customer complaints in public services. J. Res. Community Serv. **4**(5), 1110–1115 (2023)
2. Biro Pengaduan Awam. https://www.pcb.gov.my/bm/infomedia/penerbitan/laporan-tahunan. Accessed 21 July 2024
3. Oguntosin, V., Oluwadurotimi, M., Adoghe, A., Abdulkareem, A., Adeyemi, G.: Development of a web-based complaint management platform for a university community. J. Eng. Sci. Technol. Rev. **14**(1), 150–159 (2021)
4. Ibitomisin, R., Oghenekaro, L.: Web-based integrated student complaint system. Int. J. Recent Contrib. Eng. Sci. IT **12**, 7–15 (2023)
5. Aziz, A., Baharudin, A.S., Karkonasasi, K.: ADUN e-community portal: e-complaints for community service. Int. J. Appl. Eng. Res. **11**(6), 4214–4218 (2016)
6. Pejabat Setiausaha Negeri Kedah. https://kedah.spab.gov.my/eApps/system/index.do. Accessed 21 July 2024

7. Alve, S., Babardesai, M.V., Bhosale, M.S., Kapadi, M.S., Yadav, A.B.: Web application for complaint tracking and resolving. Int. Res. J. Eng. Technol. (IRJET) **4**, 1396–1401 (2017)
8. Cai, Q., Wang, H., Li, Z., Liu, X.: A survey on multimodal data-driven smart healthcare systems: approaches and applications. IEEE Access **7**, 133583–133599 (2019)
9. Saksono, H.: Innovation hub: Media kolaborasi menuju pemerintahan daerah inovatif. Nakhoda: Jurnal Ilmu Pemerintahan **19**(1) (2020)
10. Alex, A.O.I., Usman, K., Ekwealor, O.U.: Design and implementation of a tertiary institution web-based student complaint management system. Int. J. Softw. Hardware Res. Eng. **9**(8) (2021)
11. Pittaway, J.J., Montazemi, A.R.: Know-how to lead digital transformation: The case of local governments. Gov. Inf. Q. **37**(4), 101474 (2020)
12. Pathmanathan, P., Poulier, R.: A comprehensive citizen engagement framework for effectively resolving public complaints in cities. In: 2017 International Conference on Smart Technologies For Smart Nation (SmartTechCon), pp. 1128–1133 (2017)
13. Ullah, A., Singha, T., Sarker, H., Pia, F., Hossain, A.: Citizen-centric complaint reporting and analyzing mechanism. J. Softw. Eng. Appl. **16**, 223–263 (2023)
14. Merang, E.K., Ibrahim, A.F., Jamaluddin, M.N.F.: Development of a web-based water supply complaint management system. J. Comput. Res. Innov. **7**(2), 211–222 (2022)
15. Afify, E.A., Kadry, M.A.: Electronic-customer complaint management system –a generic approach. Int. J. Adv. Netw. Appl. **11**(1), 4125–4141 (2019)
16. Sudhir, B., Regina, B.: Electronic complaint management system for municipal corporation. Commun. Appl. Electron. **2**(8), 8–11 (2015)
17. Kous, K., Pušnik, M., Heričko, M., Polančič, G.: Usability evaluation of a library website with different end user groups. J. Librarianship Inf. Sci. **52**(3), 096100061877313 (2018)
18. Hajesmaeel-Gohari, S., Khordastan, F., Fatehi, F., Samzadeh, H., Bahaadinbeigy, K.: The most used questionnaires for evaluating satisfaction, usability, acceptance, and quality outcomes of mobile health. BMC Med. Inf. Decis. Making **22**(1) (2022)
19. ISO Standards - usability partners. https://www.usabilitypartners.se/about-usability/iso-standards. Accessed 21 July 2024
20. Lund, A.M.: Measuring usability with the USE questionnaire. Usability Interface **8**(2), 3–6 (2001)
21. Suryasari, S., Wiratama, J., Desanti, R.I.: The Development of web-based sales reporting information systems using rapid application development method. Ultima InfoSys: Jurnal Ilmu Sistem Informasi **13**(2), 110–116 (2022)
22. Top 7 software development methodologies: Pros & cons, insights - web and mobile development services and solutions. https://www.netsolutions.com/insights/top-3-software-development-methodologies-pros-cons/. Accessed 21 July 2024
23. Ahmad, W.F.W., Sulaiman, S., Johari, F.S.: Usability Management System (USEMATE): A web-based automated system for managing usability testing systematically. In: 2010 International Conference on User Science and Engineering (i-USEr), pp. 110–115 (2010)

Hey Anak Malaysia! Jom Bermain: A Culture Cultivation Knowledge Educational Game

Nor Farzana Abd Ghani(✉) and Muhammad Ihsan Akif Anuar

Universiti Utara Malaysia, Sintok, 06010 Kedah, Malaysia
farzana@uum.edu.my

Abstract. Raising awareness about racial tension from a young age is crucial to preventing future issues. A practical method to address this is through educational games that teach racial identities. This paper explores the development of Hey Anak Malaysia! Jom Bermain is an educational game designed for school students in Malaysia to enhance their understanding of different ethnicities and promote cultivating cultural knowledge. The game uses interactive and immersive experiences to facilitate communication, reduce stereotypes, and foster empathy among diverse players. It creates a neutral space where players can collaborate and compete, helping to bridge ethnic divides. The study investigates the game's system requirements, evaluates its usability, and provides a reference model for future development. Planned future enhancements include adding a chat function for cultural interaction, introducing additional questions and game types to maintain engagement, and implementing a level feature to increase game difficulty and provide a sense of achievement. Games are shown to be a powerful tool for knowledge management, making learning engaging and interactive through elements such as rewards, achievements, and progress tracking.

Keywords: Educational Game · Culture Cultivation Knowledge · System Requirements for the Educational Game

1 Introduction

Malaysia, a multi-ethnic, multi-religious, and multi-lingual country, is a rich tapestry of cultural diversity. However, this diversity has also given rise to racial tensions that have occasionally disrupted social harmony [1]. Addressing and mitigating these tensions requires innovative approaches beyond traditional interventions, such as policy reforms and education [2–4]. One promising avenue is the development of games as tools for promoting social cohesion and reducing racial tensions [5].

Games, particularly those designed for educational or social purposes, can be powerful mediums for changing perceptions, building empathy, and fostering cross-cultural understanding [5]. Games can engage players in a way that fosters deeper reflection and personal growth by simulating real-world scenarios, promoting collaboration, and encouraging dialogue. In Malaysia, where stereotypes and misunderstandings sometimes perpetuate ethnic divides, a well-designed game could serve as a bridge for intercultural knowledge and communication.

Z. Mat Aji et al. (Eds.): KMICe 2024, ISEM 49, pp. 286–297, 2025.
https://doi.org/10.1007/978-3-031-91485-0_22

Racial tension awareness must begin young to prevent this issue from happening again. To overcome this problem, an effective method plays a significant role in teaching young or school students racial identities. For example, an educational game about teaching racial identities in Malaysia could be helpful in teaching and testing their knowledge of different ethnicities in Malaysia [4]. Games, as interactive and immersive experiences, offer unique opportunities to influence social dynamics. Through structured play and shared goals, games can facilitate communication, reduce stereotypes, and foster empathy among participants from different racial backgrounds. By creating a safe and neutral environment where players collaborate or compete, games encourage interactions that transcend ethnic divisions and build common ground.

This paper explores the potential of game development to reduce racial tensions in Malaysia by addressing cultural and social issues through strategic design. It focuses on the creation of Hey Anak Malaysia! Jom Bermain is an educational game prototype developed for students to enhance their knowledge of Malaysian ethnicities. The study investigates system requirements, evaluates the game's usability, and is a reference model for future developers and researchers. Following background research, the paper details the game's design and development, presents the usability testing results, and concludes with recommendations for future work.

2 Background and Related Studies

This section describes the background of the educational games, related studies investigating the effects of educational games on school students, and past projects with the same scope as Hey Anak Malaysia! Jom Bermain. An educational game is a software with a precise instructional aim and content that assists learners in completing learning objectives by engaging them in games [5]. Rather than rote memory or simple comprehension, educational games need planning, hypothesis testing, or problem-solving. A system of incentives and goals that drive players, a narrative framework that situates action and defines norms of participation, learning content relevant to the narrative plot, and interactive signals that stimulate learning and offer feedback are all features of such games [6]. Several studies have found that video games have a favourable impact on education compared to traditional teaching techniques. This is true in terms of motivation, attitudes toward learning, and knowledge gained [5].

Dynamic computer-based educational games have received widespread recognition as an efficient means of utilizing technology to build interactive and constructivist learning environments. Gamification can be used to encourage students' active involvement and investment in anything other than the academic goals of a class. Digital games are gaining popularity as a powerful tool for creating socially engaged and constructivist learning environments. Games can provide elements of interactivity, thus increasing knowledge and gaining motivation [6–8]. Integrating game elements such as rewards, achievements, and progress tracking makes knowledge management more engaging and motivating [6, 7].

Few studies and past projects with educational game elements have been evaluated, such as Keragaman Suku Bangsa dan Agama di Negeriku. The game was based in Indonesia (see Fig. 1). In this game, the player will learn about Indonesian culture, religion, and traditional dress. The game has three sections: a learning section, a game section, and an evaluation question section, as shown in Figs. 2, 3 and 4. This game features both a learning section and a game section. However, the learning section in this game is text-heavy, which may lead players to skip it and move directly to the game section. In contrast, it is more engaging when any game incorporates a mix of texts, images, animations, and videos in its learning section. It encourages players to absorb all the information provided [5]. In the Keragaman Suku Bangsa dan Agama di Negeriku game, the learning section involves interacting with images to advance, as shown in Figs. 2 and 3. Following this, players face evaluation questions or trivia quizzes, a game's central feature, as illustrated in Fig. 4.

Fig. 1. Keragaman suku bangsa dan agama di negeriku.

Another game, Umbara si Anak Indonesia, also focuses on Indonesian culture. The art and design of this game are much more appealing. This game has various game types that players can play; however, it lacks information or learning sections. The game has many levels for every kind of game to give players more challenges if the level goes beyond. There are three variety of types of games available in Umbara si Anak Indonesia (see Fig. 5). Each level of each type of game in Umbara si Anak Indonesia is shown using a familiar graphic, the Indonesia country map, with intended cultural information related to its distinctive provinces (see Fig. 6).

3 Methodology of the Study

Based on the insights gained from previous projects, we incorporated relevant elements and characteristics to align with the design and development of Hey Anak Malaysia! Jom Bermain. The study followed the rapid prototyping methodology, an iterative approach to

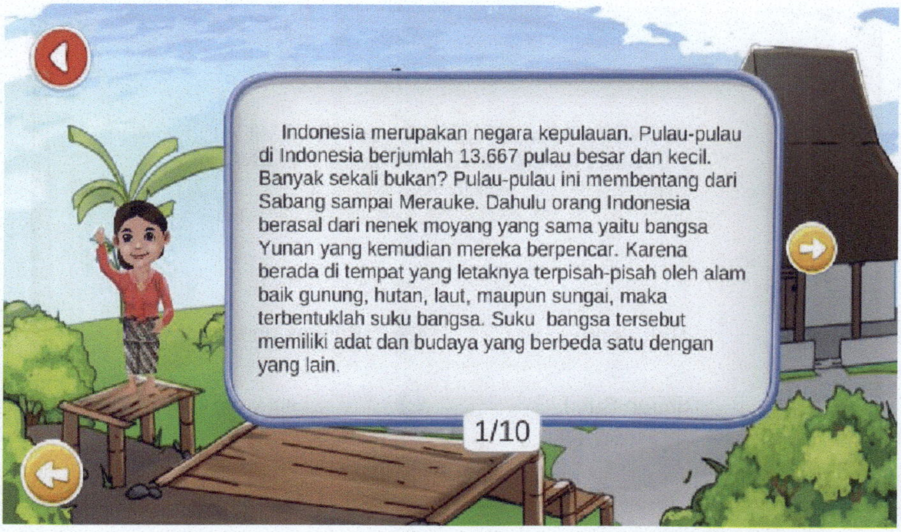

Fig. 2. Learning section in keragaman suku bangsa dan agama di negeriku.

Fig. 3. Game section in keragaman suku bangsa dan agama di negeriku.

software design that emphasizes quick improvements using regularly updated prototypes and short development cycles. The process involves four phases: requirements planning, prototype building, prototype testing, and prototype deployment. This cycle repeats until a high-fidelity prototype is achieved. In the requirements planning phase, research is conducted to gather the necessary content and requirements for the game. This stage also involves defining goals, creating storyboards, and establishing milestones and deadlines. The next phase, building the prototype, focuses on the game's actual development. The prototype was built for a desktop platform using the Godot game engine. This stage

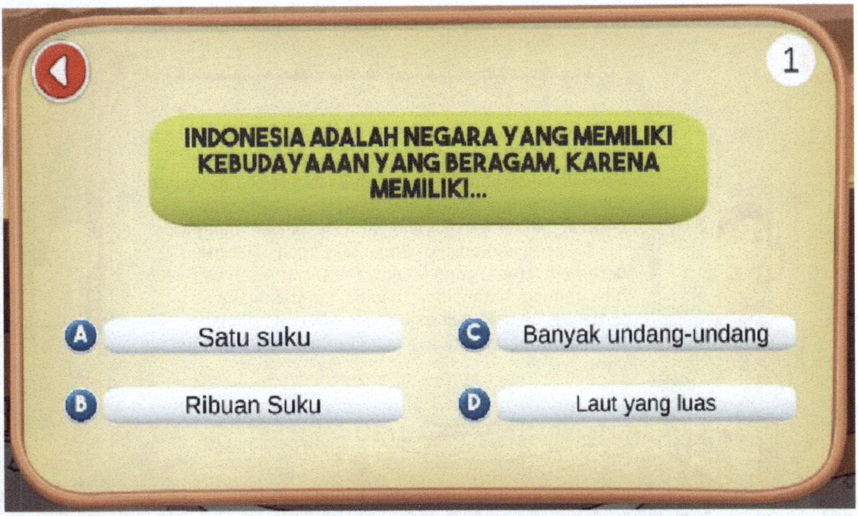

Fig. 4. Evaluation section in keragaman suku bangsa dan agama di negeriku.

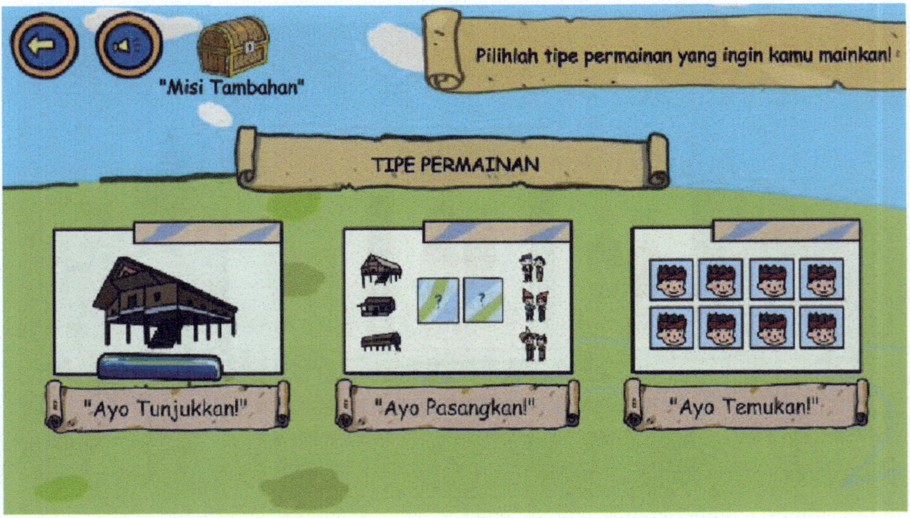

Fig. 5. Types of games in Umbara Si Anak Indonesia.

translates the design specifications from the planning phase into a visual form, ranging from low to high fidelity and may be interactive or non-interactive. This phase also develops critical elements such as game mechanics, the graphical user interface (GUI), animations, and sound. In the testing phase, the prototype is evaluated to identify and fix any bugs, ensuring the design aligns with the target users' needs. Selected users are chosen to test the game's playability and design. Upon successful testing, the prototype is exported as a web-based game. Finally, the deploying phase involves releasing the

Fig. 6. Levels of each type of game in Umbara Si Anak Indonesia.

prototype to gather feedback from players. This feedback is critical for improving the game and enhancing the user experience. The prototype is deployed on the website itch.io, where player feedback is collected to inform further game iterations. This iterative process continues until the prototype reaches a high-fidelity state, requiring no additional changes.

4 Design and Development of Hey Anak Malaysia! Jom Bermain

This section outlines the design and development of Hey Anak Malaysia! Jom Bermain, following the initial four phases of the rapid prototyping methodology. It is divided into two subsections: (1) the game's requirements and (2) the prototype development. Hey Anak Malaysia! Jom Bermain is an educational game created to reflect and demonstrate the requirements gathered during the process's earlier stages.

The requirement-gathering process was conducted through a questionnaire distributed via Google Forms. The form was shared with primary and secondary school students, and 20 students responded. The questionnaire gathered input on the types of games they would like to see in the game, the cultural aspects they would like to learn, desired game features, and preferred art styles. The feedback provided insight into the cultural knowledge elements they expected the game to incorporate. The students' opinions were recorded, and these inputs helped elicit the game's requirements. Additionally, the respondents were actively involved throughout the development phase, providing feedback and critiques as the prototype's interface was presented.

The requirements depicted in Table 1 were translated into system functionalities, followed by the visualization and modelling of the game using appropriate techniques and tools. This study employed the Unified Modeling Language (UML) to visualize and represent the system requirements. The UML diagram included a use case diagram,

Table 1. Requirements for Hey Anak Malaysia! Jom Bermain.

ID	Requirement Description	Priority
1	USER INTERFACE (UI)	
1.1	The game shall have a user interface as a medium of interaction between the player and the game. It consists of a set of screens and visual elements in the game	High
2	ENTER LEARNING SECTION	
2.1	The game shall have a learning section so the player can enter and learn new information	High
2.2	The learning section shall have image, video, and text information	High
2.3	The game will enable the player to skip to the next screen and exit the learning section	Medium
3	ENTER TRIVIA QUESTION	
3.1	The game shall have trivia questions to test the player's knowledge	High
3.2	The game will enable the players to choose the answers	High
3.3	The game shall have an exit function if the player wants to quit the trivia question	Medium
4	ENTER GAME SECTION	
4.1	The game shall have a section to make the players enjoy the game	High
4.2	The game shall have two game types: the Drag and Drop game and the Crossword game	High
4.3	The game will enable the players to choose the type of games available	High
4.4	The game shall have an exit function if the player wants to quit	Medium
5	ENTER LEADERBOARD	
5.1	The game shall have a leaderboard to show the player's names and scores	High
6	ENTER SETTINGS	
6.1	The game shall have settings to give players an option on whether to turn off or on the music and sound	Medium

which later explained the structural components of the game. The diagram was created using StarUML software. The use case diagram for the game represents the interactions between the actors and the various use cases (see Fig. 7). There are five primary use cases: Enter the Learning Section, Enter the Trivia Question Section, Enter the Game Section, View Leaderboard, and Enter the Settings. In the Enter Game Section use case, players can choose between two types of games: a Drag and Drop game and a Crossword puzzle. The Enter Trivia Question Section use case enables players to perform additional subfunctions, such as Save Name and Score, which can also be performed during the drag-and-drop and crossword games. Players can toggle sound and music on or off for the Enter Settings use case. By default, sound and music are turned on when the game starts.

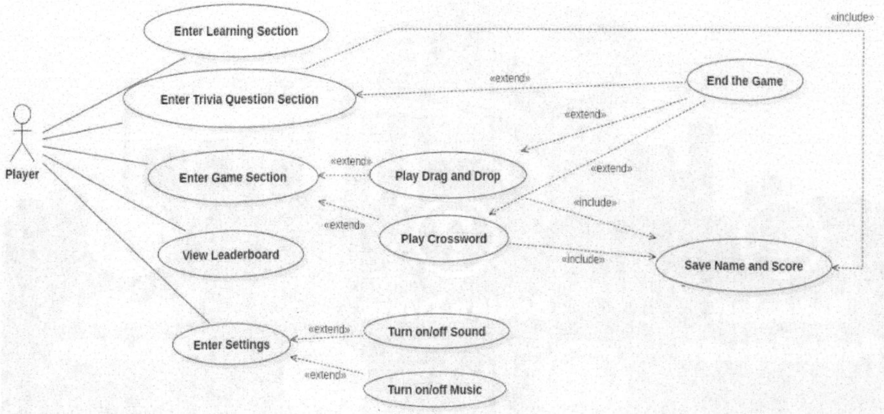

Fig. 7. The use case diagram of Hey Anak Malaysia! Jom Bermain.

The structural components of the game are later identified. In this study, 15 primary classes were identified, including Main Menu Interface, Learning Section Controller, Learning Section Interface, Trivia Question Controller, Trivia Question Interface, Congratulations Interface, Game Section Controller, Game Section Interface, Drag and Drop Interface, Crossword Interface, Leaderboard Controller, Leaderboard Interface, Leaderboard Table, Setting Controller, and Setting Interface.

5 The Prototype of Hey Anak Malaysia! Jom Bermain

A prototype of an educational game for school students was developed, incorporating the requirements outlined in the previous section. Software prototyping is a widely used method for showcasing software requirements, allowing developers to gather user feedback and recommendations based on their interactions with the prototype [6]. The Godot game engine served as the primary Integrated Development Environment (IDE), while Adobe Illustrator and Adobe Photoshop were utilized for level design. Screenshots of selected interfaces from Hey Anak Malaysia! Jom Bermain is depicted in Fig. 8, 9, 10 and 11.

5.1 The Evaluation Setting

The recruited participants are primary and secondary school students, as the game scope is for school students. A total of 30 school students were involved in this field testing. Fifteen students are from primary school (Sekolah Kebangsaan Kampung Baru), and another 15 are from secondary school (Sekolah Menengah Kebangsaan Alma). The evaluation method used in this field testing is an online usability test. Participants were given a Google Form to evaluate the game. Then, they must play the game according to the test cases included in the Google Form before doing the evaluation or post-task questionnaire. The type of evaluation used to evaluate the game is usability evaluation.

Fig. 8. The main menu.

Fig. 9. The trivia question section.

Usability Evaluation focuses on how healthy players can learn and play the game to achieve their goals. It also evaluates how satisfied players are with the game. Players must perform all test cases to proceed with the evaluation. There are 6 test cases that the participants need to perform to achieve the expected result. Then, there are two sections on a post-task questionnaire: Section A, Demographic and Background Information, and Section B, Usability Evaluation. Usability Evaluation included in this testing is the usefulness, the ease of use, the user interface, and the satisfaction of the Hey Anak Malaysia! Jom Bermain. There are a total of 23 questions of usability evaluation that

Fig. 10. Drag and drop game.

Fig. 11. Crossword game.

the participants need to answer based on a 5-point Likert scale (Strongly Agree, Agree, Neutral, Disagree, and Strongly Disagree). The players performed the following step-by-step procedure for the evaluation: (1) read and agree with the consent form, (2) open the game with the link given, (3) read and perform the test cases, (4) give feedback on each test cases, (5) leave a comment but it is optional, (6) answered the post-task questionnaire.

5.2 The Respondents' Demographic Information

Analysis of the respondents' demographic information revealed that 70% are male and 30% are female. Regarding age group, 50% of the respondents are between 7 and 12 years old, while another 50% are 13 to 17. No respondents are above 18 years old due to the field testing focusing on primary and secondary school students. For race, 70% of the respondents are Malay, while 23% are Indian. Only 7% of the respondents were Chinese recruited for this field testing. Regarding type of school, 50% of respondents are from primary school, and another 50% are from secondary school. 73% of respondents have not played educational video games before, and only 27% stated that they have played them. This is due to the participant's interest in playing a video game other than the educational one. This is for respondents who have experience or played a video game like Hey Anak Malaysia! Jom Bermain, most respondents stated no to this question at 83%. Only 17% of respondents have experience playing video games like the game we developed.

5.3 The Usability

An analysis was conducted on the respondents' responses in Section B, Usability Evaluation. The section measures the respondents' perception of Hey Anak Malaysia! Jom Bermain usefulness, ease of use, user interface, and satisfaction. The respondents rated four or five post-task scales for the four usability aspects. None of the respondents rated one or two and only a few rated neutral.

The evaluation results demonstrate that the game prototype is practical and user-friendly in knowledge management [7, 8]. Most respondents reported high satisfaction with the game and were willing to recommend it to others. Feedback analysis shows that the Learning, Trivia Question, and Game Sections were straightforward, easy to use, and beneficial for learning about Malaysian culture and ethnicities. This validates the study's objective of using the game as a tool for cultural knowledge cultivation and is supported by recommendations through findings related to recent research [7, 8].

Regarding knowledge management, the game's interface was praised for being appealing and well-designed, contributing to an engaging user experience. Respondents appreciated the game's capacity to enhance their understanding of cultural diversity, and elements such as rewards (Congratulations Interface), achievements (Leaderboard Interface), and progress tracking were noted to make the learning process more motivating and interactive. Thus, the game effectively utilizes game elements to manage and disseminate knowledge, making the learning experience enjoyable and impactful.

6 Conclusion and Future Works

This paper outlines the design and development of an educational game and illustrates how cultural knowledge cultivation can be effectively conveyed through gaming. Moving forward, several enhancements are planned for Hey Anak Malaysia! Jom Bermain. Future updates will include a chat function, allowing players to interact and learn about each other's cultures. New questions and game types will also be introduced to keep players engaged and prevent monotony. A level feature will also be added to escalate the game's difficulty as players progress, offering more significant challenges and a sense of achievement. Games are a powerful tool for knowledge management, enriching the learning experience, facilitating information exchange, and applying knowledge in practical contexts. Integrating game elements such as rewards, achievements, and progress tracking makes knowledge management more engaging and motivating, transforming learning into an interactive and compelling experience.

References

1. New Straits Times, https://www.nst.com.my/news/nation/2020/09/628320/pm-racial-tension-still-major-issue-malaysia. Accessed 22 Nov 2024
2. Muzaffar, C.: Ethnicity, ethnic conflict and human rights in Malaysia. In: Asian Perspectives on Human Rights, pp. 107–141, Routledge (2021)
3. Pauzi, N., Ghazali, S.R., Chen, Y.Y., Majani, A.F., Adenan, F., Manogaran, K.: School-aged bullying history, post-traumatic stress disorder (PTSD), and depressive symptoms: a study of university students in Malaysia. Asia-Pacific Soc. Sci. Rev. **23**(3) (2023)
4. Berns, A., Ruiz-Rube, I.: Use of digital game creation tools in computer assisted language learning: a systematic literature review. Digital Games Lang. Learn., 35–51 (2022)
5. Adarsh, K.S., Baby, G.S., Kumar, H., CK, H., Prasad, V.: Educational and immersive game development. J. Adv. Zool. **44** (2023)
6. El Mawas, N., Truchly, P., Podhradský, P., Muntean, C.: The effect of educational games on children's learning experience in a Slovakian school. In: 11th International Conference on Computer Supported Education. (2019)
7. Peterson, M., Yamazaki, K., Thomas, M., ÇEKİÇ, A.: Digital games and language learning (2021)
8. Friedrich, J., Becker,M., Kramer, F., Wirth, M., Schneider, M.: Incentive design and gamification for knowledge management J. Bus. Res. **106**, 341–352 (2020)

A Systematic Literature Review on Waste Data in Heterogeneous Data Integration

Muhammad Noor Syarifuddin[1,2](✉), Fauziah Baharom[1], and Haslina Mohd[1]

[1] Universiti Utara Malaysia, Sintok, 06010 Kedah, Malaysia
m_noor@ahsgs.uum.edu.my

[2] Politeknik Kesehatan Banjarmasin, Banjarbaru 70714, Kalimantan Selatan, Indonesia

Abstract. In the era of big data, integrating heterogeneous data is a critical process where diverse data sources from various structures and forms need to be combined into a centralized data source for further analysis. However, this process often poses data quality problems such as duplication, inconsistencies, missing values, and incomplete and unpleasant data. All these problems are considered waste in heterogeneous data integration. Waste is defined as any activity that consumes resources but creates no value for customers. When dealing with heterogeneous data sources, the presence of waste can be particularly problematic as it can significantly impact the effectiveness and reliability of the integration process and also produce a huge mistake in data analysis. Hence, it is crucial to understand the concept of waste in heterogeneous data integration to determine strategies to enhance data quality through the reduction of waste. Therefore, this paper systematically reviews the literature on waste in the data integration processes. The main steps in conducting this paper include planning, conducting, and reporting the results. The review underscores the importance of strategies and practices for handling waste in data integration. By synthesizing findings from multiple studies, this paper contributes to the broader discourse on data management, notably to support educational administrators, data scientists, and IT professionals in optimizing data management, ultimately promoting more accurate and actionable insights within heterogeneous data integration environments.

Keywords: Waste Identification · Data Quality · Data Integration · Heterogeneous Data · Systematic Literature Review

1 Introduction

Since the beginning of the 21st century, numerous technological advancements have occurred in the information technology sector. The widespread adoption of digital technologies has recently significantly impacted personal lifestyles and business operations. This trend increased in the mid-2000s, with many companies restructuring their organizations to meet the needs of the changing digital environment. This process is known as digital transformation. Many businesses and organizations have developed information systems leveraging these technological advancements to enhance their operations. Consequently, these technologies have been continuously generating vast amounts of data

Z. Mat Aji et al. (Eds.): KMICe 2024, ISEM 49, pp. 298–310, 2025.
https://doi.org/10.1007/978-3-031-91485-0_23

over the internet, business enterprises, or government sector at an unprecedented rate. The production of the data unit is no longer the GB and TB, but the PB (1 PB = 210 TB), EB (1 EB = 210 PB), and ZB (1 ZB = 210 EB). For example, it is estimated that society created 463 Exabytes of data daily in 2025 [1, 2].

Data is considered as one of the most valuable assets where almost all business strategic decisions are made based on insights from data [3]. Hence, data must be well organized and managed to be exploited. However, data is complicated. It is available in many structured, semi-structured, and unstructured forms. It originates from diverse sources involving cleaning, filtering, processing, integration, merging, partitioning, transportation, sketching, and storage procedures [4]. In the preprocessing stage, conducting a quality assessment is essential to refine, evaluate, and ensure the quality of raw data [5]. Completeness, uniqueness, timeliness, accessibility, conformity, consistency, volatility, relevancy, readability, security, ease of manipulation, and integrity are the essential characteristics to be considered in data quality [6]. When the data quality degrades, the consequences are unpredictable and can lead to completely wrong insights [7]. Besides, low-quality data can lead to inaccurate predictions or wrong decisions, take longer analysis time, decrease customers' satisfaction, and generate unclear information [8]. It is well-known that, in contrast, high-quality data leads to enhanced predictive capability and comprehensive understanding [9].

Data also plays a pivotal role in higher education institutions (HEI) [10], serving as a fundamental tool for decision-making [11], improving educational quality [12], and enhancing student outcomes [13]. Data in HEI come from various sources, which include data about students, lecturers and faculty staff, courses, research and publications [14–16]. Moreover, this data must be integrated if HEI wants to conduct an in-depth analysis. Data integration is a process that involves extraction, transformation, and loading. Data extraction is the first and most important task in the data integration process that involves gathering voluminous data from diverse sources and make it available in a centralized data source [17]. Based on the integrated data, decision-makers or top management can make decisions and improve the business process more effectively.

The main challenge in the data integration environment is its heterogeneity nature, in which data is distributed across heterogeneous and diverse sources, necessitating the identification and management of duplicate entries [18] due to differences in terms of structures, formats, and semantics [19]. Therefore, data integration might introduce some quality issues such as data duplication, inaccuracy, inconsistency, missing values, and incomplete data [20–22]. Moreover, Terol et al. [23] Stated that another issue in data quality is related to unpleasant data, which can be defined as out-of-context data such as spam. All these issues are considered waste in data management [24]. These wastes occur due to a lack of data governance in handling waste and improving quality [25]. To address this issue, researchers have focused on understanding the concept of waste in heterogeneous data integration and determining the challenges and strategies for handling waste.

Waste refers to any activity that consumes resources but creates no value for customers [26]. Waste is often hidden by bureaucratic processes, multitasking, inadequate prioritization, and invisible cognitive activities [27]. Reducing waste in heterogeneous data integration is not easy because identifying waste is difficult. Furthermore, waste in data management is not frequently discussed and published by researchers. In the context of this, we have conducted a systematic literature review using the approach Kitchenham and Charters [28]. As such, this literature review aims to clearly understand the concept of waste and how it can occur during heterogeneous data integration. Therefore, the main contribution is to provide a comprehensive understanding of the concept of waste data and the emergence of waste data during heterogeneous data integration. It is expected to define various types of waste data and explore waste data detection and identification approaches using multiple methods.

The subsequent sections of this paper are organized as follows: Sect. 2 describes the review protocol. Section 3 discusses the analysis of the literature and answers the research questions. Section 4 provides a discourse on the outcomes. Finally, Sect. 5 concludes the article, offers insights for researchers, and clarifies the limitations of our study.

2 Research Method

In our research process, the applied review protocol is proposed based on the guidelines proposed by Kitchenham and Charters [28]. Establishing the review protocol is necessary to ensure the literature review has been done systematically and minimize researcher bias. Hence, the review process is focused on the constructed research questions derived from this review's aims. Following this, we detailed the primary steps of the systematic review process, explicitly "planning, conducting, and reporting" the review results [28].

2.1 Planning the Review

As mentioned earlier, this systematic literature review aims to clearly understand the concept of waste and how it can occur during heterogeneous data integration. Therefore, this phase involves defining research questions that align with the purpose of the study, followed by describing the search strategy, search string, and criteria for inclusion and exclusion. More detail is presented below.

Research Questions: As discussed in the previous section, data is an important asset and needs to be of enough quality. Data quality is referred to as fitness for use. Poor data quality will introduce impacts such as low company reputation, poor decisions, decreased customer confidence, and low productivity. Usually, data is stored at different sources with different formats and structures. Data integration is a necessary process to improve analysis and decision-making. However, the process might introduce waste which needs to be managed. Generally, waste is any activity that consumes resources but creates no value for customers [26]. In project management, waste refers to any activity, process, condition, property, or object that consumes resources without benefiting any project stakeholder [27]. As been discussed, reducing waste can improve quality and productivity. Therefore, it is essential to understand the concept of waste and determine the strategy used to handle waste in heterogeneous data integration.

Moreover, waste is often hidden by bureaucracy and lack of governance. Hence, handling waste in heterogeneous data integration might introduce some challenges. Therefore, identifying the challenges is vital to planning data-driven activities to increase data value and manage waste.

Hence, we formulated the research questions in order to fulfill these objectives:

RQ1. What are the strategies used to handle waste in heterogeneous data integration?
RQ2. What are the challenges of handling waste in heterogeneous data integration?

Search Strategy: After defining our research questions, we began by formulating a formal search strategy to analyze all available empirical materials relevant to this review's objective. This approach covered the search scope consisting of electronic databases and academic papers, as shown in Table 1. The initial studies were gathered from electronic databases, and relevant literature was identified by exploring references. Furthermore, three distinct rounds applied inclusion and exclusion criteria to the collected studies.

Table 1. Search sources.

Electronic databases	ACM Digital Library, IEEE Xplore, Science Direct
Searched items	Journal, conference, or proceeding papers
Search applied on	Full text
Language	English
Publication period	From January 2014 to February 2024

Search Criteria: The search criteria used for this review consist of two parts search criteria (C1 and C2), defined as follows:

- C1 is a string of keywords related to waste methods such as "waste", "issue", and "problem";
- C2 is a string of keywords related to heterogeneous data integration, such as "data integration" and "heterogeneous data integration".

Equation (1). Boolean expression search criteria

$$C1 \text{ AND } C2 \tag{1}$$

An example of a search done in the electronic databases is shown below:

("waste" OR "issue" OR "problem") AND ("data integration" OR "heterogeneous data integration").

The search string for each database was manually constructed utilizing the search features provided by the respective database. Each database was approached as an analytical process of exploration and experimentation.

Inclusion and Exclusion Criteria: This study used inclusion and exclusion criteria to determine the eligibility of the study:

Inclusion criteria:

(I1) The study is published in a peer-reviewed journal;
(I2) The study is written in English;
(I3) The study should be relevant to the search criteria;

(I4) The study should be an empirical research paper, an experience report, or a workshop paper; and
(I5) The study was published between January 2014 to February 2024.

Exclusion criteria:

(E1) The study does not discuss Waste Management in Data Integration;
(E2) The study is inconsistent with inclusion criteria; and
(E3) Opinion, editorials, discussions, comments, viewpoints, keynote, tutorials, prefaces, anecdote papers, and presentations in slide formats without any associated papers.

2.2 Conducting the Review

This section presents the results obtained from our systematic search and extracts information from relevant sources and databases. The studies were searched and retrieved from the selected electronic databases using the systematic search and selection procedure to identify primary studies. Subsequently, appropriate databases were chosen to explore relevant literature within the software engineering domain. The most suitable databases for conducting research are ACM Digital Library, IEEE Xplore, and Science Direct. At the beginning of the research process, relevant keyword strings were inputted to the search field in the databases. Moreover, the search string was configured with (OR, AND) operators to maximize search results. The total of retrieved papers is 1,613. The result is presented in Table 2 and explained in the following subsections.

Table 2. The number of studies identified during systematic search process.

Database	Retrieved	Round 1		Round 2		Round 3	
		Included	Excluded	Included	Excluded	Included	Excluded
ACM	569	175	394	62	113	11	51
IEEE Xplore	164	93	71	14	79	2	12
Science Direct	880	268	612	65	65	7	58
Total	1,613	536	1,077	141	395	20	121

Round 1: in this stage, researchers screened the title, abstract, and keyword by applying inclusion and exclusion criteria from each paper retrieved from databases. Thus, the papers selected in round 1 to process in round 2 were reduced to 536 papers.

Round 2: in this stage, researchers read the abstract carefully to select papers addressing the waste in heterogeneous data integration that will be discussed in the paper. The papers selected were reduced to 141 papers.

Round 3: this stage involves reading the full text carefully from Round 2 to select the paper that discusses the kind of waste, how to identify waste, and challenges and strategies to handle such waste from heterogeneous data integration.

Finally, the papers selected for this study from round 3 are 20 papers. The challenges faced during the review process are that not much research discusses the waste data comprehensively, including the types of waste, strategies for identifying that waste data, and challenges in handling the waste data.

3 Findings

This section presents the outcomes of our review based on the research questions. As noted earlier, a total of 20 studies were identified in this review. Table 3 shows how these studies are distributed across different publication sources. Of these 20 studies, about 65% (13) were published in journals, while 35% (7) appeared in conference proceedings.

Additionally, Table 3 indicates the distribution of studies across publication venues. This indicates that no single publication source is preferred among data integration paper authors.

Table 3. Distribution of studies based on the publication sources.

Publication Source	Type	Number of Publication
ACM	Journal	Six studies
ACM	Proceeding	Five studies
IEEE	Journal	Two studies
ScienceDirect	Proceeding	Two studies
ScienceDirect	Journal	Five studies

Regarding the years of publication, Fig. 1. Presents the distribution of reviewed papers published from 2015 to 2024. This research has selected articles from the past 10 years because researchers do not frequently discuss and publish this topic. In addition, this research also wants to explore the development of research related to this topic along with technology advancements. Out of the 20 studies, three studies were conducted between 2015 and 2016, two studies were conducted from 2017 to 2018, and another three studies were conducted between 2019 and 2020. The following year possess an increase with five studies were conducted from 2021 to 2022, and the last seven studies were completed between 2023 and 2024. Furthermore, it shows a phenomenon where an increasing number of researchers discuss waste identification in heterogeneous data integration in their papers. Thus, it indicates that researchers realize the importance of research to identify waste in data integration.

This study extracted information, including the waste type, the strategy to handle the waste, and the challenges of handling that waste from 20 studies selected after study search and selection.

Table 4 summarizes various studies investigating the types of waste data, such as duplication, incomplete data, inconsistencies, missing values, unpleasant data, erroneous data, and data anomaly. Moreover, researchers use many strategies to address these issues. While the studies cover a range of methods and issues, but there is still not

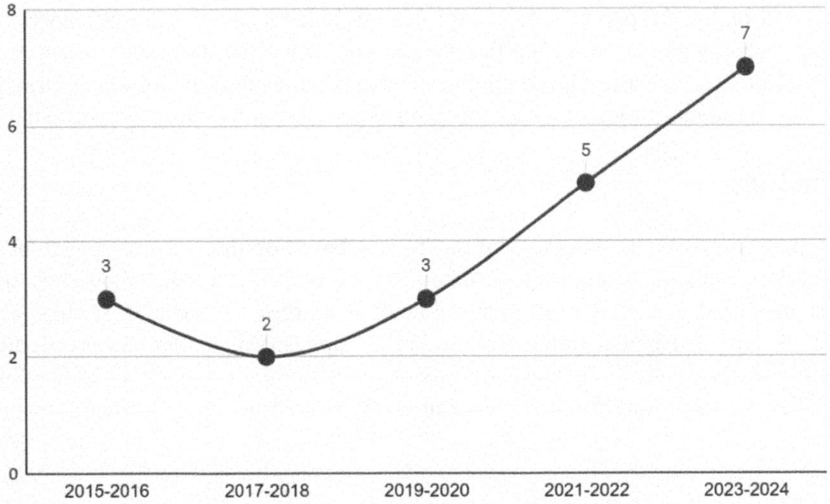

Fig. 1. Distribution of selected studies by year.

Table 4. Summary of strategies to handle waste from studies that investigated them.

Study	Waste	Strategy	Challenges
Dinh, 2023 [29]	Duplication, Erroneous data	Entity Matching	Identification, Inconclusive match
Li et al., 2021 [30]	Duplication	Entity Matching	Identification
Ahmed et al., 2022 [31]	Duplication	Entity Resolution	De-duplication
Choi et al., 2018 [32]	Duplication	Entity Resolution	Prevention
Genossar et al., 2023 [33]	Duplication	Entity Resolution	Detection
Azeroual et al., 2022 [34]	Duplication, Data anomaly	Extract Transform Load, Data wrangling	Identification
Brown et al., 2024 [35]	Duplication	Extract Transform Load	Reduce Possibility
Yu et al., 2019 [36]	Duplication	Extract Transform Load	Detection
Abboura et al., 2016 [37]	Duplication	Matching Dependencies	Detection
Schirmer et al., 2020 [38]	Duplication	Matching Dependencies	Detection
Azzalini et al., 2022 [39]	Incomplete	Functional Dependencies	Detection

(*continued*)

Table 4. (*continued*)

Study	Waste	Strategy	Challenges
Miao et al., 2016 [40]	Incomplete	Query Answering	Detection
Song et al., 2015 [41]	Incomplete	Record Matching	Data Imputation
Barbella & Tortora, 2023 [42]	Incomplete	Schema Matching	Identification
Alfano et al., 2023 [43]	Inconsistencies	Query Answering	Repairing Data
Heidari et al., 2024 [44]	Inconsistencies	Record Fusion	Identification
Mazilu et al., 2022 [45]	Inconsistencies	Schema Mapping	Identification
Berti-Équille et al., 2018 [46]	Missing Values	Functional Dependencies	Data Imputation
Youngmann et al., 2023 [47]	Missing Values	Knowledge Graph	Data Imputation
Terol et al., 2020 [23]	Unpleasant data	Extract Transform Load	Unpleasant data checking

much research on addressing and handling waste data comprehensively. Hence, there is a need for further research on scalable and automated solutions using technological advancements to address and handle waste data.

4 Discussion of the Results

This study presents a systematic literature review that addresses solutions and challenges to handle waste on data integration. 1,613 articles were retrieved from selected databases after applying keywords while searching. Moreover, after a meticulous multi-stage screening process using inclusion and exclusion criteria, researchers extracted information from 20 relevant studies that mainly focused on identifying challenges to handling waste and offered solutions to handle that waste in data integration. Out of 20 articles, about 65% (13) were published in journals, and the rest were published in proceedings. The years of publication show an increased number of publications since 2021, indicating that researchers realize that understanding waste and determining the strategy to handle waste in data integration becomes a more meaningful and interesting topic for further investigation.

RQ1. What are the strategies used to handle waste in heterogeneous data integration?

As shown in Table 4, our review found that waste in data integration can be divided into seven (7) types: duplicate data, erroneous data, data anomaly, incomplete data, inconsistent data, missing value, and unpleasant data. According to Dinh [29], these problems arise due to the lack of standardized protocols for systematically storing these entities and their unique identifiers.

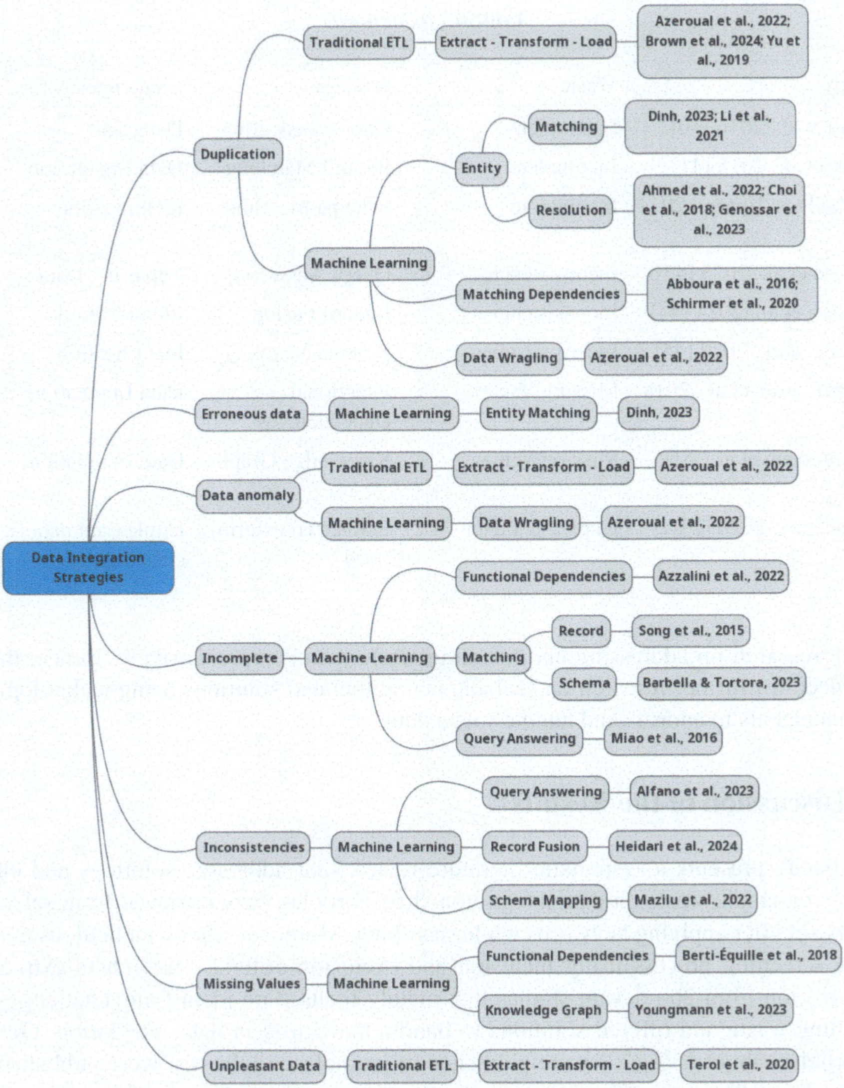

Fig. 2. Data integration strategies distribution of selected studies.

Figure 2 shows a summary of strategies to handle the waste from selected studies in response to RQ1. The strategies extracted from studies are divided into two groups: some using traditional extract transform load (ETL) and others using machine learning.

Traditional ETL handles duplication and unpleasant data because detection can be done manually. This involves human-aided waste handling, consisting of three primary stages: data extraction from various sources, data transformation into a suitable format and structure, and data loading into a target repository. However, traditional ETL has weaknesses, such as time-consuming, high cost, inconsistency, etc.

In contrast, machine learning has become more popular along with technological advances due to several advantages, such as efficient, low-cost, fast, and accurate responses. Recently, researchers are more interested in developing machine learning methods to address the issue of waste data in heterogeneous data integration.

RQ2. What are the challenges of handling waste in heterogeneous data integration?

In this study, researchers extracted challenges to handling waste from studies to respond to RQ2. RQ2 is closely related to RQ1 and explains the challenges faced by researchers in implementing the strategies (Fig. 3).

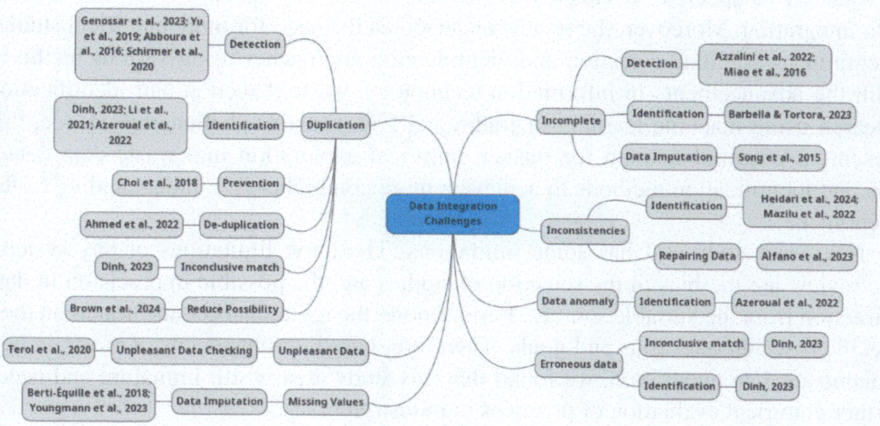

Fig. 3. Data integration challenges of selected studies.

Moreover, researchers extracted challenges to handling waste from studies to respond to RQ2. RQ2 has a close relationship with RQ1. RQ2 explains the challenges faced by researchers in implementing the strategies.

Challenges in duplication are divided into five types: detection, identification, inclusive matching, reducing possibility/ prevention, and de-duplication. Moreover, missing data has challenges in data imputation. It involves selecting the imputation method, which can significantly affect the result. For example, data imputation using simple methods with the mean or median can reduce data variability and introduce bias. In contrast, complex methods like machine learning are computationally expensive and risk overfitting. Inconsistency data has two types of challenges: inconsistent data and data repair. Data repairing is challenging because inconsistent data comes from various sources, such as human errors, system glitches, or integration issues, making it difficult to detect and fix accurately. In addition, the massive volume of datasets complicates this issue, as manual cleaning is impractical, and automated tools might have wrong information or make mistakes. Data anomaly has challenges in identification. Erroneous data has challenges in inclusive matching. Incomplete data has in detection, identification or discovering data, and data imputation. Last, unpleasant data has challenges in unpleasant data checking.

5 Conclusions and Future Works

In this study, researchers conducted a systematic literature review to explore strategies and challenges to handle waste data in heterogeneous data integration of higher education institutions. Following the established guidelines for conducting Systematic Literature Reviews (SLRs) as proposed by Kitchenham and Charters [28], an extensive search was conducted across various source databases, resulting in the initial identification of 1,613 papers. Subsequently, through a meticulous multi-stage screening process, 20 relevant papers were selected.

This review presents several implications for both practitioners and researchers. In research perspective, it enables researchers to identify specific types of waste in data integration. Moreover, the review underscores the need for more empirical studies integrating waste data detection and identification approaches using various methods. With the advancements in information technology, waste detection and identification research using machine learning or traditional ETL remains challenging. Hence, this presents a potential domain for further empirical exploration into waste data detection and identification methods to achieve a more comprehensive understanding of the phenomena.

Next, our study still has some limitations. The basic limitations of any systematic review are the bias in the selection of studies and the possible imprecision in data extraction from the variable sources. Furthermore, the researchers concentrated on their specific research questions and goals. Therefore, based on our review of waste in heterogeneous data integration, we found that this study area is still immature and needs further empirical evaluation of practices in industrial cases.

Acknowledgement. This research was supported by the Ministry of Higher Education (MoHE) of Malaysia through the Fundamental Research Grant Scheme (FRGS/1/2022/ICT01/UUM/02/2).

References

1. Anand, A.: AI driven data governance for the enterprise intelligence. SSRN Journal (2024)
2. Makhdoom, I., Abolhasan, M., Lipman, J., Piccardi, M., Franklin, D.: PrivySeC: a secure and privacy-compliant distributed framework for personal data sharing in IoT ecosystems. Blockchain: Res. Appl. **5**, 100220 (2024)
3. Saggi, M.K., Jain, S.: A survey towards an integration of big data analytics to big insights for value-creation. Inf. Process. Manage. **54**, 758–790 (2018)
4. Taleb, I., Dssouli, R., Serhani, M.A.: Big data pre-processing: a quality framework. In: 2015 IEEE International Congress on Big Data, pp. 191–198. IEEE, New York City, NY, USA (2015)
5. Taleb, I., Serhani, M.A., Bouhaddioui, C., Dssouli, R.: Big data quality framework: a holistic approach to continuous quality management. J. Big Data. **8**, 76 (2021)
6. Elouataoui, W., El Alaoui, I., El Mendili, S., Gahi, Y.: An advanced big data quality framework based on weighted metrics. BDCC. **6**, 153 (2022)
7. Taleb, I., Kassabi, H.T.E., Serhani, M.A., Dssouli, R., Bouhaddioui, C.: Big data quality: a quality dimensions evaluation. In: 2016 Intl IEEE Conferences on Ubiquitous Intelligence & Computing. Advanced and Trusted Computing, Scalable Computing and Communications, Cloud and Big Data Computing, Internet of People, and Smart World Congress (UIC/ATC/ScalCom/CBDCom/IoP/SmartWorld), pp. 759–765. IEEE, Toulouse (2016)

8. Lai, S.-T., Leu, F.-Y.: Data preprocessing quality management procedure for improving big data applications efficiency and practicality. In: Barolli, L., Xhafa, F., Yim, K. (eds.) Advances on Broad-Band Wireless Computing, Communication and Applications, pp. 731–738. Springer International Publishing, Cham (2017)
9. Taleb, I., Serhani, M.A., Dssouli, R.: Big data quality: a survey. In: 2018 IEEE International Congress on Big Data (BigData Congress), pp. 166–173. IEEE, San Francisco, CA, USA (2018)
10. Tulasi, B.: Significance of big data and analytics in higher education. IJCA **68**, 21–23 (2013)
11. Gaftandzhieva, S., Hussain, S., Hilcenko, S., Doneva, R., Boykova, K.: Data-driven decision making in higher education institutions: state-of-play. IJACSA. **14**, (2023)
12. Ashaari, M.A., Singh, K.S.D., Abbasi, G.A., Amran, A., Liebana-Cabanillas, F.J.: Big data analytics capability for improved performance of higher education institutions in the Era of IR 4.0: A multi-analytical SEM & ANN perspective. Technol. Forecast. Soc. Change **173**, 121119 (2021)
13. Sarker, F.: Linked data technologies to support higher education challenges: student retention, progression and completion (2014)
14. Akande, A.O., Van Belle, J.-P.: Cloud computing in higher education: a snapshot of software as a service. In: 2014 IEEE 6th International Conference on Adaptive Science & Technology (ICAST), pp. 1–5. IEEE, Ota, Nigeria (2014)
15. Alblawi, A.S., Alhamed, A.A.: Big data and learning analytics in higher education: demystifying variety, acquisition, storage, NLP and analytics. In: 2017 IEEE Conference on Big Data and Analytics (ICBDA), pp. 124–129. IEEE, Kuching (2017)
16. Kardan, A.A., Sadeghi, H., Ghidary, S.S., Sani, M.R.F.: Prediction of student course selection in online higher education institutes using neural network. Comput. Educ. **65**, 1–11 (2013)
17. Ma, B., Jiang, T., Zhou, X., Zhao, F., Yang, Y.: A novel data integration framework based on unified concept model. IEEE Access. **5**, 5713–5722 (2017)
18. Schildhauer, M.: Data integration: principles and practice. In: Recknagel, F., Michener, W.K. (eds.) Ecological Informatics, pp. 129–157. Springer International Publishing, Cham (2018)
19. Chromiak, M., Stencel, K.: A data model for heterogeneous data integration architecture. In: Kozielski, S., Mrozek, D., Kasprowski, P., Małysiak-Mrozek, B., Kostrzewa, D. (eds.) Beyond Databases, Architectures, and Structures, pp. 547–556. Springer International Publishing, Cham (2014)
20. Allen, G.I.: Statistical data integration: challenges and opportunities. Stat. Modell. **17**, 332–337 (2017). 21.Haneem, F., Ali, R., Kama, N., Basri, S.: Resolving data duplication, inaccuracy and inconsistency issues using Master Data Management. In: 2017 International Conference on Research and Innovation in Information Systems (ICRIIS), pp. 1–6. IEEE, Langkawi, Malaysia (2017)
21. Almutairi, M.M., Yamin, M., Halikias, G.: An analysis of data integration challenges from heterogeneous databases. In: 2021 8th International Conference on Computing for Sustainable Global Development (INDIACom)
22. Terol, R.M., Reina, A.R., Ziaei, S., Gil, D.: A machine learning approach to reduce dimensional space in large datasets. IEEE Access. **8**, 148181–148192 (2020)
23. Kadadi, A., Agrawal, R., Nyamful, C., Atiq, R.: Challenges of data integration and interoperability in big data. In: 2014 IEEE International Conference on Big Data (Big Data), pp. 38–40. IEEE, Washington, DC, USA (2014)
24. Al-Badi, A., Tarhini, A., Khan, A.I.: Exploring big data governance frameworks. Procedia Comput. Sci. **141**, 271–277 (2018)
25. Sedano, T., Ralph, P., Peraire, C.: Software development waste. In: 2017 IEEE/ACM 39th International Conference on Software Engineering (ICSE), pp. 130–140. IEEE, Buenos Aires (2017)

26. Sedano, T., Ralph, P., Péraire, C.: Removing Software Development Waste to Improve Productivity. In: Sadowski, C., Zimmermann, T. (eds.) Rethinking Productivity in Software Engineering, pp. 221–240. Apress, Berkeley (2019)
27. Kitchenham, B., Charters, S.: Guidelines for performing systematic literature reviews in software engineering (2007)
28. Dinh, X.T.: Name2Vec: name matching using character-based with deep learning. Procedia Comput. Sci. **230**, 305–315 (2023)
29. Li, Y., Li, J., Suhara, Y., Wang, J., Hirota, W., Tan, W.-C.: Deep entity matching: challenges and opportunities. J. Data Inf. Q. **13**, 1–17 (2021)
30. Ahmed, S., Gentili, M., Sierra-Sosa, D., Elmaghraby, A.S.: Multi-layer data integration technique for combining heterogeneous crime data. Inf. Process. Manage. **59**, 102879 (2022)
31. Choi, H., Gim, J., Seo, Y.-D., Baik, D.-K.: VPL-based big data analysis system: UDAS. IEEE Access. **6**, 40883–40897 (2018)
32. Genossar, B., Shraga, R., Gal, A.: FlexER: flexible entity resolution for multiple intents. Proc. ACM Manag. Data. **1**, 1–27 (2023)
33. Azeroual, O., Schöpfel, J., Ivanovic, D., Nikiforova, A.: Combining data lake and data wrangling for ensuring data quality in CRIS. Procedia Comput. Sci. **211**, 3–16 (2022)
34. Brown, N., Gruber, S., Pulsifer, P., Hayes, A.: A prototype field-to-publication data system for a multi-variable permafrost observation network. Environ Model Softw. **175**, 106006 (2024)
35. Yu, Y., et al.: ADEpedia-on-OHDSI: a next generation pharmacovigilance signal detection platform using the OHDSI common data model. J. Biomed. Inform. **91**, 103119 (2019)
36. Abboura, A., Sahri, S., Baba-Hamed, L., Ouziri, M., Benbernou, S.: Quality-based online data reconciliation. ACM Trans. Internet Technol. **16**, 1–21 (2016)
37. Schirmer, P., Papenbrock, T., Koumarelas, I., Naumann, F.: Efficient discovery of matching dependencies. ACM Trans. Database Syst. **45**, 1–33 (2020)
38. Azzalini, F., Criscuolo, C., Tanca, L.: E-FAIR-DB: functional dependencies to discover data bias and enhance data equity. J. Data Inf. Q. **14**, 1–26 (2022)
39. Miao, X., Gao, Y., Chen, G., Cui, H., Guo, C., Pan, W.: SI2P: a restaurant recommendation system using preference queries over incomplete information. Proc. VLDB Endow. **9**, 1509–1512 (2016)
40. Song, S., Zhang, A., Chen, L., Wang, J.: Enriching data imputation with extensive similarity neighbors. Proc. VLDB Endow. **8**, 1286–1297 (2015)
41. Barbella, M., Tortora, G.: A semi-automatic data integration process of heterogeneous databases. Pattern Recogn. Lett. **166**, 134–142 (2023)
42. Alfano, G., Greco, S., Molinaro, C., Trubitsyna, I.: An approximation algorithm for querying inconsistent knowledge bases. Intell. Syst. Appl. **17**, 200146 (2023)
43. Heidari, A., Michalopoulos, G., Ilyas, I.F., Rekatsinas, T.: Record fusion via inference and data augmentation. ACM/IMS J. Data Sci. **1**, 1–23 (2024)
44. Mazilu, L., Paton, N.W., Konstantinou, N., Fernandes, A.A.A.: Fairness-aware data integration. J. Data Inf. Q. **14**, 1–26 (2022)
45. Berti-Équille, L., Harmouch, H., Naumann, F., Novelli, N., Thirumuruganathan, S.: Discovery of genuine functional dependencies from relational data with missing values. Proc. VLDB Endow. **11**(8), 880–892 (2018)
46. Youngmann, B., Cafarella, M., Salimi, B., Zeng, A.: Causal data integration. Proc. VLDB Endow. **16**, 2659–2665 (2023)

Assessing the Effectiveness of System for Managing Integrated Data of Higher Public Institutions

Rahayu Ahmad[1](✉), Suzilah Ismail[1], Fauziah Baharom[1], Nur Haryani Zakaria[1], Haslina Mohd[1], Rohaida Romli[1], Yuhanis Yusof[1], Mohamed Ali Saip[1], Juhaida Abu Bakar[1], Wan Hussain Wan Ishak[1], Mohd Hasbullah Omar[1], and Salwati Badroddin[2]

[1] Universiti Utara Malaysia, 06010 Sintok, Kedah, Malaysia
rahayu@uum.edu.my

[2] Strategic Planning Division, Ministry of Higher Education, 62200 Putrajaya, Malaysia

Abstract. Effective data management in higher education institutions is essential for maintaining national higher education quality. Centralizing and integrating this data within a unified database enhances accuracy in estimates, improves decision-making, and aids in developing better strategies. To support this, a system was developed to collect and integrate data from public higher education institutions, and it has undergone continuous refinement to ensure reliability and efficiency. This study evaluates the effectiveness of the integrated system and assesses user satisfaction among public and private higher education institutions. Two data collection methods were employed for this study: surveys and interviews. The primary method involved surveys with 77 respondents, while the supplementary approach included interviews with eight officers involved in data management and strategic planning across various public higher education institutions. Overall, participants expressed satisfaction with the integrated system, noting that it complements internal institutional data and is a valuable resource for strategic planning. However, several areas for improvement were identified, including the need for better integration with other systems, the development of a data dictionary, and automated data mapping to address issues in the data transmission process.

Keywords: Data Integration · Portal · Strategic Planning · Data Quality

1 Introduction

Managing data in higher education institutions is crucial to ensuring national higher education quality is consistently upheld, particularly in fostering competitive human capital. There is a significant amount of academic data in higher education institutions. This includes data about students, staff, graduates, research and publications, and institutional information, among other things. Higher education data is heterogeneous since it comes from various institutions and is kept in distinct locations on a range of platforms. To enable a comprehensive and methodical examination, it is essential that this

Z. Mat Aji et al. (Eds.): KMICe 2024, ISEM 49, pp. 311–320, 2025.
https://doi.org/10.1007/978-3-031-91485-0_24

data be maintained and integrated into a centralized database. Integrated data analysis can enhance accuracy in estimates, more precise decision-making, and formulation of better strategies [1]. Among the challenges during data integration are missing, erroneous, or incomplete data, even with the abundance of support tools available to aid with data integration [2]. To ensure data quality, a developed system has been developed which collects data from public higher education institutions. Since its creation in 2006, the system has continuously improved to guarantee reliable and efficient functioning. Therefore, it is imperative to evaluate the effectiveness of the system and how the system is used in strategic planning of the Public Higher Education Institutions.

2 Literature Review

2.1 Heterogeneous Data Management and Integration

Nowadays, most organizations focus on data because it is one of their valuable assets. This perception of data as an asset requires organizations to commit to improving data quality. If the data managed is of good quality, more benefits will be obtained [3]. Generally, data can be processed to produce value that can be used by stakeholders. Therefore, data needs to be managed well and effectively so that the value obtained from the data can benefit the organization.

Poor data quality can affect the productivity of an organization. However, in general, the management of large-sized data is very complicated and complex. Large data processing can be processed in batch processing and stream processing. Batch processing is performed based on data analysis in a certain period of time and is adopted when there are no constraints involving response time. On the other hand, streaming processing is suitable for real-time feedback applications [2]. In addition, one of the challenges faced in managing large-sized data is that if the data involved is heterogeneous, the data is stored in various platforms separately. The data needs to be integrated to enable a holistic analysis.

However, integrating heterogeneous data is challenging and requires an iterative process [4]. Among the challenges is when performing data extraction, which involves the process of pulling data from data sources. This process is the first step that needs to be taken when extracting data. It is a complicated and time-consuming process because the data source uses different formats, structures, and types. Therefore, after the data is extracted and before it is integrated into the destination system, the data needs to be transformed to make it consistent and match [5].

In addition, data integration also involves managing unstructured data. Therefore, the data lake approach (data pool) must be implemented for this purpose. According to [4], a data lake (data pool) is a central storage repository supporting large quantities and data types, which involves structured and unstructured data. Nevertheless, metadata development is necessary to avoid ambiguity to ensure the data lake implementation is done effectively. Without any metadata, dumping all the data into a data lake will result in the data being unable to be used to its best potential. In addition, according to [4], when collecting data from heterogeneous systems, there are three sources of data errors: data entry errors, data type incompatibility, and semantic incompatibility in business entity definitions.

Furthermore, according to [6], data quality is vital in continuously managing and maintaining data. If the data is not managed correctly, some data errors, such as missing values, spelling errors, inaccurate data, incorrect formats, inconsistencies, etc., may occur. These small mistakes can cause data to become unusable and create poor data quality issues, leading to more severe problems. It is well known that poor-quality data can cause problems such as increased cost and time required to repair the data.

In addition, good data management regarding data security must also be emphasized, especially when sharing data with third parties [7]. This is followed by the need for comprehensive data governance, data collection, and data usage that continues to grow. Therefore, excellent and efficient data management by using certain technologies or methods, such as through a data management system that takes data security into account and using data analytics methods, can help the management of an organization to ensure data security [7], obtain helpful information, helping them manage data management better, make transparent, and accurate decisions, also perform data management.

Apart from data security issues, data governance needs to emphasize critical success factors (Critical Success Factors (CSF)) that can determine the direction of data governance in an organization. [8] through their research related to the financial data governance model, has identified seven main core categories that are CSFs for data governance: 1) employee data efficiency, 2) transparent data processes and procedures, 3) data tools and technologies that are flexible, 4) standardized data policy, 5) developed/acquired data roles and responsibilities, 6) precise inclusive data needs, 7) focused and visible data strategy. They also found that creating successful data governance requires high attention from stakeholders and the need to establish data roles and responsibilities. They were followed by focusing on actions related to the underlying data-focused strategy to create a successful data governance program.

An Integrated System for Managing Public Higher Institutions Data

In an attempt to address the issues regarding data management and integration, an integrated system was developed. The integrated system for integrating and improving data quality from Public Higher Education Institutions (PHEI) was established in 2006. It is an application system that runs on the web and consists of multiple parts: The e-Lounge System, the Web Analytics System, and the Information Portal.

The e-Lounge System component was created to integrate and improve data quality over time. This integrated system also helps prevent potential disagreements over the legitimacy and quality of the data given. This system facilitates data quality validation after data collection to guarantee no errors before the data is stored.

Another component of the Web Analytics System enables the preparation of reports for specific modules in higher education institutions. Four (4) modules are included: the Student Module, Staff Module, Research, Development and Commercialization (R&D&C) Module, and Institutional Information Module. Since its development, this system has undergone several revisions. Hence, it is imperative to evaluate the efficacy of the integrated system and user satisfaction among public and private higher education institutions.

3 Methodology

For this study, two approaches to data collection were employed. The primary data collection uses surveys, while the complementary approach executes interviews.

3.1 Survey

The survey was distributed to all officers in PHEI who were experienced in using the integrated system. The target respondents (population) consist of 161 coordinators of all PHEI. Using a purposive sampling approach, the survey link was distributed to a social media group of all coordinators and officers in PHEI. To increase the response rate, reminders were sent occasionally to the group, and 77 respondents participated in the survey.

The survey is composed of several parts: Part A: Demographics. This section contains four questions: the institution, the respondent's role in sending data to the integrated system, the modules involved by the respondent, and the duration of experience using the system. Part B: Infrastructure. This section also contains questions about respondents' infrastructure problems when using the integrated system. These include Internet access problems, hardware/device problems, application/software problems, and technology/computer skills issues. Part C: Usability of the Integrated System. This section has several questions adapted from [9] that measure system quality, service quality, information quality, and satisfaction.

3.2 Interviews

An interview was conducted with eight officers involved in managing data and strategic planning in various Public Higher Education Institutions to determine the system's effectiveness. Each interview session lasted from 20–30 min. The Interview guide questions are as follows:

1. How do higher education institutions perceive the usefulness and function of the integrated system?
2. How do higher education institutions utilize data from the system in strategic planning or policy formulation and framework of action plans?
3. What is the level of support from senior management towards the implementation and management of the system?
4. Are there any suggestions for improvement?

3.3 Data Analysis for Survey

A descriptive SPSS software study was conducted to determine user satisfaction with the integrated systems. The analytical report is included in the upcoming section.

Based on Fig. 1, a large portion of respondents (65%) were technical officers and coordinators of the integrated system at each PHEI. Specifically, the participants consisted of 30% coordinators, while 35.56% were technical officers. Around 28.8% of respondents are data owners. Hence, the respondents were relevant to this study since they manage data transmissions and use the integrated system.

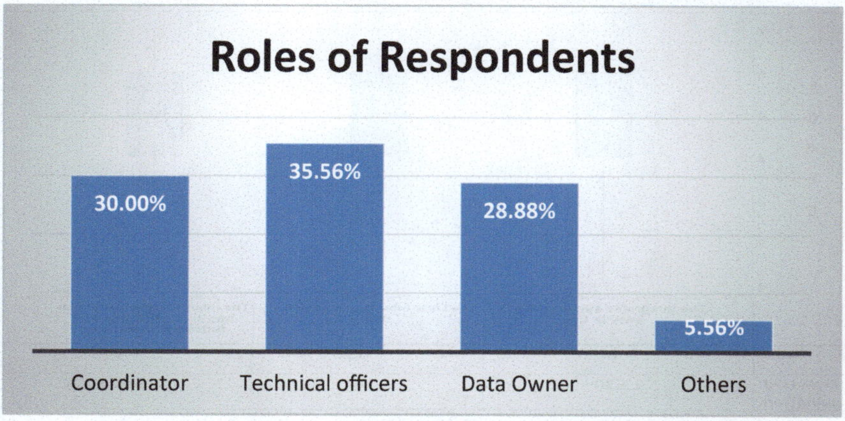

Fig. 1. Distribution of respondents based on their roles.

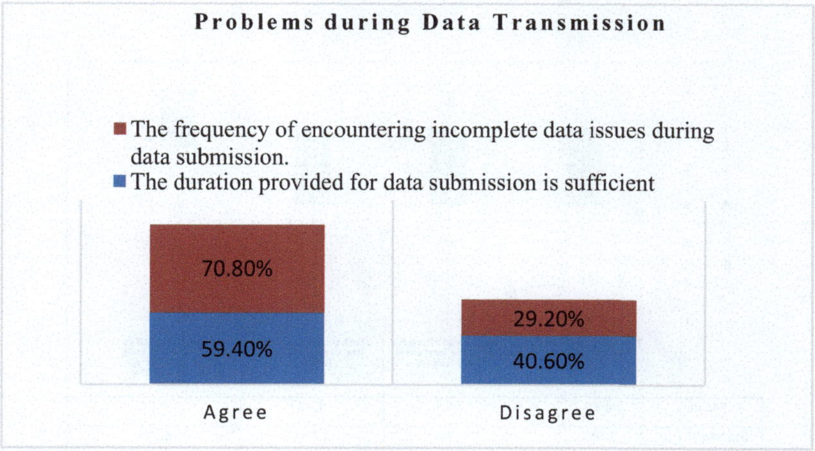

Fig. 2. Problems during Data Transmission.

The study results indicate that 70.8% of respondents encountered issues with incomplete data during data submission. This problem is partly related to the insufficient data submission period (40.6%) (Fig. 2).

Perception of System Quality. Most respondents strongly agreed that the integrated system is easy to use and has all the required features, as reflected by the mean score in Fig. 3.

Perception of Service Quality. Similarly, most respondents agreed that the integrated system provided support through manual, workshop, documentation and online help (Fig. 4).

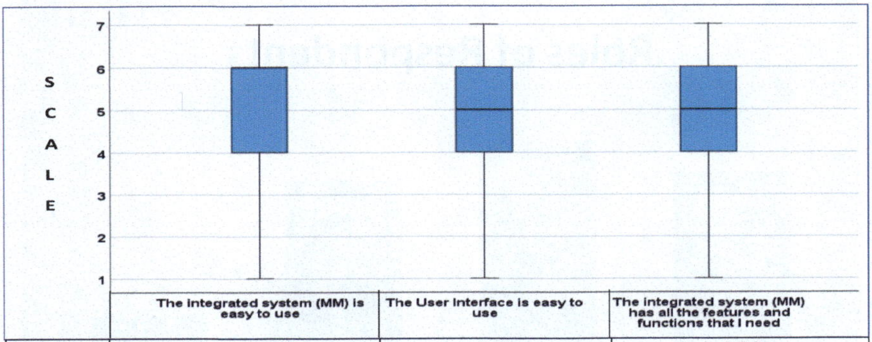

	The integrated system (MM) is easy to use	The User Interface is easy to use	The integrated system (MM) has all the features and functions that I need
Mean	5.14	5.06	4.94
Standard Deviation	1.570	1.344	1.321

Strongly Disagree:1.00-2.49 Disagree: 2.50-3.99 Agree 4.00-5.49 Strongly Agree:5.50-7.00

Fig. 3. Perception of System Quality.

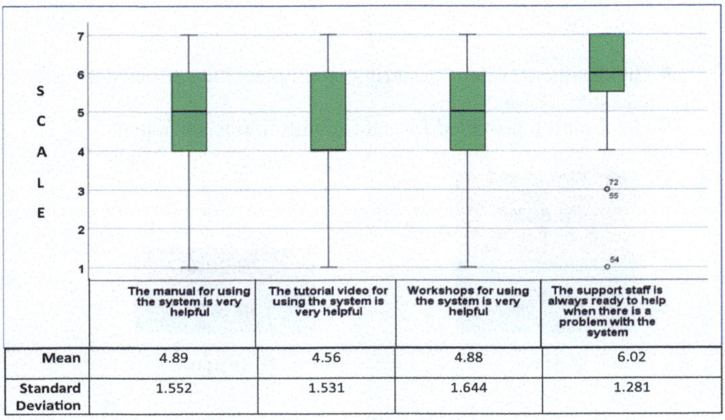

	The manual for using the system is very helpful	The tutorial video for using the system is very helpful	Workshops for using the system is very helpful	The support staff is always ready to help when there is a problem with the system
Mean	4.89	4.56	4.88	6.02
Standard Deviation	1.552	1.531	1.644	1.281

Strongly Disagree:1.00-2.49 Disagree: 2.50-3.99 Agree 4.00-5.49 Strongly Agree:5.50-7.00

Fig. 4. Perception of Service Quality.

Perception of Information Quality. Many respondents had positive perceptions about the completeness and accuracy of the data provided by the integrated system, as demonstrated in the figure below. Most of them strongly agree that they understand the data format required by the system and that the data format provided is complete and updated. They also strongly agree that they know the problematic data displayed by the system (Fig. 5).

Satisfaction with the Integrated System. In conclusion, most respondents were satisfied with using the integrated system and perceived it to be easy to use. The system

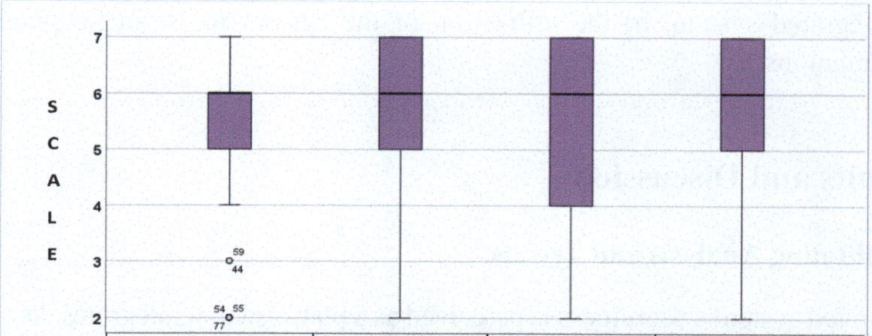

	I am very clear with the data format required by the system	The data format document provided is complete and updated	I really understand about the problematic data displayed by the system	The system is helpful for me to review the data quality through the Real Time Checking (RTC) function
Mean	5.40	5.62	5.33	5.52
Standard Deviation	1.465	1.408	1.459	1.533

Strongly Disagree:1.00-2.49 Disagree: 2.50-3.99 Agree:4.00-5.49 Strongly Agree:5.50-7

Fig. 5. Perception of Information Quality.

provided the required documentation and help using it, and it provided complete and accurate information (Fig. 6).

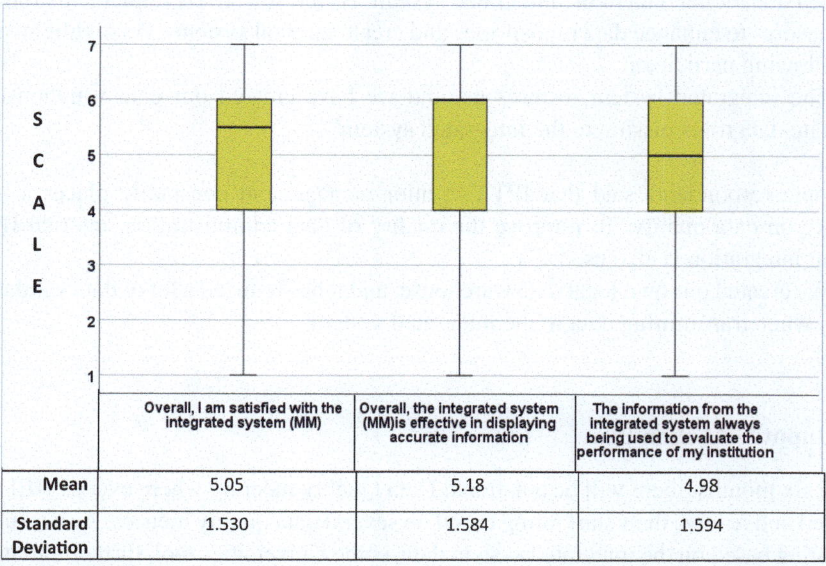

	Overall, I am satisfied with the integrated system (MM)	Overall, the integrated system (MM)is effective in displaying accurate information	The information from the integrated system always being used to evaluate the performance of my institution
Mean	5.05	5.18	4.98
Standard Deviation	1.530	1.584	1.594

Strongly Disagree:1.00-2.49 Disagree: 2.50-3.99 Agree:4.00-5.49 Strongly Agree:5.50-7

Fig. 6. Satisfaction towards the Integrated System.

Data Analysis for Interview. Data analysis was conducted with two coders. Open coding analysis was undertaken to identify emergent themes on the a) usefulness of the integrated system, b) the utilization of the system for strategic planning, c) recommendations

4 Results and Discussion

4.1 Facilitating Analysis and Trends

The integrated system's usefulness is perceived as a platform that integrates data sources for all Public Higher Education Institutions (PHEI). The data allow a more profound study to spot trends and deal with graduate and student-related problems. For example, a participant commented:

"In our university, we combined data from this system and our internal database to make analysis on graduate employability"

P3

4.2 Catalyst for Development of Local Data Warehouse

The existence of this integrated system also serves as a catalyst for the development of a local data warehouse with structures that align with the integrated system. Awareness of data quality and data management systems inside PHEI improves when all are mandated to send educational data to the integrated system. As a result, IPTA implements follow-up measures to enhance data repositories and create internal systems. As highlighted by the following participant:

"The integrated system inspires us, and we have created our data warehouse to facilitate data transmission to the integrated system"

P1

Some respondents said that IPTA's senior management constantly places a high priority on data quality. To improve the quality of data administration, internal IPTA system integration is also used:

"We created our own local data warehouse, and it has reduced a lot of data validation issues when transmitting data to the integrated system"

P4

4.3 Input for Action Plan

Every six months, there will be a national Data Quality meeting where every PHEI will be rated and receive their start rating based on several data quality metrics. As a result of this rating based on the integrated system data, some Universities took further initiatives to devise their action and strategic plan as commented below:

"After each data quality meeting and the release of star rating, we will hold a meeting led by the vice-chancellor, with participation from data owners, to discuss the performance and the quality of data provided to the integrated system"

P1

Action plans were devised based on the rating performance of data supplied to the integrated system. These plans include enhancing data governance procedures and devising methods to improve data quality. Certain PHEIs have created integration systems and implemented Real-Time Checking (RTC) features in line with the integrated system's requirements to detect erroneous data early on before it is submitted to the integrated system.

4.4 Minimal Utilization

However not every PHEI optimized the data and report provided by the integrated system. Certain PHEIs only supply data to the integrated system to fulfil the reporting requirements. These IPTAs don't use this information for internal strategy planning. Additionally, any needs related to strategic planning depend increasingly on extensive internal IPTA data sources. Moreover, insufficient knowledge among senior management of the significance and accessibility of data provided by the integrated system has led to inadequate utilization of the information. Additionally, any needs related to strategic planning depend increasingly on extensive internal IPTA data sources. Moreover, insufficient knowledge among senior management of the significance and accessibility of MyMohes and SKPG data has led to inadequate use of the information.

4.5 Recommendations for Improvement

Some of the respondents highlighted the issue of frequent changes in data format that need to be supplied to the integrated system. To address this issue, a standardized data dictionary will be provided for all the PHEIs. A few of them also commented that they lacked technical competencies since they were new users. Hence, responsible officers must attend technical workshops periodically and consistently communicate with the system's technical team. There were also suggestions for creating integration with other systems and developing a standardized data model. Additionally, users requested access to dynamic data and fields in real time. An automated operational definition mapping between PHEI and the data in the integrated system needs to be implemented to facilitate this demand. This will lead to more efficient data transmission processes.

5 Conclusion

The integrated system is generally beneficial in providing data about students, staff, research, publications, and commercialization as inputs for creating institutional action plans by most Public Higher Education Institutions (PHEI). Institutions have initiated the development of internal systems and data warehouses that align with the integrated system format to facilitate more efficient data management and transmission. It is believed that the integrated system's data complements the internal institutional data and becomes an important resource for developing strategic planning. However, several improvements in data dictionaries and automated data mapping can resolve some issues during data transmission.

References

1. Zipkin, E.F., Inouye, B.D., Beissinger, S.R.: Innovations in data integration for modelling populations. Ecology **100**(6), 1–3 (2019)
2. Varadharajan, C., et al.: Challenges in building an end-to-end system for acquisition, management, and integration of diverse data from sensor networks in watersheds: lessons from a mountainous community observatory in East River. Colorado. IEEE Access **7**, 182796–182813 (2019)
3. Mahanti, R.: Data quality: Dimensions, measurement, strategy, management, and governance. Quality Press (2019)
4. Wang, L.: Heterogeneous data and big data analytics. Autom. Control Inf. Sci. **3**(1), 8–15 (2017)
5. Le Sueur, H., Bruce, I.N., Geifman, N.: Masterplans Consortium: the challenges in data integration–heterogeneity and complexity in clinical trials and patient registries of Systemic Lupus Erythematosus. BMC Med. Res. Methodol. **20**, 1–5 (2020)
6. Azeroual, O., Schöpfel, J.: Quality issues of CRIS data: an exploratory investigation with universities from twelve countries. Publications **7**(1), 14 (2019)
7. Gupta, R., Gupta, I., Singh, A.K., Saxena, D., Lee, C.N.: An IoT-centric data protection method for preserving security and privacy in cloud. IEEE Syst. J. **17**(2), 2445–2454 (2022)
8. Karkošková, S.: Data governance model to enhance data quality in financial institutions. Inf. Syst. Manag. **40**(1), 90–110 (2023)
9. DeLone, W.H., McLean, E.R.: Information systems success: the quest for the dependent variable. Inf. Syst. Res. **3**(1), 60–95 (1992)
10. DeLone, W.H., McLean, E.R.: The DeLone and McLean model of information systems success: a ten-year update. J. Manag. Inf. Syst. **19**(4), 9–30 (2003)

An Exploratory Study on Data Integration of Heterogeneous Higher Education Towards Big Data Preparation

Fauziah Baharom[1(✉)], Nurul Asshura Mohamad Suyutey[1], Muhammad Noor Syarifuddin[1], Haslina Mohd[1], Yuhanis Yusof[1], Shafinah Farvin Packeer Mohamed[1], and Salwati Badroddin[2]

[1] Universiti Utara Malaysia, 06010 Sintok, Kedah, Malaysia
fauziah@uum.edu.my
[2] Ministry of Higher Education, 62200 Putrajaya, Malaysia

Abstract. Nowadays, higher education institutions generate vast amounts of data integrated from diverse data sources, ranging from academic information to research achievements. These data are gathered from several systems implemented at each higher education institution in Malaysia. These data are often heterogeneous in nature and have different formats and structures. Hence, the data should be integrated into a central warehouse to perform a holistic analysis. Currently, data about students and graduates are stored in two different warehouses, referred to in this article as WarehouseA and WarehouseB. Unfortunately, heterogeneity poses significant challenges to data integration efforts that require extensive data checking and preprocessing to ensure data quality. Since these warehouses contain some similar attributes to students' profiles, the integration of these warehouses is necessary to reduce the challenges. However, the integration of these warehouses still introduces some quality issues. Therefore, this exploratory study was carried out to investigate the challenges of data integration of higher education data in Malaysia using an action research approach. Findings from the investigation highlighted the data quality issues faced during the integration process and outlined suggestions to address these issues. This contribution is pivotal for improving data governance frameworks to ensure the quality of integrated data for higher education institutions.

Keywords: Data Integration · Heterogeneous Higher Education Data · Data Quality · Big Data

1 Introduction

In the digital transformation era, the world is changing rapidly due to the emergence of innovative technologies [1]. Every industry is using a large number of technological devices, and these devices are constantly producing data, including the education sector. Besides, data is growing continuously with the rapid development of data driven platforms such as social networks, Internet of Things, Cloud computing, and other technologies. Hence, increasing data has brought tremendous value to organizations to introduce the big data age.

Z. Mat Aji et al. (Eds.): KMICe 2024, ISEM 49, pp. 321–333, 2025.
https://doi.org/10.1007/978-3-031-91485-0_25

In the higher education environment, data are heterogeneous due to vast amounts of data are generated from many institutions that stored at diverse platforms with different structures and formats [2]. These data are large volumes of structured and unstructured data of academic data, including data about students, enrolment, graduates, employees, research and publications, institutions, teaching materials, etc. With the construction of the digital and smart campus, education-related data have become increasingly sophisticated and show a much more obvious shape of big data. Hence, transforming these data into valuable information for improving the core tasks of higher education is an essential challenge for the future of this sector. Moreover, there is an increasing demand for rapid and innovative insights to forecast performance, identify risks, and seize opportunities related to higher education.

Brown [3] highlighted that as more organizations concentrate on digital transformation initiatives, data integration becomes more crucial when everything from operations to customer loyalty is dependent on their ability to integrate disparate datasets and extract value. Data integration involves merging two or more data sets from multi-location data sources, which might be stored in various formats, then compiling them into one accessible storage [3], and providing users with a unified view [4]. Systematic analyses of the integrated data may improve the precision of estimates, more accurate decisions as well as the formulation of anticipated strategies [5].

Data integration involves an ingestion process to bring various types of data sources from its source into a central database or unified information system where it is easy to analyze and store [2, 6]. Data ingestion is defined as the transportation of data from various sources and transferring it to a storage medium where it can be accessed, used, and analyzed by an organization. The destination is typically a data warehouse, data mart, database, or document store [3]. Data ingestion is critical and should be emphasized for any big data project, as the volume of data is becoming very large and consists of various formats [2]. Although there are many support tools available for data integration, this process still has some challenges that may be encountered while implementing data integration. Amare and Simonova [2] highlighted several key challenges in data ingestion, one of the challenges is data quality. Data ingestion faces data quality issues because of several factors which are heterogeneous data, poor quality, and security [3]. Moreover, as the higher education data are generated from the abundance of data from diverse sources within these institutions, it frequently leads to data quality issues such as inconsistent data formats, data repetition, and missing values which are related to data veracity issues [2, 7].

In current practice, data about higher education students and graduates are stored in two different warehouses, namely, referred to in this article as WarehouseA and WarehouseB. These data are gathered from several systems which are implemented at each higher education institution in Malaysia. These data are often heterogeneous in nature and have different formats and structures. Unfortunately, the heterogeneity poses significant challenges to data integration efforts that require extensive data checking and preprocessing to ensure data quality. Since these warehouses contain some similar attributes to students' profiles, the integration of these warehouses is necessary to reduce the challenges. However, the integration of these two warehouses still introduce some quality issues. Therefore, this exploratory study was carried out to investigate the challenges of data integration of higher education data in Malaysia using an action research approach. Findings from the investigation highlighted the data quality issues faced during the integration process and outlined suggestions to address these issues.

2 Literature Review

In the digital age, higher education institutions (HEIs) increasingly rely on data-driven approaches to improve decision-making, optimize resource allocation, enhance institutional effectiveness, and streamline operations. However, the data generated and used within these institutions is often heterogeneous and fraught with quality issues, complicating the processes of data integration and analysis. This literature review examines recent research on higher education data, the challenges of heterogeneous data, data quality issues, and approaches to heterogeneous data integration.

2.1 Heterogeneous Data in Higher Education Institutions

Higher education institutions collect and maintain a wide range of data, ranging from structured student information to unstructured research data and multimedia content [8]. These vast amounts of data include student records, financial information, research data, course materials, etc. Unfortunately, these data are inherently heterogeneous, originating from multiple sources such as learning management systems (LMS), student information systems (SIS), and external databases. Hence, this heterogeneity is not limited to the source but extends to the format, structure, and semantics of the data, which presents significant challenges for data integration and analysis [9]. For example, data may be stored in relational databases, flat files, or even unstructured formats like text or multimedia. The lack of standardization across these sources complicates data integration and analysis.

Higher education data is crucial for various purposes, such as tracking student progress, optimizing resource allocation, and conducting institutional research. Higher education decision-makers are keen to utilize these data to produce better decisions. According to Webber and Zheng [10], data analytics in higher education can significantly enhance institutional effectiveness by enabling data-driven decisions [10]. For example, Nguyen et al. [11] discuss the role of data analytics in personalizing learning experiences and improving institutional decision-making [11]. Similarly, studies by Umer et al. [12] highlight the potential of predictive analytics to identify at-risk students and tailor interventions accordingly. Despite these advances, the effective use of data in higher education is often hindered by challenges related to data heterogeneity and quality. However, the potential of this data is often underutilized due to challenges in managing and integrating disparate data sources.

2.2 Data Quality Issues

Big Data has become an essential research area for governments, institutions, and private agencies to support their analytics decisions. Big Data refers to all about data and how the data is collected, processed, and analyzed to generate value-added data-driven insights and decisions. In the era of big data, data quality has become a critical concern across various domains, including business, healthcare, and education. The quality of data directly influences the outcomes of data-driven decision-making processes, making it a central focus for researchers and practitioners [13]. In the Big Data context, owning a huge amount of data does not often lead to valuable insights and decisions. Data characteristics such as volume, multi-heterogeneous data sources, and fast data generation, increase the risk of quality degradation and require efficient mechanisms to check data quality. However, ensuring data quality is a costly and time-consuming process since excessive computing resources are needed.

Data quality refers to the concept of understanding data of "fitness of use" or being fit for its intended use from a consumer viewpoint [14]. According to Wang and Strong [14], data quality can be evaluated across several dimensions, including accuracy, completeness, consistency, timeliness, and relevance. These dimensions serve as benchmarks for assessing whether data meets the requirements of its intended purpose. Recent studies underscore the importance of addressing data quality to ensure accurate and reliable outcomes. The quality of data in higher education is a critical concern, as poor data quality can lead to erroneous, inaccurate analysis, misguided decisions and also loss of confidence and worthiness in the data and its sources [15]. In fact, Laney in Nagle et al. [16] argued that poor data quality can have economic and social impacts, from strategic decisions which can lead to the death of a business to operation decisions that can lead to the death of individuals.

Azeroual et al. [17] highlighted common data quality issues that introduced data errors, such as missing values, duplicates, spelling mistakes, incorrect formatting, and inconsistencies. Moreover, data quality issues can arise from various factors, including incomplete data, data entry errors, outdated information, and inconsistent format across data sources [18]. In the context of higher education, these issues can have serious implications, such as misidentifying at-risk students or misallocating resources. As a result, ensuring high data quality is paramount for effective data utilization. In a related study, Wijayanti et al. [19] conducted a study to analyze some problems related to data quality on higher education in Indonesia. The issues found include data with empty values (null value), invalid data entry, data duplication, and unavailability of data or columns that match the data requirements. Additionally, Hartwell [20] highlighted the need for effective data quality management practices in higher education.

Improving data quality requires a multi-faceted approach involving technical and organizational interventions. There are studies propose various strategies for enhancing data quality:

- Data Governance: Effective data governance frameworks are essential for maintaining high data quality. As Karkošková [21] discusses, data governance involves establishing clear policies and procedures for data management that are required to enhance data quality in financial institutions.
- Automated Data Cleaning: Advances in machine learning have led to the development of automated data cleaning tools that can identify and correct errors in large datasets. Gudivada et al. [18] highlighted manual and even semi-automated approaches are impractical in the big data context for data cleaning. They believe machine learning and other advances in computer science offer unprecedented opportunities to automate

data cleaning, assessment, and monitoring tasks and propose a data governance-driven framework for the data quality lifecycle [18].

- Continuous Monitoring: Continuous monitoring of data quality is critical to identifying and addressing data quality issues. As highlighted by Pol et al. [22], dealing with large amounts of data generated from various sources in real time is a critical factor in big data management, particularly due to the constant repetition of previously generated data. Even though the existing big data processing technologies, such as Hadoop and Spark frameworks, have been developed for handling data replication across multiple clusters, these technologies are still inadequate in addressing the challenges related to data redundancy, data quality, inconsistency, and cost of maintaining storage. Hence, it becomes essential to design a framework that is capable of addressing and minimizing redundancy issues aimed at catering to present and future needs.

2.3 Integrating Heterogeneous Higher Education Data

Higher education institutions increasingly rely on diverse and large datasets. The data generated and used in these institutions are often heterogeneous, originating from various sources with different formats, structures, and semantics. The integration of such heterogeneous data is crucial for leveraging its full potential, yet it presents significant challenges. Data quality is a pervasive issue in higher education, where inconsistent or incomplete data can hinder integration efforts. As highlighted by Daraio and Glänzel [23], poor data quality can lead to inaccurate analyses and misinformed decisions, underscoring the need for robust data quality management practices in the integration process.

According to Maltese [24], the diversity of data types and sources in higher education creates challenges in data consistency, interoperability, and semantic alignment. They stated that without effective integration strategies, institutions may struggle to obtain a holistic view of their data, which is essential for informed decision-making. The tasks are complex and require sophisticated approaches. Traditional data integration approaches, such as Extract, Transform, Load (ETL) processes, often struggle with the complexity and diversity of educational data. More advanced methods, including schema matching, ontology-based integration, and machine learning techniques, have been proposed to address these challenges [25]. Recent research has focused on developing and refining techniques to manage this complexity. For example, Gu et al. [26] propose a machine learning-based framework for integrating heterogeneous data, which improves data accuracy and consistency.

For instance, a study by Tapia-León et al. [27] identified the various applications of ontologies in higher education, such as student modelling, curriculum management, and resource sharing. Similarly, Putrama and Martinek [28] highlighted the use of ontology-based approaches or semantic association, and graph-based solutions to facilitate data integration in overcoming the challenges posed by heterogeneous data sources. Another study proposed a cloud-based architecture to integrate and manage heterogeneous data in colleges and universities, highlighting the need for improved hardware utilization and resource sharing [26]. These methods aim to harmonize data from disparate sources, allowing for more comprehensive and accurate analysis.

3 Research Method

The study was conducted using action research as the primary methodology to perform the integration of higher education data. This method is ideal for addressing real-world challenges, such as the complex process of integrating data across various higher education systems, while allowing for continuous reflection and improvement. The iterative, participatory nature of action research is well-suited for both generating actionable knowledge and implementing practical solutions. There are four key stages of Action Research: Planning, Action, Observation, and Reflection. This section will elaborate on how each stage was performed in this study.

This exploratory study was conducted to check data quality during the integration of higher education data from two warehouses. It focuses on four (4) data quality criteria, which are data completeness, data accuracy, data consistency and data currentness. Table 1 provides a definition of these criteria based on ISO/IEC25012.

Table 1. Data quality criteria.

Criteria	Definition
Accuracy	The degree to which data has attributes that correctly represent the actual value of the intended attribute of a concept or event in a specific context of use. Example: incorrect program codes where a student may be recorded under the wrong academic program, such as being listed as enrolled in "Computer Science" instead of "Information Technology."
Completeness	The degree to which subject data associated with an entity has values for all expected attributes and related entity instances in a specific context of use. Example: In some cases, fields such as data related to the National Education Code are missing from the dataset, leaving crucial information incomplete
Consistency	The degree to which data has attributes free from contradiction and coherent with other data in a specific context of use. It can be either or both among data regarding one entity and across similar data for comparable entities. Example: mismatched of National Education Code. A program may be labelled with different codes in the two warehouses
Currentness	The degree to which data has attributes of the right age in a specific context of use. Example: During the integration, some graduates' records might not appear in the integrated dataset because the status of these students is not being updated. Instead of being updated as graduated, they are still considered as enrolment in warehouses

4 The Application of Action Research in Data Integration of Higher Education

This section discusses how action research was adapted to integrate higher education data from two warehouses. The four main steps, which are Planning, Action, Observation, and Reflection, were implemented as follows:

Step 1: Planning

The aim of this step is to understand the problem and define the solution to solve the problem. The Ministry of Higher Education in Malaysia (MoHE) continuously monitors and analyses HEIs performance such as students' intake, enrolment and graduates, academic achievements related to publications, researches, and collaborations. In order to have better management and holistic view of the analysis, data from every HEIs are integrated and stored in a centralized warehouse, referred in this article as WarehouseA. The WarehouseA is used to collect, manage and integrate data from 20 public universities, 36 polytechnics and 108 college community. It provides information related to data student (register, enrolment and graduation), data staff (academic and administration), data research, publication and commercialization, and also data institution. Apart from these information, MoHE also collect and manage data about graduates and their working status. These data are integrated from 20 public universities, 36 polytechnics, 108 college communities, 306 private universities and 419 other higher education institutions. These data are stored in another warehouse, referred to as Warehouse B. Two separate technical teams do the processes of collecting and managing data of these two warehouses.

The main problem arises during this process is related to data quality, where data that submitted by HEIs are commonly incomplete, have incorrect values, are inconsistent, etc. The process of data checking and cleaning is complex and time consuming. Since these two warehouses have 31 similar attributes about student profiles, therefore, data integration from these two warehouses is planned to solve this obstacle. This study focuses on integrating two data warehouses that manage higher education data in Malaysia, aiming to identify the quality issues that arise during the integration process.

Step 2: Action

This step begins with analyzing all the data formats related to student profiles. Based on our analysis, we decided to focus on integrating data from WarehouseA to WarehouseB. We identified 31 attributes in these two warehouses that share similar formats, such as name, gender, identification number, address, nationality, study program, faculty, and others. Since these formats are similar, they are considered feasible for integration and are expected to have matching values. However, for the purpose of this testing process, we decided to focus on 19 attributes only. The structure of the data integration is shown in Fig. 1.

As shown in Fig. 1, before data from WarehouseA are integrated into WarehouseB, a testing process was conducted by mapping the data in WarehouseA and WarehouseB. This testing involves data from four (4) universities. The mapping process is performed based on the flow, as shown in Fig. 2.

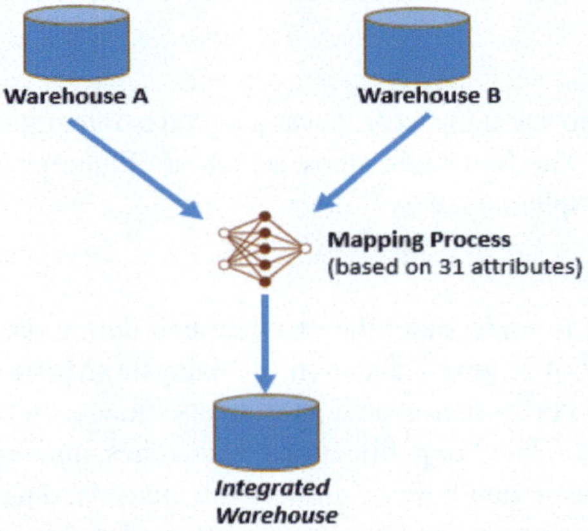

Fig. 1. Action plan for the data integration of higher education data.

As shown in Fig. 2, the process begins with the data admin of WarehouseB will prepare the data. In the current practice, every year, each university submits the list of students with their profiles to the data admin of WarehouseB. The data admin will check to ensure that the data are free from errors or any quality issues. Once the data are ready, these data will be stored in a temporary table known as Table_B. Then, the data admin of WarehouseB will send the list of students' matric numbers together with their institution ID to the data admin of WarehouseA. Based on this data, the data admin of WarehouseA will prepare the data accordingly and store the data in a temporary table named Table_A. Initial data mappings were tested to ensure accuracy, completeness, consistency and currentness between the two warehouses.

Step 3: Observation
This step focuses on analyzing the report produced from the mapping process. The process of data mapping is performed by comparing the data in Table_A and Table_B. During the comparison process, if the data is found mismatched, then a report on the data quality issues will be generated. Otherwise, the integration system will produce a report related to the integration process. After the mapping was completed, data flow and system performance were continuously monitored. Any discrepancies, such as missing data, incorrect transformations, or performance issues, were identified and logged. Regular reviews were conducted to analyze the mapped data.

Step 4: Reflection
The aim of this step is to assess the results and determine the success of the mapping, as well as plan further improvements. The outcomes of the data mapping were evaluated in terms of data quality and system efficiency. The results of the test surprised the researchers because they contradicted the assumptions that had been made. From the testing process, we found that our earlier assumption about the data of these 19 attributes is wrong. These data are supposed to have the same values because their formats are the same. Unfortunately, the findings indicate that some of the values in these attributes are

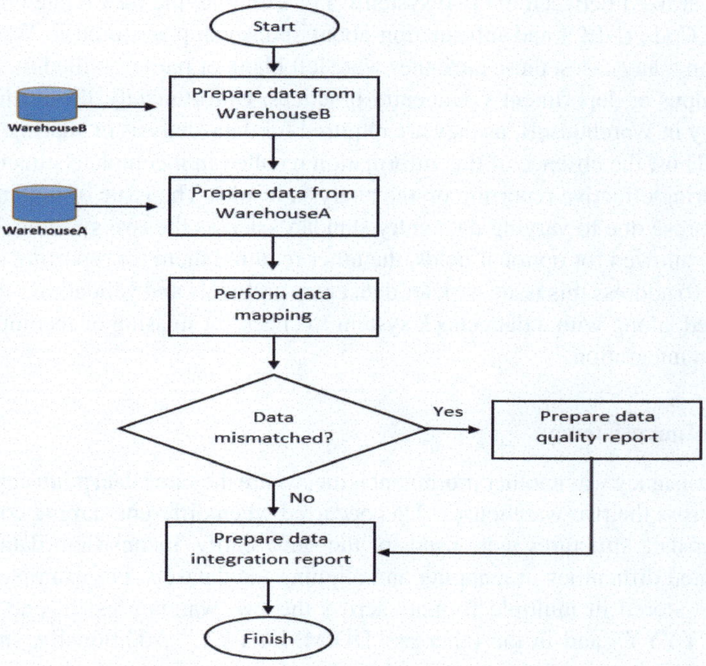

Fig. 2. The Flow of the mapping process.

mismatched and have some quality problems. Even some of the attributes considered sensitive data, such as name, identical number and gender also still contain mismatched values. Therefore, from the test results, several brainstorming sessions were held. As a result of the discussion, the researchers agreed to appoint several experts who have experience in higher education data management. Through discussion sessions with experts, proposed solutions were designed and presented.

5 Findings and Discussion

The mapping process between the two data warehouses revealed several key quality issues that affected the effectiveness and efficiency of the data integration process. The three most prominent issues identified during the mapping process were data incompleteness, data inconsistency, and data inaccuracy. Each of these challenges is discussed below, along with specific examples.

5.1 Data Incompleteness

One of the significant issues that surfaced during the data mapping process was the incompleteness of data transferred between the two systems. Certain data fields from one warehouse were not adequately populated in the other, resulting in gaps in the integrated dataset. This was especially prevalent in the case of optional fields or fields that had not been standardized between the two systems. For example, the data related to National Education Code (NEC) and information about sponsorship is stored in WarehouseA. However, in many cases, these attributes were left blank or partially filled in, depending on the campus or department's data entry practices. Unfortunately, these attributes are compulsory in WarehouseB, as they are required for data analysis or sharing with other agencies. Thus, the absence of this information resulted in incomplete student profiles, thus hindering effective reporting on the analysis results. The issue of incomplete data primarily arose due to varying data entry standards across the two systems. While one warehouse allowed for optional fields, the other required them for reporting or analysis purposes. To address this issue, stricter data entry protocols and validation rules need to be proposed, along with a data check system to check for missing or incomplete fields before data integration.

5.2 Data Inconsistency

Data inconsistency was another prominent issue, where the same data point appeared differently across the two warehouses. This occurred when different naming conventions, formats, or data structures were used for the same entity. Inconsistent data representation created difficulties in mapping and aligning the datasets. For example, "Date of Birth" was stored in multiple formats across the two warehouses: in one system as "MM/DD/YYYY" and in the other as "DD/MM/YYYY." Additionally, student race was inconsistently recorded because of these two systems using the lookup, which is not standardized. These inconsistencies complicated the process of mapping data correctly and caused errors in merged reports, such as duplication or misaligned entries for the same individuals. The inconsistency in data format and representation arose due to the lack of common data standards across the two systems. This highlighted the need for standardized formats and the use of data validation rules during data entry. To mitigate this issue, transformation rules need to be implemented to convert related attributes into a common format. Future efforts should include the development of a shared data dictionary that specifies naming conventions, data types, and formats, ensuring uniformity across both systems.

5.3 Data Inaccuracy

Data inaccuracy was also identified as a major issue during the mapping process. Certain data points were found to be incorrect or not updated, which impacted the missing dataset. This issue was particularly problematic for fields that were manually updated, leading to human error or delays in data synchronization between systems. For example, final-year student status was immediately updated; hence, this data might be missing. Another example is about the CGPA of the students. Some universities might not have a centralized system yet so any changes in this information might not be updated. When this inaccurate data was mapped to the second warehouse, it caused discrepancies in reporting and leading to incorrect information about student outcomes and enrolment. The issue of data inaccuracy underscored the importance of data quality controls and real-time validation during data entry. Manual processes are prone to human error, and the lack of synchronization between the two systems led to the propagation of inaccuracies across the integrated dataset. To address this, automated validation rules is required to detect outlier data points, flagging them for review before integration. Moreover, ensuring more frequent synchronization and updates between the two data warehouses can help minimize the occurrence of outdated or inaccurate information in the future.

6 Conclusion

The study adopted an Action Research approach, focusing primarily on integrating two data warehouses. It emphasized the importance of data quality, the painstaking process of ensuring it, the critical need for data integration, and the challenges inherent in integrating two warehouses. High data quality is paramount in any data integration project, as accurate, consistent, and complete data forms the basis for reliable decision-making, reporting, and analysis in higher education institutions. In this study, data quality issues such as incompleteness, inconsistency, and inaccuracy significantly impacted the integration process. Poor data quality led to unreliable reports and diminished trust in insights. This highlights the necessity of implementing rigorous data quality checks at every stage of the integration process.

However, the study also revealed the tedious and labour-intensive nature of checking data quality. Ensuring correct alignment between the two warehouses required detailed inspection, ongoing validation, and the correction of mismatches. While essential, this process was both time-consuming and resource-intensive. Data inconsistencies, such as differing formats or naming conventions, had to be identified and addressed, requiring repeated cycles of testing and adjustment. Despite automation efforts, significant human intervention was still necessary to oversee and resolve these issues.

Despite the challenges, data integration is critically important for institutions relying on multiple data warehouses. Integrating two warehouses provides a unified view of data, enhances accessibility, and supports better decision-making by consolidating disparate data into a cohesive system. However, this process is not straightforward. Challenges such as aligning data structures, addressing quality issues, and ensuring data synchronization across platforms present significant obstacles. In this study, issues like schema mismatches, varying data formats, and outdated information complicated the mapping process, necessitating ongoing troubleshooting.

In conclusion, while integrating heterogeneous data is essential for enhancing institutional efficiency and providing comprehensive data access, the process is fraught with challenges, particularly in ensuring data quality. The tedious tasks of validating and cleaning data must be balanced with the need to establish a system that is both accurate and reliable. Moving forward, developing a framework for Big Data Quality Governance is crucial for addressing the challenges of data integration, especially in higher education institutions where data complexity and volume are significant.

The framework will clearly define the roles and responsibilities of a cross-functional governance council, including IT officers, data owners, data stewards, and legal officers, who will oversee data quality governance policies and standards to ensure accountability for data quality and management. It also needs to be supported by automated data profiling and quality monitoring tools that enforce key dimensions such as accuracy, consistency, and completeness. Moreover, developing automated AI-driven tools for cleansing, transforming, and integrating data using AI and machine learning is required to identify quality issues and predict future inconsistencies, thereby reducing manual intervention while ensuring adherence to data quality standards. Additionally, the framework must focus on data privacy and security to ensure compliance with data protection regulations that meet legal, regulatory, and institutional standards.

Acknowledgement. This research was supported by the Ministry of Higher Education (MoHE) of Malaysia through the Fundamental Research Grant Scheme (FRGS/1/2022/ICT01/UUM/02/2).

References

1. Chae, B.K.: A general framework for studying the evolution of the digital innovation ecosystem: the case of big data. Int. J. Inf. Manage. **45**, 83–94 (2019)
2. Amare, M.Y., Simonova, S.: Learning analytics for higher education: proposal of big data ingestion architecture. SHS Web Conf. **92** (2021)
3. Brown, T.: The 3 biggest data integration challenges (how to solve them), https://itchronicles.com/big-data/3-biggest-challenges-for-data-integration/. Accessed 01 Aug 2024
4. Jovanovic, P., Nadal, S., Romero, O., Abelló, A., Bilalli, B.: Quarry: a user-centered big data integration platform. Inf. Syst. Front. **23**(1), 9–33 (2021)
5. Zipkin, E.F., Inouye, B.D., Beissinger, S.R.: Innovations in data integration for modeling populations. Ecology **100**(6), 1–3 (2019)
6. Tehreem, N.: 3 Challenges of Integrating Heterogeneous Data Sources - DZone, https://dzone.com/articles/3-challenges-of-integrating-heterogeneous-data-sou. Accessed 08 Sep 2024
7. Pansara, R.: Cultivating data quality to strategies, challenges, and impact on decision-making. Int. J. Manag. Educ. Sustain. Dev. **6**(6), 24–33 (2023)
8. Ang, K.L.M., Ge, F.L., Seng, K.P.: Big educational data & analytics: survey, architecture and challenges. IEEE Access **8**, 116392–116414 (2020)
9. Campos, V.V.D.S., Brancher, J.D., Farias, F.P., Mioni, J.L.V.M., Brahim, P.L.G.: Review and comparison of works on heterogeneous data and semantic analysis in big data. Semina: Ciências Exatas e Tecnológicas **42**(1) (2021)
10. Webber, K.L., Zheng, H.Y.: Data analytics and the imperatives for data-informed decision making in higher education. Big Data on Campus: Data Analytics and Decision Making in Higher Education, pp. 3–29 (2020)
11. Nguyen, A., Gardner, L., Sheridan, D.: Data analytics in higher education: an integrated view. J. Inf. Syst. Educ. **31**(1), 61 (2020)
12. Umer, R.T. Susnjak, Mathrani, A., Suriadi, L.: Current stance on predictive analytics in higher education: opportunities, challenges and future directions. Interact. Learn. Environ. **31**(6), 3503–3528 (2023)
13. Simard, V., Rönnqvist, M., Lebel, L., Lehoux, N.: A method to classify data quality for decision making under uncertainty. ACM J. Data Inf. Qual. **15**(2), 1–27 (2023)
14. Wang, R.Y., Strong, D.M.: Beyond accuracy: what data quality means to data consumers. J. Manag. Inf. Syst. **12**(4), 5–33 (1996)
15. Taleb, I., Serhani, M.A., Bouhaddioui, C., Dssouli, R.: Big data quality framework: a holistic approach to continuous quality management. J. Big Data **8**(1), 76 (2021)
16. Nagle, T., Redman, T., Sammon, D.: Assessing data quality: a managerial call to action. Bus. Horiz. **63**(3), 325–337 (2020)
17. Azeroual, O., Saake, G., Wastl, J.: Data measurement in research information systems: metrics for the evaluation of data quality. Scientometrics **115**(3), 1271–1290 (2018)

18. Gudivada, V., Apon, A., Ding, J.: Data quality considerations for big data and machine learning: Going beyond data cleaning and transformations. Int. J. Adv. Softw. **10**(1), 1–20 (2017)
19. Wijayanti, W., Hidayanto, A.N., Wilantika, N., Adawati, I.R., Yudhoatmojo, S.B.: Data quality assessment on higher education: a case study of Institute of Statistics. In 2018 International Seminar on Research of Information Technology and Intelligent Systems (ISRITI), pp. 231–236, Yogyakarta, Indonesia (2018)
20. Hartwell, R.H.: Assessing data quality in higher education: a study of information management practice, Ed.D., Delaware Valley University, United States -- Pennsylvania, https://www.proquest.com/docview/2488530863/abstract/27B1173BFAB7401FPQ/1. Accessed 08 Sep 2024
21. Karkošková, S.: Data governance model to enhance data quality in financial institutions. Inf. Syst. Manag. **40**(1), 90–110 (2023)
22. Pol, A.A., Cerminara, G., Germain, C., Pierini, M.: Data quality monitoring anomaly detection. Artif. Intell. High Energy Phys. World Sci. 115–149 (2022)
23. Daraio, C., Glänzel, W.: Grand challenges in data integration—state of the art and future perspectives: an introduction. Scientometrics **108**(1), 391–400 (2016)
24. Maltese, V.: Digital transformation challenges for universities: ensuring information consistency across digital services. Cataloging Classif. Q. **56**(7), 592–606 (2018)
25. Alkhamisi, A.O., Saleh, M.: Ontology opportunities and challenges: discussions from semantic data integration perspectives. In: 6th Conference on Data Science and Machine Learning Applications (CDMA), pp. 134–140. IEEE, Riyadh, Saudi Arabia (2020)
26. Gu, C., Dai, C., Shi, X., Wu, Z., Chen, C.: A cloud-based deep learning model in heterogeneous data integration system for lung cancer detection in medical industry 4.0. J. Ind. Inf. Integr. **30**, 100386 (2022)
27. Tapia-Leon, M., Rivera, A.C., Chicaiza, J., Luján-Mora, S.: Application of ontologies in higher education: a systematic mapping study. In: 2018 IEEE Global Engineering Education Conference (EDUCON), pp. 1344–1353, Tenerife: IEEE (2018)
28. Putrama, I.M., Martinek, P.: Heterogeneous data integration: challenges and opportunities. Data Brief **56**, 110853 (2024)

Understanding Cognitive Load's Effect on Mobile Application Usability: A Review

Ain Syamira Che Jailani(✉), Ridha Omar, Sharul Azim Sharudin, Masyura Ahmad Faudzi, and Zaihisma Che Cob

Universiti Tenaga Nasional, Kajang, Selangor, Malaysia
ain.syamira@uniten.edu.my

Abstract. Mobile applications are deemed to be an important tool with a significant rise, making their usability crucial for user satisfaction and successful interaction. Understanding cognitive load is essential for designing mobile applications that are easy to use. This paper examines the interaction between cognitive load and mobile application usability by analyzing factors influencing cognitive load. The primary objective is to understand further how cognitive load affects mobile application usability and to provide insights for designing user-friendly applications. It is based on a systematic literature review of articles published between 2019 and 2024 and includes definitions from existing literature. The review covers intrinsic, extraneous, and germane cognitive load and their interaction during mobile application use. It also explores how cognitive load impacts learnability, efficiency, memorability, errors, and satisfaction. In conclusion, this paper emphasizes the importance of considering cognitive load factors during the design and implementation of mobile applications.

Keywords: Cognitive Load · Usability · Mobile Application

1 Introduction

In today's digital landscape, mobile applications have become ubiquitous platforms for communication and entertainment. The usability of these applications is a critical factor that influences user experience and satisfaction [1]. Usability depends on how easy, efficient, and learnable the application is, as well as how satisfied users are with their experience [2] Previous works have acknowledged the issue of mobile applications' usability and its impact on users' satisfaction; few works have aimed to look at the correlation between cognitive load and usability in mobile applications [3–5].

Cognitive load refers to the demands placed on a user's working memory during interaction with an application, which also significantly impacts usability [6]. Measuring the cognitive load while using a mobile app provides insights into how challenging the app is for users to understand, navigate, and complete tasks [7].

Understanding how cognitive load affects mobile application usability is crucial for developers seeking to create more usable and user-friendly interfaces [8]. An increase

Z. Mat Aji et al. (Eds.): KMICe 2024, ISEM 49, pp. 334–345, 2025.
https://doi.org/10.1007/978-3-031-91485-0_26

in the number of interrelated components within an app raises the mental load required to complete tasks [9].

Primarily, this literature aims to explore the effects of cognitive load on the usability of mobile applications. To achieve this, the paper also examines the methods used to measure the user ability of mobile applications. Designing apps with optimal cognitive load can enhance usability and improve the overall user experience [7]. These findings may be helpful for designers in practice to consider cognitive load factors during the designing stage and implementation of mobile applications.

The rest of this literature is structured as follows: Sect. 2 discusses related work on cognitive load and mobile application usability. Section 3 explains the methodology used in the review process. Section 4 presents the output with the results and the ensuing discussion. Finally, Sect. 5 provides conclusions and suggestions for the upcoming work.

2 Literature and Related Works

The concept of cognitive load, developed by Sweller in the 1980s, refers to the requirements assigned to an individual's functional memory during information storage and processing [10]. The depth of cognitive load experienced by a person depends on the task's complexity, their registered knowledge and the educative materials involved [11]. Cognitive load has three main components: Intrinsic, Extrinsic, and Germane.

Intrinsic cognitive load originates from the inherent difficulty of the task and the learner's prior knowledge. Extraneous cognitive load is related to the directional design's complexity and impact on experience and skill set. Germane cognitive load involves the user's involvement in the education process and how it contributes to learning achievement [12].

Cognitive load can be measured indirectly through mental load, effort, and performance. The mental load is the severity, and the depth of a task, which is influenced by the quantity of objects that comprise the task. For instance, learning new vocabulary in a second language can be challenging due to the numerous independent elements involved [9]. Escalation of the quantity of irrelevant objects raises the mental load required to complete the task.

Mental effort pertains to the correlation between the assigned object and the individual's behaviour. It varies among individuals based on their prior knowledge, which can influence the difficulty of the assigned task [9]. Performance measures the number of errors and successes in a learning activity, like mental effort, as the user's intrinsic knowledge and behaviour influence the output [9].

In the context of cognitive load in mobile app usability, various techniques can be used to measure it, including eye tracking and questionnaires. One standard method is using NASA's Task Load Index (TLX) questionnaire [13], which assesses the perceived task load. This questionnaire is particularly effective when participants have not previously completed the task, providing more accurate results [14].

Usability examines the interconnection between system and users, activities and attributes within the use framework. Many system applications have been found inferior in meeting user needs, leading to extensive literature under the term "usability" to understand better and measure relevant factors, aiming to encapsulate all significant phenomena within a single framework or model [15].

Nielsen studied five critical features of usability: efficiency, satisfaction, learnability, memorability, and errors [16][17]. Efficiency refers to extending resources relative to the precision with which users achieve their goals. Satisfaction is the absence of negativities and the presence of positive attributes towards to object [18]. Learnability indicates how easily users can quickly start using the system effectively. Memorability means that casual user revisits idle applications without having to initiate the re-education process. Errors refer to the system's ability to maintain a low error rate, allowing users to make few mistakes and quickly recover from them [17].

Usability evaluations measure a product's utility in terms of effectiveness, efficiency, and satisfaction. Usability testing assesses efficiency, learning-free and the ability to recall interactions after an idle period. Four features of usability are measured: success rate, time-based efficiency, error rate and system usability. Based on Nielsen, usability testing involves directly testing the system with target users to determine its usability level [19]. The major point of usability testing is to ensure that the produced object effectively meets the users' needs.

Researchers have made significant progress in finding ways to address the unique challenges of mobile computing, using human-centred design principles to improve user experiences [20]. Studies show the importance of creating and using the proper methods to evaluate mobile application usability [21]. While traditional usability models provide a base, the complexities of mobile apps require a more detailed approach [22].

A vital part of this research is identifying important usability factors that make mobile applications successful. Researchers also stress the importance of human-centred design in developing mobile apps, highlighting the need to understand target users and their specific needs deeply [23].

3 Methodology

The Systematic Literature Review (SLR) is implemented applying the Preferred Reporting Items for Systematic Reviews and Meta-Analyses (PRISMA) framework [24]. PRISMA improves the output of system, assessment and analysis. The PRISMA statement consists of 27 checklist includes when systematic review reporting (with or without meta-analysis) and four phases of processes. Checklist serves as a tool for authors to ensure that all relevant information is included in their systematic review or meta-analysis manuscript. It covers various aspects of the research process, including the identification, screening, eligibility criteria, data extraction, and synthesis of studies included in the review.

The PRISMA diagram provides a visual representation of the flow of information through various phases of systematic review reporting, including the identification, screening, eligibility assessment, and inclusion of studies [24]. It helps readers understand how studies were identified, screened, included, or excluded at each review stage. The flow of the process is shown in Fig. 1.

Firstly, the development phase is to formulate research questions and the Boolean string applied and identify the number of literature for this SLR. 3 references were the

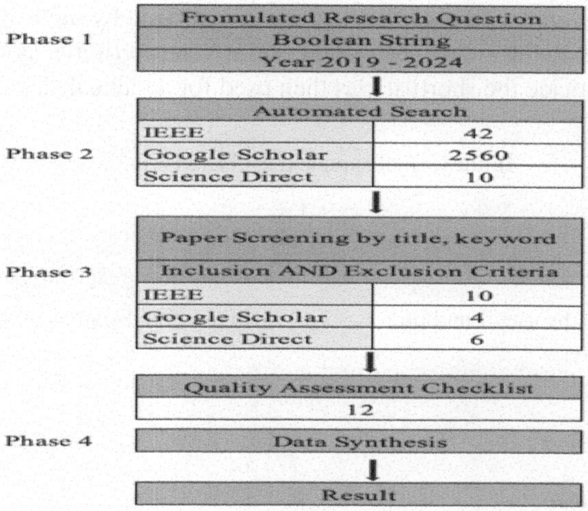

Fig. 1. Methodology of Systematic Literature Review.

sources for this report: IEEE Explore, Science Direct, and Google Scholar. It is based on publications from 2019 to 2024 and the selected relevant articles. Secondly, digital references were conducted to 2612 literature. In the subsequent phase, literatures were segregated by definition to highlight inclusiveness and exclusiveness. Literatures were then categorized by title, keyword and abstract. The ensuing phase is determined by the number of literatures. Finally, the literature was then reviewed using Quality Assessment Checklist (QAC) questions. Only 12 articles were included for the data synthesis in Phase 4.

3.1 Research Questions (RQ)

SLR aimed to understand how cognitive load affects mobile application usability. This SLR aims to explore the correlation between cognitive load and the usability of mobile applications. To gain the purpose, four RQs were formed:

RQ1: What are the measurement methods and criteria for cognitive load measuring in Mobile Applications?

RQ2: What are the measurement methods and criteria of usability in mobile applications?

RQ3: What are the implications of cognitive load on mobile application design?

RQ4: How does cognitive load affect usability in mobile applications?

3.2 Search Strategies

Search strategies were carried out in the SLR, as shown Fig. 1. The search strategies started with digital reference sources using IEEE Explore, Science Direct, and Google Scholar. The primary terms used in the digital reference sources were "Cognitive load," "Usability," and "Mobile Application." The full search strings used were "Usability," "Mobile Application," "Cognitive load," and "Usability."

The output of digital reference sources was categorized by their title, keyword, and abstract, together with the inclusiveness and exclusiveness criteria, as shown in Table 1. The literature that made the shortlist was then used for a manual search.

Table 1. Inclusion and exclusion criteria.

Inclusion Criteria	Exclusion Criteria
1. English-language article	1. Non-English language article
2. Article related to the use of mobile application	2. Article presented studies of AI/VR and AI

3.3 Study Selection

Originally, 2612 literatures were collected during the digital references sources process. The literature was then categorized by inclusiveness and exclusiveness criteria, as shown in Table 1. Then, it was categorized by its title, keyword, and abstract, together with the inclusiveness and exclusiveness criteria. This procedure involved shortlisting the article.

3.4 Quality Assessment Checklist (QAC) Questions

Twenty articles were found following the filtering procedure and the manual search of all available articles in the database. The obtained articles were then checked against the QAC questions to ensure that each piece of literature addressed the RQ. In each QAC question, each literature was classified either "Yes" if the article addressed the question adequately, "Partially" if it provided some level of response to the question, or "No" if it did not respond to the question at all. Each answer to the question was given a score of either Yes = 1 mark or Partially = 0.5 mark. From the overall score derived from these five questions, only articles that obtained 2.5 marks were approved. The QAC questions are shown in Table. Only 12 pieces of literature were produced by this technique (Table 2).

3.5 Data Synthesis

In detail, the primary goal of data synthesis was to present and provide findings through 12 selected pieces of literature that may aid in responding to the outlined research questions. This was applied to determine, synchronize, and analyze data to obtain data that would directly address the study questions. The information collected to answer RQ1, RQ2, RQ3 and RQ4 were well arranged, and the outputs were compiled under well-arranged tables and figures.

4 Result and Discussion

Each of the 12 articles that satisfied every QAC was examined separately. In this part, the literature's output was interpreted based on RQs.

Table 2. Quality Assessment Checklist (QAC) questions.

No	QAC Questions
1	Does the article clearly mention the measurement method for cognitive load in mobile applications?
2	Does the article clearly mention the criteria for cognitive load in mobile applications?
3	Does the article clearly mention measurement methods for usability in mobile applications?
4	Does the article clearly mention the criteria for usability in mobile applications?
5	How does cognitive load affect usability in mobile applications?

4.1 RQ1: What Are the Measurement Methods and Criteria for Cognitive Load Measuring in Mobile Applications?

The study identifies the cognitive load criteria reviewed in earlier studies as falling within the scope of the research. This exercise utilized extraneous, intrinsic, and germane cognitive load classifications reviewed within the cognitive load.

Table 3 represents the criteria for each cognitive load. Of the 12 articles, two explained extraneous load, two explained extraneous and intrinsic loads, and four explained all intrinsic, extraneous, and germane loads. Different cognitive load measurements are also being used to quantify each cognitive load. As represented by 12 articles, there were a variety of cognitive load measurements to quantify cognitive load criteria.

Table 3. Article cognitive load criteria and cognitive load measurement in each publication.

No	Cognitive Load Criteria	Cognitive load Measurement	Article
1	Extraneous	Suboptimal learning design	[25]
		Multimedia learning	[26]
2	Extraneous and intrinsic	Instructional teaching	[27]
		prospective, randomized, double-blind, control study	[28]
3	Intrinsic, Extraneous and Germane	Response time, mental effort ratings	[29]
		Cognitive load scale	[30]
		Time estimation	[31]
		Eye movement and Head movement	[32]

4.2 RQ2: What Are the Measurement Methods and Criteria of Usability in Mobile Applications?

Other than the five attributes of usability identified by Nielsen, success rate, time-based efficiency, error rate and system usability scale were also included as the usability measurement method. Another researcher found effectiveness, cognitive load and ease of use (Table 4). The frequency criteria of mobile applications have been mentioned in the previous article.

Table 4. Article usability criteria of the mobile application for each publication.

Usability Criteria	Number of Articles	List of Article
Efficiency	8	[2][33][34][35][36] [37][38][39]
Effectiveness	12	[2][26][27][33][34][35][36][37][38][39][40]
Satisfaction	8	[2][26][31][33][35][36][37][39]
Learnability	4	[33][35][36][38]
Memorability	4	[33][35][36][38]
Error	5	[33][34][35][36][38]
Cognitive Load	3	[33][36][38]
Ease of Use	2	[30][35]

Measurement usability in mobile applications involves various methods to evaluate the usability criteria in mobile applications. As shown in Table 5, most researchers use respondents from the Center of Mobile Application Development (PACMAD) model to measure mobile usability criteria. Other methods include eye tracking, interviews, questionnaires, and observation tests.

Table 5. Article usability measurement method of a mobile application for each publication.

Usability Measurement Method	Number of Articles	List of Article
Eye Tracking	1	[35]
Interview	1	[35]
PACMAD usability model	3	[33][36][38]
Questionnaire	2	[37][39]
Observation Test	2	[27][39]

4.3 RQ3: What Are the implications of Cognitive Load on Mobile Application Design?

The implications of cognitive load findings on designing mobile applications can hardly be overestimated. Knowledge of cognitive load effects on usability will allow optimization of system applications and make them more comprehensible and easier to use. Table 6 shows conclusions that need to be considered for mobile application design.

Table 6. Article implication of cognitive load on mobile application design.

Implication of cognitive load on mobile application design	Number of Articles	List of Article
Simplicity in feature design	8	[14][34][35][36][37][38][39][41]
Feedback mechanism	6	[14][34][37][38][39][41]
Cognitive load reduction	3	[34][28][35]
Customization Options	3	[14][28][34]
Design Guideline	3	[37][38][39]

4.4 RQ4: How Does Cognitive Load Affect Usability in Mobile Applications?

Most studies show a correlation between cognitive load and the usability of a mobile application through the user interface design. Table depicts the relationship between cognitive load and usability of a mobile system application (Table 7).

Table 7. Article relationship cognitive load and usability in mobile application.

No	Relationship Cognitive Load and Usability in Mobile Application	Reference
1	Types of design choices used led to a low amount of cognitive load	[29]
2	Interface design guidelines emphasize reducing the cognitive load in older adults, which will improve their quality of life	[41]
3	The relationship between cognitive load, usability and the subjective experience of time established. Cognitive load increases with poor design of the product interfaces	[31]
4	Instructional designers reduces the intrinsic and/or extraneous load; learning can be increased without overutilization the functional memory	[27]
5	Cognitive load can be enhanced by increasing the capabilities of the learners	[26]

4.5 Discussion

The assessment identified the methods for measuring cognitive load (RQ1) and usability (RQ2). As for RQ1, the extraneous load is the expected cognitive load that is being measured regarding the usability of mobile applications. Extraneous load is caused by how the information is presented [12], it will affect the usability of mobile applications through efficiency and effectiveness [8].

Two of Nielsen's attributes, effectiveness and satisfaction, were among the most selected methods by most articles when measuring a mobile application's usability (RQ2). All the articles use effectiveness as their method of evaluating usability. These three mainly used methods are aligned with the ISO 9241-11:2018 general usability guideline [42].

Based on the result of (RQ3), we can tailor specific programs for applications in mobile interfaces that are easier to use, faster, and best suited for the user's cognition, thus increasing overall satisfaction and usability.

The result of (RQ4) shows that usability is affected by cognitive load, particularly the extraneous load, through user interface design. Bad user interface design places a high demand on visual attention [43, 44], which later leads to a higher cognitive load. Figure 2 shows the relationship between cognitive load and the usability of mobile applications.

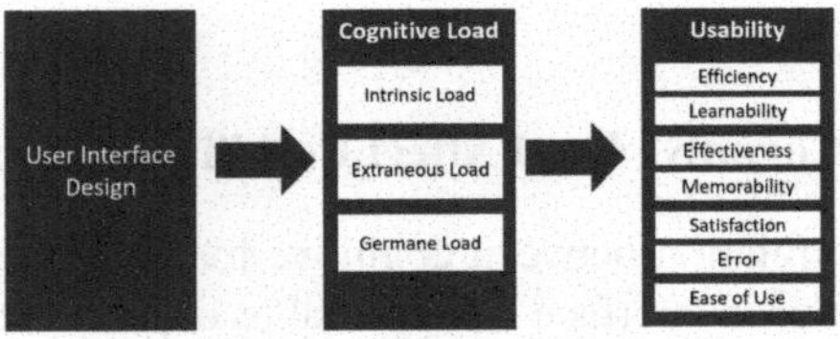

Fig. 2. The relationship between cognitive load and usability.

5 Conclusion

Researching how cognitive load affects mobile application usability can enhance our understanding in several ways. Theoretically, it can deepen our comprehension of cognitive load in mobile applications by defining its aspects and differentiating between intrinsic, extraneous, and germane types. Data from experimental setups can provide concrete statistics on how cognitive load impacts user performance, satisfaction, and efficiency.

This research can offer guidelines and recommendations for mobile application development to reduce cognitive load and improve usability. Using (PACMAD) model to measure the usability criteria of a mobile application can reduce the cognitive load and improve the practical efficiency and satisfaction of mobile application usability. These insights can benefit the industry and inform the creation of educational materials for UX/UI designers and developers.

In conclusion, this research addresses gaps in the existing literature and highlights areas for future exploration, such as the impact of AR/VR and AI on cognitive load and usability. As these emerging technologies evolve, investigating their effects on cognitive load and usability could provide valuable insights for their application in mobile design.

Acknowledgement. This study, funded by the Malaysian Ministry of Higher Education (MOHE) under the Fundamental Research Grant Scheme (FRGS) No. FRGS/1/2023/SSI07/UNITEN/03/1.

References

1. Subiyakto, A., Adhiazni, V., Nurmiati, E., Hasanati, N., Sumarsono, S., Irfan, M.: Redesigning user interface based on user experience using goal-directed design method. In: 2020 8th International Conference on Cyber and IT Service Management (CITSM), pp. 1–6. IEEE (2020)
2. Desak, G.F.P.: List of most usability evaluation in mobile application: A systematic literature review. In: 2020 International Conference on Information Management and Technology (ICIMTech), pp. 283–287. IEEE (2020)
3. Brangier, E., Hammes-Adelé, S., Bastien, J.M.C.: User experience and the impact of cognitive load: a new methodology for evaluating the mental load of user experiences in interactive systems. Behav. Inf. Technol. **39**(6), 637–657 (2020)
4. Saravani, S., Hassanzadeh, A., Gohari, M.: Cognitive load measurement in usability evaluation: a review of recent trends and future research. Univ. Access Inf. Soc. **20**, 883–897 (2021)
5. Huang, K., Liu, Y., Gao, X., Yang, J.: A model of cognitive load for complex HCI systems: application to mobile applications. Int. J. Hum. Comput. Stud. **140**, 102443 (2020)
6. Faudzi, M., Cob, Z., Sharudin, S., Omar, R., Ghazali, M.: The effects of user interface design for mobile learning application on learner's extraneous cognitive load: a conceptual framework. In: AsianChi'23 (2023)
7. Darejeh, A., Marcus, M., Mohammadi, G., Sweller, J.: A critical analysis of cognitive load measurement methods for evaluating the usability of different types of interfaces: Guidelines and framework for Human-Computer Interaction. eprint arXiv:2402.11820 (2024)
8. Alasmari, T.: The effect of screen size on students' cognitive load in mobile learning. J. Educ. Teach. Learn. **5**(2), 280–295 (2020)
9. Klepsch, M., Schmitz, F., Seufert, T.: Development and validation of two instruments measuring intrinsic, extraneous, and germane cognitive load. Front. Psychol. **8**, 1997 (2017)
10. Sweller, J.: Cognitive load during problem-solving: effects on learning. Cogn. Sci. **12**(2), 257–285 (1988)
11. Ayoob, A., Owen, J., Lee, J.: Managing cognitive load in multimedia presentations: reducing extraneous processing. Radiographics **40**(1), 151–152 (2020)
12. Sweller, J.: Cognitive load theory. In: Psychology of Learning and Motivation, vol. 55, pp. 37–76. Academic Press (2011)
13. Sandra, G., Lowell, E.: Development of NASA-TLX (Task Load Index): results of empirical and theoretical research. Adv. Psychol. **52**, 139–183 (1988)
14. Ali, W., Riaz, O., Mumtaz, S., Khan, A., Saba, T., Bahaj, S.: Mobile application usability evaluation: a study based on demography. IEEE Access **10**, 41512–41524 (2022)
15. Winter, S., Wagner, S., Deissenboeck, F.: (2008). A comprehensive model of usability. In: Engineering Interactive Systems: EIS 2007 Joint Working Conferences. EHCI 2007, DSV-IS 2007, HCSE 2007, Salamanca, Spain, March 22–24, pp. 106–122. Springer, Berlin Heidelberg (2007)
16. Nielsen. J.: Usability Engineering. Morgan Kaufmann (1994)

17. Harrison, R., Flood, D., Duce, D.: Usability of mobile applications: literature review and rationale for a new usability model. J. Interact. Sci. **1**, 1–16 (2013)
18. Ahmad, N., Hamid, N., Lokman, A.: Performing usability evaluation on multi-platform based application for efficiency, effectiveness and satisfaction enhancement. Int. J. Interact. Mob. Technol. **15**(10), 103–117 (2021)
19. Nielsen, J.: Enhancing the explanatory power of usability heuristics. In: Proceedings of the SIGCHI Conference on Human Factors in Computing Systems, pp. 152–158 (1994)
20. Alturki, R., Gay, V.: Usability attributes for mobile applications: a systematic review. In: Recent Trends and Advances in Wireless and IoT-Enabled Networks, pp. 53–62 (2019)
21. Zhang, D., Adipat, B.: Challenges, methodologies, and issues in the usability testing of mobile applications. Int. J. Hum. Comput. Interact. **18**(3), 293–308 (2005)
22. Zahra, F., Hussain, A., Mohd, H.: Usability evaluation of mobile applications; Where do we stand? In: AIP Conference Proceedings, vol. 1891, no. 1. AIP publishing (2017)
23. Eshet, E.: Human-centered design in mobile application development. Int. J. Mob. Hum. Comput. Interact. **4**(4), 1–21 (2012)
24. Moher, D., Liberati, A., Tetzlaff, J., Altman, G.: Preferred reporting items for systematic reviews and meta-analyses: the PRISMA statement. Int. J. Surg. **8**(5), 336–341 (2010)
25. Mutlu-Bayraktar, D., Cosgun, V., Altan, T.: Cognitive load in multimedia learning environments: a systematic review. Comput. Educ. **141**, 103618 (2019)
26. Banu, A., Al Siyabi, R., Al Minje, Y.: A conceptual review on integration of cognitive load theory and human-computer interaction. In: International Conference on Software Engineering & Computer Systems and 4th International Conference on Computational Science and Information Management (ICSECS-ICOCSIM), pp. 667–672, IEEE (2021)
27. McMullan, M.: Evaluation of a medication calculation mobile app using a cognitive load instructional design. Int. J. Med. Informatics **118**, 72–77 (2018)
28. Chang, H., Wu, H., Chang, Y., Tseng, Y., Wang, Y.: The effects of a virtual simulation-based, mobile technology application on nursing students' learning achievement and cognitive load: randomized controlled trial. Int. J. Nurs. Stud. **120**, 103948 (2021)
29. Englund, M., Lorenzo, S.: Investigating how design of a mobile application affects cognitive load: UI development of a mobile application for the Volvo Group IT. Umeå University, Sweden (2021)
30. Ibili, E., Billinghurst, M.: Assessing the relationship between cognitive load and the usability of a mobile augmented reality tutorial system: A study of gender effects. Int. J. Assess. Tools Educ. **6**(3), 378–395 (2019)
31. Anamalamudi, K., Surampudi, B., Peddapalyam, P., Modin, R.: Subjective time estimation to measure the cognitive load of interactive mobile user interfaces. In: International Conference on Intelligent Systems for Communication, IoT and Security (ICISCoIS), pp. 427–431, IEEE (2023)
32. Zhong, X., Hou, W.: A method for classifying cognitive load of visual tasks based on eye tracking features. In: 9th International Conference on Virtual Reality (ICVR), pp. 131–138, IEEE (2023)
33. Ali, W., Riaz, O., Mumtaz, S., Khan, A., Saba, T., Bahaj, S.: Mobile application usability evaluation: a study based on demography. IEEE Access **10**, 41512–41524 (2022)
34. Ejaz, A., Rahim, M., Khoja, S.: The effect of cognitive load on gesture acceptability of older adults in mobile application. In: IEEE 10th Annual Ubiquitous Computing, Electronics & Mobile Communication Conference (UEMCON), IEEE, pp. 0979–0986 (2019)
35. Weichbroth, P.: Usability of mobile applications: a systematic literature study. IEEE Access **8**, 55563–55577 (2020)
36. Afif, M.: Evaluating PSAU mobile application based on people at the center of mobile application development (PACMAD) usability model: empirical investigation. J. Comput. Sci. **17**(3), 275–283 (2021)

37. Bokingkito, P., Caparida, L.: Usability evaluation of a real-time water quality monitoring mobile application. Procedia Comput. Sci. **97**, 642–649 (2022)
38. Rey, W.: Assessing MABIS mobile app based on people at the center of mobile application development (PACMAD) usability model: Empirical investigation. In: 13th International Conference on Software Technology and Engineering (ICSTE), IEEE, pp. 37–43 (2023)
39. Abbasi, M., Sarmento, G., Mota, M., Martins, P., Sa, F., Cardoso, F.: Comparing user experience: an analysis of usability testing methods for mobile applications. In: 18th Iberian Conference on Information Systems and Technologies (CISTI), IEEE, pp. 1–10 (2023)
40. Tarafdar, P., Leung, A., Yue, E., Bose, I.: Understanding the impact of augmented reality product presentation on diagnosticity, cognitive load, and product sales. Int. J. Inf. Manage. **75**, 102744 (2024)
41. Chirayus, K., Nanthaamornphong, A.: Cognitive mobile design guidelines for the elderly: a preliminary study. In 17th International Conference on Electrical Engineering/Electronics, Computer, Telecommunications, and Information Technology (ECTI-CON), pp. 673–678, IEEE (2020)
42. ISO 9241-11:2018 (en), Ergonomics of human-system interaction — Part 11: Usability: Definitions and concepts. https://www.iso.org/obp/ui/#iso:std:iso:9241:-11:ed-2:v1:en. Accessed 23 Nov 2021
43. Kumar, B., Goundar, M., Chand, S.: A framework for heuristic evaluation of mobile learning applications. Educ. Inf. Technol. (Dordr) **25**(4), 3189–3204 (2020)
44. Punchoojit, L., Hongwarittorrn, N.: Usability studies on mobile user interface design patterns: a systematic literature review. Adv. Hum. Comput. Interact. **2017**(1), 6787504 (2017)

Security Requirement Engineering Practices: Sharing the Insights from Software Practitioners in a Packaging Company

Shafinah Farvin Packeer Mohamed[1] (✉), Sathiyaroobaa Vasuthevan[2], Fauziah Baharom[1], and Haslina Mohd[1]

[1] Universiti Utara Malaysia, 06010 Sintok, Kedah, Malaysia
shafinah@uum.edu.my

[2] Jabil Sdn. Bhd., 11900 Bayan Lepas, Pulau Pinang, Malaysia

Abstract. Security requirement engineering (SRE) is becoming increasingly important because the requirements are the groundwork of software which can ensure that the software produced addresses the security element. Although its significance has been disclosed, there still exist complaints received by a packaging company about security breaches in the systems they are using. The most critical issues the users face are authentication problems in accessing their internal systems, password cracking and software malfunctions. Despite rapid advances in security research, there remains a substantial knowledge gap between software practitioners in the industry and academic research. Similarly, current practices of SRE being performed by software practitioners in the packaging company are unknown. Thus, a survey was conducted among 32 software practitioners in the abovementioned company to examine current practices of SRE in real-world projects. The study was conducted using a quantitative approach through an online survey. The data was analysed by using descriptive statistical analysis. The findings of this study indicate that software practitioners in the firm are becoming more aware of the necessity of security requirements engineering practices. Unfortunately, some are yet to implement security standards and guidelines, although the company enforces security standards during software development. This study enhances the body of knowledge by uncovering the current practices of SRE among software practitioners.

Keywords: Security Requirement Engineering Practices · Survey · Software Practitioners · Packaging Company

1 Introduction

Secured software is software that can react against malicious attacks and other hacker risks and continues to function correctly under such potential risks. Every software product requests contracted security installed within it to refrain from being misused. Security is an instinctual property of software, which a large portion of the applications are lacking today, leading to software vulnerabilities persisting and impacting millions

Z. Mat Aji et al. (Eds.): KMICe 2024, ISEM 49, pp. 346–360, 2025.
https://doi.org/10.1007/978-3-031-91485-0_27

of users. Creating secure software is becoming necessary in today's environment since many unsecured software involves vulnerabilities humans can exploit. A large number of users all over the world have become victims of cybercrimes and are also easily exposed to malicious attacks, such as website crashes, password cracking, and security breaches. Deputy Communications and Digital Minister Teo Nie Ching mentioned that 3,837 cyber security incidents were reported in Malaysia from January to August 2023. The top four incidents were fraud (2,400), intrusion (396), content-related (378) and malicious code (359) [1]. The reported incidents significantly impact the software industry since they could lead to physical or financial harm. Conventional security practices are no more effective, for instance, anti-virus software firewalls, intrusion detection and intrusion prevention. Thus, researchers agree that the security practices should be addressed early during software development, which is during the requirement engineering phase [2]. The obstacle toward secure software development is due to constant pressure on software development teams to produce workable software at a quick pace. Therefore, the development team's focus has consistently been on the functional requirements. In that capacity, identifying and implementing non-functional requirements, mainly security, which is generally considered time-consuming, have consistently been dismissed or not correctly performed [3, 4]. However, as mentioned by Khan et al. [5], starting security requirement analysis at the early stage of software development can cut the cost of software development and maintenance by 12–21%.

Security requirements engineering (SRE) is the procedure of eliciting, specifying, and examining the security requirements for system fundamental ideas like "what" security requirements are [4]. If SRE is neglected, it may expose the software produced to threats. Despite its importance, the current practice of SRE among software practitioners is inadequate in tending to security issues in early development phases [3]. Researchers have conducted studies from the software practitioners' insights [6–13]. Most of the findings reveal that software practitioners are still adapting security practices despite their awareness of its needs. Some do not even implement security-related practices during software development.

Similar to the existing studies, a packaging company faces security issues with its software. This packaging company was established in 1953, and the packaging market is its core business. It has a large number of employees in numerous countries all over the world. This company has developed software systems that users use to develop imaging graphics. Comparable to other companies, this company aims to produce high quality software free from malicious attacks. Therefore, the company has defined a clear security policy for software development. Referring to the policy, software practitioners have effective response plans to prevent and mitigate cyber-attacks. Nevertheless, recently, this company has received complaints from its customers regarding the threat of cyber security breaches, such as authentication issues in accessing the system, password cracking, and software malfunction. These issues may lead to revenue loss for the company. Therefore, the company needs to investigate the current practice of SRE among its software practitioners. This chapter elaborates on the outcomes of a survey carried out in the company.

The chapter is structured as follows: first, the related works are discussed. Then, the methodology is explained and continued with the result. Subsequently, the discussion is provided. Finally, the chapter concludes with a conclusion.

2 Related Works

This section discusses the background of the study by including explanations about security requirements engineering and existing empirical studies.

2.1 Security Requirements Engineering

Security Requirement Engineering is a crucial software engineering activity that seeks to integrate security early in the software development lifecycle. Ensuring that software systems are resilient against threats and vulnerabilities is crucial. Given that security is occasionally viewed as a non-functional need, security assessments are typically performed at the final stage of the software development life cycle [14]. Yet, researchers and software practitioners agree that the security requirements should be addressed from the beginning of software development. Addressing security requirements will produce software that protects its confidentiality, integrity and availability [14]. SRE consists of four main activities: requirement elicitation, requirement specification, verification and validation, and requirement management.

Researchers have developed software lifecycle models that facilitate security efforts throughout software development. Among them are Microsoft's Security Development Lifecycle [15], Comprehensive, Lightweight Application Security Project (CLASP) by Open Web Application Security Project (OWASP) [16] and Cigital's Touchpoints [17]. Several researchers also focused on the SRE process itself, namely Secure Tropos [18, 19], Security Quality Requirements Engineering (SQUARE) [20] and CORAS [21]. Secure Tropos is an extension of the goal-driven requirements engineering technique that is security-focused [18, 19]. Mead et al. [20] proposed the SQUARE technique to gather and document security requirements. It facilitates the elicitation, categorization, and arrangement of security requirements. Conversely, CORAS is a method to guide security risk analysis [21]. It focuses on three primary artefacts: language, tools and method. The Requirements Engineering Security Maturity Model (RESMM) proposed by Niazi et al. [22] is a framework for secure requirement engineering. Chatterjee et al. [23] focused on converting security requirements and threats into designs to prevent possible threats. Ivan et al. [24] introduced the AEGIS approach to building trustworthy systems, which includes identifying security requirements based on usage context, modelling, and risk analysis. This study referred to six leading research to investigate the SRE practices: SQUARE, CORAS, SecureTropos, CLASP, Cigital Touchpoints, and Microsoft SDL.

2.2 Existing Empirical Studies

Many research has been conducted to explore software practitioners' practices for incorporating security throughout software development. Janisar et al. [6] investigated the software practitioners' awareness of secure software development. They found that the practitioners still lack awareness and must collaborate with the academicians to enhance their awareness and understanding. Furthermore, Bodin and Golberg [7] conducted interviews among software practitioners to investigate their software security culture. Three main factors were examined: individual, organizational, and material. They found that the practitioners need to be educated, security roles should exist in the organization, and the correct use of tools should be enforced. Other researchers focused on secure software development in the Agile software development environment, among them is Rindell et al. [8]. Mihelič et al. [9] investigated the Agile development of secure software for small and medium-sized enterprises, while Nagele, Korn and Matthes [10] researched information security practices in large-scale Agile software development.

Furthermore, Khan et al. [25] investigated the SRE best practices from the global software development context and obtained 70 security best practices. Assal and Chiasson [11] conducted an online survey among 123 software developers working on different software applications. Their study reveals that the software developers are well-motivated and aware of security needs. However, the secured software could not be produced without the organization's support. Moreover, Venson, Alfayez, Gomes, Figueiredo and Boehm [12] analysed the implementation of industrial software security practices. The result shows that security is extensively implemented in the projects and that high variance in secure software development activities is demonstrated throughout the projects. Assal and Chiasson [11] explored the real-life software security practices during each development lifecycle stage that highlight steps made by groups to guarantee the security of their applications. Additionally, the experiences and practices of the secure software process among software practitioners were examined in another study [13]. The researchers concluded that software practitioners were progressively becoming aware of the secure software practices importance. However, they still do not emphasize security procedures throughout software development; they just consider security requirements verifiably.

In parallel with the literature, a packaging company is receiving complaints from users regarding security issues in its software system. The company has enforced the security policy that software practitioners must follow. However, the statistical evidence of their current practices is still unknown. Khan [3] also mentioned this and stated that empirical evidence is needed for this domain. Therefore, this study examined the current practices of SRE among software practitioners in the packaging company.

3 Research Methodology

This study utilized a quantitative approach, and the data was collected through a survey. It was conducted among software practitioners in a packaging company. It comprises theoretical research, instrument design, sampling, a pilot study, data collection, and analysis. The literature was reviewed as part of the theoretical investigation to determine the issue related to the research field. The problem and scope of the study were established from these existing studies. Besides, the SRE practices were mainly identified from the literature. In this study, the software practitioners in the packaging company are the target population, and there are 35 of them. Purposive sampling was used since it is appropriate for the study. Using this sampling technique, the sample is guaranteed to meet the objectives because it is selected based on specific characteristics [26].

The instrument was constructed by employing the theoretical findings and adapting several existing instruments [11, 13, 27]. The instrument has two main sections: 1) demographic information and 2) security requirement engineering practices. A seven-point semantic differential scale ranging from Extremely Not Important to Extremely Important was used, besides yes/no answers. The instrument was validated through a pilot study to ensure the instrument's validity and readability, as well as to avoid misinterpretation of questions. The instrument was distributed via Facebook groups, involving software practitioners with at least 3 years of working experience in the related field. The respondents suggested reducing the number of questions and simplifying the questions to be more apparent. Accordingly, an amendment was made. Cronbach's Alpha test was obtained for the pilot test, which is 0.908. This value is acceptable based on Nunnally and Bernstein [28]. For the real survey, 35 potential respondents were invited to join the survey, who are software practitioners in the packaging company. The survey was conducted for two weeks. Of the 35 potential respondents, 32 participated in the study, and all their responses were usable. This number of responses is considered sufficient, as suggested by Krejcie and Morgan [29]. Data collected from the survey was analyzed using the Software Package for Social Science (SPSS) Version 23.0.

4 Results

This section details the results of the survey that was conducted.

4.1 Respondents' Background

The respondents' background is illustrated in Table 1. The majority of those respondents are programmers (53.1%), followed by system analysts (12.5%) and team leaders (9.4%). 6.3% are the quality assurance/testers/auditor and project manager. 12.5% stated the'other' position. Among the stated positions are application support executive and system engineer. Furthermore, most respondents are from the Computer Science field (65.6%). Another 15.6% are from Computer/Electrical Engineering, while 18.8% stated 'other' fields, including Information Technology and Electronics and Communication Engineering. The highest level of education for the respondents is the Bachelor's degree level (68.8%). Next, 15.6% hold a master's degree; the'other' level is a Master of Computer Applications (MCA). All of them are aware of the SRE practices, which indicates that they have prior knowledge and experience of SRE practices. They believe the security mechanism and its implementation from the beginning phases of software development are critical. Cross-tabulation analysis is used to identify respondents based on their experience and position, as demonstrated in Table 2. Only five (5) respondents have more than 10 years of experience. Most have 1 to 5 years of experience, and eight (8) are programmers.

4.2 Common Attacks Prevention Technique

The respondents were asked how they avoid introducing common attacks that occurred previously. Many of them do not consider attacks that happened in the past (43.8%).

Table 1. Respondent's position in company.

Description	Frequency	Percentage
Position		
Programmer	17	53.1
System Analyst	4	12.5
Other	4	12.5
Team Leader	3	9.4
Quality Assurance/Tester/Auditor	2	6.3
Project Manager	2	6.3
Education Field		
Computer Science	21	65.6
Computer/Electrical Engineering	5	15.6
Other	6	18.8
Education Level		
Bachelor's Degree	22	68.8
Master's Degree	5	15.6
Other	5	15.6

Table 2. Respondents' experience.

Positions level	<1 year	1–5 years	6–10 years	11–20 years	Total
Project Manager	0	1	0	1	2
Programmer	1	8	6	2	17
Quality Assurance /Tester /Auditor	0	2	0	0	2
System Analyst	0	3	1	0	4
Team Leader	0	0	2	1	3
Other	0	2	1	1	4
Total	**1**	**16**	**10**	**5**	**32**

28.1% of them consult with security experts to prevent common attacks. The other respondents refer to the document which records the security attacks that have occurred previously (15.6%), while 12.5% of respondents look for well-known common security attacks and vulnerability databases. The results are shown in Table 3.

4.3 Security Training

Respondents were asked about the percentage of security training provided. A cross-tabulation analysis is used to classify the respondents based on their position and the rate

Table 3. Prevention techniques from common attacks.

Prevention Techniques	Frequency	Percentage
Exclude past attacks from consideration	14	43.8
Consult with security experts to prevent common attacks	9	28.1
Refer to documentation that records previous security incidents	5	15.6
Check attack and vulnerability databases for widely recognized common security threats	4	12.5
Total	**32**	**100**

of security training they receive annually. Most respondents (31.3%) receive between 50% and 75% security training annually. Only 28.1% are provided with security training, more than 75% annually. On the other hand, 21.9% attended security training, between 25% and 50% in a year. Besides, only 3.1% are not provided with any security training. Meanwhile, the training is provided primarily for programmers and system analysts, 53.1% and 12.5%, respectively, as listed in Table 4.

Table 4. Percentages of security training provided.

Positions	Percentages of Trainings per Year					Total
	None	25% or less	Between 25% to 50%	Between 50% to 75%	Greater than 75%	
Project Manager	0 (0%)	0 (0%)	0 (0%)	2 (6.3%)	0 (0%)	**2 (6.3%)**
Programmer	1 (3.1%)	3 (9.4%)	3 (9.4%)	4 (12.5%)	6 (18.6%)	**17 (53.1%)**
Quality Assurance/ Tester/Auditor	0 (0%)	1 (3.1%)	0 (0%)	0 (0%)	1 (3.1%)	**2 (6.3%)**
System Analyst	0 (0%)	0 (0%)	1 (3.1%)	2 (6.3%)	1 (3.1%)	**4 (12.5%)**
Team Leader	0 (0%)	1 (3.1%)	0 (0%)	1 (3.1%)	1 (3.1%)	**3 (9.4%)**
Other	0 (0%)	0 (0%)	3 (9.4%)	1 (3.1%)	0 (0%)	**4 (12.5%)**
Total	**1 (3.1%)**	**5 (15.6%)**	**7 (21.9%)**	**10 (31.3%)**	**9 (28.1%)**	**32 (100%)**

4.4 The Use of Security Standards

Respondents were asked whether they employ security standards, guidelines, or checklists when building software. Most of them do not use security standards and guidelines but are aware that they should be used in their organization (37.5%). However, 34.3% utilize Common Criteria for Information Technology Security Evaluation guidelines. Meanwhile, only a small percentage (9.4%) use the Certified Information Systems Security Professional (CISSP) common body of knowledge in software development, and 6.3% employ the ISO 17799 Security Standard. The rest (9.4%) do not use security standards or guidelines. Table 5 presents the analysis result.

Table 5. The use of security standards.

Responses	Frequency	Percentage
No, but I am aware that security standards and guidelines are used in our organization	12	37.5
Common Criteria for Information Technology Security Evaluation	11	34.3
Do not use any security standards and guidelines	4	12.5
Certified Information Systems Security Professional (CISSP) common body of knowledge	3	9.4
ISO 17799 Security Standard	2	6.3
Total	**32**	**100**

4.5 Security Incidents Faced

The respondents were enquired about the security incidents that they face. As shown in Fig. 1, the respondents experienced many security incidents. The most popular security incidents faced by the respondents are malicious code (37.5%), followed by SQL injection (34.4%) and spamming (31.3%). Other security incidents include spoofing, denial of service, eavesdropping, and password cracking. However, 37.5% claimed they never faced any security incidents.

4.6 Software Security Procedures

The respondents were asked how they regulate IT and software security procedures. 78% of the respondents agreed that the procedure is standardized across the organization. However, 16% of them defined the procedure within each project. Only 6% answered that it is up to the individuals. Figure 2 presents the analysis result.

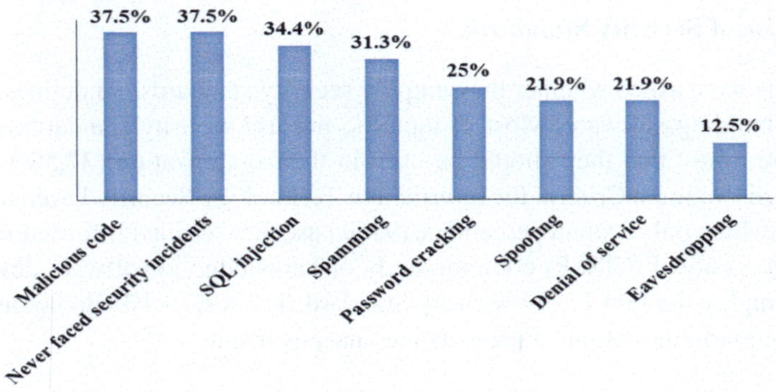

Fig. 1. The security incidents faced.

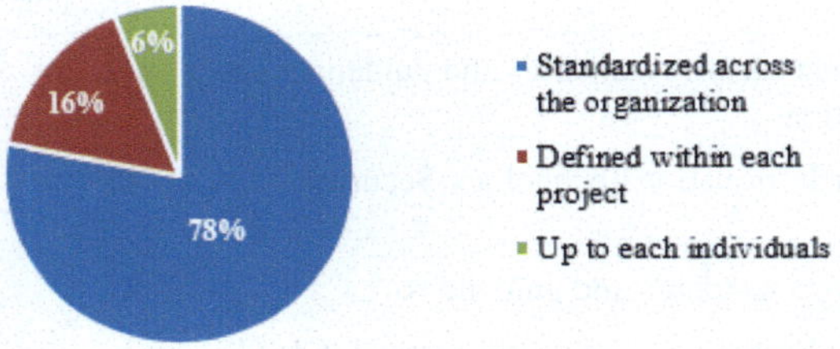

Fig. 2. The software security procedures.

4.7 Security Requirement Engineering Tools

Of the 32 respondents, only 47% have knowledge about the SRE tools. Most were not exposed to any SRE tools (53%). Among the tools that they are familiar with are STS-TOOL (25%), Magic Draw tool (10%), CORAS (3%), and SecTro2 (3%), while another 6% stated other tools like CSRF and Symantec End Point Protection. The result of the analysis is shown in Fig. 3.

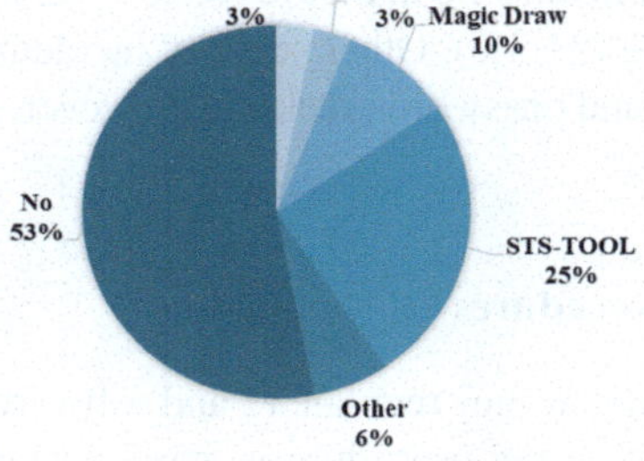

Fig. 3. The SRE tools.

4.8 Security Requirement Elicitation Practice

Next, the respondents were asked whether they had explicitly identified and documented security requirements at an early stage. 21.9% of the respondents address the security requirement from an early stage, but the requirements are not documented. Nonetheless, 25% of them gather and document the security requirements explicitly during requirement gathering. On the other hand, 12.5% of respondents only discuss security issues during the implementation process or after the system has been developed. Most commonly, 40.6% of the respondents gather and document the security requirements explicitly throughout the software development stages. Table 6 presents the analysis result.

Table 6. Eliciting security requirements explicitly during requirement gathering.

Responses	Frequency	Percentage
Security requirements are obtained and documented throughout the software development stages of the projects	13	40.6
Security requirements are obtained and documented in the early stages of the projects before the development begins	8	25
Security requirements are discussed early on but not documented	7	21.9
Security issues are only dealt with during the implementation phase or after the system is developed	4	12.5
Total	**32**	**100**

4.9 Opinion on the Security Requirement Engineering Practices

The respondents were asked their opinions on the SRE practices needed to develop high-quality software. The seven-point numerical scale is used for this question, ranging from Extremely Not Important to Extremely Important. Each practice's mean score is mapped to interval values and the degree of importance. The interval values between 4.45 to 5.30 are considered as important (I), while 5.31 to 6.16 are very important (VI) practices, as summarized in Table 7.

Table 7. The mean values for opinion on the security requirement engineering practices.

Activities	Practices	Mean	DI
Security Requirements Elicitation	Considering attackers' perspective while eliciting security requirements	5.5	(VI)
	Eliciting security requirements by considering common attacks that happened before	5.47	(VI)
	Referring to the latest lists of common attack patterns, vulnerabilities and threats to keep up-to-date with current trends	5.69	(VI)
	Obtaining security requirements explicitly	5.16	(I)
Security Requirements Specification	Security requirements and security mechanisms lead to trade-offs that must sacrifice some other requirements to obtain better security	5.5	(VI)
	Documenting security requirements in a particular notation (e.g., misuse case, attack tree)	5.62	(VI)
Security Requirements Verification and Validation	Evaluating security risks associated with the system that has been developed	5.87	(VI)
	Analysing more than one alternative security solution (e.g. security mechanisms for authentication are biometrics and passwords)	5.97	(VI)
Security Requirements Management	Producing and revising security policy in software development and implementation in organization	5.94	(VI)
	Planning and documenting the security plan	5.88	(VI)
	Documenting and maintaining a set of well-defined security requirements to prevent from introducing common attacks	5.84	(VI)
	Monitoring the identified threats, vulnerabilities and impacts continuously	5.84	(VI)
	Updating security requirements iteratively	5.78	(VI)

(*continued*)

Table 7. (*continued*)

Activities	Practices	Mean	DI
	Ensuring the newly identified risks are reported and mitigated as soon as possible	5.75	(VI)
	Performing risk analysis iteratively throughout the software development (to identify the possible threats, vulnerabilities and impacts of the application)	5.75	(VI)
	Planning mitigation strategy to countermeasure the identified threats, vulnerabilities and impacts	5.56	(VI)

DI: Degree of Importance

5 Discussion

The survey results reveal the software practitioners' current SRE practices in a packaging company. This company is currently facing security issues with its software. Aligned with previous studies [2, 6, 8, 11, 13], the respondents agreed that security mechanisms and implementation from the early stages of software development are extremely important (100%) and can lead to better quality software. They also agreed that the security mechanism is necessary to protect software code from being exploited in today's business environment. SRE is an essential phase in software development that emphasizes non-functional requirements, particularly security. Software security has been neglected or addressed towards the end of software development [3]. However, this feature is essential in producing high-quality, secure software resistant to attacks, especially in today's environment which is exposed to threats.

Software practitioners should refer to references that guide handling security issues to elicit security requirements. Aligned with the previous findings [13], the respondents in this study did not make any security references while evoking security requirements. Only a small number of them referred to the documents that record the previous attacks, consistent with the findings of Assal et al. [11] and Elahi et al. [27] studies. This shows that only some software practitioners are aware of the industry's current threats, attacks, countermeasures, and risks from previous projects. Thus, there should be a mechanism which forces them to make reference, as a real threat helps prevent the negative distinctions and can lead to improved software practitioners' attitudes and behaviors towards security. Furthermore, training has been acknowledged as one of the significant approaches to creating awareness of security issues among software practitioners [7]. The majority of the respondents attended the security training (31.3%). This outcome is in line with the findings in the study of [27], whereby most of their respondents were provided security training. However, the outcomes from the previous studies [7, 13] are contradictory, where the respondents received less security training. Only a small percentage of them went to security training. Lack of training may contribute to inappropriate implementation of the secure software system, as it does not provide a clear guideline for its execution.

Moreover, security requirements should be elicited at the beginning of software development to develop high-quality software. Only a small percentage of the respondents considered security requirements from the beginning of software development. Some respondents only managed security requirements during the implementation stage or after the system was created. This result is aligned with the findings of the existing study [13]. Oppositely, as Elahi et al. [27] indicated, most of their respondents left the security requirements undocumented and only considered them implicitly. However, in this study, the respondents' feedback is not consistent. It shows that the security requirements elicitation practices are not standardized among the software practitioners in the company. Next, the most popular security incident faced by most respondents is malicious code, which aligns with the top four security incidents reported by Bernama [1]. However, some claimed they do not confront any security incidents. This outcome is controverted with the previous study's findings [13], which showed that the more significant part of their respondents faced numerous security incidents such as password cracking, malicious code attacks and SQL injection.

Security standards are very significant in developing secured software since they provide a basis to reduce differences in security processes within the company. Most respondents did not utilize security standards and guidelines, even though the company enforces security standards. This might be why the security requirement elicitation practices are not standardized among the software practitioners in the company. Conversely, Rindel [8] and Elahi et al. [27] stated that their respondents utilized security standards or guidelines. Additionally, SRE tools help elicit security requirements, which are a comprehensive and effective way to elicit them. An example of a recently developed tool is the E-store [30]. However, the majority of the respondents of this study were not familiar with the security requirement engineering tools. Furthermore, the respondents expressed that the SRE practices included in this study are essential and must be performed. Security requirement elicitation focuses on gathering system security requirements and goals from different perspectives. Eliciting security requirements explicitly is one of the most essential practices [8].

Moreover, most researchers also emphasized that security requirements should be established from an attacker's perspective [15, 16, 20]. This task will be more effective when referring to the latest lists of familiar patterns, vulnerabilities, and threats to keep current trends up to date. Once the security requirements have been identified, they must be documented to make them official. This is to ensure that the documentation is transparent, reliable and traceable. These security requirements need to be documented in particular notation [20, 25]. Next, they must be verified to ensure the security requirements are appropriate and responsive to possible threats. The potential solutions should be analyzed so that threats can be reduced at the required levels; for example, biometrics and passwords are alternative protection mechanisms for authentication. Besides, security requirement management is vital to monitoring and recording the security requirements, the relationships between requirements, and the interactions between requirements documents. The security policy should be in place and revised regularly to ensure that the security requirements are appropriately managed. A good security policy document identifies security objectives, more concrete requirements, and design features. In addition, a security plan could be utilized to distinguish and document security requirements [15]. The company should provide support with security policy and strategy to ensure they are well-executed. Thus, security awareness is essential for software practitioners. Overall, similar to a study by Janisar et al. [6], the respondents of this study are aware of SRE and its importance; however, they lack practical aspects.

6 Conclusion

This chapter discusses the security requirement engineering practices among software practitioners in a packaging company. The outcome reveals that they are increasingly aware of the importance of security requirements engineering practices. However, some still have not implemented security standards and guidelines, although the company enforces security standards during software development. This could cause them to be unaware of the industry's current threats. They could also replicate the same risks they encountered in previous projects. This situation highlights the need for better enforcement of SRE practices among software practitioners within the company. This study has contributed to the understanding of SRE practices. It has revealed the current practices implemented by software practitioners in the real-world setting. This study can be enhanced by targeting more respondents from various software development companies to generalize the results.

References

1. Bernama, https://www.malaymail.com/news/malaysia/2023/09/13/teo-nie-ching-3837-cyber-security-incidents-reported-from-january-to-august-this-year-fraud-intrusion-among-top-four/90696, last accessed 2024/7/17
2. Khan, R.A., Khan, S.U., Akbar, M.A., Alzahrani, M.: Security risks of global software development life cycle: Industry practitioner's perspective. Journal of Software: Evolution and Process **36**(3), e2521 (2024)
3. Khan, R.A., Khan, S.U., Khan, H.U., Ilyas, M.: Systematic mapping study on security approaches in secure software engineering. IEEE Access **9**, 19139–19160 (2021)
4. Salini, P., Kanmani, S.: Survey and analysis on security requirements engineering. Comput. Electr. Eng. **38**(6), 85–97 (2012)
5. Khan, R.A., Khan, S.U., Khan, H.U., Ilyas, M.: Systematic literature review on security risks and its practices in secure software development. IEEE Access **10**, 5456–5481 (2022)
6. Janisar, A., Shafee, K., Sarlan, A., Maiwada, U., Salameh, A.A.: Securing Software Development: A Holistic Exploration of Security Awareness in Software Development Teams. International Journal of Academic Research in Business and Social Sciences **14**(1), 1326–1338 (2024)
7. Bodin, N., Golberg, H. K. B.: Software Security Culture in Development Teams: An Empirical Study, Master's Thesis, NTNU (2021)
8. Rindell, K., Ruohonen, J., Holvitie, J., Hyrynsalmi, S., Leppänen, V.: Security in agile software development: A practitioner survey. Information and Software Technology (2021)
9. Mihelič, A., Vrhovec, S., Hovelja, T.: Agile development of secure software for small and medium-sized enterprises. Sustainability **15**(1), 801 (2023)

10. Nägele, S., Korn, L., Matthes, F.: Adoption of information security practices in large-scale agile software development: A case study in the finance industry. 18th International Conference on Availability, Reliability and Security, pp. 1–12. ACM, Benevento (2023)
11. Assal, H., Chiasson, S.: Think secure from the beginning: A survey with software developers. Conference on Human Factors in Computing Systems, pp. 289–303 Glasgow (2019)
12. Venson, E., Alfayez, R., Gomes, M. M., Figueiredo, R. M., Boehm, B.: The impact of software security practices on development effort: An initial survey. ACM/IEEE International Symposium on Empirical Software Engineering and Measurement, pp. 1–12 (2019)
13. Mohamed, S.F.P.: Fauziah Baharom XE "Baharom, Fauziah", Aziz Deraman, Jamaiah Yahya, Haslina Mohd: An exploratory study on secure software practices among software practitioners in Malaysia. JTEC **8**(8), 39–45 (2016)
14. Mufti, M., Niazi, M., Alshayeb, S.: Mahmood: A readiness model for security requirements engineering. IEEE Access **6**, 28611–28631 (2018)
15. Microsoft, https://www.microsoft.com/en-us/securityengineering/sdl/practices, last accessed 2024/7/20
16. OWASP, https://owasp.org/, last accessed 2024/7/20
17. McGraw, G.: Software security: Building security in. Datenschutz und Datensicherheit-DuD **36**(9), 662–665 (2012)
18. Omilab Europe, https://austria.omilab.org/psm/content/sectro/info?view=home, last accessed 2024/7/20
19. Mouratidis, H., Giorgini, P., Manson, G.: When security meets software engineering: a case of modelling secure IS. J. Inf. Syst. **30**, 609–629 (2005)
20. Mead, N. R.: Identifying security requirements using the security quality requirements engineering (SQUARE) method. Integrating Security and Software Engineering: Advances and Future Visions, 43–69, IGI Global (2007)
21. Lund, M.S., Solhaug, B., Stølen, K.: Model-driven risk analysis: the CORAS approach. Springer-Verlag, Berlin HeidelbergSpringer Science & Business Media (2014)
22. Niazi, M., Saeed, A.M., Alshayeb, M., Mahmood, S., Zafar, S.: A maturity model for secure requirements engineering. Comput. Secur. **95**, 101852 (2020)
23. Chatterjee, K., Gupta, D., De, A.: A framework for development of secure software. CSI Transactions on ICT **1**, 143–157 (2013)
24. Ivan, F., Angela, M. Sasse, Hailes, S. M. V.: Bringing Security Home: A process for developing secure and usable systems, New Security Paradigms, Switzerland, 49–57 (2003)
25. Khan, R.A., Akbar, M.A., Rafi, S., Almagrabi, A.O., Alzahrani, M.: Evaluation of requirement engineering best practices for secure software development in GSD: an ISM analysis. Journal of Software: Evolution and Process **36**(5), e2594 (2024)
26. Zikmund, W.G., Babin, B.J., Carr, J.C., Griffin, M.: Business research methods (ninth edit). Cengage Learning, Erin Joyner (2013)
27. Elahi, G., Yu, E., Tong, L., Lin, L.: Security requirements engineering in the wild: a survey of common practice. IEEE Annual Computer Software and Applications Conference (2011)
28. Nunnally, J. C., Bernstein, I. H.: Psychometric theory. McGraw-Hill. (1994)
29. Krejcie, R.V., Morgan, D.W.: Determining sample size for research activities. Educ. Psychol. Measur. **30**(3), 607–610 (1970)
30. Hnaini, H., Mazo, R., Champeau, J., Vallejo, P., Galindo, J.: E-SCORE: A web-based tool for security requirements engineering. SoftwareX **26**, 101704 (2024)

Sharing Insights on Learning Difficulties in Introductory Programming Courses: A Preliminary Study Among Undergraduate Students

Asmalinda Adnan, Nur Hafiza Jamaludin(✉), and Rohaida Romli

Data Management and Software Solution, Research Lab, School of Computing, Universiti Utara Malaysia, 06010 UUM Sintok, Kedah, Malaysia

za5828@gmail.com

Abstract. Teaching and learning programming can be challenging within ecosystems that incorporate various techniques for engaging both educators and students with technology. Despite this, insufficient studies have been conducted to assess the current state of learning programming techniques in education. To address this gap, a preliminary study was conducted to collect data on introductory programming courses among students in tertiary education. This article presents an analysis of the results pertaining to undergraduate students, drawn from the data collected during the preliminary study, with a focus on the challenges they encounter in learning programming. A total of ninety-five undergraduate students from various public educational institutions in northern Malaysia participated in this preliminary study. The data sets were analysed using SPSS version 29.0 and interpreted using a 7-point Likert scale. The results indicated that students rated positively in aspects such as the system development environment, internet access, programming structure, program design, and problem-solving at the basic difficulty level. Students preferred self-directed learning over conventional lectures when faced with learning challenges. They identified specific concepts as the most difficult in programming and preferred examples, exercises, and visualisations over lectures. These findings are significant for determining practical teaching tools, approaches, and techniques for future programming education and for enhancing computational thinking skills as part of the fifth Industrial Revolution agenda.

Keywords: Teaching And Learning Programming · Programming Difficulties

1 Introduction

Computer programming is a fundamental skill for undergraduate students in Science, Mathematics, and Engineering [1]. However, learning programming can pose significant challenges for novice programmers [2, 3]. For instructors, teaching programming effectively requires ensuring students understand complex concepts and develop the necessary skills for solving real-world problems. This involves various processes that necessitate different ways of thinking [4]. Existing research has continued to address

Z. Mat Aji et al. (Eds.): KMICe 2024, ISEM 49, pp. 361–372, 2025.
https://doi.org/10.1007/978-3-031-91485-0_28

these difficulties, aiming to enhance teaching and learning systems by proposing effective methods, supporting digital educational tools, and developing relevant programming courses. Despite these efforts, introductory programming courses still encounter high withdrawal [5] and failure rates [6]. The rapid advancement of technology in everyday life has led to an increased demand for skilled computer scientists with strong programming capabilities and a solid foundation in computer programming.

Several researchers have identified various factors contributing to teaching and learning programming challenges. A study by Rahmat et al. [4] identified five key factors affecting programming difficulties: problem-solving ability, learning resources, teaching and learning approaches, time management, and confidence levels. Similarly, Xinogalos [7] highlighted five challenges from the student's perspective, constructing algorithms, translating algorithms into programming languages, writing code, modularisation, and testing and debugging. Gomes et al. [8] identified seven aspects that lead to programming difficulties and poor performance in programming courses. These aspects encompass students' prior knowledge, behaviours, teaching and learning methods, social backgrounds, inadequate problem-solving skills, insufficient mathematical understanding, and limited exposure to question-and-answer sessions during programming classes. Khaleel et al. [5] pointed out that challenges arise from ineffective learning experiences, lack of interest in programming courses, and low motivation. Finally, Cheah [9] identified four primary factors contributing to these difficulties: inadequate problem-solving abilities, low student efficacy, ineffective teaching methods, and personal traits and attitudes.

Although several factors have been identified as contributing to difficulties in learning programming, such as programming structures, students' characteristics, instructors' effectiveness, and learning environments, previous literature emphasises the importance of employing a combination of two or more effective teaching techniques to ensure successful educational delivery. Techniques for teaching programming should be adjusted to suit students' different levels, as an appropriate approach can help foster their motivation and engagement [10]. Moreover, studies on integrating learning techniques have revealed that 23.2% of students fail semester exams due to weaknesses in problem-solving skills [11]. Additionally, Computer Science educators often encounter challenges in assisting students with their development in programming skills. Educators can implement effective strategies and tools to support students' learning by identifying the factors contributing to these difficulties, better journeys [12].

Currently, there is a dearth of empirical data assessing the integration of programming techniques into introductory courses for students in secondary and higher education. Due to this reason, a preliminary study was conducted to gather data on these courses among secondary and tertiary education students. A portion of this data focused on the challenges associated with learning programming. This article presents an analysis and results specifically related to this aspect, involving only undergraduate students.

The remaining sections of this article are organised as follows: Sect. 2 reviews related works that address the challenges of learning programming. Section 3 outlines the methodology, and the analysis and results are presented in Sect. 4. Finally, Sect. 5 provides the article's conclusion.

2 Related Works

The challenges of teaching and learning introductory programming have been recognized since the introduction of the first programming language [13]. High failure and dropout rates among beginners have prompted researchers to investigate the factors contributing to these difficulties and seek practical solutions for enhancing students' performance and motivation in programming.

Various studies have been conducted to explore the challenges in learning programming. Scholars believe that the inherent complexities of programming contribute significantly to the difficulties faced in introductory courses. These challenges often manifest as misconceptions related to syntactic knowledge, conceptual knowledge, and strategic knowledge [3, 7, 14–16]. One of the significant issues stems from a lack of understanding of basic programming structures, which is tied to students' conceptual knowledge. This gap can lead to misconceptions and inaccurate mental models [1, 17].

Students themselves are a critical factor in the difficulties of learning programming. Interpreting problem descriptions into technical solutions is particularly challenging, as it requires students to develop appropriate mental models that enable critical and systematic thinking when solving problems. According to Bosse and Gerosa [18], tackling complex real-world problems necessitates comprehension beyond the programming language's syntax and flow. Previous studies have reported that many students struggle with problem-solving approaches and have difficulty interpreting problems accurately [1, 3, 8, 9, 16, 19].

Al Rekhawi [20] observed that instructors frequently find managing large groups of students in introductory programming courses challenging. Gomes et al. [8] suggested that the ideal class size for effectively identifying students' weaknesses should be fewer than twenty. Furthermore, instructors who lack experience in preparing course materials, employing effective teaching strategies, and using appropriate teaching tools may unintentionally contribute to misconceptions in the teaching and learning of programming. Cheah [9] highlighted the importance of aligning students' learning methods with adequate teaching resources to tackle the challenges in computer programming. Currently, reference materials and teaching resources are often static [3], outdated, and ineffective, which can lead to extended learning times and student demotivation. Additionally, ageing hardware and inadequate access to both hardware and software exacerbate the difficulties associated with learning programming [1].

In addition to addressing programming difficulties, several researchers have proposed solutions to enhance the learning experience in programming. Milne and Rowe [19] suggest using a visualisation-based approach to help students improve their problem-solving skills. Jenkins [3] provides guidelines for conducting programming courses effectively, while Brennan and Resnick [21] recommend design-based learning activities that incorporate interactive media. Khaleel et al. [5] advocate for programming learning requirements that create enjoyable environments, such as gamification. Tan [22] proposes using visualisation tools, and Susanti [13] suggests tailored approaches to address individual students' struggles with basic programming. Additionally, research by Tim [23] indicated that block-based programming offers significant advantages over text-based programming in terms of students' interest and perceived difficulty in learning.

3 Methodology

The preliminary study was conducted using a quantitative approach to collect relevant data through a survey technique. The instrument used is a questionnaire. Respondents were selected through stratified purposive sampling from various public institutions in northern Malaysia. Since this article focuses specifically on undergraduate students, the data was gathered from 95 respondents. The data was analysed using SPSS v29 to determine the minimum, maximum, median, and standard deviation.

The questionnaire used in the preliminary study was adapted from existing research [22]. This article concentrates on the challenges of learning programming, covering four primary constructs: Basic Difficulties Level (BDL), Learning Activity Difficulties (LAD), Difficult Concepts (DC), and Programming Teaching Tools (PTT). Each construct consists of different items. The BDL construct encompasses eight items, including the system development environment, internet access, programming structure, syntax, program design, bug identification, time management, and problem solving. The LAD construct includes four items, such as lecture, tutorial, practical, and self-learning. The DC construct comprises twelve items, including variables, selection, loops, recursion, arrays, pointers, parameters, data types, abstract data types, input/output handling, error handling, and library usage. The PTT construct contains six items: digital e-books, lecture notes, exercises, program examples, images, and visualisations. In total, 30 items were measured on a 7-point Likert scale adapted from another study [22], where 1 represents "very bad" and 7 represents "very good." Table 1 presents the constructs and items related to teaching and learning programming difficulties, while Table 2 details the 7-point Likert scale measurement.

Table 1. Constructs and items related to difficulties in learning programming.

B1. BASIC DIFFICULTIES LEVEL
B1 1.1 System development environment
B1 1.2 Internet access
B1 1.3 Programming structure
B1 1.4 Syntax
B1 1.5 Program design
B1 1.6 Bug identification
B1 1.7 Time management
B1 1.8 Problem solving
B2. LEARNING ACTIVITY DIFFICULTIES

(*continued*)

Table 1. (*continued*)

B1. BASIC DIFFICULTIES LEVEL
B2 2.1 In lecture
B2 2.2 In tutorial
B2 2.3 In practical
B2 2.4 In self-learning
B3. DIFFICULT CONCEPTS
B3 3.1 Variables
B3 3.2 Selection
B3 3.3 Loops
B3 3.4 Recursion
B3 3.5 Arrays
B3 3.6 Pointers
B3 3.7 Parameters
B3 3.8 Data types
B3 3.9 Abstract data types
B3 3.10 Input/output handling
B3 3.11 Error handling
B3 3.12 Library usage
B4. PROGRAMMING TEACHING TOOLS
B4 4.1 Digital e-books
B4 4.2 Lecture notes
B4 4.3 Exercises
B4 4.4 Program examples
B4 4.5 Images
B4 4.6 Visualisations

4 Analysis and Result

This section outlines the analysis and results of the preliminary study that examined the challenges faced in learning programming among undergraduate students. The descriptive statistics for each construct are detailed in the following paragraphs.

In the constructs of Basic Difficulties Level (BDL), Learning Activity Difficulties (LAD), Difficult Concept (DC), and Programming Teaching Tools (PTT), the ratings ranged from a minimum of 1 (very bad) to a maximum of 7 (very good) for all items. Within the BDL construct, six out of eight items were rated as good, while two items were assessed as neither good nor bad. For the LAD construct, all items received neutral ratings, falling within the range of 3.58 to 4.43. In the DC construct, eight items were

Table 2. Detailed measurement of the 7-point likert scale.

Likert Scale	Mean Value Interval	Interpretation
1	1.00–1.85	Very bad
2	1.86–2.71	Rather bad
3	2.72–3.57	Bad
4	3.58–4.43	Neither good nor bad
5	4.44–5.29	Good
6	5.30–6.15	Rather good
7	6.16–7.00	Very good

rated as good, two were considered neither good nor bad, and two were rated as relatively good. Finally, in the PTT construct, all items—including digital e-books, lecture notes, exercises, program examples, images, and visualisations—were rated as neither good nor bad. Tables 3, 4, 5, and 6 present the results for each learning programming construct, displaying the minimum, maximum, and mean values.

Table 3. Analysis and findings of the basic difficulties level.

Construct	Descriptive Statistic			
B1. BASIC DIFFICULTIES LEVEL	Min	Max	Mean	Interpretation
B1 1.1 System development environment	1	7	4.80	Good
B1 1.2 Internet access	1	7	4.58	Good
B1 1.3 Programming structure	1	7	4.64	Good
B1 1.4 Syntax	1	7	4.60	Good
B1 1.5 Program design	1	7	4.55	Good
B1 1.6 Bug identification	1	7	4.29	Neither good nor bad
B1 1.7 Time management	1	7	4.19	Neither good nor bad
B1 1.8 Problem solving	1	7	4.52	Good

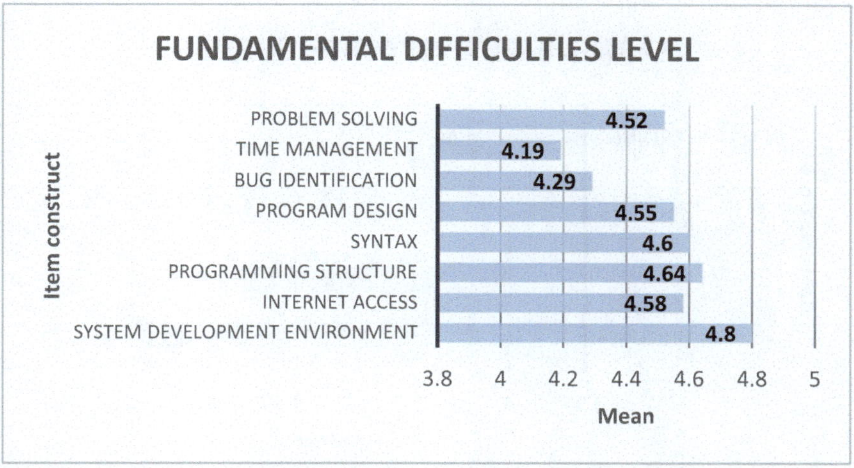

Fig. 1. Graphical analysis of means for the basic difficulties level.

According to Fig. 1, respondents identified the system development environment, internet access, programming structure, program design, and problem-solving as significant challenges in learning programming, with positive interpretations. In contrast, respondents expressed moderate agreement regarding these issues related to bug identification and time management, rating these aspects as neither good nor bad.

Table 4. Analysis and findings of learning activity difficulties.

Construct	Descriptive Statistic			
B2. LEARNING ACTIVITY DIFFICULTIES	Min	Max	Mean	Interpretation
B2 2.1 In lecture	1	7	4.25	Neither good nor bad
B2 2.2 In tutorial	1	7	4.19	Neither good nor bad
B2 2.3 In practical	1	7	4.20	Neither good nor bad
B2 2.4 In self-learning	1	7	4.06	Neither good nor bad

The analysis in Fig. 2 shows that most participants prefer self-learning over other programming-related activities such as practical exercises, tutorials, and lectures. They view self-learning as an effective method for acquiring programming skills. Additionally, respondents favour small-group discussions, such as those conducted in tutorial classes, over practical work in small classes without discussion. Overall, they appear less inclined to learn programming in large class settings.

Figure 3 summarises the difficulties associated with various programming concepts for undergraduate students. The means indicate that students face challenges with concepts such as selection, loops, pointers, parameters, data types, input/output handling, error handling, and library usage. Additionally, students experience moderate

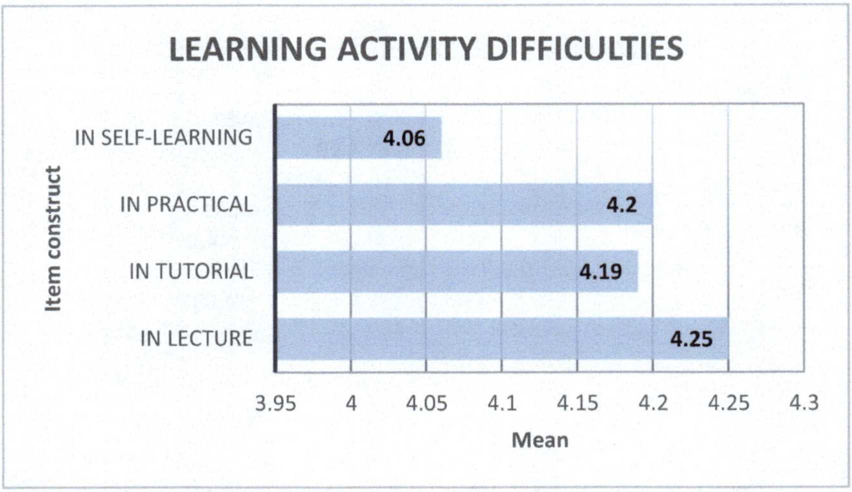

Fig. 2. Graphical analysis of means for the learning activity difficulties.

Table 5. Analysis and findings of difficult concepts.

Construct	Descriptive Statistic			
B3. DIFFICULT CONCEPTS	Min	Max	Mean	Interpretation
B3 3.1 Variables	1	7	5.47	Rather good
B3 3.2 Selection	1	7	5.24	Good
B3 3.3 Loops	1	7	4.97	Good
B3 3.4 Recursion	1	7	5.70	Rather good
B3 3.5 Arrays	1	7	4.33	Neither good nor bad
B3 3.6 Pointers	1	7	4.56	Good
B3 3.7 Parameters	1	7	5.21	Good
B3 3.8 Data types	1	7	4.97	Good
B3 3.9 Abstract data types	1	7	4.31	Neither good nor bad
B3 3.10 Input/output handling	1	7	5.29	Good
B3 3.11 Error handling	1	7	4.56	Good
B3 3.12 Library usage	1	7	4.48	Good

difficulty with the concept of variables, which is foundational to the problem-solving process. Recursion also proved challenging for students, as it is introduced later in the sub-sequence level of the programming course.

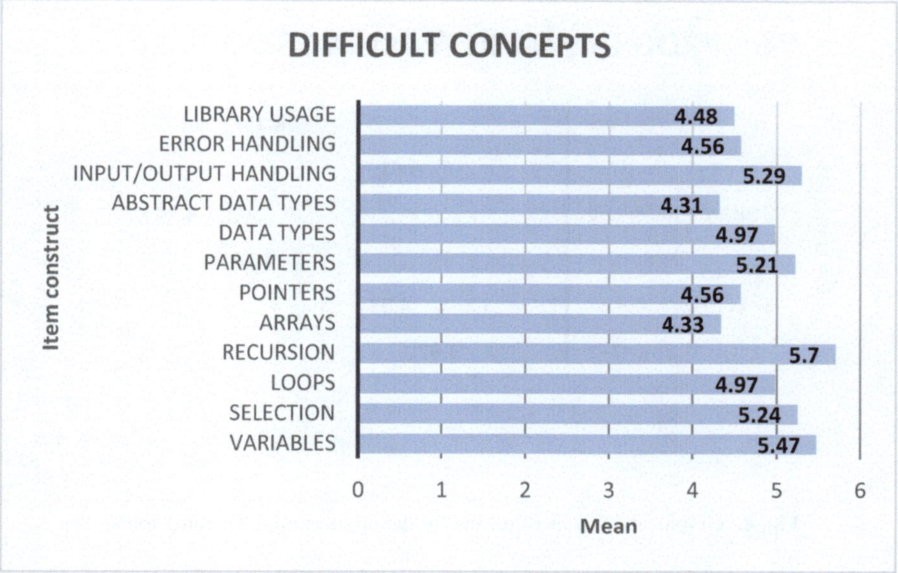

Fig. 3. Graphical analysis of means for the difficult concepts.

Table 6. Analysis and findings of programming teaching tools.

Construct	Descriptive Statistic			
B4. PROGRAMMING TEACHING TOOLS	Min	Max	Mean	Interpretation
B4 4.1 Digital e-books	1	7	3.97	Neither good nor bad
B4 4.2 Lecture notes	1	7	4.02	Neither good nor bad
B4 4.3 Exercises	1	7	4.16	Neither good nor bad
B4 4.4 Program examples	1	7	4.24	Neither good nor bad
B4 4.5 Images	1	7	4.01	Neither good nor bad
B4 4.6 Visualisations	1	7	4.15	Neither good nor bad

The analysis in Fig. 4 shows that respondents chose program examples as the most valuable teaching tools for learning programming, followed by exercises and visualisations. In contrast, the use of digital e-books was rated as less beneficial, possibly due to their less familiar context in teaching programming than lecture notes and images.

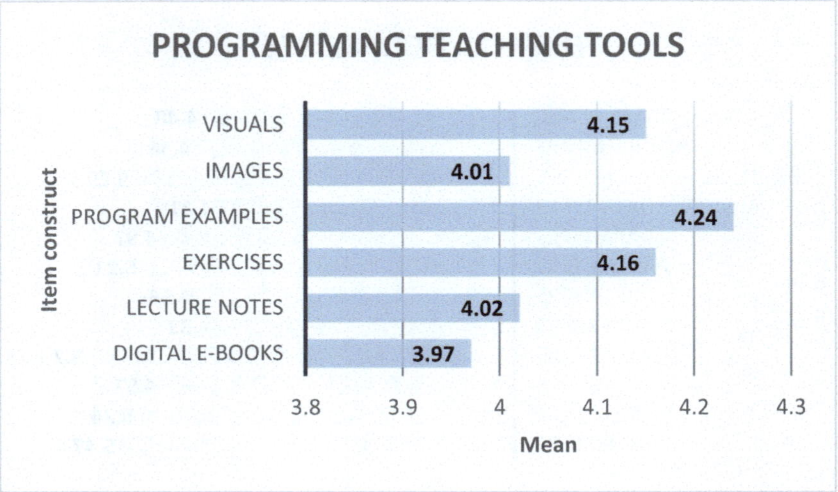

Fig. 4. Graphical analysis of means for the programming teaching tools.

5 Discussion and Conclusion

This article presents the analysis and results of the constructs related to learning programming difficulties as part of the preliminary study that has been conducted. The results indicate that students rated the system development environment, internet access, programming structure, program design, and problem-solving positively at the fundamental difficulties level. Additionally, students expressed a preference for self-directed learning over lectures in terms of managing learning activity difficulties. Students identified variables as the most challenging concepts for learning programming; they preferred program examples, exercises, and visualisations over conventional lectures.

The findings indicate that most items align with existing key conclusions. Respondents agreed that programming structure, syntax, program design, and problem-solving are challenging, which supports previous research highlighting difficulties stemming from the nature of programming [7, 13, 14, 26], and problem-solving [3, 4, 8, 9, 16, 21]. Additionally, Jenkins [3] and Lahtinen et al. [26] emphasise that programming competence can be effectively acquired through practice in practical and tutorial classes, and through self-learning. This aligns with the results of the preliminary study presented in this article, which extensively evaluated practical, tutorial, and self-study methods as effective strategies for overcoming learning activity difficulties. Furthermore, research by Qian et al. [14] and Xinogalos [7] indicates that students struggle with syntactic, conceptual, and strategic knowledge, which corresponds to the complex concepts often rated as relatively excellent or challenging in learning programming. Finally, the results regarding programming teaching tools show that respondents prefer visualisations, program examples, and exercises to enhance their learning experience in programming. This finding aligns with other studies suggesting that reference materials should not be static [3, 15], and that effective teaching materials should be considered in conjunction with learning methods to overcome programming difficulties [9].

Overall, the findings from this preliminary study are crucial for identifying suitable teaching tools, methods, and techniques for future programming education. The new generation of teaching and learning tools should support computational thinking (CT), a vital skill for the Industrial Revolution 5.0 (Industry 5.0). By emphasising human centricity, resilience, and sustainability in Industrial Revolution 5.0 [27], future programmers must develop strong programming abilities and problem-solving strategies closely related to CT skills.

References

1. Kadar, R., Wahab, N.A., Othman, J., Shamsuddin, M., Mahlon, S.B.: A study of difficulties in teaching and learning programming: a systematic literature review. Int. J. Acad. Res. Progressive Educ. Dev. **10**(3), 591–605 (2021)
2. Demilie, W.B.: Causes of failure of university students in computer programming courses: The case of Wachemo University. Int. J. Sci. Res. Comput. Sci. Eng. Inf. Technol. **5**(5), 123–132 (2019)
3. Jenkins, T.: On the difficulty of learning to program. In: Proceedings of the 3rd Annual Conference of the LTSN Centre for Information and Computer Sciences, vol. 4, pp. 53–58 (2002)
4. Rahmat, M., Shahrani, S., Latih, R., Yatim, N.F.M., Zainal, N.F.A., Ab Rahman, R.: Major problems in basic programming that influence student performance. Procedia Soc. Behav. Sci. **59**, 287–296 (2012)
5. Khaleel, F.L., Ashaari, N.S., Tengku, T.S.M., Ismail, A.: Programming learning requirements based on multi perspectives. Int. J. Electr. Comput. Eng. **7**(3), 1299–1307 (2017)
6. Bennedsen, J., Caspersen, M.E.: Failure rates in introductory programming: 12 years later. Assoc. Comput. Mach. **10**(2), 30–36 (2019)
7. Xinogalos, S.: Designing and deploying programming courses: strategies, tools, difficulties and pedagogy. Educ. Inf. Technol. **21**(3), 559–588 (2016)
8. Gomes, A.J., Santos, A.N., Mendes, A.J.: A study on students' behaviours and attitudes towards learning to program. In: Proceedings of the 17th ACM Annual Conference on Innovation and Technology in Computer Science Education, pp. 132–137 (2012)
9. Cheah, C.S.: Factors contributing to the difficulties in teaching and learning of computer programming: a literature review. Contemp. Educ. Technol. **12**(2), 1–2 (2020)
10. Attard, L., Busuttil, L.: Teacher perspectives on introducing programming constructs through coding mobile-based games to secondary school students. Inform. Educ. **19**(4), 543–568 (2020)
11. Sa'adah Binti, N., et al.: Persepsi pelajar dan pensyarah terhadap kegagalan pelajar dalam kursus pengaturcaraan (2020)
12. Singh, S.: Identifying learning challenges faced by novice/beginner computer programming students: an action research approach. In: 6th Proceedings of Software Engineering Education Workshop (SEED), pp. 63–71 (2022)
13. Susanti, W.: An overview of the teaching and learning process basic programming in algorithm and programming courses. Turk. J. Comput. Math. Educ. (TURCOMAT) **12**(2), 2934–2944 (2021)
14. Qian, Y., Lehman, J.: Students' misconceptions and other difficulties in introductory programming: a literature review. ACM Trans. Comput. Educ. (TOCE) **18**(1), 1–24 (2017)
15. Gomes, A., Mendes, A.J.: Learning to program-difficulties and solutions. In: International Conference on Engineering Education–ICEE, pp. 1–5 (2007)

16. Medeiros, R.P., Ramalho, G.L., Falcão, T.P.: A systematic literature review on teaching and learning introductory programming in higher education. IEEE Trans. Educ. **62**(2), 77–90 (2018)
17. Sirkiä, T., Sorva, J.: Exploring programming misconceptions: analysing of student mistakes in visual program simulation exercises. In: 12th International Proceedings of the Koli Calling International Conference on Computing Education Research, pp. 19–28 (2012)
18. Bosse, Y., Gerosa, M.A.: Why is programming so difficult to learn? Patterns of difficulties related to programming learning mid-stage. ACM SIGSOFT Softw. Eng. Notes **41**(6), 1–6 (2017)
19. Milne, I., Rowe, G.: Difficulties in learning and teaching programming—Views of students and tutors. Educ. Inf. Technol. **7**, 55–66 (2002)
20. Al Rekhawi, H.A.: Android applications development intelligent tutoring system. Int. J. Acad. Inf. Syst. Res. (IJAISR) **4**(5), 33–58 (2020)
21. Brennan, K., Resnick, M.: New frameworks for studying and assessing the development of computational thinking. In: Proceedings of the 2012 Annual Meeting of the American Educational Research Association, pp. 1–25. Vancouver, Canada (2012)
22. Tan, P. H., Ting, C. Y., Ling, S. W: Learning difficulties in programming courses: Undergraduates' perspective and perception. In: 2009 International Conference on Computer Technology and Development, pp. 42–46, IEEE (2009)
23. Tim, W.C.: Comparing programming interest, perceived programming difficulty, and computational thinking skills in block-based and text-based programming learning in Minecraft in secondary school students (2022)
24. Dimopoulos, D.I., Pantis, J.D.: Knowledge and attitudes regarding sea turtles in elementary students on Zakynthos Greece. J. Environ. Educ. **34**(3), 30–38 (2003)
25. Pimentel, J.L., Pimentel, J.L.: Some biases in likert scaling usage and its correction. Int. J. Sci. Basic Appl. Res. (IJSBAR) **45**(1), 183–191 (2019)
26. Lahtinen, E., Ala-Mutka, K., Järvinen, H.M.: A study of the difficulties of novice programmers. ACM SIGCSE Bull. **37**(3), 14–18 (2005)
27. Maddikunta, P.K.R., et al.: Industry 5.0: a survey on enabling technologies and potential applications. J. Ind. Inf. Integr. **26**, 1–19 (2022)

Advancing the Effectiveness of Automated Programming Assessment Systems: A Systematic Mapping Study

Abdulsamad Umar[1](✉), Rohaida Romli[2], and Shafinah Farvin Packeer Mohamed[2]

[1] School of Computing, Universiti Utara Malaysia, 06010 Sintok, Kedah, Malaysia
s.abdullah097@yahoo.com

[2] Data Management & Software Solution, Research Laboratory, School of Computing, Universiti Utara Malaysia, UUM, 06010 Sintok, Kedah, Malaysia

Abstract. Assessing students' programming assignments, which measure their understanding and proficiency with coding concepts, is among the most important aspects of Computer Science education. The rising number of students enrolling in programming courses has historically made this process more tedious, subjective, and time-consuming. The Development of automated programming assessment systems (APAS), which offer quick, consistent, fair assessments, is one way to handle these problems. APAS enhances the learning environment by simplifying the grading system and supporting fairness and transparency, freeing instructors to devote more time to teaching. Seeking to enlighten educators and researchers about the state of APAS and pinpoint areas for further research, this systematic review has been done to evaluate the present approaches and technological breakthroughs in APAS fully. The main focus is standard APAS techniques and their effectiveness in offering timely, helpful feedback in learning environments. The findings indicate that the primary grading metrics of APAS include correctness, readability, maintainability, and documentation, while its testing strategies are categorized into dynamic testing and static analysis. It has also appeared that most APAS possess feedback features. However, they face limitations in providing comprehensive feedback essential for improving students' programming skills. Future research should focus on improving the feedback systems by providing timely and personalized student feedback, which may help improve learning outcomes and enhance programming skills. Furthermore, APAS could also benefit from integrating some artificial intelligence and machine learning techniques into its testing strategies.

Keywords: Automatic Programming Assessment (APA) · Programming Assessment Techniques · Assessment Feedback Criteria

1 Introduction

Evaluating students' programming assignments is crucial component in Computer Science education as it examines their comprehension and proficiency with coding principles. This has historically been a labor-intensive procedure where instructors have

Z. Mat Aji et al. (Eds.): KMICe 2024, ISEM 49, pp. 373–384, 2025.
https://doi.org/10.1007/978-3-031-91485-0_29

had to assess manually and grade each assignment, which can be subjective and time-consuming [1, 2]. The growing number of students enrolling in programming courses is a substantial problem for educational institutions in effectively handling and promptly responding to these assignments [2, 3].

An answer to these problems has been found in Automated Programming Assessment Systems (APAS). Ala-Mutka [4] and Douce et al. [5] describe how APAS are made to automatically assess programming assignments and give students timely feedback based on preset criteria and processes. According to Wang et al. [6], these methods support fairness and openness in educational evaluations by streamlining the grading process and providing consistent, objective assessments.

In addition, by providing students with timely feedback on their programming abilities, APAS helps to create a more productive learning environment by encouraging iterative learning and progress [5]. Automating routine grading activities allows teachers to concentrate more on giving pupils personalized guidance and mentorship, improving overall education quality [4].

APAS uses various software testing methods, including static analysis and dynamic testing, to assess code efficiency, correctness, programming style, and other things without requiring human involvement. By ensuring that programming assignments are regularly scored based on established criteria, these systems lessen the possibility of bias and human error in the assessment process.

However, to fully understand their effectiveness, identify advantages and disadvantages, and highlight areas for further study, it is necessary to thoroughly evaluate the present techniques or approaches, and technological advancements employed in APAS. By offering a thorough summary of the state-of-the-art in APAS, this systematic review seeks to close this knowledge gap and educate researchers and educators on the state of the field and areas for future improvement in ongoing research.

The remaining content of this paper is structured as follows: Sect. 2 discusses the related work; Sect. 3 highlights the detailed discussion of the methodology; Sect. 4 provides the analysis and result of review; and Sect. 5 concludes this paper and provides suggestions for future research direction.

2 Related Work

Many systematic reviews have been carried out on Automated Programming Assessment Systems (APAS) to investigate different facets of its creation, application, and effectiveness. The following subsections provide an overview of recent systematic review studies that contribute to our understanding of APAS and guide our current systematic review.

2.1 Overview of Automated Programming Assessment Techniques

Ala-Mutka [4] conducted an in-depth analysis of automated assessment methods, focusing on programming assignments in particular. The review grouped several APAS techniques, including automatic code compilation, feedback systems, and syntax analysis. Ala-Mutka also compiled the available body of knowledge and concentrated on developing APAS technologies and their uses in learning environments.

Douce, Livingstone, and Orwell [5] also investigated APAS techniques within the field of programming education. In this paper, they underlined the part automated testing systems do in the objective and efficient assessment of programming tasks. The study covered some of the assessment policies and procedures APAS follows to guarantee correct grading and prompt feedback transmission. Their study delved deeply into the several automated techniques accessible for assessing programming assignments from a static analysis and dynamic testing standpoint.

2.2 Advances and Challenges in APAS Technology

Much research has been done on the challenges and new developments in APAS. Ihantola et al. [7], Ismail and Lakulu [8], and Mekterović et al. [9] all looked at market trends and technical advances in APAS by looking at adaptive algorithms, interactive feedback mechanisms, and machine learning techniques. Ismail and Lakulu [8] said that there needs to be fairness, openness, and structured learning support to improve educational result. It was also noted that there needs to be more pedagogical integration. Meterovic et al. [9] looked at the most current APAS and found that it couldn't fully meet the needs of teaching several different programming languages. They developed Edgar, an open-source APAS that combines assessment instruments and educational materials for several programming languages.

Paiva et al. [10] investigated the present state of automated assessment in Computer Science education. They evaluated the exercises provided, the testing methods employed, the security measures implemented, the types of feedback given, and the educational benefits these systems offer educators. Additionally, they discussed technological advancements such as encapsulation and static analysis and outlined future research goals in automated assessment. Moreover, Messer et al. [11] analyzed 121 academic papers about feedback systems and automated grading for programming education. They categorized these tools based on the methodology used, supported language paradigms, tested skills, automation levels, and evaluation techniques. While they acknowledged the lack of methods for evaluating readability, maintainability, and source code documentation, they stressed the significance of assessing correctness in object-oriented languages.

2.3 Features of Tools and Feedback Mechanisms

Souza et al. [12] thoroughly examined several tools used in programming assignment assessment. Their review noted various features these tools provide, such as automated feedback, multi-language support, and interaction with educational platforms. The study also filled in knowledge gaps regarding how these tools scale and adapt to various learning environments. Also, Keuning et al. [13] conducted a comprehensive investigation of studies regarding the automatic feedback generation in programming exercises. Their study focused on how well the feedback mechanisms offered by APAS enhance learning outcomes and student engagement. The review identified problems with how feedback was delivered and provided solutions to improve the standard and suitability of automated feedback for teaching programming. Ulah et al [14] also comprehensively analyzed automatic assessment systems (AAS) for novice programmers, emphasizing the value of instant feedback and requiring the least amount of work from teachers. The assessment addressed dynamic, static, and hybrid analysis approaches in AAS and underlined the need for standardized systems that can handle a range of educational settings.

These studies provide unique insights that improve the current APAS systematic review. To understand the APAS ecosystem comprehensively, the current study aims to integrate the findings of previous studies that have focused on technological advancements, system capabilities, pedagogical integration, and evaluation techniques. It emphasizes evaluating efficacy, outlining potential areas for future research, and highlighting benefits and drawbacks. By combining and analyzing multiple points of view, this study seeks to provide a thorough understanding of APAS that influences research and practical applications in educational contexts.

3 Methodology

This review study employs the Systematic Mapping Studies (SMS) technique, as outlined by Petersen, Feldt, Mujtaba, and Mattsson [15]. The SMS process involves the following five steps:

3.1 Definition of Research Question

Defining the research question(s) is pivotal in any systematic review, including SMS. The identified research questions are as follows:

RQ1: What techniques are utilized in Automated Programming Assessment Systems (APAS)?
RQ2: What are the feedback criteria in Automated Programming Assessment Systems (APAS)?

3.2 Conduct Search

Searching is a crucial step in selecting relevant studies [15]. This process involves two steps: (1) using search strings to retrieve information from electronic resources, and (2) selecting databases. The keywords and search strings used to search for relevant studies in electronic databases include ("automatic" OR "automated") AND ("assessment" OR "evaluation" OR "grading") AND ("programming" OR "program" OR "code"), "automated programming assessment systems OR "APAS" OR "automated grading" AND "programming education".

After designing the search string, the relevant databases have been chosen. Seven electronic databases were selected including ACM Digital Library, Google Scholar, Scopus, and ScienceDirect, which are known as the most relevant to scientific sources in primary studies are likely to be contained [16] and some of the selected databases are among the primary databases in the state of Computer Science.

3.3 Screening of the Papers

This stage is about screening the primary studies for the topic of this study such that the studies that are not related to answering the RQs can be excluded [15]. The inclusion and exclusion criteria applied in this study are depicted in Table 1.

Table 1. Inclusion and exclusion criteria.

Inclusion Criteria	Exclusion Criteria
• Abstracts and keywords are written in English • Studies that focused on automated systems for programming assignment evaluation or APA • Studies that presented APAS-related empirical results, theoretical insights, or creative techniques • Studies published in academic books, conference proceedings, or peer-reviewed journals	• Studies that did not focus on programming assessment • Papers where the primary language is not English • Duplicated papers • Studies that do not indicate issues related to programming assessment • Papers that are not in the field of programming education

3.4 Keywording Using Abstracts

Petersen et al. [15] describe a two-step process for keywording. Initially, abstracts are examined to identify keywords and concepts that represent the paper's contribution. These keywords are then used to cluster and categorize mapping studies. Following the categorization of papers, selected articles are thoroughly read. As each paper is reviewed, categories are updated, or new ones are created to accommodate novel findings. This systematic approach results in a systematic map comprising clustered categories derived from all relevant papers on specific topics.

3.5 Data Extraction

Data Extraction is the final stage of the SMS process. The data extraction procedure was conducted in one stage. The stage was to collect the information about the paper to address the RQs of this mapping study by presenting them in tabulation form as a summary. In this stage, nine categories or dimensions were classified by using a classification scheme for the selected studies. The classification scheme is shown in Table 2.

Table 2. Classification scheme of selected study.

Categories	Description
Feedback Criteria	Feedback criteria used or proposed in the selected studies
Strength	Importance of having the feedback criteria
Limitations	Limitation (s) on the proposed study and feedback criteria
Technique (s) of testing applied	The type (s) of the testing applied in the assessment, could be static analysis, dynamic testing or both
Feedback details	The detailed descriptions of feedback provided via the performed assessment

4 Analysis and Results

Automated Programming Assessment Systems (APAS) have changed how Computer Science educators grade their students' programming tasks. These systems use both static analysis and dynamic testing to check that syntax, logic, and how they work are all up to standards that have already been set. By automating the assessment process, APAS removes the burden of marking assignments by hand and ensures that all grades are fair and consistent. One of the best things about APAS is that they can give students quick, helpful feedback that helps them improve at coding over time. This section discusses the techniques of APAS (RQ1) and feedback criteria (RQ2) employed in the reviewed studies, emphasizing their significance in contemporary education.

4.1 The Techniques Utilized in Automated Programming Assessment Systems (APAS)

APAS uses a variety of strategies or techniques to quickly and accurately grade programming tasks. These strategies have two main types: dynamic testing and static analysis. Different types of dynamic testing include black box testing, white box testing, and grey box testing. Each type uses a different set of testing methods, such as equivalence partitioning (EP), boundary value analysis (BVA), and decision table testing (DTT) for black box testing; code coverage analysis (CCA), path coverage analysis (PCA), unit testing (UT), and data flow testing (DFT) for white box testing. Pattern matching (PM), style check (SC), source code metrics (SCM), and complexity metrics (CM) are some of the techniques used in static analysis. Each method has its own advantages and disadvantages, as demonstrated by a review of various studies and research findings. These assessment techniques can evaluate computer skills such as correctness, readability, maintainability, documentation, etc. New studies also show that AI and machine learning are widely used as ways to test computer skills.

Black box testing techniques evaluate the software without access to the internal code structure, focusing on input and output behaviour. Standard methods include EP, which reduces test cases by grouping inputs into partitions that should behave similarly [17]; BVA, which tests the edges of input ranges to detect edge cases, and off-by-one errors [18]; and decision table testing, which uses tables to ensure all input combinations and corresponding behaviors are tested, useful for complex business rules [19]. Conversely, white box testing techniques, such as unit testing, data flow testing, path coverage analysis, and condition coverage testing, focus on evaluating the internal workings of the code, ensuring comprehensive testing of code logic and structure [9, 20–22].

Static analysis involves examining code without executing it to detect errors, ensure coding standards compliance, and assess other quality metrics. Research by Ihantola et al. [7] and Liu and Petersen [23] underscores its role in identifying logical errors and syntax mistakes early in development, thereby enhancing code quality. Style check (SC) focuses on adherence to coding style rules such as naming conventions and indentation, promoting code readability, as highlighted by Brown and Parnin [24] and Chen et al. [25]. SCM provides quantitative assessments of code complexity, length, and maintainability, with Keuning et al. [13] and Liu and Petersen [23] noting their importance in objectively evaluating code quality. Complexity metrics, such as cyclomatic complexity, measure the number of linearly independent paths through the program [26], while pattern matching tests expressions for specific characteristics. However, static analysis may not fully capture runtime errors and dynamic behaviours, SC might seem restrictive to creativity, and high complexity or length doesn't always imply poor quality.

Machine Learning. The machine learning algorithms used by APAS could make the processes of grading and giving comments automatically. Machine learning algorithms can look at huge amounts of programming submissions to find common mistakes, trends in the quality of the code, and ways to give each student customized feedback based on their needs [27]. Model interpretability and adaptability to multiple programming languages are still important areas for improvement, but these methods use data-driven insights to make assessments more accurate and scalable [28–30]. As APAS continues to grow, adding machine learning could improve how well students learn code and speed up the assessment process.

Grading Metrics. APAS usually grades computer skills based on a number of factors, such as correctness, readability, maintainability, and documentation. Unit testing and output matching are the main ways that correctness is checked to make sure that the code works as it should [9, 20, 21]. Enforcing writing standards and best practices and using tools like Style Check and Static Analysis to make code more transparent [24, 25] makes it easier to read. Static analysis and source code metrics measure maintainability by looking at how easy it is to understand, change, and add to code [13, 23]. Documentation is essential for communication in software development, but it's not reviewed very often. Automated tools give feedback on the quality and quantity of comments [31].

4.2 The Criteria of Feedback in Automated Programming Assessment Systems (APAS)

This research is based on a thorough review of several studies examining different aspects of APAS (see Table 3). The studies are grouped into five main types of feedback: Feedback quality, timeliness, adaptability, feedback types, and feedback delivery.

Feedback Quality. The capacity of APAS to offer thorough and timely feedback is one of its main advantages since it dramatically improves learning results. Research by [4, 7, 11–14] demonstrates how APAS uses interactive feedback mechanisms to sustain high levels of student engagement. Additionally, these systems guarantee consistent feedback for various students and tasks and support a variety of programming languages. High-quality feedback is essential for enhancing learning results since it enables students to grasp complex concepts better and make performance changes. Thanks to its support of several languages, all students will get clear and reasonable counsel regardless of their language background. Furthermore, insightful and interactive comments let students participate more, enhancing the learning process's exciting and motivating aspects. The absence of thorough comments can, however, have clear negative consequences. If students do not get high-quality feedback, they could get inconsistent and unreliable comments which could hinder their learning process by not knowing what is expected of them in their programming tasks. When students receive too much information without clear guidance on how to use or prioritize it, they may experience feedback

Table 3. Analysis of studies based on feedback criteria in APAS.

Feedback Criteria	Studies	Strength	Limitation
Feedback quality	[4, 7, 11, 12, 13, 14]	-Improves outcomes of learning -Multiple language support -High student engagement through interactive feedback -Automated feedback consistency	-Inconsistent quality of feedback -Potential for feedback overload -Limited ability to address complex, higher-order thinking skills -Feedback may lack depth
Timeliness	[5, 8, 10, 12, 14]	-Quick turnaround time for feedback -Reduces instructor workload -Enables frequent practice and iterative improvement -Supports real-time feedback	- May not provide in-depth feedback -High reliance on uptime and system efficiency -Possible delays during times of high traffic
Adaptability	[7, 10, 13]	-Adapts feedback based on student performance -Personalized learning experience - Supports various programming languages and exercise types -Configurable models	-Difficulty in maintaining consistency across different programming languages -Adaptive mechanisms can be complex to implement -May require extensive customization
Feedback Types	[11, 13, 14]	-Supports diverse feedback types (textual hints, code annotations, etc.) - Can provide qualitative feedback based on standard errors -Immediate, detailed feedback	-Feedback may be generic and not tailored to individual student's needs - Limited by predefined feedback templates -Can lead to dependency on system-generated feedback

(*continued*)

Table 3. (*continued*)

Feedback Criteria	Studies	Strength	Limitation
Feedback Delivery	[5, 13]	-Effective delivery methods (e.g., immediate on submission, integrated with tutorials) - Enhances student motivation and engagement - Multiple delivery channels	-May require significant setup and maintenance -Potential technical issues in feedback delivery can disrupt the learning process - Dependence on robust infrastructure

overload, which reduces the input's overall usefulness. Students are lack the critical problem-solving skills and in-depth understanding necessary to succeed in increasingly complex programming environments when they receive inadequate feedback, which frequently ignores the sophisticated, higher-order thinking skills required for complicated programming jobs. General and shallow has little value in helping students learn from their mistakes and become better programmers. As a result, students tend to repeat the same mistakes, and their progress is inhibited.

Timeliness. APAS are very good at giving students their feedback quickly, which cuts down on their waiting time. Research by [5, 8, 10, 12, 14] provides evidence for this criterion. Giving students rapid feedback makes grading easier for teachers and motivates them to practice more often, which helps them develop their skills incrementally. Feedback inside APAS must be provided promptly for several reasons. Students can apply suggestions quickly, which leads to faster skill development and a more dynamic learning process. This helps students learn and improve more quickly. This timely feedback also increases instructor efficiency by simplifying their workload and allowing them to concentrate on more strategic areas of instruction. Additionally, it also encourages continual learning cycles, making it possible to practice frequently and enhance concepts through iterative improvement. Furthermore, timely feedback enables real-time adjustments, allowing students to make immediate corrections and stay on track while fully understanding concepts in the moment. Some systems allow for real-time feedback while coding. On the other hand, rapid response may sacrifice depth. The input is timely, although it might not be as detailed as is necessary. Furthermore, system availability and efficiency are critical to APAS's efficacy, and delays may occur during high traffic.

Adaptability. Studies by [7, 10, 13] show how APAS is in terms of providing customized feedback based on student performance/assessment. These systems offer a customized learning experience, support multiple programming languages and exercise forms, and have changeable models fit for different educational requirements. Still, it might be challenging to maintain consistency in feedback across many customizing languages. Using adaptive mechanisms can be technically difficult and usually requires major customizing.

Types of Feedback. APAS can handle different kinds of feedback, such as textual comments and code annotations, as seen by [11, 13], and [14]. These systems can immediately give clear, and complete feedback based on common mistakes. Adding different types of feedback to APAS makes a complete support system that works with all learning styles. This makes the learning environment more effective and interesting. On the other hand, without different kinds of comments in the APAS, students will probably get generic answers that neglect their own learning needs and particular strengths or shortcomings. Insufficient personalized feedback can compromise the efficacy of the educational process by depriving pupils of the subtle direction required for their individual growth. The problem is exacerbated by the system's reliance on a small number of pre-established feedback templates, which limits the feedback to a one-size-fits-all strategy that may miss important aspects of individual performance. As a result, students may become overly dependent on the system-generated feedback, which can diminish their ability to engage in self-assessment and critical thinking.

Feedback Delivery. Two studies show the efficacy of APAS's feedback delivery methods [5, 13]. These systems efficiently provide feedback, either in real-time upon submission or in conjunction with tutorials, increasing the motivation and involvement of students. Feedback can be sent in a variety of ways, giving you freedom. On the other hand, substantial setup and upkeep are needed to guarantee efficient feedback supply. The learning process may be interfered with by potential technical problems. Hence, the system must have a robust infrastructure.

5 Conclusion and Future Work

Static analysis is one of the best strategies for evaluating code quality. It can discover logical and syntax errors, and whether the code meets specific standards, all without executing the code. However, it fails to accurately depict the code's performance during execution, omitting information about runtime errors and dynamic behaviours. Additionally, it has the potential to enforce strict coding conventions, which may restrict the student's creativity. Dynamic testing is extensively used in APAS because it examines the code's behaviour during execution, providing a thorough insight into its performance in real-world scenarios. Black-box and white-box testing thoroughly validate Code functionality and robustness, which provide comprehensive insights into program operations under different scenarios. Dynamic testing is a crucial and commonly used method in automated programming assessment systems. It helps identify runtime flaws and evaluate code logic in various situations, improving evaluations' reliability and accuracy.

The limitations of APAS can be seen from its deficiency in providing comprehensive feedback. Comprehensive feedback is crucial in programming education since it improves learning by addressing errors/faults and fundamental coding concepts. Giving feedback on correctness, readability, maintainability, and documentation helps students understand the intricacies of effective coding practices. It facilitates incremental enhancement, aiding students in comprehending and implementing programming topics proficiently. APAS plays a crucial role in evaluating programming skills, so further development in this area of research is required.

With the emergence of Artificial Intelligence (AI) in recent times, integrating AI with APAS may potentially improve some of the limitations present in the current APAS. Furthermore, future studies could focus on improving the feedback features of APAS by providing more responsive, real-time feedback, which could significantly improve student code and enhance their programming skills.

Acknowledgment. The authors acknowledge the Malaysian Ministry of Higher Education (MoHE) for funding this research through the Fundamental Research Grant Scheme (FRGS/1/2022/ICT01/UUM02/1).

References

1. Romli, R., Sarker, S., Omar, M., Mahmod, M.: Automated test cases and test data generation for dynamic structural testing in automatic programming assessment using MC/DC. Int. J. Adv. Sci. Eng. Inform. Technol. **10**(1), 120–127 (2020)
2. Denny, P., Luxton-Reilly, A., Simon, B.: Evaluating a new exam question: parsons problems. In: Proceedings of the Fourth International Workshop on Computing Education Research, pp. 113–124 (2011)
3. Romli, R., Sulaiman, S., Zamli, K.Z.: Improving automated programming assessments: user experience evaluation using FaSt-generator. Procedia Comput. Sci. **72**, 186–193 (2015)
4. Ala-Mutka, K.M.: A survey of automated assessment approaches for programming assignments. Comput. Sci. Educ. **15**(2), 83–102 (2005)
5. Douce, C., Livingstone, D., Orwell, J.: Automatic test-based assessment of programming: a review. J. Educ. Res. Comput. (JERIC) **5**(3), 4 (2005)
6. Wang, T.H., Su, Y.S., Cheung, S.K., Kwok, L.F.: Evaluating the effectiveness of an automated feedback system for computer programming courses. J. Educ. Comput. Res. **56**(7), 1345–1366 (2018)
7. Ihantola, P., Ahoniemi, T., Karavirta, V., Seppälä, O.: Review of recent systems for automatic assessment of programming assignments. In: Proceedings of the 10th Koli Calling International Conference on Computing Education Research, pp. 86–93 (2015)
8. Ismail, M.H., Lakulu, M.M.: A critical review on recent proposed automated programming assessment tool. Psychol. Educ. J. **57**(8), 1049–1060 (2020)
9. Mekterović, I., Brkić, L., Milašinović, B., Baranović, M.: We are building a comprehensive automated programming assessment system. IEEE access **8**, 81154–81172 (2020)
10. Paiva, J.C., Leal, J.P., Figueira, Á.: Automated assessment in computer science education: a state-of-the-art review. ACM Trans. Comput. Educ. (TOCE) **22**(3), 1–40 (2022)
11. Messer, M., Brown, N.C., Kölling, M., Shi, M.: Automated grading and feedback tools for programming education: a systematic review. ACM Trans. Comput. Educ. **24**(1), 1–43 (2024)
12. Souza, D.M., Felizardo, K.R., Barbosa, E.F.: A systematic literature review of assessment tools for programming assignments. In: 2016 IEEE 29th International Conference on Software Engineering Education and Training (CSEET), pp. 147–156 (2016)
13. Keuning, H., Heeren, B., Jeuring, J.: A systematic literature review of automated feedback generation for programming exercises. ACM Trans. Comput. Educ. (TOCE) **19**(1), 1–43 (2018)
14. Ullah, Z., Lajis, A., Jamjoom, M., Altalhi, A., Al-Ghamdi, A., Saleem, F.: The effect of automatic assessment on novice programming: strengths and limitations of existing systems. Comput. Appl. Eng. Educ. **26**(6), 2328–2341 (2018)

15. Petersen, K., Feldt, R., Mujtaba, S., Mattsson, M.: Systematic mapping studies in software engineering. In: Proceedings of the 12th International Conference on Evaluation and Assessment in Software Engineering, pp. 68–77 (2008)
16. Souza, F.C.M., Papadakis, M., Durelli, V., Delamaro, M. E.: Test data generation techniques for mutation testing: a systematic mapping. In: Conference on Software Engineering, (17), pp. 419–432 (2014)
17. Myers, G.J., Sandler, C., Badgett, T.: The art of software testing. John Wiley & Sons (2011)
18. Kaner, C., Falk, J., Nguyen, H.Q.: Testing computer software. John Wiley & Sons (1999)
19. Copeland, B.J. (Ed.): The essential Turing. Clarendon Press (2004)
20. Arifi, S.M., Abbou, R.B., Zahi, A.: Assisted learning of C programming through automated program repair and feedback generation. Indonesian J. Electr. Eng. Comput. Sci. **20**(1), 454–464 (2020)
21. Almeida, J.B., Cunha, A., Macedo, N., Pacheco, H., Proença, J.: Teaching how to program using automated assessment and functional glossy games (experience report). In: Proceedings of the ACM on Programming Languages, **2**(ICFP), pp. 1–17 (2018)
22. Rapps, S., Weyuker, E.J.: Selecting software test data using data flow information. IEEE Trans. Software Eng. **4**, 367–375 (1985)
23. Liu, D., Petersen, A.: Static analyses in Python programming courses. In: Proceedings of the 50th ACM Technical Symposium on Computer Science Education, pp. 666–671 (2019)
24. Brown, C., Parnin, C.: Nudging students toward better software engineering behaviours. In: 2021 IEEE/ACM Third International Workshop on Bots in Software Engineering (BotSE), pp. 11–15 (2021)
25. Chen, H.M., Chen, W.H., Lee, C.C.: An automated assessment system for analysis of coding convention violations in Java programming assignments. J. Inf. Sci. Eng. **34**(5), 1203–1221 (2018)
26. McCabe, T.J.: A complexity measure. IEEE Trans. Software Eng. **4**, 308–320 (1976)
27. Bey, A., Jermann, P., Dillenbourg, P.: An empirical study comparing two automatic graders for programming. MOOCs context. In: Proceedings of Data Driven Approaches in Digital Education: 12th European Conference on Technology Enhanced Learning, EC-TEL, pp. 537–540. Springer International Publishing (2017)
28. Dong, Y., Hou, J., Lu, X.: An intelligent online judge system for programming training. In: Proceedings of Database Systems for Advanced Applications: 25th International Conference, DASFAA 2020, Part III 25, pp. 785–789. Springer International Publishing (2020)
29. Carmichael, T., Blink, M.J., Stamper, J., Gieske, E.: Linkage objects for generalized instruction in coding (LOGIC). In: Proceedings of The Thirty-First International Flairs Conference (2018)
30. Day, M., Penumala, M.R., Gonzalez-Sanchez, J.: Annette: an intelligent tutoring companion embedded into the Eclipse IDE. In: Proceedings of 2019 IEEE First International Conference on Cognitive Machine Intelligence (CogMI), pp. 71–80 (2019)
31. Gerdes, J.H.: Developing applications to grade introductory visual basic courses automatically, In: Proceedings of 23rd Americas Conference on Information Systems, AMCIS 2017 (2017)

Enhancing Intrinsic Motivation Through Gamification in Computer Programming Fundamental Course for Higher Education Students

Fazli Azzali[1](✉), Azizi Abas[1], Noraini Mohd Banua[2], and Nor Hazlyna Harun[3]

[1] InterNetWorks Research Lab, School of Computing, Universiti Utara Malaysia, Sintok 06010, Kedah, Malaysia
fazli@uum.edu.my

[2] Politeknik Sultan Abdul Halim Muadzam Shah, Bandar Darulaman, Jitra 06000, Kedah, Malaysia

[3] Data Science Research Lab, School of Computing, Universiti Utara Malaysia, Sintok 06010, Kedah, Malaysia

Abstract. Gamification is increasingly recognized as an innovative educational approach involving students more actively in their learning journey. It holds the potential to enhance the learning environment by making it more enjoyable and captivating, thereby benefiting education. This research explores enhancing student intrinsic motivation in Computer Programming Fundamental courses through gamification. The problem statement addresses maintaining student motivation and interest in technical subjects like programming. The study's objectives include identifying practical gamification elements, the usability of the gamification, and evaluating their impact on student intrinsic motivation. This research followed a three-phase approach: analysis, design and integration, and evaluation. The study employs a quantitative methodology, utilizing surveys to measure various constructs of motivation: attention, confidence, relevance, and satisfaction. Results show high motivation levels across all constructs, with mean scores translating to percentages of 85.6% for attention, 86.2% for confidence, 86.8% for relevance, and 86.8% for satisfaction. The study's results highlight the positive influence of gamification on student intrinsic motivation in Computer Programming Fundamental courses. By integrating game elements effectively, students showed increased attention, relevance, confidence, and satisfaction in their learning experience. This research contributes to theoretical advancements in educational gamification, innovative applications in teaching programming, and the promotion of self-directed learning.

Keywords: Gamification · Student Intrinsic Motivation · Computer Programming Fundamental · Usability · Self-Directed Learning

Z. Mat Aji et al. (Eds.): KMICe 2024, ISEM 49, pp. 385–398, 2025.
https://doi.org/10.1007/978-3-031-91485-0_30

1 Introduction

In this era of rapid development of science and technology information, the fields of computer science and information technology continue to rise in prominence, and we are increasingly reliant on software in many facets of our daily lives. As a result, the excitement that the development of new technologies has generated in the public consciousness is luring more and more individuals into computer-related fields of study. According to the standard for Malaysia's higher education institutions, students must take programming as a prerequisite for any computer science and information technology degree. Yet, although the need for software engineers is skyrocketing in worldwide markets, the proportion of computer science and information technology undergrads who withdraw from school is shockingly high [1].

Most computer science and information technology beginners find it challenging to learn how to code. Many students report that they are intimidated to tackle these difficulties because they perceive learning computer programming as difficult and time-consuming and needs constant endeavor. Several researches have highlighted some challenges associated with learning programming, such as the learning environment in programming courses affecting students' motivation and performance [2]. Conceptual comprehension, insufficient experience, poorly planned lab activities, a lack of enthusiasm, intellectual effort, logical thinking, and cognitive and problem-solving abilities are some other challenges students confront [1]. All the difficulties mentioned above might highly reduce the student's motivation and performance while learning programming [2]. As a result, most students decided to drop out of college [1].

Learning programming has always been challenging, particularly for novice learners. Thus, one study suggested a semantic approach whereby the gaming environment is integrated to emphasize programming structures and abstract notions of programming, which may enhance the learning process regardless of the programming language and syntax [1]. Furthermore, several studies show how students' motivation and engagement may be improved by making lessons more engaging. The use of gamification may help with this issue [2]. Although online learning approaches have been highly popularized in recent years, especially during a pandemic, there has shown a decline situation in student interest or engagement as the complexity of the programming content increases; for example, despite 81% of students enrolled in the online Python course detailed in passed the course's first half in the 2018–2019 winter semester, just 18% finished the course in its entirety [3]. In contrast, several studies have shown that using gamification features boosts student motivation to learn [2]. The term "gamification" describes incorporating game elements and strategies into non-game contexts such as educational and business domains where fun and amusement are not the primary objectives. [2]. The goal of incorporating game design ideas and mechanics into educational settings is to increase student engagement and motivation.

Additionally, research shows that using games as a teaching tool is more effective than conventional methods in capturing and holding students' attention and motivating them to learn [1, 1]. The gamification model contains three parts, including dynamics, mechanics, and components, as stated by Werbach and Hunter [2]. The dynamics include emotion, narrative progression, and social, fostering teammates' cooperation. At the same time, the mechanics contain challenges and opportunities, while badges, points, leader boards and avatars are categorized as components [2]. Besides, gamification, visual presentation, animations and simulation of algorithms are all inseparable, which attempt to make learning more dynamic, visible, and interactive; plus, it contributes to better comprehending intrinsically dynamic concepts than the paper form. Thus, this project is implemented to develop a web-based gamification for programming course content in the highest institute learning to increase the university's student engagement, attention, and motivation while learning programming content.

Gamification, defined as "attracting attention, making students more interested in learning, motivating students to be more motivated, encouraging self-learning, problem-solving with game mechanics, and game thinking," has proven effective in raising student engagement across various educational contexts in several case studies [4]. Gamification is another teaching and learning approach that can attract students' interest and boost their motivation in computer programming classes, which are typically considered challenging by most students [5]. This approach helps to improve students' motivation and involvement and achieve learning outcomes [6]. By incorporating game elements into educational activities, such as badges, points, levels, leaderboards, and challenges, outside traditional gaming contexts, like classrooms and online learning environments, the gamification approach intends to increase student engagement and motivation by making the learning process more enjoyable, interactive, and immersive [7].

Gamification has been demonstrated to improve students' academic performance [6]. This approach makes learning more enjoyable, reduces cognitive load and improves psychological and behavioral characteristics in students [7]. Digital-based games have proven effective in improving learning outcomes and increasing student enthusiasm [7]. Furthermore, gamification promotes a spirit of competitiveness, teamwork and success, which can encourage students to take an active part in their educational pursuits [8].

Enhancing higher learning students' intrinsic motivation to learn computer programming courses through gamification is a topic of increasing interest in education. Since they are essential for IT development and in many other businesses, programming abilities are becoming more valuable in the real world. Therefore, students should enhance their computational thinking by learning computer programming languages, a crucial component of multidisciplinary skills in contemporary education [7]. Gamification in educational settings turns learning into an exciting journey that provides quick feedback and builds a sense of achievement. Incorporating features such as leaderboards, awards, challenges, and badges makes learning more engaging and motivational for students [9]. Compared to other programming subjects, there are notable distinctions in each gamification. More specialized gamification features that align with learning objectives are necessary due to the technical complexity of programming. Programmers must combine gamification features with coding challenges and problem-solving activities for maximum impact, unlike non-technical disciplines where more straightforward elements are sufficient.

Thus, this study is implemented to develop a web-based gamification for programming course content in the highest institute learning to increase the university's student engagement, attention, and motivation while learning programming content. Here are all identified objectives of this project: To determine the student in learning programming content through an accessible gamified learning website. To develop the gamification content of programming courses in the high-level institution. To evaluate the motivation for gamification content in programming courses at high-level institutions. The targeted crowd involved the students of the high-learning institution who were required to take the programming course. This website offers programming content enriched with gamification elements to engage students, increase motivation, and improve learning outcomes while helping to lower university dropout rates.

2 Related Works

2.1 Gamification Overview

The term "gamification" was initially used in 2008 in digital learning, but it wasn't until 2010 that it gained widespread acceptance [10]. To facilitate learning, gamification in education aims to accurately identify the qualities that make games enjoyable and encourage players to keep playing. The objective is to increase students' motivation to study more thoroughly and meticulously to become more competitive and driven to enhance their academic standing [11]. Beyond education, gamification is becoming more popular across various fields, such as business and health. It has recently attracted much interest from the public and commercial sectors [12]. According to Morschheuser et al. [13] and Shafie and Abdullah [14], that different kinds of game mechanics and game dynamics elements are shown in Table 1. These elements include points, levels, badges/achievements, virtual goods, leaderboards, virtual gifts, and a progress bar.

Table 1. Type of game elements.

Game Mechanics	Game Dynamics
Points	Rewards
Levels	Status
Badges/Achievements	Achievements
Virtual goods	Self-expression
Leaderboards	Competition
Virtual gifts	Altruism
Progress bar	Timers & Pressure

Gamification in education was implemented many years ago to encourage pupils to put in more practice and effort by giving them badges. Furthermore, these approaches can encourage students to become more actively involved in their education and achieve additional favourable outcomes [15]. Table 2 shows the differences between traditional methods and gamification approaches.

Students recognized that playing games is more enjoyable than using conventional calibration techniques. As mentioned by Shafie and Abdullah in [14], the assessment of the efficacy of students' attitudes and interests showed that gamification, in contrast to

Table 2. Traditional method vs. gamified approach.

Traditional Method	Gamified Approach
Teacher teaches students	Self-learning
Marks to evaluate students	Points to evaluate students
Text (books and board)	Design (more attractive)
Topics to define course contents	Levels (more competition)
Increased complexity	Stages (more complexity)
Test	Master level
Grade	Ranking

conventional techniques, can influence students' perspectives favourably. This is making them more receptive and promoting delightful learning experiences. The main differences between game-based learning, gamification, and gaming are outlined in Table 3 [16].

Table 3. Difference between game, game-based learning, and gamification distinctions.

Game	Game-based Learning	Gamification
Games are played solely for fun; they may or may not have established guidelines and objectives	In games, learning objectives are stated	It depends on doing tasks that have rewards or points
Both winning and losing are inherent to the game	The goal is to motivate students to take action and learn	The goal is to spur students to take some sort of action
Playing games comes first, with rewards afterwards	In essence, winning the game is gratifying	Existence is fundamentally voluntary and pleasurable
In general, games are safe and expensive to design	The building often costs a lot of money	In general, gamification is more straightforward and less costly
Scenes and stories are included in the game	Content is frequently changed to fit the game's narrative	Game-like topographies rather than content are typically added to the learning management system (LMS)

2.2 Gamification in Computer Programming Courses

Gamification is a technique that boosts motivation and engagement by incorporating elements of game design into non-gaming environments, like education. Programming classes become more engaging, motivating, and committed when gamification is implemented [17]. Gamification raises students' interest in computer programming and enhances learning outcomes [2, 17] Click or tap here to enter text. However, not all students will benefit from this method, which may not work in all situations [18]. Furthermore, developing suitable assessment techniques and identifying pertinent factors utilized are necessary for gauging the success of gamification. Since gamification can increase student motivation and engagement, it is imperative to consider these obstacles. Inadequately crafted gamification components might not inspire or involve developers [19].

This study highlights how different gamification components can be used to track how they affect students' increased intrinsic motivation. Every kind of game discussed also has components intended to pique students' interest in playing while boosting their drive and academic performance. Table 4 shows the game elements used in earlier research. The most used gamification elements are badges, leaderboards, points, and levels; feedback, progress, challenge, achievement, and story/theme are followed in order of use. Since they constitute a component of the student's intrinsic motivation, this study suggests using these elements: points, leaderboards, rankings, and progress. Based on studies examining how Keller's produced ARCS Model affects students' intrinsic motivation, these components were chosen [20].

Based on 28 previous research studies in Table 4, the element of badges is the type of success and recognition employed in 23 (82.1%) studies as the most popular gamification element. This demonstrates how well-liked they are for rewarding pupils when they accomplish particular goals. Leaderboards and points are also commonly utilized; 19 research (67.9%) included both. These components are essential to building a competitive atmosphere that may motivate students and offer quick performance feedback. Thirteen studies (46.4%) use levels to assist students grow in an organized manner. This highlights their responsibility to sustain students' interest by giving them definite objectives and a feeling of progress. Nine (9) studies (32.1%) incorporate feedback, a crucial component for mentoring students and bolstering learning, demonstrating the significance of feedback in the learning process. However, despite their relevance, achievement aspects are employed less frequently, only referenced and used in 4 research, or 14.3% of all studies. Story/themes are the least used element, appearing in only 1 study (3.6%). This may result from the challenge of creating captivating stories that complement instructional materials. In summary, the table presents the various ways in which gamification components can be combined to produce an exciting learning environment that improves student performance and motivation. This study's proposed elements will focus on choosing the four (4) elements in the gamification, such as point, leaderboard, level, and progress, as highlighted in Table 4.

2.3 Intrinsic Motivation and ARCS Model

Intrinsic motivation is essential to support learning and academic achievement [21]. Students with intrinsic motivation are likelier to be involved, tenacious, and imaginative in their learning [22]. Teachers can encourage students' intrinsic motivation in several ways, including giving them opportunities for autonomy and choice, assigning challenging but meaningful work, giving them feedback and acknowledgement, and fostering a welcoming and helpful learning environment [23]. By encouraging intrinsic motivation,

Table 4. Summary of game elements in existing study.

Previous Study	Points	Leaderboards	Badges	Achievements	Levels	Story/themes	Feedback	Rewards	Progress	Challenges
Romero-Rodriguez et al., 2019		/	/							/
Facey-Shaw et al., 2020			/							
Dikcius et al., 2021								/		
Khaleel et al., 2019	/	/	/		/					
Khaleel et al., 2020	/	/			/					
Uz Bilgin & Gul, 2020	/	/	/		/	/	/			/
Sanchez et al., 2020							/		/	
Asiksoy & Canbolat, 2021	/	/	/		/		/			
Adams & Du Preez, 2022	/	/	/		/		/	/		/
Garnett & Button, 2018			/							
Castro & Gonçalves, 2018		/	/		/					
Coleman, 2018			/							
Gündüz & Akkoyunlu, 2020	/		/	/	/					
Milenković et al., 2019	/	/	/							
Pakinee & Puritat, 2021	/	/			/				/	/
Van Roy & Zaman, 2019		/	/							/
Ahmed & Asiksoy, 2021	/	/	/		/		/			
Marín et al., 2019	/	/						/		/
De-Marcos et al., 2020	/	/	/	/						
Donnermann et al., 2021	/		/							
Ozdamli, 2018	/		/				/			
Marín et al., 2019		/	/		/		/		/	
Cuervo-Cely et al., 2022	/	/	/	/	/			/	/	/
Rogers et al., 2021	/	/	/	/			/			
Heryadi et al., 2020	/	/	/		/			/	/	/
Costa & Aparicio, 2017	/	/	/		/				/	
Smiderle et al., 2020	/		/							
Dichev & Dicheva, 2017	/	/	/				/	/		/
Proposed Elements of Study	**19**	**19**	23	4	**13**	1	9	6	**6**	9

teachers can assist students in realising their full potential and cultivating a lifetime love of learning [24]. This study will explore gamification's effectiveness in enhancing students' intrinsic motivation. The goal is to understand students' intrinsic motivation when learning computer programming courses.

The ARCS model, created by Keller in [20], is a framework designed to enhance motivation and engagement in educational and instructional contexts [25]. The model is organized into four key categories: attention, relevance, confidence, and satisfaction, as illustrated in Fig. 1.

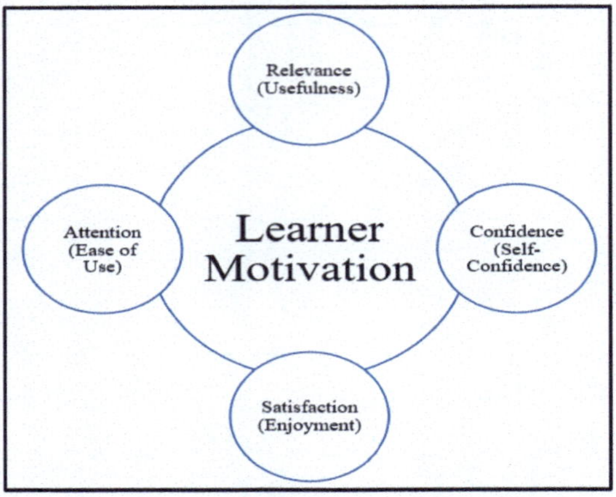

Fig. 1. The ARCS Model (adapted from [20]).

Table 5 shows the definitions for each category and examples of relevant process questions. This gamification prototype model will also examine intrinsic motivation models to address research gaps and challenges. A carefully chosen combination of elements must be considered for gamification strategies to enhance students' motivation to learn programming and maintain their enthusiasm.

Table 5. ARCS model categories, definitions, and process questions [26].

Major Categories	Definitions	Process Questions
Attention	Capturing the interest of learners, stimulating the curiosity to learn	How can I make this learning experience stimulating?
Relevance	Fulfil the personal needs/goals of the learner to cultivate a positive attitude	In what ways will this learning experience be valuable for my students?
Confidence	Helping the learners to believe/feel that they will succeed and control their success	How can I help the students to succeed and allow them to control their success via instruction?
Satisfaction	Reinforcing accomplishment with rewards (internal and external)	How can I help the students to feel good about their experience and desire to continue learning?

3 Methodology

3.1 Research Flowchart Process

Figure 2 illustrates the research flow for this study, which is divided into three phases: (i) Analysis, (ii) Design and Integration, and (iii) Evaluation.

Phase I: Analysis was based on a literature review to identify potential game elements for gamification. The findings from this phase also support the research problem statement. The literature review focused on computer programming, gamification approaches, and student motivation, covering studies from 2018 to 2023.

Phase II: Design and Integration involved designing and incorporating the game elements identified in the initial phase into a gamification tool. This phase also included usability tests and pilot testing of the instruments.

Phase III: Evaluation involved implementing the proposed gamification and assessing its impact on student motivation through questionnaires to achieve the research objectives. The questionnaire survey served as the primary data collection method, leading to a thorough analysis and discussion, culminating in a comprehensive report concluding the study.

3.2 Respondents Sampling and Research Procedure for the Motivation Evaluation

The evaluation of student motivation was conducted at three community colleges in the Northern Zone: Kolej Komuniti Bandar Darul Aman (KKBDA), Kolej Komuniti Pandang Terap (KKPT), and Kolej Komuniti Arau (KKA0. The study involved 81 students enrolled in the Computer Programming Fundamentals (STM20263) course. The participants were selected using the target population technique, where the researcher chose the entire population as the study sample, as stated by Fraenkel and Norman in 1996. This questionnaire, adapted from the Instructional Materials Motivation Survey by Keller in [20] and modified by Alcasoda and Balaoro in [27], aims to evaluate how gamification in the Computer Programming Fundamentals course influences intrinsic motivation. The researchers surveyed people across three community colleges. They first coordinated with the lecturers teaching the Programming Fundamentals course to schedule the study date, day, and time. Once the logistics were finalized, the researchers collected the data in the classroom. The respondents were provided with an explanation of the gamification to ensure that the data collected would align with the research objectives. After engaging with the gamification, the respondents completed a set of questions. All the collected data was then analysed using SPSS statistical software.

4 Data Finding and Discussion

This section presents the results of students' motivation to use gamification based on the questions from the distributed survey questionnaire. This study's 23-item survey on student motivation is divided into five sections. Section A covers the demographics of

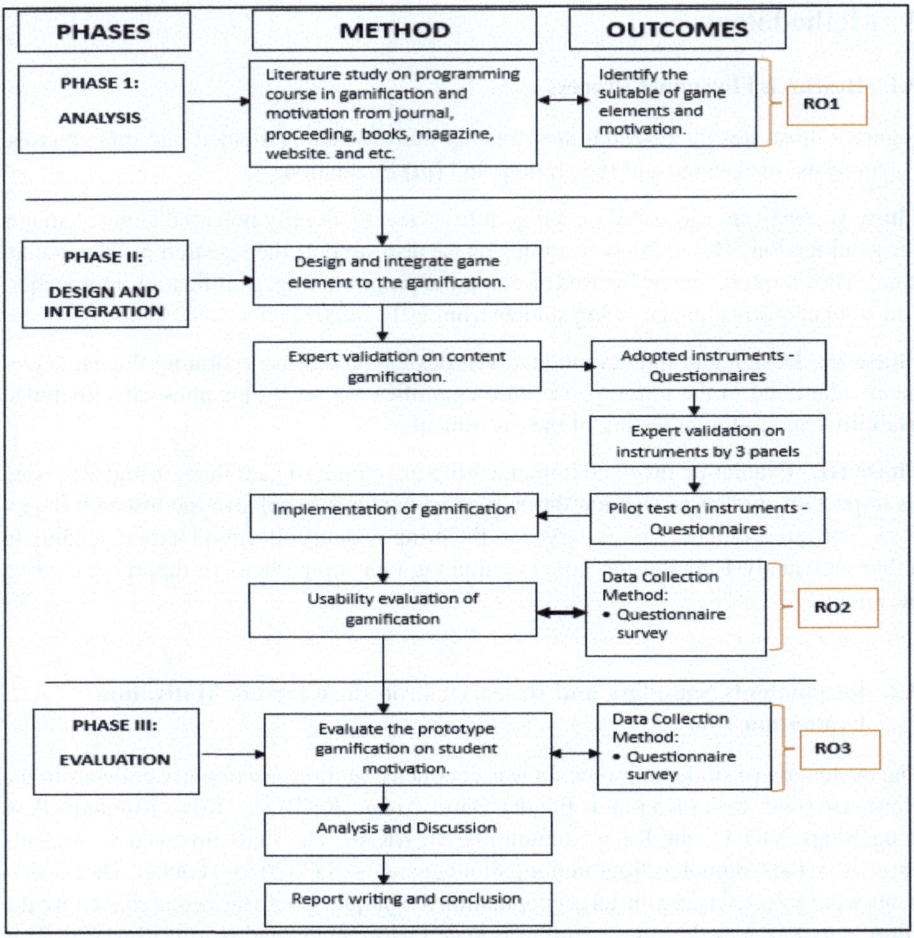

Fig. 2. Research flowchart process.

the respondents. Section B contains five items of Attention, while Section C, Section D and Section E contain six items covering Relevance, Confidence, and Satisfaction, respectively.

4.1 The Reliability of the Pilot Test Study

For the reliability of the item instruments, a pilot test was conducted to ensure the research process would follow the procedure correctly. Thirty KKA, KKBDA, and KKPT respondents were selected for the pilot test. The results of the pilot test are presented in Table 6.

Table 6 presents the pilot test results, showing the Cronbach's Alpha Coefficient Value (α) for each construct. The reliability analysis for the Attention, Relevance, Confidence, and Satisfaction construct produced Cronbach's Alpha around values of more than 0.900, indicating a high level of reliability, as the alpha value exceeded 0.8.

Table 6. Reliability test (Cronbach's alpha) of the pilot test instrument (n = 30) for student motivation questionnaire.

Construct	Number of Items	Cronbach's Alpha Coefficient Value (α)
Attention	5	.943
Relevance	6	.938
Confidence	6	.958
Satisfaction	6	.967

4.2 The Summary Result of the Student Motivation

The overall results for student motivation in 81 respondents, as data sampling encompassing the four dimensions of motivation, attention, relevance, confidence, and satisfaction, are described in this subsection. As demonstrated in Table 7, the high scores obtained for each dimension highlight how well the gamification method improves an engaging and inspiring learning environment.

Table 7. The overall result for student motivation.

Construct	N	Mean	Std. Deviation	Level
Attention	81	4.3259	0.60535	High
Relevance	81	4.2654	0.63123	High
Confidence	81	4.3189	0.65468	High
Satisfaction	81	4.3416	0.60490	High

Table 7 shows high mean scores and reflects students' strong positive responses and motivation. The attention has a mean score of 4.3259 (std = 0.60535), indicating a high level of focus and engagement with the instructional material. The relevance, with a mean of 4.2654 (std = 0.63123), suggests that students find the material pertinent and aligned with their learning needs and interests. The confidence is rated at a mean of 4.3189 (std = 0.65468), showing that students strongly believe in their ability to learn and succeed with the provided materials. Lastly, satisfaction has the highest mean score of 4.3416 (std = 0.60490), demonstrating that students find the learning experience rewarding and enjoyable. Figure 3 shows the graph of the overall result for student intrinsic motivation.

According to the findings and analysis, the overall mean score for all four categories (Attention, Confidence, Relevance, and Satisfaction) is 4.3130, or 86.26%. This indicates that the respondents were highly motivated and satisfied overall. In conclusion, respondents rated all aspects—attention, relevance, confidence, and satisfaction very highly, as stated by Rudzi Munap in [28]. The highest mean score in satisfaction indicates that respondents are particularly pleased with using gamification to learn computer programming courses.

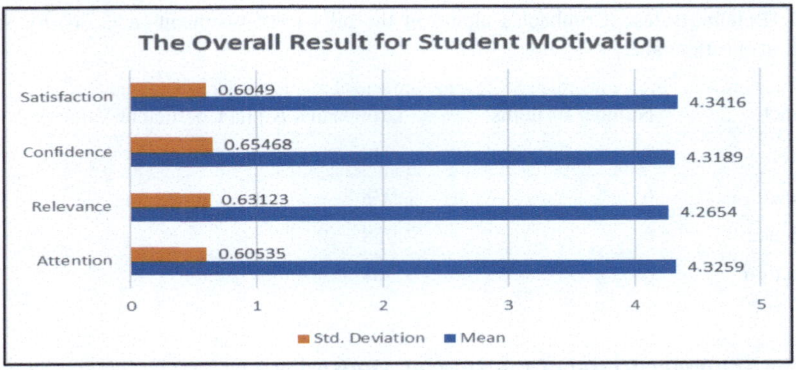

Fig. 3. The overall result for student motivation.

5 Conclusion

To summarize this section, the findings suggest that implementing gamification in the learning of computer programming courses is well-suited for community colleges. The interactive teaching methods of gamification offer flexibility, allowing students to enjoy learning without time or location constraints. However, limited exposure to this educational technology needs to be addressed by providing related skill courses. The benefits of gamification can serve as a powerful motivator for enabling this transformation in teaching. Applying these positive effects within higher education institutions in Malaysia, particularly in community colleges, could significantly enhance student achievement across various aspects. Additionally, students can view and utilize the gamification approach from a broader perspective. This study not only identifies the research gap related to the effectiveness of gamification in computer programming courses but also aims to fill this gap by conducting a comprehensive evaluation of how integrating game elements impacts student motivation. By addressing this gap in the existing literature, the study contributes valuable insights into educational technology and gamification, demonstrating how innovative approaches can enhance the learning process and promote intrinsic motivation among students.

Acknowledgement. Universiti Utara Malaysia of Malaysia supported this research through the Research Grant Scheme Universiti phase 2–2022 (Grant/2/2022/SO Code: 21435/UUM).

References

1. Shahid, M., Wajid, A., Haq, K., Saleem, I., Shujja, A.H.: A review of gamification for learning programming fundamental. Presented at the (2019)
2. Maryono, D.: Others: implementation of gamification in programming learning: literature review. Int. J. Inform. Educ. Technol. **12**, 1448–1457 (2022)
3. Swacha, J., Queiros, R., Paiva, J.C.: Towards a framework for gamified programming education. In: 2019 International Symposium on Educational Technology (ISET) (2019)

4. Krath, J., Schürmann, L., von Korflesch, H.F.O.: Revealing the theoretical basis of gamification: a systematic review and analysis of theory in research on gamification, serious games and game-based learning. Comput. Hum. Behav. **125**, 106963 (2021)
5. Ghai, A., Tandon, U.: Integrating gamification and instructional design to enhance usability of online learning. Educ. Inform. Technol. **28**, 2187–2206 (2023)
6. Khaldi, A., Bouzidi, R., Nader, F.: Gamification of e-learning in higher education: a systematic literature review. Smart Learn. Environ. **10**, 1–17 (2023)
7. Cuervo-Cely, K.D., Restrepo-Calle, F., Ramírez-Echeverry, J.J.: Effect of gamification on the motivation of computer programming students. Journal of Information Technology Education: Research. 21, (2022)
8. Heryadi, Y., Saleh Abbas, B., Trisetyarso, A., Suparta, W., Gaol, F.L.: Gamification framework for programming course in higher education. In: Proceedings of the International Conference on Information Technology and Digital Applications. pp. 54–57. Springer (2020)
9. Paez-Quinde, C., Morocho-Lara, D., Culqui, C.P.C., Escalante, M.: Gamification as a strategy in collaborative learning against virtual education in times of pandemic. In: IEEE Global Engineering Education Conference (EDUCON). pp. 752–756 (2022)
10. Deterding, S., Dixon, D., Khaled, R., Nacke, L.: From game design elements to gamefulness: defining "gamification. In: Proceedings of the 15th International Academic MindTrek Conference: Envisioning Future Media Environments. pp. 9–15. ACM (2011)
11. Khaleel, F.L., Ashaari, N.S., Wook, T.S.M.T.: An empirical study on gamification for learning programming language website. Jurnal Teknologi. **81**, 151–162 (2019)
12. Behl, A., Jayawardena, N., Pereira, V., Islam, N., Del Giudice, M., Choudrie, J.: Gamification and e-learning for young learners: a systematic literature review, bibliometric analysis, and future research agenda. Technol. Forecast. Soc. Chang. 176 (2022)
13. Morschheuser, B., Hassan, L., Werder, K., Hamari, J.: How to design gamification? a method for engineering gamified software. Inf. Softw. Technol. **95**, 219–237 (2018)
14. Shafie, A., Abdullah, Z.: Gamification in learning programming language. In: International Innovation Technology Exhibition & Conferences 2019 (Itec 19). pp. 181–187 (2020)
15. Khaleel, F.L., Ashaari, N.S., Wook, T.S.M.T., Ismail, A.: Gamification-based learning framework for a programming course. In: Proceedings of the 2017 6th International Conference on Electrical Engineering and Informatics: Sustainable Society Through Digital Innovation ({ICEEI} 2017). pp. 1–6 (2018)
16. Yamani, H.A.: A conceptual framework for integrating gamification in e-learning systems based on instructional design model. Int. J. Emerg. Technol. Learn. **16**, 14–33 (2021)
17. Shorn, S.P.: Teaching computer programming using gamification. In: Proceedings of the 14th International CDIO Conference. pp. 1–10 (2018)
18. Smiderle, R., Rigo, S.J., Marques, L.B., de Miranda Coelho, J.A., Jaques, P.A.: The impact of gamification on students' learning, engagement and behavior based on their personality traits. Smart Learn. Environ. **7**, 1–11 (2020)
19. Kabilan, M.K., Annamalai, N., Chuah, K.M.: Practices, purposes and challenges in integrating gamification using technology: a mixed-methods study on university academics. Springer, US (2023)
20. Keller, J.M.: Development and use of the ARCS model of instructional design. J. Instr. Dev. **10**, 2–10 (1987)
21. Cherry, K.: What is intrinsic motivation? https://static1.squarespace.com/static/5245a9c6e4b038b5cbe9a684/t/5990decd893fc0867d8aba46/1502666445285/What+Does+Intrinsic+Motivation+Mean%3F.pdf Accessed 1 May 2024
22. Ortiz, M., Chiluiza, K., Valcke, M., Ortiz-Rojas, M.: Gamification in computer programming: effects on learning, engagement, self-efficacy, and intrinsic motivation. In: European Conference on Game-Based Learning (ECGBL). pp. 507–514 (2017)

23. Facey-Shaw, L., Specht, M., van Rosmalen, P., Bartley-Bryan, J.: Do badges affect intrinsic motivation in introductory programming students? Simultation Gaming. **51**, 33–54 (2020)
24. Mekler, E.D., Brühlmann, F., Tuch, A.N., Opwis, K.: Towards understanding the effects of individual gamification elements on intrinsic motivation and performance. Comput. Hum. Behav. **71**, 525–534 (2017)
25. Vagianou, M., Paraskeva, F., Karampa, V., Bouta, H.: Applying motivational techniques and gamified elements on instructional design models for effective instruction in secondary education. In: Uden, L. and Liberona, D. (eds.) Learning Technology for Education Challenges. pp. 111–123. Springer International Publishing, Cham (2021).
26. Keller, J.M.: The ARCS model of motivational design. In: Motivational Design for Learning and Performance. pp. 43–74. Springer (2010)
27. Alcasoda, R.M., Balaoro, J.M.: Increasing learners' performance and intrinsic motivation using gamified instructional materials: the Arcs Model approach. Int. J. Adv. Res. Innovative Ideas Educ. **8**, 2022 (2022)
28. Munap, R.: Penilaian Program Diploma Kesetiausahaan Eksekutif di Universiti Teknologi Mara, Universiti Kebangsaan Malaysia, Bangi (2003)

Scalable Big Data Architecture: Improving Data Management at MoHE

Zhamri Che Ani(✉), Fauziah Baharom, Haslina Mohd, Yuhanis Yusof, and Mohamed Ali Saip

Data Management and Software Solution Lab, School of Computing, University Utara Malaysia, Sintok 06010, Kedah, Malaysia
zhamri@uum.edu.my

Abstract. Higher education institutions generate large amounts of data from various sources, including student information systems, learning management systems, administrative records, etc. For this reason, systematic data management and use are essential in decision-making and strategic planning. However, data management practices are currently struggling to handle the growing volume and complexity of data, making it challenging to obtain useful information for quick decision-making. Furthermore, integrating heterogeneous data sources and processing data in real-time while maintaining data quality remains challenging. Therefore, in the context of Malaysian education, this study proposes a large-scale big-data architecture designed specifically for the Malaysian Ministry of Higher Education (MoHE). It uses Apache technologies such as Apache Hadoop, Apache Spark, Apache Kafka, Apache NiFi, and Apache Cassandra to provide comprehensive solutions for data integration, real-time processing, and data quality management. This architecture aims to improve the efficiency and effectiveness of higher education data management in making better decisions. This innovative approach enable MoHE to can make more accurate decisions based on multiple data sources, enable real-time analysis, and ensure robust data governance, leading to a more responsive and data-driven educational environment.

Keywords: Big Data Architecture · Data Integration · Data Quality · Data Management

1 Introduction

Higher education institutions, from small colleges to large universities, generate and collect large amounts of data from various domains [1]. These domains include student information systems, staff information systems, learning management systems, research and publication databases, administrative records and financial systems. Data from these sources are usually heterogeneous, different not only in terms of their format, whether structured, semi-structured, or unstructured, but also in technology such as various database management systems, file storage formats and cloud services [2].

Integrating these diverse data sources to support decision-making and strategic planning is essential. However, integrating data from various heterogeneous sources in

Z. Mat Aji et al. (Eds.): KMICe 2024, ISEM 49, pp. 399–413, 2025.
https://doi.org/10.1007/978-3-031-91485-0_31

higher education institutions poses several challenges. These challenges include handling diverse data formats, ensuring real-time processing capabilities and managing data quality [3].

In the context of Malaysian education, the Ministry of Higher Education (MoHE) oversees various higher education institutions, including 20 public universities, 36 polytechnics, 105 community colleges, and 390 private universities [3]. Currently, MoHE manages 26 million data, which will increase every year. Due to the scale and complexity of the data, there is an urgent need for a scalable big data framework to manage and use the data effectively.

This paper proposes a scalable big data architecture to improve data management in MoHE Malaysia. The proposed architecture uses Apache big data technology to address data integration challenges, real-time processing and data quality management. Indirectly, it will improve the overall efficiency and effectiveness of higher education data management in Malaysia.

2 Current Situation

The Ministry of Higher Education (MoHE) Malaysia faces several challenges related to data management in the higher education environment. Currently, MoHE manages a large amount of data, estimated at 26 million. This data is stored in two central data warehouses, namely MyMoheS and SKPG (Sistem Kajian Pengesanan Graduan). However, having large amounts of data does not always lead to better decision-making because the data is not yet ready to be processed. It should go through many stages, including data cleaning and pre-processing [4]. In addition, each university, polytechnic and community college under MoHE also maintains its data warehouse, using various database management systems (DBMS) such as Oracle, Microsoft SQL Server, MySQL, PostgreSQL, and Ingres.

The diversity in DBMS across institutions creates many challenges in data integration and standardization. MoHE provides an API for periodic data submission, but many institutions, particularly young universities with limited staff and expertise, struggle to meet the requirements. This problem leads to delays and difficulties in data submission, affecting the overall quality and timeliness of the data [5].

Public universities in Malaysia use various tools, such as Pentaho, NiFi, and Talend, to manage data extraction, transformation, and loading (ETL). In some cases, data is still sent using Excel files, further complicating the process. Using multiple programming languages for data submission adds further complexity where it requires diverse technical expertise to effectively manage different systems.

The process of checking data quality is also time-consuming and inefficient. In most cases, the quality of the data always depends on the quality of the data source [4]. Furthermore, different data sources might also provide conflicting data due to incomplete, incorrect and outdated data [6]. Currently, SQL queries are used for data quality checks. Unfortunately, the same queries might not be executed across all DBMS platforms. As a result, each institution must develop its own data-checking program, which increases the risk of inconsistencies and errors. This decentralized approach to data quality management makes it difficult to ensure uniformity across all institutions and also difficult to improve the process of data submission.

Additionally, the level of expertise varies among universities, polytechnics and community colleges, which contributes to inconsistencies in data management practices. Some institutions may have well-established IT departments, while others, particularly young or smaller institutions, may require more resources and expertise to effectively manage their data.

Apart from structured data, MoHE must also manage unstructured data from social media platforms. This type of data adds another challenge in terms of storage, processing and analysis, which require advanced tools and techniques that are not uniformly available across all institutions.

Due to these challenges, there is a clear need for a scalable big data architecture to address the diverse and growing data of MoHE and all higher institutions in Malaysia. The architecture will streamline data integration, improve data quality, and provide tools and frameworks to manage structured and unstructured data effectively. By implementing the architecture, MoHE can ensure more consistent and reliable data practices across all institutions, ultimately supporting better decision-making.

3 Methodology

Since this architecture is built on existing projects, the methodology is divided into three main phases. The 4-layer Data Analytics Framework [5] and Data Management Body of Knowledge [7] were used as guidelines to propose this architecture.

This process starts with defining objectives and scope, identifying business goals and involving stakeholders, especially MOHE. Current system and data assessments were conducted to analyze the data landscape, identify gaps, and understand data governance requirements.

The second phase involves gathering functional and non-functional requirements, followed by evaluating and selecting appropriate technologies for data ingestion, storage, processing, integration, and analysis, as well as identifying security and compliance requirements. The architecture is then designed by focusing on four main layers: the Data Connection Layer, the Data Management Layer, the Analytics Layer, and the Presentation Layer [5].

The third phase involves developing prototypes to identify issues in implementing the architecture. It includes pilot implementations to test the architecture's capabilities with both real-time and batch data processing systems as performance benchmarks to assess scalability and efficiency, as well as rigorous data quality and integration testing. However, this phase is an ongoing process, and some prototypes must be done using actual test data.

4 Proposed Scalable Big Data Architecture

The scalable big data architecture for the Ministry of Higher Education (MOHE) in Malaysia is a comprehensive system that integrates Apache technologies such as Apache Hadoop, Spark, Kafka, NiFi and Cassandra to manage large volumes of structured, semi-structured and unstructured data. It enables real-time processing, efficient data integration, and ensures data quality to support decision-making and strategic planning in higher education. As depicted in Fig. 1, the proposed architecture is developed to handle the complete data life cycle from ingestion to consumption. At the same time, it will also ensure scalability, flexibility and efficiency in data management. The architecture consists of six main components: data sources, data ingestion, data storage and processing, data usage, data analysis and data management.

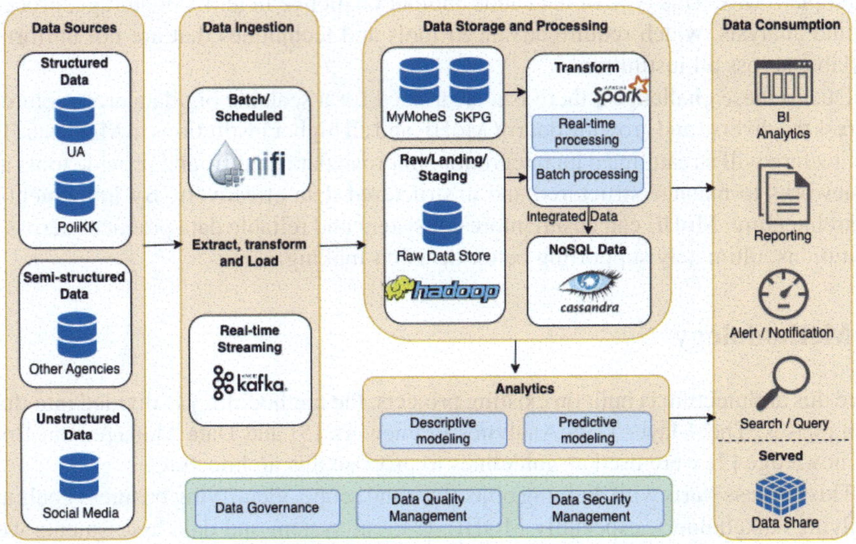

Fig. 1. Scalable big data architecture.

4.1 Data Sources

The proposed architecture is designed to support three main data types: structured, semi-structured and unstructured. This is important to achieve better decision-making and strategic planning within MoHE and the institutions. Each type of data source is described in detail below.

Structured Data. Structured data is highly organized and easily searchable, usually in a well-defined relational database or spreadsheet format. This data source can be identified by the existence of a fixed schema which means the data is stored in a table with rows and columns. Each field has a predefined data type, which enables simple and complex querying and reporting using Structured Query Language (SQL) [8]. In the context of public universities, polytechnics and community colleges in Malaysia, most structured data is stored in relational databases such as MySQL, Oracle, PostgreSQL, Microsoft SQL Server and Ingres. Structured data includes student records, course registrations, grades and financial data.

Semi-Structured Data. Semi-structured data such as JavaScript Object Notation (JSON) and Extensible Markup Language (XML) do not have rigid schemas but still contain some properties that make them easier to analyze than unstructured data. This data type usually includes tags or markers that separate semantic elements and enforce a hierarchy of records and fields [9]. JSON is commonly used to send data between servers and web applications as text. In Malaysian higher education, JSON might be used in Application Programming Interfaces (APIs) to transfer data between different systems from other agencies, such as the Department of Statistics Malaysia (DoSM) and the National Higher Education Fund Corporation (PTPTN).

Unstructured Data. Unstructured data has no predetermined format or structure, making it more challenging to process and analyze. It includes audio, video, image files, emails and social media data [10, 11]. Higher education institutions increasingly engage with students and the public through social media platforms such as Facebook, Twitter and Instagram. Data from social media is valuable for sentiment analysis, marketing, understanding student engagement and public perception [12]. MoHE can use this data to understand student interests and concerns, which can help predict future enrollment numbers and graduates' employability. It is also helpful to understand the needs and interests of the student population to develop or adapt academic programs accordingly.

4.2 Data Ingestion

Efficient data ingestion is a critical part of the proposed big data architecture. It involves collecting and integrating data from various sources into a centralized system for processing and analysis [13]. This architecture uses Apache tools and technologies to handle batch and real-time data ingestion to ensure data flows smoothly and efficiently into the system. The leading technologies used for this project are Apache NiFi and Apache Kafka. Each of them has a different purpose and complements each other to provide a comprehensive data ingestion solution.

Apache NiFi. Apache NiFi is widely used across multiple industries. It is a user-friendly, powerful, and reliable system for processing and distributing data. NiFi automates cybersecurity, observability, event streams, and generative AI data pipelines and distribution [14]. It is designed to manage and automate complex data flows, making it an ideal solution for batch-processing workflows [15].

NiFi can also automate data ingestion from various sources, including databases, file systems, and cloud services. It supports extensive data transformation capabilities, like filtering, enriching, aggregating, and format conversion [16]. NiFi is highly scalable, able to handle large volumes of data in a distributed environment and can scale horizontally as data grows [17]. Additionally, NiFi ensures reliable data flow with features such as back pressure to manage flow rates and data buffering to handle volume fluctuations to improve overall reliability [18, 19].

Apache Kafka. Apache Kafka is an open-source distributed event streaming platform used by thousands of companies to manage high-performance data pipelines, streaming analytics, data integration, and mission-critical applications [20]. It is designed to handle high-throughput and low-latency data ingestion, which make it an excellent choice for data processing and analysis.

Kafka enables real-time data streaming [21] from sources such as applications and social media platforms. It can handle millions of messages per second with low latency to ensure real-time data for processing [22, 23]. It ensures data durability and reliability by replicating data across multiple nodes to prevent data loss during hardware failure. This Kafka's distributed architecture allows it to scale horizontally by adding more brokers and handling growing data volumes without compromising performance [24, 25]. Additionally, Kafka can also be integrated with data processing and storage systems such as Apache Spark, Hadoop, and Cassandra, making it an essential tool for building end-to-end data pipelines [11]. Therefore, Kafka will allow MoHE to process and analyse data as it is generated to get quick results.

4.3 Data Storage and Processing

Adequate data storage and processing are also critical parts of the proposed architecture to ensure that large volumes of diverse data can be quickly managed and accessed. This section discusses the technologies used for data storage and processing in the architecture, focusing on their capabilities and discussing how they can contribute to the overall system.

Apache Hadoop. Apache Hadoop is a framework that facilitates distributed processing of large data sets across clusters of computers using a simple programming model. It is designed to scale from a single server to thousands of machines where each of them can provide local computation and storage. It is also designed to detect and manage failures at the application layer to ensure highly available services on a cluster of computers, even when failures occur on specific machines [26].

Basically, Hadoop has two main components: Hadoop's distributed file system (HDFS) and MapReduce. HDFS offers scalable storage across commodity hardware clusters [27], while MapReduce is an easily written application capable of processing vast amounts of data in parallel on large clusters (thousands of nodes) of commodity hardware in a reliable, fault-tolerant manner [28]. Hadoop provides a cost-effective solution using inexpensive hardware. This is the best option for MoHE to manage large amounts of growing data with a limited budget. HDFS's flexibility in handling structured, semi-structured, and unstructured data allows MoHE to store diverse datasets in a single repository, such as student records, multimedia files, and research data. Additionally, Hadoop is optimized for high-throughput data access, supporting large-scale batch-processing tasks that are essential for data-intensive operations such as reporting and analytics [21].

Apache Spark. Apache Spark is a multi-language engine designed to perform data engineering, data science, and machine learning tasks on a single-node machine or cluster. It can process data in batch and real-time streaming. It is supported by many languages, such as Python, SQL, Scala, Java, and R. Spark allows users to train machine learning algorithms on laptops and execute the same code on thousands of other machines [29].

Apache Spark also offers in-memory computing capabilities to increase processing speed by storing data in memory [30–33]. Furthermore, this beneficial Spark ecosystem includes libraries for SQL, streaming, machine learning, and graph processing to provide a comprehensive toolkit for diverse data operations on a single platform [21] to enable MoHE to manage and analyze data efficiently.

Apache Cassandra. Apache Cassandra is an open-source NoSQL distributed database that provides scalable and high-availability data storage [34]. It is particularly suitable for applications that require continuous availability and high throughput. In this architecture, Cassandra handles large amounts of integrated data resulting from batch processing by Apache Spark. Like Hadoop and Spark, Cassandra is also a distributed architecture ensuring high reliability by eliminating a single point of failure where data will be automatically replicated across multiple nodes for high availability and fault tolerance [35]. It can also scale horizontally by adding nodes to handle increased data loads and smoothly handle intensive read/write operations. Compared to other DBMSs, Cassandra excels in environments that require high throughput and low latency, which makes it ideal for real-time applications and time-series data. It is also a flexible and schema-free data model which allows easy adaptation to changing data requirements, accommodating diverse data sets without rigid schema constraints [36].

MyMoheS and SKPG Data Warehouses. MyMoheS and SKPG (Sistem Kajian Pengesanan Graduan) are existing data warehouses that use Microsoft SQL Server to perform complex queries and reporting. These databases offer robust support for relational data, facilitating efficient storage and retrieval of structured data. They effectively manage transaction data and ensure data integrity [27]. The systems developed based on these databases also integrate with Power BI to enable the creation of interactive dashboards and visualizations [37]. By incorporating Apache Hadoop, Apache Spark and Apache Cassandra, along with MyMoheS and SKPG, the proposed architecture provides a comprehensive and scalable solution for data storage and processing [34]. This integration will significantly improve MoHE's ability to produce related statistics and make data-driven decisions.

4.4 Data Analytics

Data analytics is a critical component of the proposed scalable big data architecture to enable the extraction of valuable information from large amounts of data. In the Ministry of Higher Education (MoHE) Malaysia context, analytics is more focused on descriptive and predictive modelling. Both types of modelling have different purposes and techniques to analyze data, support decision-making, and improve the educational ecosystem [10].

Descriptive Modeling. Descriptive modelling summarizes and interprets historical data to understand past behaviour and trends [38]. It aggregates and summarizes data to generate comprehensive reports, including student enrollment numbers, graduation rates, faculty performance metrics, financial expenditures, etc. Statistical techniques such as mean, median, mode, variance, and standard deviation are used to analyze data distribution and identify patterns [39]. For example, analyzing student grade averages across different departments can highlight areas that need academic support.

Data visualization tools such as bar charts, line graphs, pie charts and heat maps present data in an easy-to-interpret format to help stakeholders quickly understand complex data relationships and trends [40]. Trend analysis examines data over time to identify significant patterns and changes, which assists the MoHE in tracking key metrics such as student retention rates, research output, and funding allocations over several years.

Predictive Modeling. Predictive modelling uses historical data and statistical algorithms to predict future outcomes to help anticipate trends, identify potential issues, and make proactive decisions [38]. It relies on machine learning algorithms such as linear regression, logistic regression, decision trees, random forests, support vector machines, and neural networks to build models that predict future events [41]. In the case of MoHE, for example, predictive models can predict student performance and identify at-risk students by analyzing factors such as attendance, grades, and extracurricular activities to enable early intervention and support [42].

Additionally, predictive analytics can predict future student enrollment trends, which is useful for resource allocation, faculty recruitment, and infrastructure development [43]. It also helps optimize resource allocation by predicting course, facility, and staff demand. Furthermore, predictive models can also be used to assess the potential impact of ongoing research projects and assist in financial planning, research grants and operational costs. By using both descriptive and predictive modelling, the proposed architecture will empower MoHE to support evidence-based decision-making, improve operational efficiency and improve educational outcomes across the Malaysian higher education sector.

4.5 Data Consumption

Data consumption is the final stage in the data lifecycle, where processed and analyzed data is used to make informed decisions, generate detailed information and drive strategic actions. Adequate data use will ensure that valuable information derived from data is accessible to stakeholders at the right time. In the Ministry of Higher Education (MoHE) Malaysia context, data consumption includes several key components: business intelligence (BI) analytics, reporting, alerts/notifications, search/query capabilities, and data sharing.

BI Analytics. Business Intelligence (BI) analytics involves using data analysis tools and technologies to support strategic and operational decisions by allowing stakeholders to visualize and interpret complex data and reveal trends [44]. BI tools such as Power BI, Tableau and QlikSense also provide interactive dashboards that allow users to visualize data through charts, graphs and maps that can be customized to display key performance indicators (KPIs) related to various stakeholders [45]. In addition, BI tools are often used to integrate with advanced analytics capabilities, including predictive modeling, statistical analysis and machine learning to enable deep analysis and forecasting before informing decision makers. As such, BI analytics will help MoHE and other users to understand complex data patterns and identify anomalies or opportunities.

Reporting. Reporting involves the generation of structured documents that provide detailed information and analysis on various aspects of the organization, which are essential for documenting performance, compliance, and strategic planning [46, 47]. In the context of MoHE, many reports, such as regular, ad-hoc, and custom reports, need to be generated. Regular reports, such as monthly, quarterly, and annual reports, are used to track progress and performance over time, including registration, financial and research reports. Ad-hoc reports will be generated on request to address specific queries or issues, while custom reports are tailored to meet the needs of different stakeholders. All these reports will automatically pull data from various sources and generate reports at determined schedules to reduce MoHE workload and ensure timely availability.

Alerts/Notifications. Alerts or notifications are proactive mechanisms that ensure stakeholders are informed of critical events, changes, or thresholds that require immediate attention [48]. In the case of MoHE, this architecture focuses more on real-time and threshold-based notifications. Real-time alerts notify users of important events as they occur. For example, when student performance drops below a certain threshold, the system will send emails and push notifications to prompt timely intervention. Threshold-based notifications alert users when predefined limits are exceeded. For example, there is budget overspending, or the possibility of API being disconnected more than three times daily. The system administrator can determine the notification frequency to ensure no user disruption. These alerts or notifications will be integrated with MyMoheS and SKPG systems to provide comprehensive monitoring and timely response.

Search/Query. Search and query capabilities allow users to quickly and efficiently obtain specific information from large data sets, which is essential for data exploration, problem-solving, and informed decision-making [49]. In the case of MoHE, this architecture focuses more on advanced search functionality, which allows searching across multiple data sources using criteria such as keywords, date ranges, and data fields. Query tools, such as Structured Query Language (SQL), Jakarta Persistence Query Language (JPQL), and Cassandra Query Language (CQL), provide powerful data retrieval capabilities for performing complex queries on relational and NoSQL databases. In addition, Natural Language Processing (NLP)-based search tools are also helpful in facilitating non-technical users' interaction with data by typing natural language queries [50].

Data Share. Data sharing involves making data accessible to various stakeholders inside and outside the organization to facilitate collaboration, analysis, and decision-making [51]. To facilitate data sharing, processed data and information will be shared using Application Programming Interfaces (APIs). APIs enable programmatic data exchange between systems and integrate with external applications such as analytics tools and dashboards. In this architecture, secure data access is critical to ensure APIs can be used securely through authentication and authorization mechanisms. By implementing this robust data-sharing mechanism, the proposed architecture ensures that data is used effectively to support decision-making, improve operational efficiency, and drive strategic initiatives in MoHE. Indirectly, it will empower stakeholders to achieve continuous improvement and innovation in higher education.

4.6 Data Management

Effective data management is essential to ensure data is accurate, secure and used appropriately. In the context of MoHE Malaysia, data management includes three critical components: data governance, data quality management and data security management. Each of these components plays an important role in maintaining data integrity and usability, which is essential for decision-making and operational efficiency.

Data Governance. Data governance involves establishing policies, procedures, and standards to effectively manage data assets to ensure that the data is consistent, reliable, and compliant with organizational rules and policies It begins by developing a comprehensive policy that outlines data management across its lifecycle, covering data ownership, monitoring, access and use. MoHE plays a key role by overseeing specific data domains, ensuring data quality, compliance and security, and acting as an intermediary among higher institutions. Maintaining a data catalogue is essential for data governance. It provides a centralized repository of metadata that includes information about data sources, definitions and uses to aid in data context and quality understanding. In addition, compliance with relevant laws and regulations, such as the General Data Protection Regulation (GDPR) and the Personal Data Protection Act (PDPA) is essential [54]. This requires robust policies and procedures by MoHE.

Data Quality Management. Data quality management includes processes and technologies to ensure data is accurate, complete, reliable, and timely, which is essential for decision-making and operational excellence [55]. In the MoHE context, the activities that need attention are data profiling, data cleaning, data standardization and data monitoring. Data profiling involves analyzing data to understand its structure, content and quality, as well as to identify anomalies, inconsistencies and missing values. Data cleaning corrects or removes inaccurate, incomplete or duplicate data to ensure reliability for analysis. Data standardization ensures consistency in format and structure to facilitate integration and comparison across datasets. This is very important for the development of APIs. Continuous data monitoring with automated tools is essential to maintaining high-quality standards and ensuring that each institution provides quality data.

Data Security Management. Data security management involves protecting data from unauthorized access, breach and other security threats to ensure confidentiality, integrity and availability [38, 56]. Critical practices include implementing robust access control mechanisms, such as role-based and attribute-based access control, encrypting data, and maintaining an audit trail to track and monitor user activity [27, 57]. Developing an incident response plan, including procedures for detecting, reporting, and responding to insecure data, is critical to mitigating the impact of a security incident. By integrating robust data governance, quality management and security management practices, the proposed architecture ensures effective and secure data management to support MoHE Malaysia in maintaining high data integrity and confidentiality.

5 Proposed Implementation

Implementing the proposed scalable big data architecture involves integrating various technologies to create a robust and efficient system for managing and analyzing data. Key steps in this implementation include setting up the data ingestion pipeline, configuring the storage solution, and implementing the processing framework and data analytics. Below are suggestions for the main programming languages used in this architecture.

5.1 Java Spring Boot for Developing API and RtcMetrics

API development using Spring Boot will facilitate communication and data exchange between different architectural components [58–63]. This API handles requests from Apache NiFi, retrieves data from the database, and sends information to the Hadoop File System (HDFS). Additionally, rtcMetrics, implemented with Spring Boot, enables real-time data inspection and data quality monitoring. Continuously scheduled checks through rtcMetrics help detect and address data quality issues to ensure big data systems operate efficiently and optimally.

5.2 Python for Data Analytics Using AI

Python is used for data analytics and artificial intelligence because of its diversity and extensive ecosystem of libraries and frameworks. For data analytics, libraries such as Pandas, NumPy and SciPy are used for data manipulation, statistical analysis and numerical computation to extract meaningful information from large datasets [64–66]. In artificial intelligence, Python machine learning libraries, including PySpark, Scikit-Learn, TensorFlow, and Keras, can be used to develop predictive models and perform advanced analysis [67, 68]. For example, by analyzing historical data, these models can predict student performance, identify at-risk students and recommend timely interventions to help improve decision-making.

5.3 .NET for MyMoheS and SKPG Systems

MyMoheS and SKPG are existing comprehensive information systems that manage various administrative and academic functions in MoHE Malaysia. The systems have been developed using the.NET framework to integrate multiple modules, including student records, faculty information, course information, financial information and more. The.NET framework was chosen because it provides a robust and scalable platform for developing secure and efficient enterprise-grade applications [69].

6 Conclusion

In this paper, we have proposed a scalable big data architecture to enhance the data management framework of the Ministry of Higher Education (MoHE) Malaysia. The architecture integrates Apache big data technologies to address challenges related to data integration, real-time processing and data quality management. Using tools such as Apache Hadoop, Apache Spark, Apache Kafka, Apache NiFi and Relational and NoSQL databases, the proposed architecture offers a comprehensive solution capable of handling multiple data formats and ensuring efficient processing, data storage and analytics.

The benefits of this architecture are many. First, it provides a robust framework for integrating heterogeneous data sources, including structured, semi-structured and unstructured data. This integration is essential to create a comprehensive view across domains, such as student information, financial, administrative records and more. Second, the architecture supports real-time data processing to enable timely decision-making. For example, using real-time analytics can improve student performance monitoring and so on.

Additionally, the architecture incorporates strong data quality management practices to ensure that the data used for analysis is accurate, complete and reliable. This is important to maintain the integrity of analytical information and support evidence-based decision-making. Implementing the data governance framework will also further improve data consistency and compliance with MoHE standards, which is essential for the integrity and security of education data.

Future research and development in this area could explore some potential issues. One key area is further integrating artificial intelligence and machine learning technologies to enhance predictive analytics and automate decision-making processes. For example, developing advanced machine learning models to predict student outcomes and optimize resource allocation can improve educational strategies and outcomes.

Another promising area for future work is improving data privacy and security measures. As educational institutions increasingly rely on big data, ensuring the confidentiality and security of sensitive information becomes a priority. Research in advanced encryption techniques, secure data-sharing protocols and privacy-preserving data analysis can provide valuable information for protecting educational data.

Finally, continuous evaluation and optimization of the architecture is necessary to adapt to the evolving needs of higher education institutions. This includes monitoring system performance, addressing emerging challenges, and incorporating institutional feedback to ensure the architecture remains practical and relevant.

In conclusion, the proposed scalable big data architecture offers a comprehensive solution to MoHE Malaysia's data management challenges. By integrating advanced technologies and adopting best practices in data management, this architecture is hoped to significantly improve the efficiency and effectiveness of higher education data management, supporting better decision-making and better educational outcomes.

Acknowledgement. This study is being conducted with support from the Ministry of Higher Education (MoHE) Malaysia.

References

1. Ashaari, M.A., Amran, A., Ahmad, N.H., Bakri, H., Nazri, S.: Big data analytics technology capability and data-driven decision making in Malaysian higher education institutions: a conceptual framework. In: IOP Conference Series: Materials Science and Engineering (2020)

2. Alsheikh, N.A.: Developing an integrated framework to utilize big data for higher education institutions in Saudi Arabia. Int. J. Comput. Sci. Inform. Technol. **11**(01) (2019)
3. Ministry of Higher Education (MoHE) Malaysia. https://www.mohe.gov.my/en Accessed 06 Aug 2024
4. Taleb, I., Serhani, M.A., Bouhaddioui, C., Dssouli, R.: Big data quality framework: a holistic approach to continuous quality management. J. Big Data **8**(1) (2021)
5. Ghavami, P.: Big data management: data governance principles for big data analytics. Berlin: Walter de Gruyter GmbH (2020)
6. Mandreoli, F., Montangero, M.: Dealing with data heterogeneity in a data fusion perspective: models, methodologies, and algorithms. Data Handling Sci. Technol. **31** (2019)
7. International, D.: DAMA-DMBOK: data management body of knowledge (2nd edn.). Technics Publications, LLC (2017)
8. Connolly, T.M., Begg, C.E.: Database systems: a practical approach to design, implementation, and management, 6th edn. (2015)
9. Friesen, J.: Java XML and JSON. Apress (2016)
10. Mago, B., Khan, N.: A proposed framework for big data analytics in higher education. Int. J. Adv. Comput. Sci. Appl., **12**(7) (2021)
11. Sebei, H., Hadj Taieb, M.A., Ben Aouicha, M.: Review of social media analytics process and big data pipeline. Soc. Netw. Anal. Min. **8**(1) (2018)
12. Daniel, B.K.: Big data in higher education: the big picture. In: Big Data and Learning Analytics in Higher Education: Current Theory and Practice, pp. 18–28. Springer (2016)
13. Ghosh, P., Biswas, A., Ghosh, S.: Fundamentals and technicalities of big data and analytics. Intelligent Systems in Healthcare and Disease Identification using Data Science, 1st Edition, Tylor & Francis Group (2023)
14. Apache Software Foundation: Apache NiFi. https://nifi.apache.org/ Accessed: 26 Jul 2024
15. Wnek, K., Boryło, P.: A data processing and distribution system based on Apache Nifi. Photonics **10**(2), 210 (2023)
16. Apache NiFi User Guide https://nifi.apache.org/docs/nifi-docs/html/user-guide.html Accessed: 06 Aug 2024.
17. Ouafiq, E.M., Raif, M., Chehri, A., Saadane, R.: Data architecture and big data analytics in smart cities. Procedia Comput. Sci. **207**, 4123–4131(2022)
18. Balan Thekkummal, N.: Methodologies for managing big data analytics pipelines for smart city applications. PhD Dissertation, Newcastle University (2023)
19. Cakir, A.: Big data management and technologies. business analytics for professionals, pp. 265–286. Springer (2022)
20. Apache Software Foundation, Apache Kafka. https://kafka.apache.org/ Accessed 26 Jul 2024
21. Kapadiya, D., Shekhawat, C., Sharma, P.: A study on large scale applications of big data in modern era. In: Proceedings of 5th International Conference on Information Management & Machine Intelligence, No. 54, pp. 1–6 (2023)
22. Badidi, E.: A fog node architecture for real-time processing of urban IoT data streams. Software Engineering Methods in Intelligent Algorithms, pp. 330–341 (2019)
23. Tantalaki, N.S.S., Roumeliotis, M.: A review on big data real-time stream processing and its scheduling techniques. Int. J. Parallel, Emergent Distrib. Syst. **35**(5) (2020)
24. Raptis T.P., Passarella, A.: A survey on networked data streaming with Apache Kafka, vol. 11. IEEE Access (2023)
25. Lekkala, C.: Designing high-performance, scalable kafka clusters for real-time data streaming. Euro. J. Adv. Eng. Technol. **8**(1), 76–82 (2021)
26. Apache software foundation: Apache Hadoop. https://hadoop.apache.org/ Accessed 26 Jul 2024

27. Chen, K., Wang, F., Wei, Z.: Hardware and software infrastructure driven by big data: challenges and prospects. In: Proceedings of 3rd International Conference on Big Data, Artificial Intelligence and Risk Management, pp. 625 – 631 (2023)
28. MapReduce Tutorial. https://hadoop.apache.org/docs/r1.2.1/mapred_tutorial.html Accessed 08 Aug 2024
29. Apache Software Foundation: Apache Spark. https://spark.apache.org/ Accessed 26 Jul 2024
30. Gao, Z., Wang, T., Wang, Q., Yang, Y.: Execution time prediction for Apache Spark. In: ACM International Conference Proceeding Series, pp. 47–51 (2018)
31. Döschl, A., Keller, M.E., Mandl, P.: Performance evaluation of Apache Hadoop and Apache Spark for parallelization of compute-intensive tasks. In: ACM International Conference Proceeding Series (2020)
32. Singh, A., Khamparia, A., Luhach, A.K.: Performance comparison of Apache Hadoop and Apache Spark. In: ACM International Conference Proceeding Series (2019)
33. Saouabi M., Ezzati, A.: A comparative between Hadoop Mapreduce and Apache Spark on HDFS. In: ACM International Conference Proceeding Series (2017)
34. Apache Software Foundation: Apache Cassandra. Accessed: Jul. 26, 2024. [Online]. Available: https://cassandra.apache.org
35. Raza, M. U., Xujian, Z.: A comprehensive overview of big data technologies: a survey. In: ACM International Conference Proceeding Series (2020)
36. Chebotko, A., Kashlev, A., Lu, S.: A big data modeling methodology for Apache Cassandra. In: Proceedings of 2015 IEEE International Congress on Big Data, BigData Congress 2015 (2015)
37. Widjaja, S., Mauritsius, T.: The development of performance dashboard visualization with power bi as platform. Int. J. Mech. Eng. Technol. (IJMET), **10**(5) (2019)
38. Davoudian, A., Liu, M.: Big data systems: a software engineering perspective. ACM Comput. Surv. **53**(5), 1–39 (2020)
39. Alem, D.D.: An overview of data analysis and interpretations in research. Int. J. Acad. Res. Educ. Rev. **8**(1), 1–27 (2020)
40. Kharakhash, O.: Data visualization: transforming complex data into actionable insights. Autom. Technol. Bus. Process. **15**(2) (2023)
41. Otoo-Arthur, D., Van Zyl, T.: A systematic review on big data analytics frameworks for higher education - tools and algorithms. In: ACM International Conference Proceeding Series, No. 15, pp. 1–9 (2019)
42. Rao, A., P.N, K. Baglodi: Role of big data in education sector: a review. Int. J. Adv. Sci. Eng. Technol. **6**(1), 1–3 (2018)
43. Picciano, A.G.: The evolution of big data and learning analytics in American higher education. J. Asynchronous Learn. Netw. **16**(3), 9–20 (2012)
44. Usman, M., Moinuddin, M., Khan, R.: Unlocking insights: harnessing the power of business intelligence for strategic growth. Int. J. Adv. Eng. Technol. Innovations **1**(4), 97–117 (2024)
45. Motamedi Nia, S.: Visualizing business production performance. Bachelor Thesis, University of Twente (2019)
46. Marmo, R., Nicolella, M., Polverino, F., Tibaut, A.: A methodology for a performance information model to support facility management. Sustainability (Switzerland), **11**(24) (2019)
47. Pera, F., Mazzoncini, R.: Sustainability in strategic planning: an Italian case study. Master Thesis, Politecnico Milano (2021)
48. Damaševičius, R., Bacanin, N., Misra, S.: From sensors to safety: internet of emergency services (IoES) for emergency response and disaster management. J. Sens. Actuator Netw. **12**(3), 41 (2023)
49. Cassavia, N., Ciampi, M., De Pietro, G., Masciari, E.: A big data approach for querying data in EHR systems. In: ACM International Conference Proceeding Series, pp. 212–217 (2016)

50. Ofori-Boateng, R., Aceves-Martins, M., Wiratunga, N., Moreno-Garcia, C.F.: Towards the automation of systematic reviews using natural language processing, machine learning, and deep learning: a comprehensive review. Artif. Intell. Rev. **57**(8), 1–60 (2024)
51. Rovolis, G., Habibipour, A.: When participatory design meets data-driven decision making: a literature review and the way forward. Manage. Sci. Lett. **14**(2), 107–126 (2024)
52. Hu, Z., Zhao, F., Zhao, X.: Research on smart education service platform based on big data. In: ACM International Conference Proceeding Series, pp. 11–15 (2020)
53. Mohammad, A., Mcheick, H., Grant, E.: Big data architecture evolution: 2014 and beyond. In: Proceedings of 4th ACM Symposium on Development and Analysis of Intelligent Vehicular Networks and Applications, pp. 134–144 (2014)
54. Feng, Y., Feng, X.: Smart data analysis and data governance. In: Proceeding of the 5th International Conference on Computer Science and Application Engineering (CSAE 2021), ACM (2021)
55. El Alaoui, I., Gahi, Y., Messoussi, R.: Big data quality metrics for sentiment analysis approaches. ACM International Conference Proceeding Series, pp. 36–43 (2019)
56. Mondal, A., Chatterjee, P.S.: A systematic literature survey on data security techniques in a cloud environment. In: Proceedings of International Conference on Information Technology (OCIT), pp. 451–456 (2022)
57. El Haourani, L., El Kalam, A.A., Ouahman, A.A.: Big data security and privacy techniques. In: ACM International Conference Proceeding Series, pp. 1–9 (2020)
58. Suryotrisongko, H., Jayanto, D.P., Tjahyanto, A.: Design and development of backend application for public complaint systems using microservice spring boot. Procedia Comput. Sci. **124**, 736–743 (2017)
59. Rozaliuk, T., Kopyl, P., Smołka, J.: Comparison of ASP.NET Core and Spring Boot ecosystems. J. Comput. Sci. Inst. **22**, 40–45 (2022)
60. Ritik, S.: Spring boot backend development. Int. J. Adv. Res. Sci. Commun. Technol. (2022)
61. Wyciślik, Ł, Latusik, Ł, Kamińska, A.M.: A comparative assessment of JVM frameworks to develop microservices. Appl. Sci. (Switzerland) **13**(3), 1343 (2023)
62. Jovanović, Ž., Jagodić, D., Vujičić, D., Ranđić, S.: Java spring boot rest web service integration with Java artificial intelligence Weka framework. In: Proceedings of Unitech 2017, Vol. II, pp. 323–327 (2017)
63. Prilsafira, T., Kunang, Y.N., Putra, M.H.: Rest API backend aplikasi e-commerce secondhand menggunakan framework Spring Boot. J. Sist. Teknol. Inform. **8**(2) (2023)
64. Álvarez Cid-Fuentes, J., Álvarez, P., Amela, R., Ishii, K., Morizawa, R.K., Badia, R.M.: Efficient development of high performance data analytics in Python. Futur. Gener. Comput. Syst. **111**, 570–581 (2020)
65. Nagpal, A., Gabrani, G.: Python for data analytics, scientific and technical applications. In: Proceedings of International Conference on Artificial Intelligence (AICAI) (2019)
66. Nelli, F.: Python data analytics: with Pandas, NumPy, and Matplotlib, 2nd edn. (2018)
67. Dr. R.G.: Real time anomaly detection techniques using PySpark framework. J. Artif. Intell. Capsule Netw. **2**(1) (2020)
68. Bandi, R., Amudhavel, J., Karthik, R.: Machine Learning with PySpark – Review. (2018)
69. Schwichtenberg, H.: Modern data access with entity framework core (2018)

Part 4: Artificial Intelligence with Knowledge Management

Designing for Persons with Disabilities Using Creative Pattern Matching Strategy for Social Innovation

Husniza Husni[1](✉), Fanindia Purnamasari[2], and Rajesh Kaluri[3]

[1] Universiti Utara Malaysia, 06010 Sintok, Kedah, Malaysia
husniza@uum.edu.my
[2] Universitas Sumatera Utara, Medan, Indonesia
[3] Vellore Institute of Technology, Vellore, India

Abstract. Persons with Disabilities (PWDs) are often left out in the designs of digital artifacts such as websites, banking apps, online learning, shopping apps, and many more, which leads to inefficient usage and denying accessibility for PWDs. Such a situation triggers the need to explore how digital artifacts can be designed for PWDs to maintain inclusivity and provide accessibility. This paper aims to extract the ideas for designing digital artifacts for PWDs, specifically for the visually impaired, towards social innovation. Using a creative pattern-matching strategy for idea generation, a number of 55 participants took part in the ideation session. The findings revealed some creative and out-of-the-box ideas for designing for PWDs who are visually impaired. The ideas are presented as design strategies for visually impaired users. They are much more concerned with non-typical design and critical issues of mental health, which is suicidal among PWD students. It is enlightening that the participants gave much concern and attention to such an alarming issue, which denotes their awareness of the PWD communities. It is thus critical to increase the accessibility and inclusivity of PWDs.

Keywords: Design Thinking Ideation · Creative Pattern Matching · Persons with Disabilities

1 Introduction

The needs and requirements of Persons with Disabilities (PWDs) are often neglected when designing and developing products for general usage, including digital artifacts. The designs of buildings, staircases, signboards, websites, and applications usually do not include the PWDs' basic needs. Around 400 visually impaired students at Indonesian universities were reported in 2017, which signals the urgent need to highlight the inclusivity of PWDs in universities, as only five universities provide support and services [1]. Hence, to ensure inclusivity and accessibility for them, some effort has to be made to give them access to education and an opportunity for a better life for the PWDs.

In Indonesia, the number of students with disabilities attending higher education has grown. However, no official report exists on the enrolled numbers [2]. Students with

Z. Mat Aji et al. (Eds.): KMICe 2024, ISEM 49, pp. 417–427, 2025.
https://doi.org/10.1007/978-3-031-91485-0_32

disabilities used computers and information technology less and invested more time to meet the demands of their studies [3, 19]. For university websites to be accessible to all users, including students with disabilities, they must be designed with their needs in mind [2]. A study evaluating the accessibility of Indonesian university websites found that most were not fully accessible to people with disabilities [2, 4]. Meanwhile, a recent study highlighted the perceptions of students with disabilities regarding reasonable learning accommodations. The study indicates that while some accommodations are in place, there is still a significant need for improvement to effectively meet students' diverse needs [5].

To ensure inclusivity and accessibility for PWDs, more effort is needed to provide equal access to education, employment, and opportunities for a better life, leading to more social innovation [19]. Legislation and policies alone are insufficient without proper implementation and enforcement. Changing societal attitudes and promoting awareness of the rights and capabilities of PWDs is crucial to achieving true inclusion by introducing and promoting the use of assistive technology for PWDs. Thus, Social innovation is significant in ensuring this effort comes to fruition, such as having entrepreneurial incubators for PWD-led startups; this shows how social innovation can empower PWDs. Nonetheless, social innovation demands more than just identifying the needs and requirements of the PWDs; instead, it means empathizing with them and creating creative ideas to make it a success.

highlighted, focusing on designing for PWDs using a creative strategy called pattern matching. This strategy can be viewed as the first step towards an innovative solution to later transform the accessibility and inclusivity of tertiary education in Indonesia by providing the necessary support and services to PWDs, specifically the visually impaired students in Indonesian universities. The following section presents some related works and is followed by the methodology, which discusses how this strategy is performed to generate ideas creatively. The following section presents the findings and discussion, and the final section concludes this paper.

2 Related Works

2.1 Persons with Disabilities and Digital Inclusion

PWDs face numerous challenges in accessing and utilizing digital artifacts due to a lack of inclusive design practices. The digital divide, exacerbated by inaccessible websites, banking apps, online learning platforms, and shopping apps, often results in inefficient usage and limited access for PWDs [6]. Digital inclusion emphasizes the importance of designing technology accessible to all users, regardless of their physical or cognitive abilities [18]. Inclusive design enhances usability for PWDs and benefits a broader type of users or general users by accommodating diverse needs and preferences [7]. Nonetheless, many websites and digital platforms remain inaccessible to PWDs, leading to digital discrimination, as reported in The World Bank's Digital Progress and Trends Report 2023 [8]. The report emphasizes the need for continuous upgrades in digital infrastructure to bridge this divide, particularly in developing countries where the gap is more pronounced [8].

2.2 Visually Impaired Students

Visually impaired university students encounter specific challenges in accessing educational resources and participating fully in academic activities. Studies have shown that traditional educational environments often fail to meet the needs of visually impaired students, leading to barriers to learning and social integration [9]. Accessible digital technologies can play a crucial role in mitigating these challenges by providing alternative formats, screen readers, and other assistive tools that support independent learning and engagement [10]. Moreover, universities are responsible for ensuring that their digital infrastructure, including learning management systems and online content, adheres to accessibility standards [11].

In Indonesia, 5.1% of the total population of PWDs are in college [28]. Although the exact statistics are scarce [2], the number is estimated to increase as 3.75 billion people in Indonesia have visual impairments [29]. Based on this condition, The Indonesian government has forbidden universities from refusing to admit students with disabilities [30]. However, only universities with inclusive management have developed it, one of them being Pamulang University. Pamulang University's inclusive university management establishes programs for visually impaired students to pursue higher education. Students in these organisations, particularly those in IT programs, must have logic and high mathematics ability despite being visually impaired. PWDs, such as the visually impaired, were facilitated by providing technology such as software to support learning materials, various media information that can convert text to speech (a project initiated by Michael Curran), and sign language instruction in facilities. Teaching techniques for developing a readiness learning program for visually impaired students received significant assistance from university management [31]. On the other hand, a survey was conducted in [32] to investigate whether PWDs have access to higher education, and it showed that acceptance in university is influenced by four aspects: availability, accessibility, acceptability, and adaptability. They collected the responses of respondents who were interested in six programs, namely Law, English, French, Japanese, and Architecture. Based on the survey, the respondents found it challenging to study and follow the lessons on campus due to the absence of sign language interpreters for communication between lecturers and students. For the Japanese program, the unavailability of materials to learn kanji in braille and the refusal of PWDs from the aspect of social status, even though the student has good grades, are among the issues. Other respondents said that education should be accepted for people from various backgrounds and cultures.

2.3 Ideation and Social Innovation

The ideation process is critical in generating innovative solutions that address the specific needs of PWDs. Creative pattern-matching strategies, such as brainstorming and design thinking, effectively generate out-of-the-box ideas that can lead to socially innovative solutions [12]. Social innovation in this context refers to the development of new products, services, or models that address social challenges and improve the quality of life for marginalized groups, including PWDs [13]. Engaging diverse stakeholders, including PWDs, in the ideation process ensures that the solutions are relevant, practical, and impactful [14].

Design thinking is a structured and creative approach to problem-solving that stimulates creativity and fosters an inclusive environment where all voices, especially PWDs, can contribute to the innovation process [20]. Design thinking involves empathizing with users, defining problems, ideating solutions, prototyping, and testing [21]. This iterative process encourages the generation of diverse ideas and ensures that the proposed solutions are user-centred and feasible. This paper focuses on the ideation phase to create creative solutions or designs for social innovation for PWDs.

Social innovation is essential for addressing the unique challenges faced by PWDs, as traditional market-driven innovations often overlook their needs. Mulgan [22, 23] defines social innovation as "innovative activities and services that are motivated by the goal of meeting a social need, " which is "predominantly diffused through organizations whose primary purposes are social." We can develop sustainable and scalable interventions for PWDs by focusing on social innovation.

2.4 Design Strategies for Visually Impaired Users

Designing for visually impaired users requires a deep understanding of their unique needs and challenges. Research has highlighted several key strategies for creating accessible digital artefacts, such as providing alternative text for images, ensuring keyboard navigability, and using high-contrast colour schemes [15]. These strategies help make digital content more accessible, allowing visually impaired users to interact with websites and applications effectively. For example, using alternative text for images ensures that screen readers can describe the content to visually impaired users. At the same time, high-contrast colour schemes improve readability for users with various visual impairments [24, 25].

Additionally, addressing critical issues such as mental health, often overlooked in the design process, is essential. Visually impaired students are at a higher risk of experiencing mental health issues, including depression and suicidal ideation, due to social isolation and academic stress [16]. Therefore, incorporating mental health support features into digital designs can significantly enhance the well-being of visually impaired users. For instance, providing easy access to mental health resources, creating online communities for support, and designing stress-relief tools can help mitigate these challenges [26].

When designing educational tools for university students with visual impairments, it is crucial to consider accessibility and usability. Ensuring that learning management systems (LMS) and other educational platforms are fully accessible can help visually impaired students navigate their academic environment more efficiently. Features like screen reader compatibility, clear navigation structures, and descriptive headings are vital. Creating inclusive content, such as providing lecture transcripts and ensuring all visual materials have descriptive captions, can significantly aid learning. Engaging with visually impaired students during the design and testing phases can provide valuable insights and ensure that the final product meets their needs [26, 27].

These strategies help make digital content more accessible, allowing visually impaired users to interact with websites and applications effectively. For example, using alternative text for images ensures that screen readers can describe the content to visually impaired users. At the same time, high-contrast colour schemes improve readability for users with various visual impairments [24, 25].

3 Methodology

3.1 Participants

Fifty-five (55) undergraduate students at Universitas Sumatera Utara, Indonesia, actively participated in this ideation session. The students are from Computer Science and Information Technology programs and are currently enrolled in the university's Human-Computer Interaction course. There are 30 males and 25 females divided into 11 groups with five members each.

3.2 Creative Pattern Matching Strategy

The Creative Pattern-Matching strategy is used to curate creative, out-of-the-box ideas quickly. The process is somewhat straightforward but rather challenging, as the participants have to think of an idea individually in under one minute. Figure 1 depicts the activities they performed to think on their feet using the creative pattern-matching strategy.

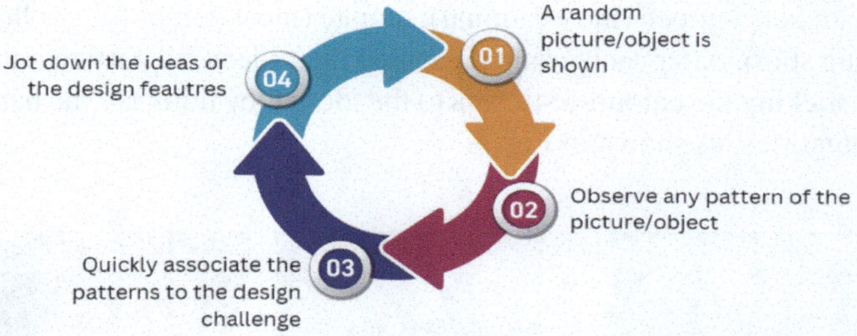

Fig. 1. Creative Pattern Matching Strategy used in the ideation session.

As shown in Fig. 1, the rapid ideation is curated by showing a number of random pictures or objects that have nothing to do with the problem at hand. What the participants must do, individually, is to try to map the patterns observed from the object to the problem or design challenge and note them on sticky notes. The process is repeated until all pictures/objects are shown (usually from 8 to 10 objects). This workshop shows eight random objects: a car key with a car as background, grapes in a basket, the San Francisco bridge, an engine, Easter eggs in a basket, a scenic view of mountains, people discussing in a library, and a flying cow. There is no particular reason why the objects were chosen; hence, pure and total randomness was maintained to avoid any prior assumption or relation.

3.3 Brainstorming for Best Possible Ideas

Once the pattern-matching strategy is completed, each participant presents their ideas to the group members and explains how they came up with them and why they are suitable or can address the issue. The task for each group is to brainstorm and improve the ideas among themselves and then select the best five to be presented to the others.

3.4 Four Categories Method and Dot Voting

After the brainstorming session, each group must come forward and paste the five best ideas on the walls. The Four Categories Method is used to further identify the best among the best ideas [17]. Figure 2 shows the four categories of ideas in this ideation method. The method involves categorizing concepts based on their varying degrees of abstraction, ranging from the most rational choice to the "long shot".

Fig. 2. The Four Categories Method for selecting the best ideas (©Teo Yu Siang and the Interaction Design Foundation).

Referring to Fig. 2, all participants are tasked to vote for the best ideas for all four categories. To do this, the Dot Voting method is used. In this workshop, we opted to use coloured stickers – green (most rational), orange (most delightful), yellow (darling), and white (long shot). After each group presented their ideas, all participants voted their favourites by sticking the coloured stickers to the ideas they think are the best, according to the four categories, as shown in Fig. 3.

Fig. 3. The Dot Voting Process.

4 Findings and Discussion

As aforementioned, the pictures or objects are chosen randomly, even before the scope was selected, i.e., visually impaired students at university, to ensure complete randomness and not have any influence on the ideas generated by the participants. Apart from having fun in the workshop, which is essential to introduce stress-free sessions to facilitate the creation of ideas, the participants have shown their ability to stay focused and try to think outside of the box. We observed that the participants hesitated at first, afraid that their ideas might be wrong or not logical; hence, emphasizing a safe and non-judgmental environment during the session is essential. After they felt they could trust the session, they started to kick off with some fantastic ideas from undergraduate students. The following are some ideas generated by the participants from the random objects shown during the session, gathered from the sticky notes collected, as shown in Table 1.

An analysis was performed to extract the themes that might transpire from the ideation session. This is to further analyse what technology lies behind the ideas generated by the participants based on the random images shown to them. Interestingly, the participants seem to have the ability to immediately relate to technology when it comes to their ideas. From low-tech to high-tech assistive technology, the participants managed to understand the needs of university students who are visually impaired. Table 2 shows the list of possible technology and innovative ideas categorized into types of AT.

Referring to Table 2, most ideas related to high-tech assistive technology were derived from the Four Categories method's 'long shot' and 'most rational' categories. These include Internet of Things (IoT)-based innovative ideas to install roadside sensors and warning systems near dangerous places for the most rational. At the same time, haptic technology was suggested and voted for the long shot' category. For the 'most rational' ideas, it is doable with our current technology – IoT, sensors, and apps to link smartphones to the IoT system. This idea is exciting and can be done with an optimal budget. What is more inspiring is that most of the ideas generated during the session concern or focus more on the safety of the visually impaired students maneuvering the campus for their daily routine or activities. It shows that the participants indeed care for the needs of their visually impaired friends.

Based on the findings and the mapping of ideas to technology and levels of AT, several key insights related to social innovation for the PWDs include the following:

Empathy-Driven Design: The emphasis and concerns on enhancing safety and accessibility for visually impaired students reflect an empathy-driven solid approach. The participants' ideas highlight a deep concern for addressing real needs and improving the quality of life for students with visual impairments. This empathy is crucial in social innovation, creating genuinely beneficial solutions for the target community.

Technology Integration: The diverse range of technologies suggested—from low-tech braille signboards to high-tech IoT-based warning systems—illustrates the potential for integrating various levels of technology to address specific needs. Social innovation often involves blending different technologies and approaches to create holistic solutions. This integration can lead to more effective and adaptable solutions.

Practical vs. Aspirational Innovation: The distinction between 'long shot' and 'most rational' ideas revealed a balance between aspirational and practical solutions. While

Table 1. Some ideas that were generated from the ideation session.

Random Objects and the Generated Ideas	

a) Movement of visually impaired students on campus; sensors to be placed at high-risk locations/positions to assist/guide them to maneuver around the campus safely; car key with an alarm system; car service for disabled students that is easier to book.	b) An Alert system for visually impaired students to prevent them from falling off bridges, roadside, stairs, and other dangerous locations.

c) e-book, a text-to-speech tool in the library to assist visually impaired students in reading the book by themselves, and sensors and voice alerts for the bookshelves so they know if they are in the correct aisle.	d) Voice feedback for food in the student café; voice-enabled pesticide detection for the grapes/fruits for food safety.

e) IoT, sensors and voice alerts to notify the students of their whereabouts on campus; tactile paving with sensors to give voice feedback to the students	f) Haptic technology can be used to label classrooms and lecture halls around the campus and Braille labels of buildings and signboards for the visually impaired.

high-tech ideas such as IoT and image recognition represent forward-thinking, 'long shot' concepts, practical ideas like braille and text-to-speech systems fall into the realm of current feasible solutions. This balance is important in social innovation as it addresses immediate and future needs.

Safety and Navigation: The focus on safety and navigation reflects a targeted approach to solving specific problems visually impaired students face. Social innovations that address critical issues such as safety improve day-to-day experiences and contribute to a more inclusive and supportive environment.

Budget Considerations: The feasibility of high-tech solutions, such as IoT-based systems, with an optimal budget highlights the importance of considering cost in social innovation. Practical social innovations should be economically viable and sustainable to ensure they can be implemented and maintained over time.

Table 2. Mapping of possible technology and types of AT.

Technology	Innovative ideas	Types of AT
Braille	a. Install braille signboards or signs around campus for more straightforward navigation and identification of buildings and landmarks in the university	Low-tech
Haptic	1 Haptic technology can be used and leveraged apart from braille for signboards, reading e-books, etc	High-tech
Text-to-Speech	a. A warning system is to be installed near dangerous places around campus, such as bridges, roadside, and stairs, to alert visually impaired students b. The TTS system is in posts where visually impaired students can get information on their location or position on campus c. TTS for navigation in the library for finding books and other materials d. TTS for reading the books in the library	High-tech
IoT	a. Installing sensors on the roadside for detection of dangerous maneuvers of passers-by especially the visually impaired b. Alerts or warnings can be voice or suitable sounds	High-tech
Image recognition	a. The visually impaired can use their phone's camera to scan labels, signs, signboards, etc., and the app will read for them	High-tech

Community Involvement: Participants' involvement in generating ideas for improving accessibility shows the value of community input in social innovation. Engaging with those directly affected by the issues ensures that the solutions developed are relevant and practical.

The findings suggest that empathy, technological integration, practical feasibility, and community involvement can drive effective social innovation. By focusing on these aspects, solutions can be developed that address immediate needs and contribute to long-term improvements in accessibility and safety for visually impaired individuals.

5 Conclusion

The idea generation session using a creative pattern-matching strategy can be potentially leveraged to facilitate idea creation to solve specific issues we face daily. Such is the case chosen for this ideation session, i.e., the visually impaired university students. With the improved education system for PWDs including the visually impaired, they can now pursue tertiary education. However, most of our buildings and campuses are yet to be fully equipped with materials and technology to support their studies and daily living. The ideation session generated many possible ideas for solving concerns for the visually impaired moving around the campus. Ideas as practical and logical implementation to the long shot ideas have been created using this creative strategy, sparking social innovation that can be taken seriously for universities to start providing assistive technology, be it low-tech, mid-tech or even high-tech. The assistive technology can benefit visually impaired students and, hence, increase accessibility where universities can provide an environment that aims to ensure inclusive and equitable quality education and promote lifelong learning opportunities for all, as mandated in the United Nations' Sustainable Development Goals, SDG 4.

References

1. Dzulfikar, L.T.: The long road for disability-inclusive education in Indonesia's universities. The Conversation. https://theconversation.com/the-long-road-for-disability-inclusive-education-in-indonesias-universities-128224
2. Arini, F.D.: How accessible the university websites in Indonesia for people with disabilities. Indonesian J. Disabil. Stud. (IJDS), 7(2), 164–169 (2020). https://doi.org/10.21776/ub.ijds.2020.007.02.05
3. Sachs, D., Schreuer, N.: Inclusion of students with disabilities in higher education: performance and participation in students' experiences. Disabil. Stud. Q. **31**(2) (2011). https://doi.org/10.18061/dsq.v31i2.1593
4. Alam, N.H.: Web accessibility of the higher education institute website based on the world wide web consortium and section 508 of the rehabilitation act (Doctoral Dissertation), http://scholarworks.uark.edu/etd/ 2319 (2014)
5. Almayeni, M., Marlina: Perceptions of students with disabilities on reasonable learning accommodations at Universitas Negeri Padang. Indonesian J. Disabil. Stud. **10**(2), 155–165 (2023). https://doi.org/10.21776/ub.ijds.2023.10.02
6. Zhou, J., Rau, P.L.P., Salvendy, G.: Use and design of handheld computers for older adults: a review and path forward. Int. J. Hum. Comput. Interact. **30**(2), 72–90 (2014)
7. Mankoff, J., Fait, H., Tran, T.: Is your web page accessible? a comparative study of methods for assessing web page accessibility for the blind. In: ACM Transactions on Computer-Human Interaction **12**(1), 41–56 (2016)
8. World bank group. Digital Progress and Trends Report 2023. https://www.worldbank.org/en/publication/digital-progress-and-trends-report (2023)
9. Sloan, D., Stratford, J.: Higher education institution responses to disability legislation. High. Educ. Res. Dev. **37**(4), 723–737 (2018)
10. Fichten, C.S., Asuncion, J.V., Barile, M., Ferraro, V., Wolforth, J.: Accessibility of eLearning in post-secondary education: from policy to practice. Educ. Inf. Technol. **24**(2), 205–227 (2019)
11. Cavender, A.C., O'Modhrain, M., Kolarik, A., Cutting, J.: Accessibility and inclusive design in higher education: strategies and guidelines. In: ACM Transactions on Accessible Computing **12**(4), 20 (2020)
12. Kolko, J.: Design thinking comes of age. Harv. Bus. Rev. **93**(9), 66–71 (2015)
13. Phills, J.A., Deiglmeier, K., Miller, D.T.: Rediscovering social innovation. Stanf. Soc. Innov. Rev. **6**(4), 34–43 (2008)
14. Mulgan, G., Tucker, S., Ali, R., Sanders, B.: Social innovation: what it is, why it matters and how it can be accelerated. London: The Young Foundation (2007)
15. Power, C., Freire, A.P., Petrie, H., Swallow, D.: Guidelines are only half of the story: accessibility problems encountered by blind users on the web. In: ACM Transactions on Accessible Computing **9**(1), 6 (2017)

16. Arslan, G., Yildirim, M., Karataş, Z.: Suicide risk, resilience, and psychological wellbeing in college students: the mediating role of social support. Arch. Suicide Res. **25**(1), 44–58 (2021)
17. Dam, R.F., Teo, Y.S.: How to select the best idea by the end of an ideation session. interaction design foundation–IxDF (2022). https://www.interaction-design.org/literature/article/how-to-select-the-best-idea-by-the-end-of-an-ideation-session
18. Petrie, H., Bevan, N.: The evaluation of accessibility, usability, and user experience. In: C. Stephanidis (Ed.), The Universal Access Handbook, pp. 61–73, FL: CRC Press, Boca Raton (2009)
19. Papadopoulos, K., Koustriava, E., Isaraj, L., Chronopoulou, E., Manganello, F., Molina-Carmona, R.: Assistive technology for higher education students with disabilities: a qualitative research. Digital **4**, 501–511 (2024). https://doi.org/10.3390/digital4020025
20. Foster, M.K.: Design thinking: a creative approach to problem solving. Manage. Teach. Rev. **6**(2), 123–140 (2021). https://doi.org/10.1177/2379298119871468
21. Brown, T.: Design thinking. Harv. Bus. Rev. **86**(6), 84–92 (2008)
22. Mulgan, G.: The process of social innovation. Innovations. Technology, Governance, Globalization **1**(2), 145–162
23. Marriott, K.: Non-visual access to graphical information on COVID-19. In: The 22nd International ACM SIGACCESS Conference on Computers and Accessibility, pp. 1–3. Association for Computing Machinery, New York (2020)
24. Siu, A.F., et al.: COVID-19 highlights the issues facing blind and visually impaired people in accessing data on the web. In: Proceedings of the 18th International Web for All Conference, pp. 1–15. Association for Computing Machinery, New York (2021)
25. Tavassoli, F., et al.: Towards an effective web-based virtual health intervention: the impact of media platforms, visual framing, and race on social presence and transportation ratings. In: Duffy, V.G. (ed.) HCII 2021. LNCS, vol. 12778, pp. 165-181. Springer, Cham (2021)
26. Jain, M., Diwakar, N., Swaminathan, M.: Smartphone usage by expert blind users. In: Proceedings of the 2021 CHI Conference on Human Factors in Computing Systems, pp. 1–15. Association for Computing Machinery, New York (2021)
27. Leporini, B., Buzzi, M., Hersh, M.: Distance meetings during the COVID-19 pandemic: are video conferencing tools accessible for blind people? In: Proceedings of the 18th International Web for All Conference, pp. 1–10. Association for Computing Machinery, New York (2021)
28. AHEAD (Association for Higher Education Access and Disability), Numbers of Students with Disabilities Studying in Higher Education in Ireland 2014/15. Huntersville, NC: AHEAD Educational Press (2016)
29. Indrawati, A.: Pertuni, persatuan tunanetra indonesia website. https://pertuni.or.id/siaran-pers-peran-strategis-pertuni-dalam-memberdayakan-tunanetra-di-indonesia/
30. Ministry of Research, Technology, and Higher Education: Regulation of the Minister of Research and Technology–Higher Education, Number 46 of (2017)
31. Wulansari, R., Baharuddin, H.: Inclusive university management with technology basis: a case study of visual impairment students at information technology study in university of pamulang (UNPAM). In: Proceedings of the 1st International Conference on Research in Social Sciences and Humanities (2020)
32. Sukirno, Premchairporn, N.: Accessibility of persons with disabilities to study in higher education institutions. Homata Int. J. Soc. Sci. **3**(1), 74–83 (2022)

Techniques for Mitigating the Vanishing Gradient Problem in Deep Neural Networks

Shakirat Oluwatosin Haroon-Sulyman[1(✉)], Siti Sakira Kamaruddin[2], Farzana Kabir Ahmad[2], Shuzlina Abdul-Rahman[3], and Nor Idalaila Aziz[4]

[1] Department of Information Technology, University of Ilorin, Ilorin, Nigeria
sulyman.sh@unilorin.edu.ng
[2] School of Computing, Universiti Utara Malaysia, Changlun, Malaysia
[3] School of Computing Sciences, College of Computing, Informatics and Mathematics, Universiti Technologi MARA, Selangor, Malaysia
[4] Politeknik Sultan Abdul Halim Mu'adzam Shah, Jitra, Kedah, Malaysia

Abstract. The Vanishing Gradient (VG) problem is a significant challenge during Deep Neural Networks (DNN) training, mainly when using the backpropagation algorithm. This issue is especially true for tasks involving sequential data and long-range dependencies, such as natural language processing (NLP). Over the years, the VG problem has received significant research attention, indicating its importance. However, the few existing studies on the VG problem that offer a subset of work in this area do not provide a structured understanding of the various methods to solve the VG problem. This study briefly explores methods proposed to mitigate the VG problem, categorizing them into three main approaches: (1) modified activation functions, such as ReLU and its variants, which help alleviate gradient shrinkage in deep layers; (2) architectural innovations, such as residual connections and LSTMs, designed to preserve gradient flow over long sequences, and (3) optimization strategies, including advanced methods like Adam, which adapt learning rate dynamically to stabilize training. A comparative analysis of these techniques highlights key insights: while ReLU variants provide simplicity and computational efficiency, they may still struggle with dying neurons; architectural approaches like residual networks offer robust gradient propagation and great potential but at the cost of increased complexity—optimization methods like Adam's balance computational overhead with improved convergence rates. We conclude by discussing potential future directions for further research in mitigating the vanishing gradient problem, instilling a sense of optimism for the future of deep learning.

Keywords: Vanishing gradient · deep neural network · backpropagation algorithm

1 Introduction

Deep Neural Networks (DNN) have transformed domains such as computer vision and text generation by enhancing the modelling, and representations of complex patterns. However, training these networks often encounters a significant obstacle known as the

Z. Mat Aji et al. (Eds.): KMICe 2024, ISEM 49, pp. 428–438, 2025.
https://doi.org/10.1007/978-3-031-91485-0_33

vanishing gradient problem [1]. This problem, a consequence of the fundamental back-propagation algorithm in training DNN, poses a critical challenge, particularly for networks dealing with high-dimensional data such as texts. Figure 1 illustrates a DNN with multiple layers of input, hidden and output layers, which shows the flow of gradients through the network.

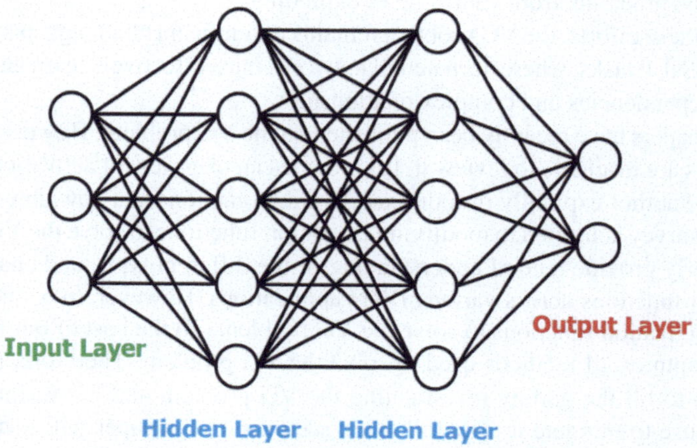

Fig. 1. Illustration of the DNN architecture

As neural networks deepen, the gradients used for updating the weights during back-propagation can become increasingly smaller or completely vanish, limiting the learning process and making it difficult for the model to preserve long-term relationships within the data [2].

The vanishing gradient is not just applicable to DNN but a particularly pronounced one in NLP with tasks such as language modelling, machine translation, and text generation, which necessitate the learning and representing intricate relationships and dependencies within textual data. The sequential nature of language and the prevalence of long-range dependencies make NLP tasks especially vulnerable to the VG problem, adding another layer of complexity to an already challenging issue [3]. However, limited research has been conducted on the VG problem. Hence, this survey is generally about VG in various DNN applications.

The vanishing gradient refers to the phenomenon where the gradients for the parameters of a neural network become very small as more layers are added, using certain activation functions like sigmoid or traditional ReLU. This can cause the network weights to be updated very slowly and ineffectively, leading to difficulty learning complex patterns in text data. The VG problem has been recognized as an issue since the early days of DNN [4]. Moreover, it has hindered the efficient training of deep architectures.

Over the years, researchers have proposed various methods to mitigate the VG problem. One approach has been to modify the activation functions, such as introducing variants of activation functions such as Leaky ReLU and parametric ReLU, which help maintain non-zero gradients even in deeper layers [1]. However, these modifications do not provide a complete solution, and their effectiveness varies across different tasks and datasets.

Another method explored is architectural innovations like residual networks and Transformer models, developed to preserve gradient flow across layers, enabling the model to learn complex long-range dependencies effectively [5]. Optimization strategies like gradient clipping and Adam optimization, further stabilize the training process by preventing the gradients from vanishing or exploding.

Despite these efforts, the VG problem remains a significant challenge, particularly in text data and NLP tasks, where deep neural networks must effectively learn and represent long-term dependencies and complex relationships.

Several papers have recently been published on the VG problem. However, few literature surveys are available to review it. [6] provides an overview of activation functions in DNN. The author explicitly introduces activation functions and how they are used in DNN; their survey is limited to modifying activation functions to solve the VG problem. [7, 8] similarly provide a good understanding of the different types and characteristics of activation functions across various DNN applications. However, they are limited to modifying activation functions to solve the VG problem. To the best of our knowledge, no survey captures all methods used to solve the VG problem. Therefore, this current review aims to fill the gap by investigating the VG problem and the various methods researchers use to mitigate it. The following sections of this paper will investigate the details of various methods, such as modification of activation functions, architectural innovation, and optimization methods, used to mitigate the VG problem. This is followed by discussions and conclusions.

2 Vanishing Gradient Concept

Deep learning is an Artificial Neural Network (ANN) that requires multiple layers of information processing stages in hierarchical architectures. These layers are trained using the backpropagation network algorithm, a widely used technique for training deep neural networks to classify and identify patterns [9]. The backpropagation algorithm is designed to modify the network weights by making incremental adjustments in the opposite direction of the error derivative. In deep networks, however, the gradients become increasingly smaller for each layer further from the output, a phenomenon known as the vanishing gradient problem [10]. The VG problem can be understood in a situation where the gradient of the objective function to the parameter approaches zero, causing minimal or no parameter updates during the network training, which leads to poor convergence and eventually stops training [7]. A significant cause of this problem is the choice of activation function [1]; a mathematical function is applied to each neuron's output in a neural network to determine whether or not the neuron should be activated. The activation function is applied in every neuron of the DNN to learn complex data patterns for accurate network predictions [1]. In the course of training Deep Neural Networks (DNNs), the weights and biases for each layer are determined by calculating the discrepancy between the predicted output and the actual label. However, in the beginning layers of the network, weights and bias updates may gradually become small due to activation functions such as sigmoid, which can lead to the VG problem and eventually stop the learning process. Figure 2 is a typical illustration of activation functions, such as the sigmoid function.

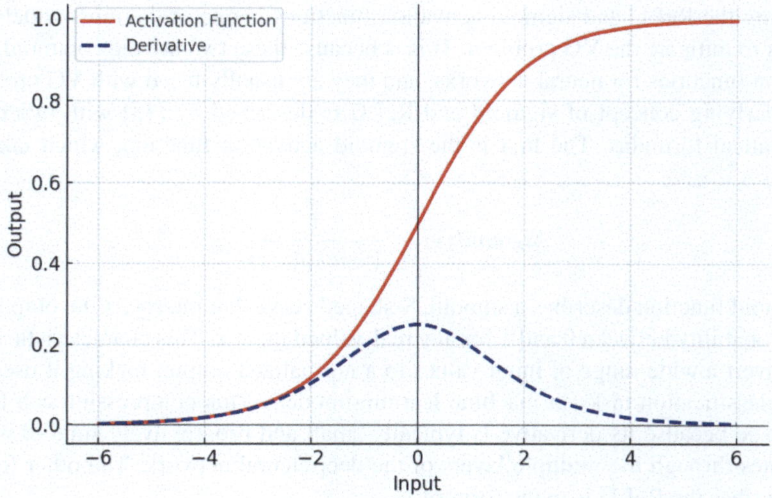

Fig. 2. Illustration of the activation function in DNN

This function compresses its input into a value between 0 and 1, which causes the output to approach 0 or 1 respectively. However, as the input becomes larger or smaller, the sigmoid function's derivative (represented by the blue curve) becomes very small. During backpropagation, gradients are calculated based on the derivative of the activation function. Since the derivative of the sigmoid function approaches zero, the gradient diminishes gradually, causing weight updates in earlier layers to become very small. This can prevent effective layer learning, slowing down the DNN training or causing it to stop completely, known as the VG problem.

As a result, DNNs such as Recurrent Neural Networks (RNNs) face this issue in their earlier layers, making it challenging to learn long-term dependencies effectively. This is because the gradients become too small to provide meaningful updates to the network weights, leading to slow or ineffective training [11].

In natural language processing (NLP) tasks, recurrent neural networks (RNNs) are commonly employed because of their capacity to examine words individually and consider the preceding state output as the current state input. However, RNNs are known to perform well only for a few text analyses and can face challenges when dealing with long-term dependencies [12].

Various modifications have been made to the activation functions and neural networks to address these issues. The following section will examine the methods employed to solve the VG problem.

3 Methods to Solve Vanishing Gradient

3.1 Modified Activation Functions

Modifying the ReLU and sigmoid activation functions is one of the most widely used methods to mitigate the VG problem. This is because these two are the commonly used activation functions for neural networks, and they are usually faced with VG problems. The underlying concept of sigmoid and ReLU is described by [13] with two crucial mathematical formulas. The first is the sigmoid activation function, which takes the form:

$$Sigmoid(x) = \frac{1}{1 + \exp(-x)} \tag{1}$$

the sigmoid function describes a smooth, S-shaped curve that interprets the outputs as a value probability between 0 and 1 for any real-valued input x. This characteristic allows it to convert a wide range of input values to a normalized output, making it useful for binary classification tasks in machine learning models. This compression can lead to a VG issue because its derivative is typically small and progressively smaller when it propagates through the multiple layers of the deep neural network. The other formula that describes the ReLU is in the form of:

$$ReLU(x) = \max(0, x) \tag{2}$$

where the ReLU function generates a constant output of 0 for any negative input. This can result in the dying ReLU phenomenon, where neurons output zero, and the gradients do not change, resulting in a VG problem variant.

Thus, changing the activation function through modification of the sigmoid [14] or reinstating the original derivative of the sigmoid to an artificial one for practical model training [13]. These modified activation functions have shown improved performance in handling VG issues.

3.2 Architectural Innovations

Recent architectural innovations have shown great potential for tackling deep neural networks' vanishing gradient (VG) issue. Unlike modifying activation functions, which focus on the individual neuron level, these architectural advancements introduce structural changes to the network architecture, facilitating more efficient gradient flow during the backpropagation process. Architectural innovations have demonstrated remarkable success in various deep-learning tasks, including natural language processing and sequential data modeling, enabling better gradient propagation.

[15] proposed a Recurrent Identity Network (RIN) to overcome the VG in the traditional recurrent network. The RIN model employs parallel computations and separates the influence of input and hidden states during the recurrent computation. This separation prevents the gradients from vanishing, as the gradients are computed independently for the input and hidden states. Crucially, the RIN architecture leverages the ReLU activation function, significantly speeding up the convergence and enabling the network to be less sensitive to the training conditions.

Another architectural innovation in RNNs is the Independently recurrent network termed (the IndRNN) model [16]. In this model, neurons are independent of each other, allowing for effective gradient propagation over different time steps and maintaining long-term memory. The IndRNN model has been successfully applied to language modelling and image classification tasks, demonstrating its ability to mitigate the VG problem. Also, BiGridLSTM architecture [5] has been proposed to alleviate the VG problem in Long Short-Term Memory (LSTM) networks. This architecture combines an LSTM grid and bidirectional architecture, enabling more effective gradient propagation and improving the performance of LSTM models on speech datasets.

The Transformer architecture [17] has effectively addressed the VG problem in tasks involving sequential data. Transformer-XL, with its self-attention mechanism, mitigates the VG problem by providing a mechanism to retain and utilize information from distant parts of the input sequence, leading to improved performance in capturing long-range dependencies in a variety of NLP tasks [18].

Therefore, architectural innovations have been crucial in mitigating the VG problem, particularly in tasks involving sequential data and long-range dependencies, such as natural language processing. By introducing structural changes that facilitate more efficient gradient flow, these architectures have significantly enhanced DNNs' representational capacity and performance in capturing complex patterns and relationships in sequential data.

3.3 Optimization

Optimization methods focus on efficiently updating the network's parameters during training and ensuring that the gradients are effectively utilized for learning. While architectural innovations and activation function modifications aim to facilitate adequate gradient flow, optimization methods have been widely used in conjunction with architectural innovations [19].

One optimization technique that has shown promise in addressing the VG issue is the Stochastic Diagonal Approximate Greatest Descent (SDAGD) algorithm. [1] demonstrated that SDAGD, when used with ReLu and sigmoid activation functions, can lead to faster training and better performance than standard stochastic gradient descent (SGD), which often struggles with slower convergence and VG problems.

Adaptive learning rate optimization algorithms, such as Adaptive Moment Estimation (Adam) and Root Mean Square Propagation (RMSProp) optimizer, have also been explored as effective methods for mitigating the VG problem and utilized the Adam optimizer in conjunction with recurrent neural network (RNN) models for text classification tasks, demonstrating its ability to stabilize the training process and overcome the VG issue [2]. Similarly, RMSProp resists the VG problem and has been successfully applied in tasks like fake news detection [20]. Another widely employed optimization method is gradient clipping [15], which involves constraining the gradients within a predetermined range to stabilize the optimization process and reduce the impact of the VG issue.

In addition to these techniques, normalization and regularisation methods have also been explored. Batch normalization and regularization techniques like dropout have been utilized in text classification tasks to help mitigate the VG issue [3].

Recent studies have shown that optimization methods can be enhanced by incorporating architectural innovations and adjusting activation functions. This can improve performance in deep neural network models and streamlined training processes, ultimately addressing the VG problem, particularly in natural language processing tasks.

4 Discussion

4.1 Modified Activation Functions

The first advantage of this method is that it requires less computational cost. Another advantage is that it makes little changes to the neural network. While these modifications have shown promising results, their effectiveness can vary across tasks and datasets. Moreover, the choice of activation function modification often needs to be customized to fit the specific needs and traits of the problem regarding aspects such as the network structure, optimization strategy, and the nature of the data. As observed from this review, studies here focus on activation functions, and the majority were tested on image datasets [13, 14]. As researchers have pointed out, this method's disadvantage is domain-specific and may lead to training instability.

4.2 Architectural Innovations

The first advantage of architectural innovation is that it improves model performance [15]. Research shows that most of the architectural innovation in DNN is focused on RNN due to its wide usage in high-dimensional data. However, it cannot memorize long-term information, leading to VG problems. Another advantage of this method is that it facilitates efficient gradient flow because it allows for structural changes to network architecture. Despite this method's flexibility to improve neural networks, it requires high computational costs and increased network complexity. However, techniques such as the transformer model prove more effective by working independently without additional objectives [18], resulting in faster training and mitigating vanishing gradient problems.

4.3 Optimization

One advantage of the optimization method is its flexibility in training neural networks and its ability to learn complex patterns and long-range dependencies in data, such as NLP tasks [15]. However, they can introduce additional hyperparameter tuning, which is time-consuming and requires high computational cost. Also, other approaches in this method, such as batch normalization, may not be practical for fewer datasets or tasks where the input distribution frequently changes. In general, while optimization can efficiently reduce the VG problem, it is essential to carefully select the most appropriate technique based on specific datasets and tasks.

Specifically, VG issues are more pronounced in NLP tasks due to their sequential nature, which limits the ability of architectures like RNNs to capture long-sequence tasks such as language modelling.

Table 1 provides a concise overview of the advantages and disadvantages of the previously mentioned methods.

Table 1. Summary of advantages and disadvantages of various methods for addressing the VG problem.

Method	Pros	Cons
Modified activation functions	-Requires less computational cost Ensures faster convergence -Can be integrated easily with little changes into an existing neural network	-It is domain-specific -Leads to training instability
Architectural innovations	- Facilitate efficient gradient flow -Improves model performance	-Effectiveness varies across tasks and datasets -Requires high computational cost
Optimization	-Flexible for hyperparameter tuning -Able to learn long-range dependencies	-Time consuming -Requires high computational cost

5 Further Research Directions

Despite substantial progress in mitigating the VG problem, critical gaps that require further research remain. While effective in some domains, current methods often fail to generalize across different tasks, mainly in NLP and other applications requiring the modelling of long-term dependencies. Future research should focus on developing more adaptive, scalable, and task-specific solutions to address this.

1. Task-Specific and Dynamic Activation Functions:
   Current activation functions, such as ReLU and its variants, address the VG problem to a degree but often lack adaptability for specific tasks, especially those involving sequential or highly nonlinear data. A promising research direction would be creating task-specific activation functions that adapt dynamically during training. These functions could vary their behaviour based on the depth of the network, the layer context, or the characteristics of the input data. These activation functions could ensure better gradient propagation by incorporating mechanisms that adjust the gradient behavior in real-time, particularly in tasks like NLP and time-series analysis. Dynamic activation functions that evolve during training could mitigate VG in deeper networks, offering a more nuanced solution than static functions like Leaky ReLU.
2. Hybrid Architectures and Optimization Techniques:
   While residual connections and Transformer-based architectures have shown promise in addressing gradient flow, there is still significant untapped potential in combining architectural innovations with adaptive optimization strategies. For instance, hybrid models that integrate self-attention mechanisms with recurrent architectures could provide short-term and long-term dependency modeling, a crucial requirement for speech recognition and reinforcement learning tasks. These hybrid architectures could also incorporate dynamic optimizers, like an enhanced version of Adam, which

adjusts learning rates and model depth or connectivity in response to gradient behaviors. Optimization strategies, such as layer-wise gradient clipping and advanced normalization techniques, could prevent vanishing gradients while maintaining efficient training.

3. Regularization Techniques Tailored to Gradient Behavior:
   Often employed to prevent overfitting, regularisation methods could also be leveraged to stabilize gradient propagation. Gradient-sensitive regularization techniques, such as layer-wise regularization penalties that adjust dynamically based on gradient magnitude, could help mitigate vanishing and exploding gradients. By selectively applying more robust regularization in deeper layers or where gradients are more likely to vanish, such methods could improve training stability in very deep networks. Techniques like variational dropout or adaptive weight regularization might also be re-imagined to directly counteract VG by enforcing constraints that maintain healthy gradient flow throughout the network.
4. Transfer Learning and Pretrained Models in VG Mitigation:
   Using transfer learning, pre-trained models, such as BERT and GPT, can circumvent some of the difficulties associated with VG. However, the role of pre-trained models in VG mitigation has not been thoroughly investigated. Future research should explore how transfer learning strategies can be optimized to alleviate VG, particularly in low-resource scenarios. Additionally, there is room to explore the fine-tuning dynamics of these models, focusing on how gradient behavior changes across different layers during fine-tuning for specific tasks. Understanding how transfer learning interacts with VG-related challenges could inform new model initialization and training techniques, making it possible to train deeper models without encountering significant gradient decay.
5. Dynamic Neural Network Architectures:
   One of the most promising future directions is the exploration of dynamic neural network architectures that adapt their depth, learning rates, or even internal structure based on gradient behavior during training. Dynamic architectures could adjust layer connections or neuron activation in real-time, reducing the likelihood of gradient vanishing in specific network parts. For example, layer-skipping mechanisms could be activated when gradients are too small, while specific layers could be expanded when gradients are too large. These architectures would evolve during training, making them more resilient to the VG problem.

6 Conclusion

This paper presented a brief review of the methods to handle VG, which can be utilized to solve the VG problem. We investigated the overview of the VG problem, its underlying causes, and various methods proposed to mitigate it. As we examined each method, we found that the modified activation functions, a popular approach to solving the VG problem, do not seem suitable for tasks with long-range dependencies, such as NLP tasks. Architectural innovations seem to be the best method, but their effectiveness has not been widely explored in literature. The optimization method looks promising. However, it is time-consuming and requires high computational requirements. Accordingly, these are some of the methods utilized to solve the VG problem. However, careful consideration is required for NLP tasks, which require a more robust method due to the nature of text data. As a result, given the current literature on VG, new methods have been researched for the VG problem. Some of these methods utilize techniques such as the Transformer model, as observed in [18]. However, these methods have not been widely explored on various tasks, such as in the context of text data. Therefore, further innovations and research are needed to address this issue, as there seems to be no specific method fit for all data types.

Acknowledgement. This research was supported by the Ministry of Higher Education (MoHE) of Malaysia through the Fundamental Research Grant Scheme (FRGS/1/2023/ICT02/UUM/02/1).

References

1. Tan, H.H., Lim, K.H.: Vanishing gradient mitigation with deep learning neural network optimization. In: 2019 7th International Conference on Smart Computing and Communications, ICSCC 2019, pp. 1–4 (2019)
2. Tarnate, K.J.M., Devaraj, M., De Goma, J.C.: Overcoming the vanishing gradient problem of recurrent neural networks in the ISO 9001 quality management audit reports classification. Int. J. Sci. Technol. Res. **9**(3), 6683–6686 (2020)
3. Hassan, A., Mahmood, A.: Deep learning approach for sentiment analysis of short texts. In: 3rd International Conference on Control, Automation and Robotics, pp. 705–710 (2017)
4. Mathew, J., Priya Sebastian, M. M.: Deep learning techniques in computational linguistics. In: Proceedings of the 2nd International Conference on Intelligent Computing and Control Systems, ICICCS 2018, pp. 576–581(2018)
5. Fei, H., Tan, F.: Bidirectional Grid Long Short-Term Memory (BiGridLSTM): a method to address context-sensitivity and vanishing gradient. Algorithms **11**(11), 172 (2018)
6. Gustineli, M.: A survey on recently proposed activation functions for Deep Learning, pp. 1–7 (2022)
7. Dubey, S.R., Singh, S.K., Chaudhuri, B.B.: Activation functions in deep learning: a comprehensive survey and benchmark. Neurocomputing **503**, 92–108 (2022)
8. Datta, L.: A Survey on Activation Functions and their relation with Xavier and He Normal Initialization, pp. 1–17 (2020)
9. Lillicrap, T.P., Santoro, A., Marris, L., Akerman, C.J., Hinton, G.: Backpropagation and the brain. Nat. Rev. Neurosci. **21**(6), 335–346 (2020)
10. Hanin, B.: Which neural net architectures give rise to exploding and vanishing gradients? Advances in Neural Information Processing Systems, pp. 582–591 (2018)
11. Noh, S.-H.: Analysis of gradient vanishing of RNNs and performance comparison. Information **12**(11), 442 (2021)
12. Li, Y., Ye, J.: Learning adversarial networks for semi-supervised text classification via policy gradient. In: Proceedings of the ACM SIGKDD International Conference on Knowledge Discovery and Data Mining, pp. 1715–1723 (2018)
13. Hu, Z., Zhang, J., Ge, Y.: Handling vanishing gradient problem using artificial derivative. IEEE Access **9**, 22371–22377 (2021)
14. Roodschild, M., Gotay Sardiñas, J., Will, A.: A new approach for the vanishing gradient problem on sigmoid activation. Prog. Artif. Intell. **9**(4), 351–360 (2020)
15. Hu, Y., Huber, A., Anumula, J., Liu, S.: Overcoming the vanishing gradient problem in plain recurrent networks 1–20 (2019)

16. Li, S., Li, W., Cook, C., Zhu, C., Gao, Y.: Independently recurrent neural network (IndRNN): building a longer and deeper RNN. In: Proceedings of the IEEE Computer Society Conference on Computer Vision and Pattern Recognition, pp. 5457–5466 (2018)
17. Vaswani, A.: Attention Is All You Need. In: Advances in Neural Information Processing Systems, Nips (2017)
18. Dai, Z., et al.: Transformer-Xl: Language Modeling with Longer-Term Dependency, pp. 1–16 (2019)
19. Haroon-Sulyman, S.O., Taiye, M.A., Kamaruddin, S.S., Ahmad, F.K.: Systematic Literature Review and Bibliometric Analysis on Addressing the Vanishing Gradient Issue in Deep Neural Networks for Text Data. In: Zakaria, N.H., Mansor, N.S., Husni, H., Mohammed, F. (eds.) Computing and informatics, pp. 168–181. Springer, Nature Singapore (2024)
20. Bahad, P., Saxena, P., Kamal, R.: Fake news detection using bi-directional LSTM-recurrent neural network. Procedia Comput. Sci.. **165**, 74–82 (2019)

Supply and Demand Analysis in Bridging the Gap of Graduate Employability in Malaysia Through Jobstreet Data and Tiktok Insights

Juhaida Abu Bakar[1,2](✉), Ruziana Mohd Rasli[3], Mohamed Ali Saip[2], Nor Hazlyna Harun[1], Chan Pui Yee[1], Chong Kai Zhi[1], Nurul Maizura Shafika Md Zaid[1], Ong Hong Kai[1], and Ho Jun Han[1]

[1] Data Science Research Lab, School of Computing, Universiti Utara Malaysia, 06010 Sintok, Kedah, Malaysia
juhaida.ab@uum.edu.my

[2] Data Management and Software Solution Research Lab, School of Computing, Universiti Utara Malaysia, 06010 Sintok, Kedah, Malaysia

[3] School of Multimedia Technology and Communication, Universiti Utara Malaysia, 06010 Sintok, Kedah, Malaysia

Abstract. Navigating the transition from academia to employment is critical for graduates, demanding a comprehensive understanding of graduate employability dynamics. However, a thorough integration of employer expectations and graduates' authentic experiences using diverse data sources is lacking. This work integrates data from JobStreet, a leading Malaysian job portal, to explore the employer's perspective and TikTok videos to capture graduates' insights. The research aims to unveil supply and demand challenges in the Malaysian job market, offering a holistic view. By analysing both employer-side insights from the JobStreet dataset and graduate perspectives through TikTok videos, the study addresses supply and demand discrepancies in the Malaysian job market for recent graduates. Employer preferences and graduate experiences are examined through the JobStreet dataset, with skills gap assessment providing a thorough employer perspective. Qualitative insights from TikTok videos offer nuanced understandings of graduates' challenges and expectations. Furthermore, prevalent financial concerns among graduates are highlighted, as indicated by the sentiment analysis of 7,661 TikTok videos. By integrating and comparing these findings, the study seeks to bridge the gap between employer demands and graduate realities, providing valuable insights for education and industry stakeholders.

Keywords: Graduate Employability · Job Market Dynamics · Employer Preferences · Supply and Demand Challenges · Skills Gap Analysis

1 Introduction

The transition from student to workforce is crucial for graduates, marking a shift from theory to practice and impacting societal progress. Graduate employability involves more than just having a degree; it includes skills, adaptability, industry alignment, and

Z. Mat Aji et al. (Eds.): KMICe 2024, ISEM 49, pp. 439–451, 2025.
https://doi.org/10.1007/978-3-031-91485-0_34

personal expectations. This work integrates data from JobStreet, a prominent Malaysian employment marketplace, and insights from TikTok videos featuring recent graduates. This combination offers a comprehensive view of employer perspectives and graduates' experiences.

Previous studies focus on specific aspects like soft skills [1], learning activities [2], and employer perceptions of employability [3]. However, there is a lack of comprehensive integration of employer expectations and graduates' authentic experiences using diverse data sources. Current research fails to provide an integrated analysis combining structured datasets like JobStreet with unfiltered perspectives from TikTok. This limits the ability to identify congruence or divergence between employer expectations and graduate reality. Additionally, the emotional and personal aspects of the job search process are often overlooked yet are crucial for understanding graduates' challenges.

In Malaysia, graduates' educational backgrounds often do not align with the skills companies seek [4]. This mismatch hinders the seamless transition from school to workforce and affects job suitability. A comprehensive strategy involving companies, graduates, and educational institutions is essential to address this issue.

To bridge these gaps, the aim is to improve graduate employability in Malaysia by integrating insights from employer-driven datasets and graduate narratives. This study seeks to develop a thorough understanding of the employability environment for new graduates by analysing supply and demand disparities in Malaysian labour market. Goals include identifying high-demand industries, determining desired skill sets, investigating graduates' difficulties, understanding factors affecting employers' decisions, and identifying skill gaps between employers and graduates. Bridging these gaps would facilitate a smooth transition from academia to the workforce, benefiting both graduates and Malaysia's economic growth.

By merging JobStreet data and TikTok anecdotes, the study aims to understand the graduate employability landscape in Malaysia. This approach seeks to uncover job market challenges from both employer and graduate perspectives, shedding light on the dynamics between employer expectations and graduate realities. This study aims to provide insights for creating more aligned and effective strategies for graduate employability, benefiting both the workforce and the economy in Malaysia.

Section 2 reviews related work on graduate employability studies. Section 3 outlines the research methodology. Section 4 presents the experiments and results, and Sect. 5 concludes the findings.

2 Related Work

Due to technological advancements, the recruitment process has transitioned towards online applications, offering increased accessibility to job vacancy data across various sectors. While online recruitment portals provide a more comprehensive range of options, there exists a potential for bias in the coverage of high and semi-skilled job vacancies, resulting in a less comprehensive representation of the overall job vacancy landscape [5]. Recognising that graduates often possess limited or no work experience, employers prioritise qualities and abilities alongside discipline-specific knowledge and skills. Thus, Graduate employability hinges on developing soft skills, constituting a harmonious blend of expertise and generic abilities that enable individuals to succeed more in the workplace [4].

2.1 Usage of Online Job Advertisement Portals for Job Profiling

Understanding academic qualifications and skill requirements for job vacancies is crucial, with job advertisements playing a pivotal role by swiftly providing valuable insights into dynamic changes in job profiles [6]. However, the core aspects of a given occupation are subject to industry influence, contributing to variations even within the same job category [7]. The recent exponential growth of specialised internet portals and services dedicated to labour market information have provided real-time insights into current and ongoing workforce needs. In this ever-changing landscape, analysis of job descriptions from numerous advertisements become crucial, offering valuable insights into graduate employability and identifying current trends in necessary competencies to enhance competitiveness and adapt to the rapidly changing era.

The employment of content analysis facilitated the examination of requisite employability skills in job advertisements through a predefined categorisation dictionary [4]. However, analysing unstructured job advertisement data presents inherent difficulties in extracting critical information. To address these challenges, implementing text mining techniques, including N-grams and regular expressions (regex), prove instrumental in mitigating content analysis limitations, especially concerning managing vast datasets [8].

2.2 Enhancing Graduate Employability in Malaysia Through Comprehensive Skill Development

Evidence from [2] shows that researchers provide a comprehensive overview of the impact of a soft skills shortage on the employability of Malaysian postgraduate students. The study used logistic regression analysis to create a predictive model, demonstrating that gender, household income, field of study, and various soft skills significantly impact graduates' employability status. The results of this study provide important insights into the complex dynamics between soft skills and employability.

Another study in [1] provides insights into the employability skills of Malaysian graduates. The study aims to identify Malaysian graduates' employability skills, prioritise each skill, and highlight the gap between the importance of these skills to employers and their satisfaction with them. The study uses importance-performance analysis (IPA) to evaluate the skills gap and its impact on graduate employability in Malaysia. The results showed that employers perceive graduates' performance in employability skills as lower than the importance assigned to those skills. The widest gap was in communication skills, especially in English. The study concludes that graduates should acquire technical and soft skills to increase their employability.

Additionally, a study in [3] investigated the trends shaping graduate employability in Malaysia based on the National Vision 2020. This multi-stage study, which began with a survey of employer perceptions, highlighted the importance of soft skills, particularly in the context of Transnational Education (TNE) students. The favourable TNE students' career prospects and our focus on language and communication skills shape our approach to understanding employer expectations and graduates' realities in the Malaysian job market.

2.3 Exploring Graduate Employability Insights in Malaysia Through Social Media Analysis

TikTok, widely recognised as a platform for influencer marketing, has rapidly gained substantial traction among millennials and Gen Z [9]. Current discussions on social media platforms offer valuable insights into evolving trends and themes related to the future of work. The rise of artificial intelligence and its applications has minimised the previously required human effort and expanded job dimensions in a scalable manner [10]. This study aims to conduct sentiment analysis to uncover potential work opportunities.

Social media data mining is defined as collecting and analysing social media data involves using advanced technologies like machine learning to discern patterns within the data. This process facilitates various applications, such as assisting businesses in evaluating employees' skills and qualifications in the workplace. Researchers have utilised LinkedIn profiles as the base for user modelling, employing techniques like scraping to extract public profiles. Employing Natural Language Processing (NLP) methods, educational backgrounds are categorised, and professional backgrounds are clustered from the gathered profiles. Other social media data sources, such as Twitter, have been leveraged in different fields using machine learning for information extraction. Moreover, techniques such as Topic modelling, exemplified by LDA, play a pivotal role in extracting insights from textual components like comments and captions in video social media, enhancing comprehension of content themes and user interactions [9]. Textual similarity is also a data mining subfield that provides information terms often used in a group of people [11].

Sentiment analysis is a crucial tool for recent graduates, allowing for the automatic identification of the emotional tone in social network text activity [12]. The interpretation of sentiment varies across different domains due to the complex nature of human emotions, leading to diverse definitions and classification algorithms in research [13]. While 'emotions' and 'sentiments' are sometimes used interchangeably, the latter refers explicitly to emotional polarity, covering positive, negative, and neutral states [9].

Sentiment analysis is dedicated to mining attitudes, emotions, and feelings rather than focusing solely on extracting informational content. This involves subtasks related to sentiment content and sentiment level. Various computational statistics and machine learning techniques, such as latent semantic analysis and bag-of-words (BoW), are commonly employed in sentiment analysis to reveal patterns and meanings within textual data [12].

In addition to addressing the complexities of circular economy investments, recent graduates can benefit from understanding sentiment analysis [14]. This valuable skill provides insights into whether discussions surrounding circular economy practices, such as those related to recycling and remanufacturing, carry a positive, negative, or neutral tone on social media platforms using relevant hashtags [13]. This knowledge can prove essential in navigating the digital landscape and gaining a nuanced understanding of how individuals express themselves online, particularly when entering the workforce.

3 Methodology

This section presents the proposed approaches to analyse graduate employability through the JobStreet dataset and TikTok insights in Malaysia. The study collected company and job advertisement data from the JobStreet Malaysia platform to represent the perspective of employers. Utilising the Bardeen scraper, a no-code visual tool, extracted data by creating scraper templates and seamlessly syncing the collected information with Google Sheets. The job advertisement dataset encompassed attributes such as job title, company name, location, job type, work type, salary, job description, requirements, job advertisement link, and company link. Similarly, the company data included details like company name, industry, company size, and state.

A total of 12,684 job advertisement records and 2,021 company records were amassed during this phase. It is important to note that job advertisements on the platform are available for only 30 days, and with over 30,000 job vacancies listed, data collection was conducted by randomly sampling 30% of the available job categories to manage time constraints and maintain the integrity of the job vacancy landscape. Moreover, company data collection was limited to firms actively hiring and featured in job advertisements with a website directory on the JobStreet Malaysia platform. This results in a partial view of the companies posting job vacancies, as not all companies have a directory listing.

In addition, TikTok insights on graduate employability in Malaysia were systematically collected. This comprehensive approach involved extracting and analysing publicly available content, discussions, and user behaviours related to the multifaceted aspects of graduate employability. A range of carefully selected hashtags, such as #jobinterviewtips, #interviewtips, #interviewtipsandtricks, #jobsearchtips, #jobsearching, #careeradvice, #CareerChoices, #JobCharacteristics, and #WorkplacePriorities, was utilised to exploit the informative potential of these hashtags for targeted exploration of the subject. The data collection phase was from December 20, 2023, to January 3, 2024.

Data was collected with a solid commitment to ethics, strictly following TikTok's rules and user privacy regulations. Furthermore, TikTok insights were extracted meticulously using Apify, a tool chosen for its efficiency and alignment with platform policies and ethical data collection procedures.

The focus of this study extended beyond merely gathering data; common challenges faced by graduates in securing employment were specifically extracted, and information concerning graduate expectations, preferences, and concerns was delved into using relevant hashtags. This holistic approach aimed to capture the diverse opinions and experiences surrounding graduate employability on the TikTok platform. The total dataset gathered comprised 11,299 instances. The dataset included columns such as "authorMeta.avatar" and "authorMeta.name," "text," "discount," "shareCount," "playCount," "commentCount," "videoMeta.duration," "musicMeta.musicName," "musicMeta.musicAuthor," "musicMeta.musicOriginal," "createTimeISO," and "webVideoUrl." These datasets reflected the richness and diversity of insights obtained through this methodical and ethical data collection process, enhanced by the strategic use of targeted hashtags.

In the data pre-processing phase, both datasets handle missing values, identify duplicates, handle incomplete data, and handle text cleaning. Moreover, several text-preprocessing techniques were explicitly deployed in the job description and requirement columns. This involved the removal of stopwords, tokenisation, and lemmatisation, essential for constructing a comprehensive dictionary conducive to practical natural language processing tasks.

On the employer side, the analytical approaches aim to unveil concealed insights, provide a nuanced understanding, and offer valuable information for an in-depth investigation into the employment landscape of graduates in Malaysia.

The analysis summarised employers' demographic characteristics by investigating their respective industries and locations. This method aimed to provide insights into the specific demands across various sectors and geographic areas. Through this process, the top 10 industries and top 4 locations were identified, exhibiting the highest job demand, thus revealing prevalent high-demand job types.

Scrutinising job types and workplace locations also analysed job vacancy distribution. This allowed us to identify the top 4 places and top 5 job types with the most frequent job vacancies and specific job types within each area to discern trends. Moreover, companies with high job demand were examined to determine if their demand aligned with the overall trends in the job market landscape.

By scrutinising employer and job vacancy distributions, correlations between employer backgrounds and job demand trends were identified, contributing to a comprehensive understanding of the dynamics within the job market. This comprehensive approach aided in identifying high-demand job types by considering factors such as company background, the overall job vacancy landscape, and the specific job demands of individual companies.

Meanwhile, on the graduate side, The TikTok video analysis involves analysing recent graduates' videos to discover common issues in their job-seeking experiences. This qualitative analysis extracts insights about graduate job-search expectations, preferences, and concerns, including a focus on the emotional aspects of the process. It also seeks to understand the elements that influence their choice of employers. This analysis indicates areas of alignment and distinction between employer goals and actual graduate realities by comparing employer expectations with graduates' experiences.

Various analytical techniques can pinpoint recent graduates' common obstacles when seeking employment. Challenges such as lack of relevant work experience, difficulty meeting company standards, and networking issues can be identified through surveys and data analysis, revealing common patterns. Job market research focuses on specific sectors or industries to identify recurring needs that graduates may lack. Sentiment analysis of online content, including social media posts and forums, provides insights into graduates' emotional experiences during their job search, complementing quantitative data. Word cloud visualisations based on sentiment analysis offers a vivid snapshot of prevalent challenges.

Qualitative insights from case studies or interviews with graduates, recruiters, or career consultants clarify uncertainties about career paths, job requirements, and networking challenges. Network analysis assesses the impact of networking opportunities on job searches, while comparisons between educational curricula and job market demand highlight skill shortages. Methodically using these diverse techniques identifies areas needing improvement, critical requirements, and recurrent obstacles. This information helps guide strategies to address fresh graduates' job search difficulties.

VADER sentiment analysis was also utilised to illuminate TikTok videos' textual and personal dimensions. This approach provided a comprehensive understanding of the diverse textual landscape and personal narratives encountered by graduates during their job search. Both positive, neutral, and negative sentiments were extracted from the text by VADER, offering essential insights into the experiences, issues, joys, and aims shared by graduates.

Additionally, word clouds were employed to visually represent the sentiments extracted by VADER, including both positive and negative textual expressions. These visualisations illustrated common sentiments and textual patterns within the videos, providing a nuanced view of the graduates' perspectives during the job search journey.

Upon completing analyses from both perspectives, synthesising insights derived from employer preferences and graduate experiences provided a comprehensive understanding of graduate employability. Integrating these two sides' perspectives was instrumental in addressing challenges related to supply and demand. The skill gap analysis served as a valuable tool to illuminate discrepancies between the skills possessed by graduates and those sought by employers. Identifying the skills highlighted by employers and the experiences and challenges shared by graduates were used to mitigate workforce shortages and foster strategic career development initiatives. The report also included the analysis results, providing a thorough and well-supported understanding of the research results.

4 Results

4.1 Employer Side

The analysis delved into the job titles, descriptions and requirements within the broader job vacancy landscape. It was noted that the job titles across various categories, such as engineer, executive, sales, technician, developer, and software-related positions, aligned closely with the observed demand in respective job categories. This observation highlights the specificity of the roles in high demand, emphasising their strong alignment with the overall distribution of job categories.

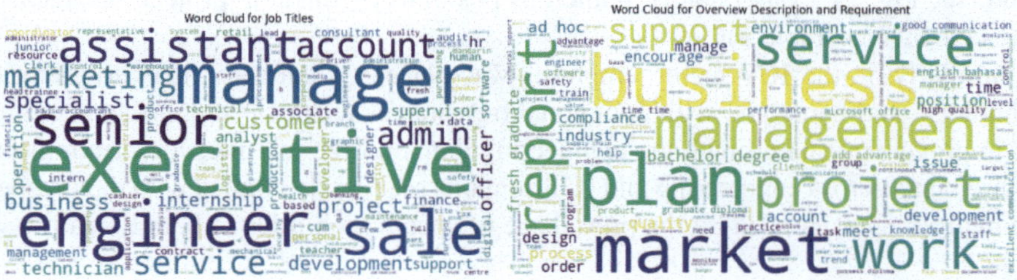

Fig. 1. Word cloud for job titles (left), overview description, and requirement (right).

The word cloud analysis of overall job descriptions and requirements (see Fig. 1) highlighted significant terms such as business, management, project, market, industry, and supply chain, particularly relevant to sales and manufacturing positions. Furthermore, the positions encompassed a wide range, from entry-level to managerial positions, indicating a diverse level of experience sought by employers. Academic qualifications typically included a minimum diploma, with a bachelor's degree recommended.

Additionally, emphasis was placed on essential soft skills such as communication, proficiency in languages including English, Bahasa, and Mandarin, and collaboration and problem-solving abilities. Proficiency in Microsoft Office was deemed a fundamental complex skill, essential for tasks like report preparation. The work environment was acknowledged to be demanding, necessitating self-motivation from potential job seekers.

The comparison between job descriptions and job titles served as a means to assess their alignment, thereby identifying any disparities between the outlined requirements and the actual job roles indicated by their titles (see Fig. 1). However, due to the impracticality of evaluating every job category, descriptive analyses played a pivotal role in identifying high-demand job types. Engineering (see Fig. 2) and ICT (see Fig. 2) emerged as notable areas of demand, prompting a focused analysis of these domains to glean valuable insights into their alignment with job descriptions (see Fig. 2).

Fig. 2. Word cloud for description and requirement of engineering positions (left) and Information & Communication Technology (ICT) positions (right).

In engineering, specific competencies such as ad hoc program management, maintenance, continuous improvement, quality control, compliance, safety health, and technical support are crucial alongside general soft and hard skills. Comparing these competencies with the job titles in engineering, a strong alignment is observed, indicating that engineering work typically follows a project-based mode, as highlighted in the word cloud (see Fig. 2). Consequently, numerous positions pertinent to project management were evident (see Table 1).

Similarly, in the ICT sector, emphasis was placed on software, network security, technology, and hardware, with commonalities such as technical support and continuous improvement shared with the engineering domain. The job titles, descriptions and requirements across engineering and ICT domains seamlessly aligned with their respective natures, as evidenced by the prevalence of specific job titles in both domains (see Table 2).

The high demand for engineering and ICT roles was not only due to their popularity but also reflected the need for adaptability in an ever-changing society. Self-learning and

Table 1. Top 10 most frequent positions in engineering.

Job Title	TF-IDF
engineer project engineer	0.234480
engineer electrical engineer	0.189817
engineer senior engineer	0.145154
mechanical design engineer	0.133988
field service engineer	0.122823
engineer mechanical engineer	0.111657
engineer production engineer	0.100491
technician project engineer	0.100491
manager project engineer	0.089326
engineer technical support	0.089326

Table 2. Top 10 most frequent positions in Information & Communication Technology (ICT).

Job Title	TF-IDF
senior software engineer	0.200384
engineer software engineer	0.133589
developer software developer	0.122457
software developer net	0.111324
full stack developer	0.100192
technical support engineer	0.100192
front end developer	0.100192
IT project manager	0.089059
software engineer senior	0.089059
software developer software	0.089059

continuous improvement were crucial to remain aligned with market trends and seize opportunities, especially in domains that drive technological advancements and change.

Additionally, an analysis was conducted to discern any variations or parallels in job demands across diverse locations within the sales sector. Across different states, sales job descriptions exhibit commonalities in both hard and soft skill requirements. Notably, they consistently sought candidates adept at devising business development strategies, analysing market trends, and cultivating robust client relationships, which were essential for success in sales roles. Despite the geographical spread of sales opportunities, the uniformity in job descriptions underscored the alignment between stated requirements and job titles (see Fig. 3).

Fig. 3. Word cloud for description and requirement of sales positions across various locations.

4.2 Graduate Side

The outcome from TikTok analysis resulted in a comprehensive set of 7661 videos, classified into three sentiment categories: 3739 videos were positive, 2958 were neutral, and 964 were negative (see Fig. 4). Due to the similar content between neutral and negative sentiments, we excluded neutral videos from further analysis. Subsequently, we then generated word clouds for negative and positive feelings to unveil prevailing themes. In Fig. 5, prominent terms such as "credit card", "debt," "pay," "loan," and "low" emerge, signifying the prevalent concerns among graduates. Including the term "low," in conjunction with discussions around "salary," underscores the challenges fresh graduates face with lower income levels, impacting their financial capacity to manage loans and debts. This observation underscores the nuanced and multifaceted nature of financial concerns within the graduate community, emphasising the complexities associated with managing financial responsibilities in the initial stages of their careers.

From the Fig. 6, the term "career" notably stands out. While numerous terms are prominent, "career" is particularly emphasised. Upon reviewing the selected TikTok videos, it is apparent that graduates express a robust desire for "skill development" and "training opportunities" provided by potential employers. This qualitative exploration aims to unveil graduates' importance to obtain training within a company context, providing valuable perspectives on how this aspect shapes their career expectations. They are eager to acquire 'new skills,' particularly those the company offers, as the word cloud indicates. The word cloud also reveals associations with "success," "opportunities," "goals," "achievements," and "new." These insights collectively signify graduates' aspirations for the opportunity to succeed in their career goals and engage

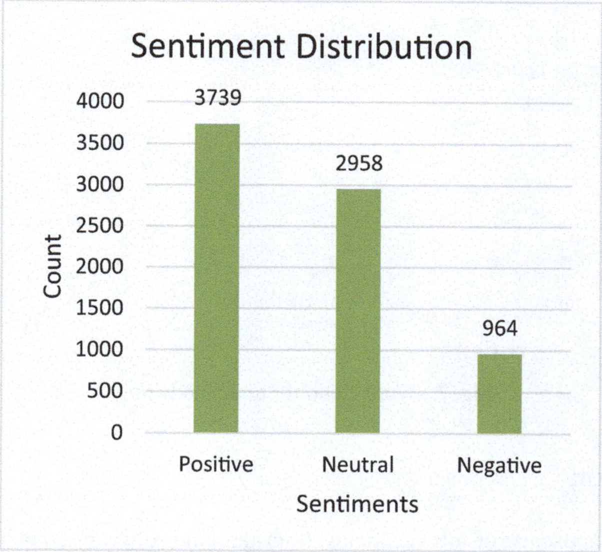

Fig. 4. Sentiment distribution

Fig. 5. Word cloud for negative sentiment analysis

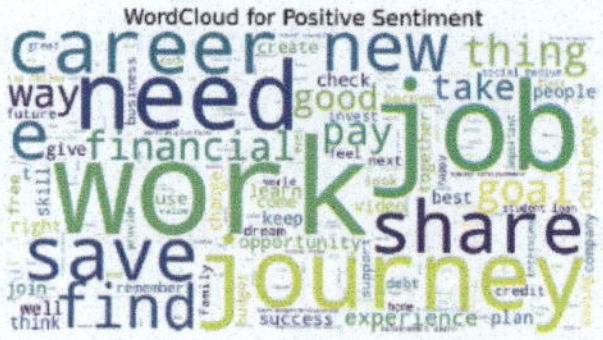

Fig. 6. Word cloud for positive sentiment analysis

in continuous learning. Additionally, graduates express a keen interest in expanding their skill set with diverse abilities that align with the company's offerings, indicating a desire for comprehensive professional development within a supportive organisational environment.

Topic Modeling. Since five topics are the highest score in the coherence test for topic modelling, five topics are included to obtain the top ten words for this analysis. This analysis was completed using TikTok insight.

Using word cloud visualisation, the data modelling step used a calculated approach to extracting and analysing the top ten words for each selected topic. The findings showed some interesting trends, with "credit" showing up as the most common theme in the first subject, "financial" in the second, "job" in the third, "student" in the fourth, and a unique fusion of "journey" and "career" in the last topic. These findings highlight the importance of financial concerns, job-related issues, and students' personal and professional journeys while shedding light on the complex facets of graduate employability (refer to Fig. 7).

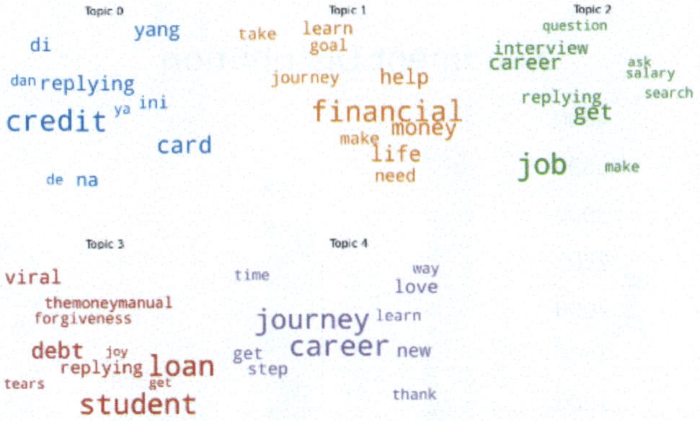

Fig. 7. Word cloud for topic modelling

5 Conclusion

In the broader landscape of job vacancies, high-demand roles are concentrated in engineering and ICT, manufacturing, transport and logistics, and sales, with more opportunities in rapidly developing regions like Selangor, Kuala Lumpur, Johor, and Penang. Analysis reveals strong alignment between job descriptions and actual positions, aiding in accurate candidate identification and consistent role responsibilities across locations. TikTok analysis offers valuable insights into recent graduates' authentic job search experiences, using VADER sentiment analysis and word clouds to uncover emotional landscapes. This approach enhances our understanding of graduates' challenges and aspirations, guiding the creation of strategies aligned with their real-world experiences. The TikTok data underscores the importance of continuous self-improvement for fresh graduates transitioning from university to the workplace. While numerous opportunities exist, adaptability and self-motivation are crucial for overcoming challenges and achieving career goals. Employers are willing to cultivate talent, but the need for self-regulation and initiative remains pivotal for a successful transition from academia to the professional world.

Acknowledgement. This research was supported by the Ministry of Higher Education (MoHE) through the Fundamental Research Grant Scheme (Ref: FRGS/1/2018/ICT02/UUM/02/3). The content of this article is solely the responsibility of the authors and does not necessarily represent the official views of MoHE, Malaysia. The authors would also like to extend their gratitude to the Data Management & Software Solution Research Lab, School of Computing, Universiti Utara Malaysia, for their generous support in sponsoring the publication of this article.

References

1. Hamid, M.S.A., Islam, R., Hazilah, A.M.N.: Malaysian graduates' employability skills enhancement: an application of the importance-performance analysis. J. Glob. Bus. Adv. **7**(3), 181–197 (2014)

2. Basir, N.M., Zubairi, Y.Z., Jani, R., Wahab, D.A.: Soft skills and graduate employability: evidence from Malaysian tracer study. Pertanika J. Soc. Sci. Humanit. **30**(4), 1975–1986 (2022)
3. Fernandez-Chung, R.M., Leong, Y.C.: TNE graduate employment study: An analysis of graduate employment trends in Malaysia (A research study in three phases) Phase III-employability of graduates in Malaysia: The perceptions of senior management and academic staff in selected higher education institutions (2018)
4. Lee, T.H., Lim, Y.M., Chithra Latha Ramalingam, T., Chew, S.H.: Job opportunities and employability skills required of business graduates in Malaysia: an investigation through online job advertisements. J. Market. Adv. Pract. **1**(1), 37–49 (2018)
5. Shabli, R.: Leveraging job vacancies advertised online to analyse the Malaysian labour market using big data analytics. International Statistical Institute (2023)
6. Pejic-Bach, M., Bertoncel, T., Meško, M., Krstić, Ž: Text mining of industry 4.0 job advertisements. Int. J. Inform. Manag. **50**, 416–431 (2020)
7. Khaouja, I., Kassou, I., Ghogho, M.: A survey on skill identification from online job ads. IEEE Access **9**, 118134–118153 (2021)
8. Ningrum, P.K., Pansombut, T., Ueranantasun, A.: Text mining of online job advertisements to identify direct discrimination during job hunting process: a case study in Indonesia. PLOS ONE **15**(6), e0233746 (2020)
9. Alaql, A.A., AlQurashi, F., Mehmood, R.: Data-driven deep journalism to discover age dynamics in multi-generational labour markets from linkedin media. J. Media **4**(1), 120–145 (2023)
10. Sarin, P., Kar, A.K., Kewat, K., Ilavarasan, P.V.: Factors affecting future of work: insights from social media analytics. Procedia Comput. Sci. **167**, 1880–1888 (2020)
11. Brambilla, M., Rahdari, B.: Event-based user profiling in social media using data mining approaches, Technical Report, School of Industrial and Information Engineering, Politecnico Di Milano (2016)
12. Batrinca, B., Treleaven, P.C.: Social media analytics: a survey of techniques, tools and platforms. AI Soc. **30**(1), 89–116 (2015)
13. Peng, S., et al.: A survey on deep learning for textual emotion analysis in social networks. Digit. Commun. Netw. **8**(5), 745–762 (2022)
14. Dınçer, H., Yüksel, S., Hacyoglu, U., Rouyendegh, B. D.: Multidimensional analysis of investment priorities for circular economy with Quantum Spherical fuzzy hybrid modelling. Int. J. Inform. Technol. Decis. Mak. (2023)

A Collaborative Knowledge Portal Model for Paddy Farming in Southeast Asia

Fadhilah Mat Yamin[1,2], Wan Hussain Wan Ishak[3(✉)], Noor Hidayah Abu[1,2], Zakirah Othman[1], Arif Imam Suroso[4], Febriantina Dewi[4], Lokita Rizky[4], Khariyah Kamilla[4], and Hansen Tandra[4]

[1] School of Technology Management and Logistics, Universiti Utara Malaysia, Changlun, Malaysia

[2] Knowledge Science Research Lab, School of Technology Management and Logistics, Universiti Utara Malaysia, Sintok, Kedah, Malaysia

[3] School of Computing, Universiti Utara Malaysia, Changlun, Malaysia
hussain@uum.edu.my

[4] Institut Pertanian Bogor Indonesia, Jawa Barat, Indonesia

Abstract. In Southeast Asia's culturally and ecologically diverse agricultural landscape, effective knowledge management has become increasingly important. This paper introduces the Paddy Knowledge Portal, a specialized platform for paddy cultivation, focusing primarily on Malaysia and Indonesia. It emphasizes the vital role of knowledge in enhancing agricultural efficiency and competitiveness in these countries, both of which have strong agricultural traditions and are major contributors to global rice production. The proposed knowledge portal as a strategic solution for promoting innovation and knowledge sharing in a competitive market. The portal framework includes an overview of user categories—administrators, agencies, and the public—and outlines key knowledge components, such as general information, tutorials, FAQs, questions, and tips, ensuring the platform is accessible and valuable to a broad audience. The paper also discusses the potential impact of the Paddy Knowledge Portal, its collaboration with government agencies like MADA and DOA, and its role in supporting sustainable agricultural practices and collaborative problem-solving. This initiative envisions the portal as an essential tool for enhancing the collective knowledge base, improving adaptability in rice farming, and strengthening the resilience and sustainability of Southeast Asia's agricultural sector.

Keywords: Knowledge Management · Paddy Knowledge Portal · Paddy Cultivation · Collaborative Model

1 Introduction

Knowledge is a valuable asset in the agricultural landscapes of nations like Malaysia and Indonesia, increasing efficiency and competitiveness. These agriculturally rich nations are increasingly understanding the need of knowledge management in their farming communities.

Z. Mat Aji et al. (Eds.): KMICe 2024, ISEM 49, pp. 452–460, 2025.
https://doi.org/10.1007/978-3-031-91485-0_35

Malaysia stands out in Southeast Asia for its rich culture and beautiful surroundings. Notably, Malaysia is one of the world's leading rice producers, with paddy production playing a critical role in the country's economy. This sector creates jobs, earns revenue, and assures food security [1, 2].

Meanwhile, with its wide archipelago, Indonesia is a leader in paddy farming. The country has a long agricultural tradition, with rice farming at the heart of its agricultural industry [3]. Indonesia's diversified agro-ecological zones and large size make it a suitable place for rice production. Rice production has enormous cultural, social, and economic importance in Indonesia, serving as a staple food for the population and adding significantly to the country's export revenues.

In today's highly competitive economy, using information as a driver of innovation is critical [4–6]. As a result, knowledge portal-based application systems are critical to the performance of organizational operations and objectives across a wide range of agricultural contexts [7, 8].

This research describes a knowledge portal approach designed for paddy agriculture. The planned portal, which has been methodically built to centralize rice production knowledge and expertise, is intended for a wide range of audiences, including farmers, researchers, agricultural specialists, and enthusiasts from Malaysia and Indonesia alike. The portal is an important instrument for distributing knowledge throughout these nations' rice farming communities, and it includes a variety of features and functionalities. These include sharing best practices, communicating research findings, demonstrating cutting-edge agricultural technologies, and offering practical assistance. Users can actively participate in discussions, share their experiences, and give insights, creating a collaborative environment that promotes learning and growth.

Over time, the Paddy Knowledge Portal has the potential to have a tremendous influence on rice farming communities in Malaysia and Indonesia. It can gradually improve the community knowledge base, promote sustainable agricultural practices, and help the rice farming business adapt to changing requirements and challenges. As more users from both countries actively contribute to and engage with the platform, it has the ability to expand and flourish, ensuring that rice farming remains resilient and capable of meeting future global demands and environmental considerations.

2 Literature Review

Knowledge portals are specialized internet platforms. It is critical for centralizing and facilitating access to vast amounts of knowledge, expertise, and resources in certain sectors or themes [9]. The portals provide access to organized and selected knowledge, appealing to a wide range of users, from novices to experts. These portals are often designed with user-friendly interfaces, extensive search capabilities, and tools to encourage knowledge exploration and exchange [10, 11].

Many agricultural knowledge portal formats have been created to encourage interactivity, smooth data access, and effective content management. These portals allow for collaborative participation among various agricultural stakeholders, with the primary goal of acquiring and sharing agricultural knowledge. This knowledge sharing supports innovation, sustainability, and informed decision-making in the agricultural domain, eventually benefiting farming communities.

A study by Van-Baalen et al. [8], for example, focuses on the formation of networks of practice via information portals. This study highlights the importance of information portals in promoting knowledge dissemination across weakly connected organizational contexts. This study emphasizes the importance of a sense of urgency, fragmented awareness, and the function of information brokers in forming networks of practice. It also provides a conceptual framework for understanding how knowledge portals work in network development.

FisOmics, developed by India's Center for Agricultural Bioinformatics, is an example of an information portal focused on agricultural bioinformatics and biotechnology [12]. FisOmics provides a diverse set of fish genomics data and related analysis, making it an invaluable resource for academics and anglers looking for information on types of fish and habitats.

Another innovation in knowledge portals is e-KrishiShiksha, developed on the e-learning platforms [13]. This e-learning platform has had a tremendous impact on agricultural education and research, attracting a large number of undergraduate students from a variety of agricultural fields. It successfully offers online courses, decreasing the need to physically distribute course materials while also offering access to carefully designed content that corresponds to the ICAR-approved syllabus.

As discussed above, the key needs for an agricultural information portal include a number of important characteristics. First, the portal should include comprehensive material [8, 9] that covers a wide range of agricultural issues, from crop cultivation to pest control and market trends. This is to ensure that it fulfills the unique demands of its visitors. Furthermore, accessibility and usability are critical. Therefore, user-centered design with an intuitive interface and adaptability to various devices is vital [12]. A search functionality is also required for rapid information retrieval [14]. The collaborative features that support discussion and active communication among members are also required [12].

3 Methodology

This study includes the following steps:

a) Literature Review

This phase consists of three components:

1) Exploration of existing literature: In this phase, the existing literature, research papers, and case studies that are related to knowledge portal design within the agriculture domain are examined.
2) Identification of key concepts, best practices, and challenges: As part of the literature review, there is an emphasis on identifying the critical conceptual underpinnings governing knowledge portal design in the agricultural domain. Furthermore, the review identifies the best practices that have demonstrated efficacy in agricultural knowledge management and dissemination.
3) Analysis of successful knowledge portals in other domains: The literature review also covers knowledge portals in diverse domains. This cross-domain perspective gives an idea of the new concepts and features that may not have been explored within the agricultural context.

b) Stakeholder Engagement

During this phase, our study focuses on identifying and engaging various stakeholders in the agriculture sector. These stakeholders include agricultural professionals and government agencies, whose input is critical due to their domain-specific knowledge and practical experience. Their involvement is critical in developing a knowledge portal that fits the agricultural sector's diverse and evolving needs. This study's key government agencies include the MUDA Agricultural Development Authority (MADA), the Kedah State Department of Agriculture (DOA), and the Lembaga Zakat Negeri Kedah (LZNK). Table 1 summarizes the roles and contributions of various agencies.

Table 1. Stakeholders.

Agency	Role
MADA	MADA is responsible for Kedah's granary region, which is a significant area for paddy farming. The granary region is distinguished by its vast rice fields and is an important part of Kedah's agricultural environment. MADA's role in this region is to provide effective water resource management, infrastructure upkeep, and farmer support
DOA	DOA oversees and nurtures Kedah's agriculture economy. Its complex responsibilities include ensuring expansion, sustainability, and adherence to agricultural practices, which contribute significantly to the state's agricultural and economic development. In terms of paddy farming, the DOA has control over Kedah's non-granary paddy fields, actively regulating, advocating for, and supporting this essential component of state agriculture
LZNK	Lembaga Zakat Negeri Kedah (LZNK) is the Kedah State Zakat Board, responsible for collecting and distributing Zakat, a form of almsgiving in Islam, within the state of Kedah, Malaysia. While the primary mission of LZNK is related to charitable and welfare activities, it also extends its role to paddy estates, particularly in supporting and uplifting paddy farmers' welfare

To better understand stakeholders' specific requirements and expectations, we employ a straightforward, multi-step methodology. This includes open, two-way discussions to identify each group's specific needs, challenges, and goals for sharing agricultural knowledge. The information gained from these talks serves two main functions. First, they supply critical information that influences the design and content of the knowledge portal. Understanding stakeholders' actual demands and expectations allows us to ensure that the portal satisfies the actual requirements of the agricultural community. Second, this technique allows stakeholders to communicate their ideas and expectations, which fosters a sense of involvement and collaboration in the portal's design and development.

c) Conceptual Model Development

The conceptual model of the knowledge portal serves as a roadmap for its design and development, providing a clear strategy for its creation. The model defines three main user categories—administrators, farmers, and agricultural agencies—each with specific roles and access levels tailored to their needs. Administrators have full access to system management, while farmers use the portal to acquire knowledge. Agricultural agencies may have access to specialized tools and resources. The model outlines the portal's core purpose of sharing agricultural knowledge, detailing the types of information and resources it will provide. This includes identifying relevant agricultural topics, selecting data sources, and organizing content to meet the diverse needs of its users.

4 The Proposed Framework for the Paddy Knowledge Portal

Figure 1 illustrates the proposed conceptual framework of the envisioned knowledge portal designed exclusively for paddy cultivation. In essence, this knowledge portal serves as a centralized, one-stop reference center that connects various stakeholders, including farmers and researchers, with agricultural agencies involved in paddy cultivation. The knowledge repository comprises tacit (experiential) and explicit (documented) knowledge addressing fundamental agricultural questions. It addresses fundamental agricultural questions, such as when to plant, which crops to select, who is involved in the process, how to implement specific practices, where to obtain resources, why certain methods are effective, and what tools or techniques to use.

The portal caters to three main user categories: administrators, agencies, and the public. A breakdown of these user categories and their subcategories, along with examples in the context of Malaysia, is visually presented in Fig. 2. Public users encompass a diverse range of individuals, including farmers, researchers, suppliers, and other interested parties. The portal is conceived as a pivotal platform where users can seamlessly interact and access valuable information.

Farmers stand to benefit significantly from this initiative. The portal offers a treasure trove of resources on best practices for paddy cultivation, covering planting, nurturing, and harvesting techniques, strategies for identifying and managing paddy diseases, comprehensive recommendations on irrigation systems, and insights into effective fertilization methods.

Government agencies, prominently MADA and DOA, play a pivotal role by contributing their expertise and knowledge to the portal. Their contributions aim to meet the needs of public users, fostering a collaborative environment for knowledge sharing.

This extensive knowledge reservoir (tacit and explicit) is thoughtfully categorized into five distinct segments. A comprehensive breakdown of the knowledge components is presented in Fig. 3.

1. General Information: Providing comprehensive insights into foundational aspects of paddy cultivation, including information on crop varieties, climate considerations, and historical context.
2. Tutorials: Offering step-by-step guides and instructional materials, walking users through various facets of paddy cultivation, from the initial planting stages to the final harvest.
3. FAQs (Frequently Asked Questions): A repository of common inquiries and their corresponding solutions, enabling quick resolutions to frequently encountered challenges.
4. Questions: Creating a platform for users to pose their unique queries and receive expert responses from the community and agricultural agencies, encouraging collaborative problem-solving.
5. Tips and Tricks: Comprising a collection of practical advice, time-tested strategies, and innovative techniques to enhance the efficiency and effectiveness of paddy cultivation.

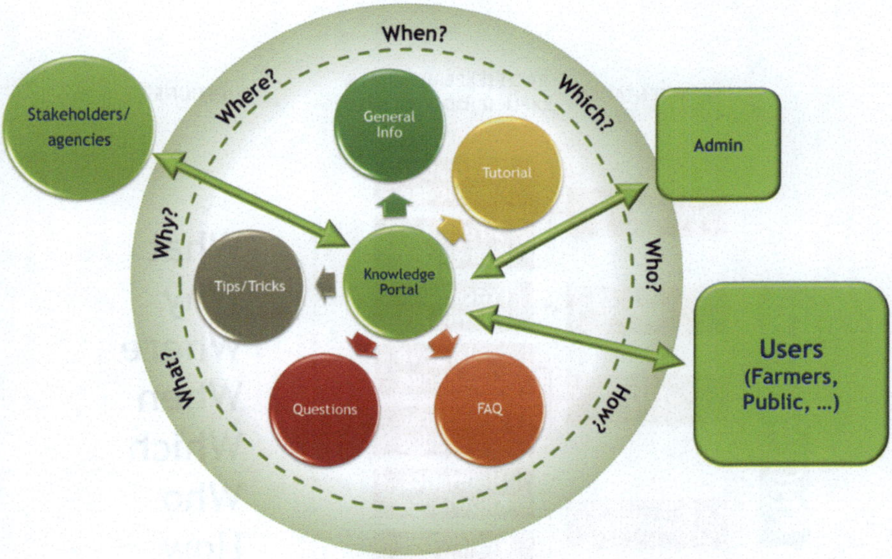

Fig. 1. Conceptual framework

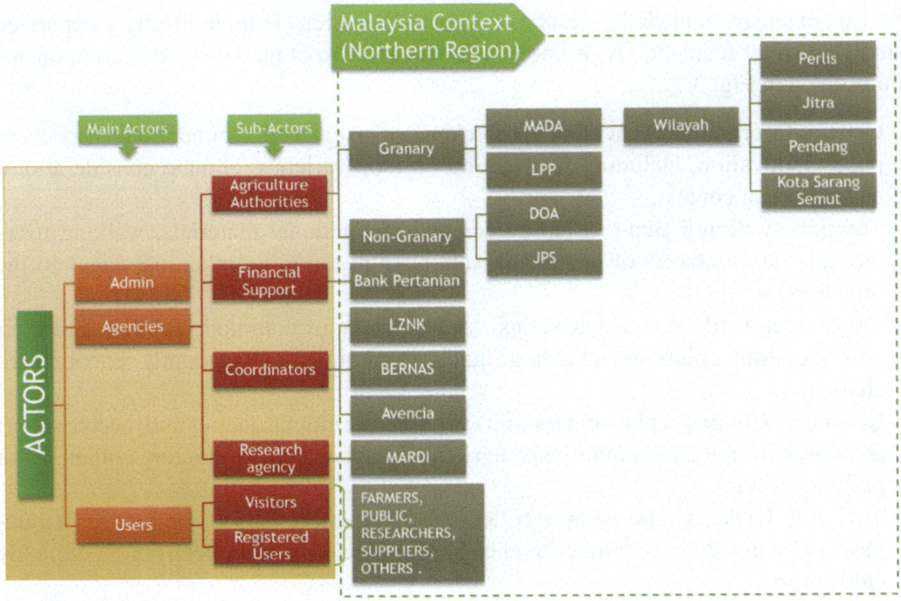

Fig. 2. Users of the knowledge portal

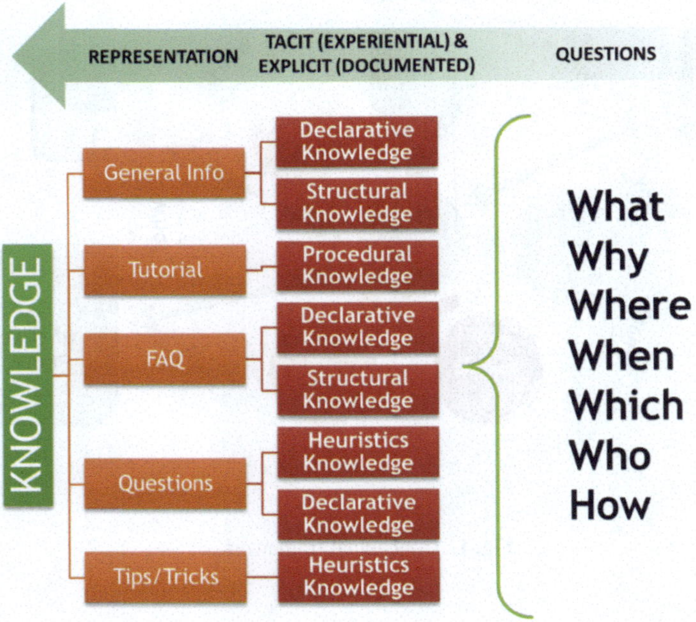

Fig. 3. Knowledge components

5 Conclusion

In conclusion, as outlined in the proposed framework, the envisioned Paddy Knowledge Portal offers a holistic and innovative solution to address the diverse needs of stakeholders involved in paddy cultivation. This knowledge portal is a centralized hub connecting farmers, researchers, and agricultural agencies, fostering a collaborative environment for knowledge sharing and promoting best practices in paddy cultivation.

The portal's user-centric design, catering to administrators, agencies, and the general public, ensures that it is accessible and valuable to a wide range of users. It significantly benefits farmers by providing a comprehensive repository of resources on paddy cultivation, covering everything from planting and nurturing to disease management, irrigation, and fertilization techniques. This wealth of knowledge is complemented by the active participation of government agencies, such as MADA and DOA, who contribute their expertise to meet the needs of the public and enhance the collaborative knowledge-sharing environment.

The thoughtful categorization of knowledge components, including general information, tutorials, FAQs, questions, and tips and tricks, ensures that users can easily access the information they require and engage in collaborative problem-solving. This approach is designed to enhance the efficiency and effectiveness of paddy cultivation while promoting sustainable agricultural practices.

The proposed framework for the Paddy Knowledge Portal represents a significant step forward in addressing the knowledge needs of the paddy cultivation community. Centralizing and sharing both experiential and documented knowledge can revolutionize how knowledge is exchanged and applied in this critical agricultural sector. As the portal continues to evolve and expand, it will serve as a vital resource for those involved in paddy cultivation, ultimately contributing to the prosperity and sustainability of this essential agricultural practice.

Acknowledgement. This article is part of a study carried out under the grant "Matching Grant Kod S/O: 21249" provided by the Universiti Utara Malaysia and Institut Pertanian Bogor, Indonesia.

References

1. Dorairaj, D., Govender, N.T.: Rice and paddy industry in Malaysia: governance and policies, research trends, technology adoption and resilience. Front. Sustain. Food Syst. **7**, 1093605 (2023)
2. Rajamoorthy, Y., Abdul Rahim, K.B., Munusamy, S.: Rice industry in Malaysia: challenges, policies and implications. Procedia Econ. Financ. **31**, 861–867 (2015)
3. Sutardi, Apriyana, Y., Rejekiningrum, P., Alifia, A.D., Ramadhani, F., Darwis, V., Setyowati, N., et al.: The transformation of rice crop technology in Indonesia: Innovation and Sustainable Food Security. Agronomy **13**(1), 1 (2022)
4. Idrees, H., Xu, J., Haider, S.A., Tehseen, S.: A systematic review of knowledge management and new product development projects: Trends, issues, and challenges. J. Innov. Knowl. **8**(2), 100350 (2023)
5. Kanyundo, A.J., Chipeta, G.T., Chawinga, W.D.: An analysis of knowledge management practices at Lilongwe University of Agriculture and Natural Resources, Malawi. Soc. Sci. Hum. Open **8**(1), 100640 (2023)

6. Orlando, F., Alali, S., Vaglia, V., Pagliarino, E., Bacenetti, J., Bocchi, S.: Participatory approach for developing knowledge on organic rice farming: Management strategies and productive performance. Agric. Syst. **178**, 102739 (2020)
7. Szilágyi, R., Lengyel, P., Herdon, M.: Portal for knowledge of agricultural informatics. In: Sideridis, B.A., Herdon, M., Várallyai, L. (eds.) Agricultural Informatics, pp. 37–42 (2010)
8. Van-Baalen, P., Bloemhof-Ruwaard, J., Van Heck, E.: Knowledge sharing in an emerging network of practice: the role of a knowledge portal. Europ. Manag. J. **23**(3), 300–314 (2005)
9. Staab, S., Maedche, A.: Knowledge portals ontologies at work. AI Mag. **22**(2), 63–75 (2001)
10. Chau, M., Huang, Z., Qin, J., Zhou, Y., Chen, H.: Building a scientific knowledge web portal: The NanoPort experience. Decis. Support Syst. **42**(2), 1216–1238 (2006)
11. Jamaludin, Z., Ishak, W.H.I.: A virtual repository approach to departmental information sharing. In: Baharom, F., Mahmuddin, M., Yusof, Y., Wan Hussain, W.I., Saip, M.A. (eds.) Proceedings of Knowledge Management International Conference (KMICe), pp. 479–484 (2010)
12. Pathak, A.K., et al.: FisOmics: A portal of fish genomic resources. Genomics **111**(6), 1923–1928 (2019)
13. Kumari, P., Rani, S.: e-KrishiShiksha: e-Learning portal on agriculture education in India. Library Philosophy and Practice (e-journal), 7814 (2023)
14. Yamin, F.M., Ramayah, T., Ishak, W.H.W.: Information searching: the impact of user knowledge on user search behavior. J. Inf. Knowl. Manag. **12**(3), 1350023 (2013)

Underwater Image Segmentation for Early Detection of White Spot Disease in Fish: Evaluating Multiple Techniques for Enhanced Aquaculture Management

Nor Hazlyna Harun[1](✉), Mohamad Ghozali Hassan[2], Juhaida Abu Bakar[1], Fazli Azzali[3], and Nur Syafiqah Suhiman[1]

[1] Data Science Research Lab, School of Computing, Universiti Utara Malaysia, 06010 Sintok, Kedah, Malaysia
hazlyna@uum.edu.my

[2] School of Technology Management & Logistic, Universiti Utara Malaysia, 06010 Sintok, Kedah, Malaysia

[3] InterNetWorks Research Lab, School of Computing, Universiti Utara Malaysia, 06010 Sintok, Kedah, Malaysia

Abstract. Image segmentation is a crucial technique in the field of computer vision, particularly for underwater imagery, which poses unique challenges due to varying lighting conditions and water clarity. Embracing digital acceleration and social innovation in knowledge management, this study focuses on image segmentation for the detection of white spot disease in fish, a common and severe issue in aquaculture. The proposed method uses 4 segmentation techniques to segment infected regions on the fish's body from underwater images. The segmented images are then analyzed to look for white spots, signs of the disease in question. That is why the possibility of early and highly accurate diagnosis of the disease using this approach will contribute to better management and prevention of diseases in aquaculture, and improvement of the health of fish stocks, and reduced losses accordingly. This study establishes the application of image segmentation as a useful feature in underwater disease detection systems and the application of advanced digital technologies and knowledge management in solving practical issues.

Keywords: Image Segmentation · Underwater Image · Fish Disease

1 Introduction

Image segmentation is a critical process in which an image is divided into multiple distinct and meaningful segments, each corresponding to a specific object or significant region within the image. The main objective of image segmentation is to partition an image into different segments based on colour, texture, intensity and other features that are relevant [1]. This method is essential for various image processing applications,

Z. Mat Aji et al. (Eds.): KMICe 2024, ISEM 49, pp. 461–472, 2025.
https://doi.org/10.1007/978-3-031-91485-0_36

including object detection, recognition, and scene analysis. Finding and grouping areas in a picture that share similar traits but can still be distinguished from nearby regions is the aim. A collection of unique, non-overlapping sections that collectively represent the full image are produced by effective image segmentation [2]. Each of these zones usually correlates to a particular area or object in the image.

Underwater images are pictures that are made or collected in aquatic surroundings, especially in water, below the water surface. Usually, underwater pictures include divers, underwater views, actions, and underwater creatures, fish, corals, plants, and other underwater organisms [3]. The protozoan *Ichthyophthirius multifiliis* is the source of "*Ichthyophthiriasis*", also referred to as "ich" or "white spot disease," which is a parasitic illness that affects fish [4]. The development of tiny, elevated white or grayish spots on the fish's body, fins, and even gills is a major sign of this illness. These patches are in fact cysts as the parasite begins to fix itself onto the fish body part.

The identification and segmentation of specific regions in underwater images especially those showing early symptoms of diseases like "Ichthyophthiriasis" (white spot disease) are difficult. New problems arise, such as lighting and the quality and saturation of water, and color change, which makes segmentation more difficult. These factors make it challenging to identify minor changes in the visual signals that define the early stages of the disease, and consequently, diagnosis is often late. Therefore, the use of traditional segmentation approaches leads to low effectiveness and inaccuracy, which contributes to the fast dissemination of the disease and large losses for fish farmers and breeders.

Therefore, this study suggests using a variety of image segmentation approaches to accurately identify the existence of white spot disease in fish [5]. The goal is to create an automated system that can identify sick fish early and stop the sickness from spreading to other aquatic life. By enabling them to efficiently monitor the health of their fish populations and take prompt action to protect the general health of aquatic ecosystems, this method can be extremely beneficial to fish farmers and breeders.

2 Related Works

An overview of earlier studies that were relevant to segmentation techniques was provided in this section. The main goals of the study were to evaluate the effectiveness of contrast-limited adaptive histogram equalization (CLAHE) and auto-level color correction. A thorough examination of each technique separately as well as their synergistic integration resulted in a substantial improvement in the quality of underwater images. The technique was created especially to enhance pictures impacted by protozoan white spot illness [6]. Several metrics, including Mean Squared Error (MSE), Peak Signal-to-Noise Ratio (PSNR), Entropy, and Structural Similarity Index (SSIM), were used for performance evaluation in this work. The article provides a thorough examination of several picture enhancement methods designed to address issues related to underwater photography.

Furthermore, an enhanced version of the K-means algorithm was developed to segment the backgrounds of underwater images. This improved algorithm aimed to address issues related to selecting the optimal K value and minimizing the influence of the initial centroid positions on the gray level quantization process, which were limitations of the original K-means algorithm [7, 8]. The experimental results demonstrated that the improved K-means algorithm effectively segmented the backgrounds of underwater images characterized by poor color, low contrast, and fuzzy edges. The segmentation results in this study indicated a significantly higher effectiveness, with the upgraded K-means algorithm achieving a value of 90.34%.

A previous study demonstrated the detection of white spot syndrome virus in shrimp (Penaeid prawn) using the K-means clustering technique. This research conducted a comparative analysis of white spot detection across different shrimp species employing the K-means clustering method. The devised approach included image capture, histogram analysis, K-means clustering, and an initial image sensing step [9]. Consequently, the K-means clustering method produced quick and accurate results, with the added potential of cost savings.

Additionally, CLAHE (Contrast Limited Adaptive Histogram Equalization) enhancement and thresholding techniques were developed for image segmentation [10]. Originally designed for medical imaging, CLAHE has proven effective in enhancing low-contrast images, such as portal films. This technique achieves significant improvements by dividing the image into contextual regions and applying histogram equalization to each segment. Thresholding, on the other hand, performed well when the image's gray-level histogram divided the background and object's pixelation into two predominant modes. The study conducted a comparative analysis of various enhancement techniques and found that the CLAHE approach not only effectively equalized the image histogram but also significantly improved contrast.

Researchers have proposed a straightforward segmentation method for color under water images of live grouper fish against a seawater backdrop. The study introduced a segmentation framework designed to isolate the complex ocean background from the fish image. Utilizing the K-means function in the MATLAB Clustering Toolbox, RGB images of the fish were used as input matrices in a clustering segmentation approach [11]. Notable contributions of the study include the division of RGB images into three separate planes and the subsequent analysis of their histograms. To segment images of live seabass fish with the specific aim of identifying the entire fish while excluding extraneous objects, a K-means clustering framework was developed.

3 Proposed Method

The proposed framework in Fig. 1 employs multiple image segmentation techniques to identify white spot disease in seabass from underwater images, which are affected by factors such as lighting and water quality and have a complex background. The process begins with the acquisition of high-resolution images using a specialized underwater camera, which serves as the input for four segmentation techniques: The techniques include; thresholding, region-based segmentation, the watershed algorithm, and K-means clustering. The contrast-based method of thresholding separates an image into the targeted region and the rest of the image while region-based segmentation separates pixels into different regions having similar behavior. The algorithms of watershed differ in segmentation of boundaries in mosaic images especially when the structures overlap, in case of white spots in this image the watershed was correct while K-means clustering captured the areas of interest by creating groups of the image. The comparison of the technique's performance is done using the Dice Coefficient, Jaccard Index, and Accuracy, Specificity, and Precision to determine the best technique for early disease diagnosis. Last, the study establishes a stand-alone MATLAB application of the most effective approach for screening white spot disease to aid fish farmers and breeders in early detection of the disease for better fish health management in aquaculture.

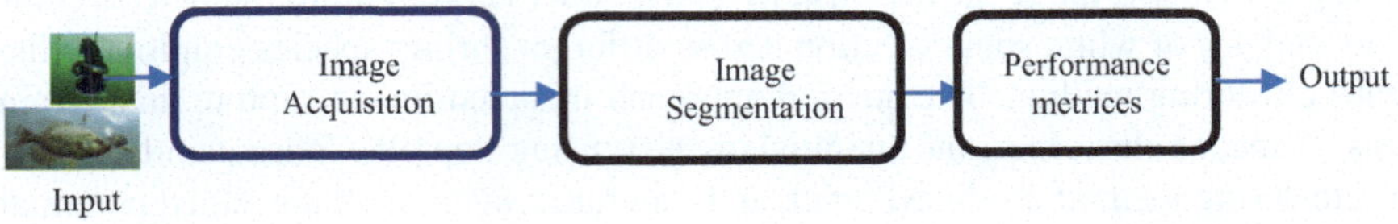

Fig. 1. Proposed framework.

3.1 Image Acquisition

In this phase, data was gathered at the National Fish Health Research Center (NaFisH) of the Fish Research Institute (FRI) in Batu Maung, Penang, Malaysia. Images of seabass fish were captured using a GoPro Hero 9 + at 4K resolution and 60 frames per second. These images represented as the input data for the segmentation technique. Figure 2 presents a representative image of an infected fish.

Fig. 2. Sample images of infected fish.

3.2 Image Segmentation

Image segmentation is a widely used technique for processing and analyzing digital images. Its primary function is to partition an image into multiple regions or segments, often based on pixel characteristics. This process involves grouping clusters of pixels based on similar color or shape features, or distinguishing foreground from background elements [12]. Previous studies have employed a variety of segmentation techniques, including edge-based, thresholding, and clustering-based methods [13]. In this study, four image segmentation approaches were applied to seabass images: thresholding, region-based segmentation, the watershed algorithm, and K-means clustering [14].

Threshold. One technique used to differentiate between foreground and background features in an image is threshold segmentation. This method is especially useful for photographs where there is a strong difference between the foreground objects and the background since it uses a grayscale intensity image to extract the foreground [15, 16]. A threshold value of 0.5 was used in this investigation. The following one-line command converts a grayscale image to a binary image based on a specified threshold value:

$$binary\ Image = imbinarize\ (input\ Image,\ threshold\ Value\ /\ 255);$$

In this command:

inputImage is the grayscale image to be *thresholded.*

0.5 is the threshold value, where pixels with intensity values greater than or equal to 0.5 will be set to 1 (white), and others will be set to 0 (black).

Region-Based. In this method, the geographical aspect is generating a single connected area of pixels that belong to the same category, while clustering is putting pixels with similar values together. To divide the image into several components with comparable pixel properties, an algorithm is used [17]. At this point, is a one-line command to label connected components and then get properties of these regions:

$$region\ Properties = regionprops(imbinarize\ (inputImage,\ 0.5))$$

This command performs the following:

Thresholds the *inputImage* with a value of 0.5 to create a binary image.

Labels the connected components in the binary image.

Measures all properties of these labeled regions.

Watershed. The watershed algorithm is employed to segment complex images when traditional methods such as contour detection and thresholding fail to yield accurate results. This algorithm begins by identifying distinct regions in the image's foreground and background. Markers are then utilized to initiate the watershed process, leading to precise boundary identification. In this study, a set of 10 markers was utilized to enhance segmentation performance by effectively separating touching and overlapping objects. This approach is particularly valuable for accurately identifying overlapping objects within an image [18, 19]. Below is the one-line command for the watershed algorithm format:

$$L = watershed\ (imgradient\ (rgb2gray\ \left(imread\ \left('inputImage.bmp'\right)\right),\ 10));$$

This command performs the following:

Reads the inputImage.

The image is converted to grayscale.

The algorithm calculates the gradient magnitude of the grayscale image using a specified value of 10.

Applies the gradient magnitudeimage's watershed transform.

K-Means.

In the context of separating regions of interest from the background without supervision, the K-Means clustering algorithm serves as a fundamental technique. It functions by partitioning the provided dataset into K clusters, each characterized by its centroid. K represents the predefined number of groups aimed at identifying distinct patterns within the data [20, 21]. Below is the one-line command for K-means clustering format:

$$K = 3; \%\, Number\ of\ clusters$$

$$[idx,\ C] = kmeans(reshape(double(rgb2gray(inputImage)),\ [],\ 1),\ K);$$

This command performs the following:

Converts *inputImage* to grayscale.

Reshapes the grayscale image into a column vector.

Applies K-means clustering with $K = 3$ clusters to segment the image. *Idx* contains the cluster indices, and *C* contains the cluster centroids.

3.3 Performance Metrics Evaluations

In order, to quantitatively measure the overall performance of the four segmentations, techniques used in this study, standard metrics was used such as the dice coefficient, Jaccard index, accuracy, specificity, and precision were employed, each calculated using its respective formula.

Dice Coefficient. The Dice coefficient is a performance metric for image segmentation that measures the similarity between two sets by calculating the ratio of their intersection to the average size of the sets, using the formula:

$$\text{Dice Coefficient} = \frac{2 \times |A \cap \mathrm{B}|}{|A| + |B|} \tag{1}$$

where:

|A ∩ B| represents the size of the intersection between sets A and B.

|A| represents the size of set A.

|B| represents the size of set B.

Jaccard Index. A popular performance statistic that assesses how similar or comparable two sets are is the Jaccard index. This index is frequently used in domains like information retrieval and picture segmentation. It is computed as the ratio of the sets' union size to the size of their intersection. **The formula for calculating the Jaccard index is:**

$$\text{Jaccard Index} = \frac{|A \cap \mathrm{B}|}{|A \cup B|} \tag{2}$$

where:

|A ∩ B| represents the size of the intersection between sets A and B.

|A ∪ B| represents the size of the union between sets A and B.

Accuracy. One of the most important performance metrics for determining whether a classification or prediction model is correct is its accuracy. It can be expressed as a percentage that represents the percentage of accurately predicted cases to all examples in a dataset. Accuracy is calculated using the following formula:

$$\text{Accuracy} = \frac{\text{Number of Correct Predictions}}{\text{Total Number of Prediction}} \tag{3}$$

Specificity. In binary classification, specificity is a critical performance metric that assesses how well a model recognizes negative examples. It is computed as the proportion of real negative cases in a dataset divided by the number of true negative predictions. The formula for calculating specificity is as follows:

$$\text{Specificity} = \frac{\text{True Negatives}}{\text{True Negatives} + \text{False Positives}} \tag{4}$$

where:

True negatives (TN) are instances that are correctly identified as belonging to the negative class by the model.

False Positives (FP) refer to instances that belong to the negative class but are incorrectly classified as positive by the model.

Precision. In binary classification, precision is a critical performance metric that assesses how well a model predicts the positive outcomes. The model determines the percentage of true positive predictions among all the cases it expected to be positive. Precision is calculated using the following formula:

$$\text{Precision} = \frac{\text{True Positives}}{\text{True Positives} + \text{False Positives}} \tag{5}$$

where:

True Positives (TP) are the instances correctly classified as the positive class.

False Positives (FP) are the instances incorrectly classified as the positive class when they belong to the negative class.

4 Design and Development

The proposed image segmentation application was developed to address issues faced by breeders and farmers. In the context of innovation management, application prototyping is a vital method for illustrating requirements and gathering feedback from user experiences. By employing an iterative process, the final outcome is progressively refined to more closely align with user requirements and expectations. The system was developed as a standalone application using MATLAB R2021a software. The program is user-friendly, requiring no additional software installations, making it accessible even for non-technical users. Figure 3(a) shows the interface of main system and (b)default folder. To start the system, click on the 'Select Image' button to choose an image from the default folder setup.

Figures 4, 5, 6 and 7 shows the results of image segmentation based on chosen method.

Figures 3, 4, 5, 6 and 7 show that each segmentation method affects the identification of white spot disease in fish images in various ways with its advantages and limitations. While using the thresholding technique works well in high contrast scenarios, it performs poorly in complex underwater scenes and therefore yields low overlap accuracy. Region-Based Segmentation is accurate for different regions but is not very effective in fine underwater changes. The Watershed algorithm provides good results in terms boundary

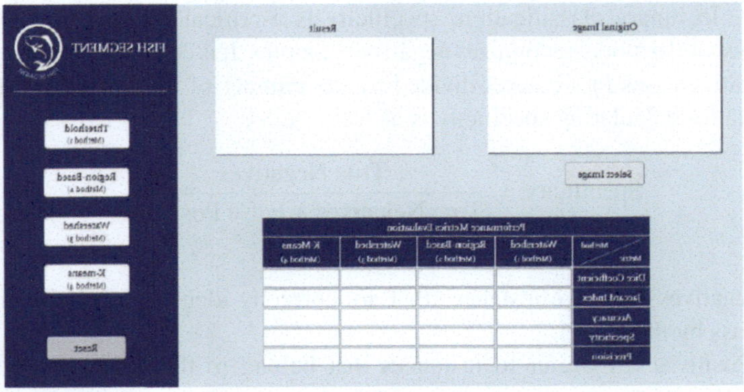

(a)

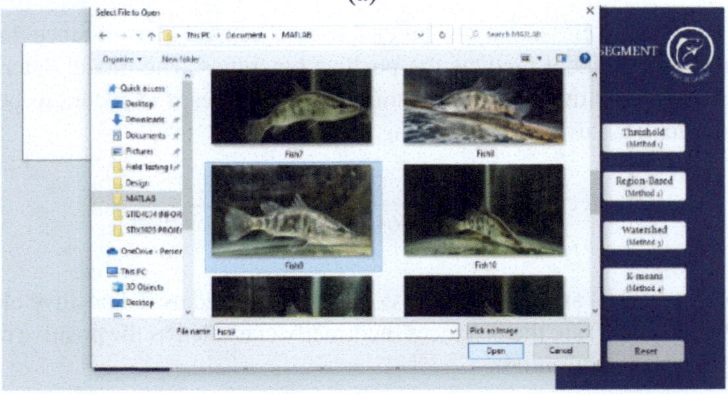

(b)

Fig. 3. Interface of image segmentation application.

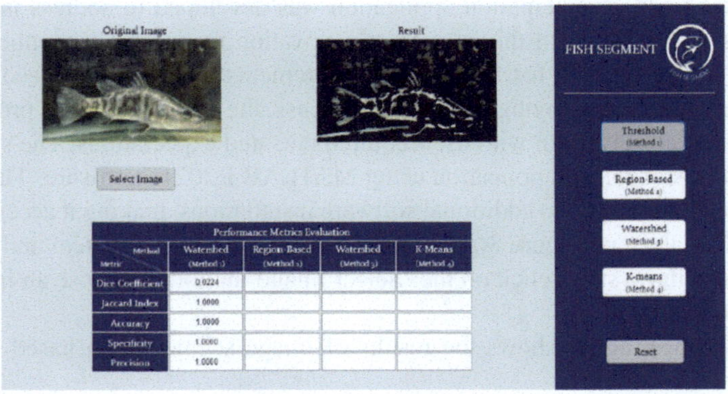

Fig. 4. Interface for Threshold technique button.

definition, a considerable degree of accuracy but over-segmentation is likely to affect the precision. K-Means Clustering yields the highest value of Dice Coefficient, but its stability, especially in regions of uniform intensity, is not very consistent. The type of

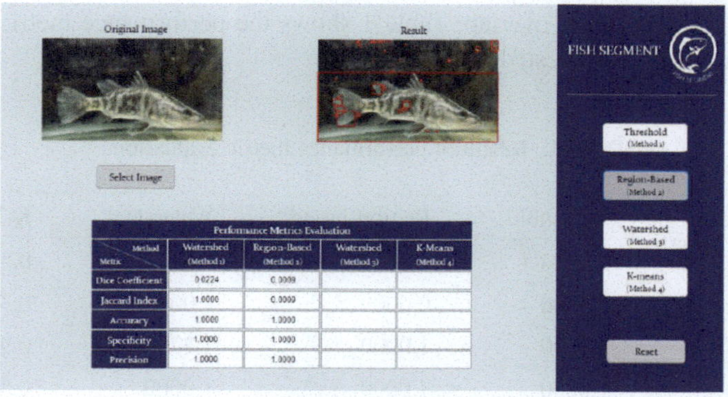

Fig. 5. Interface for Region-Based technique button.

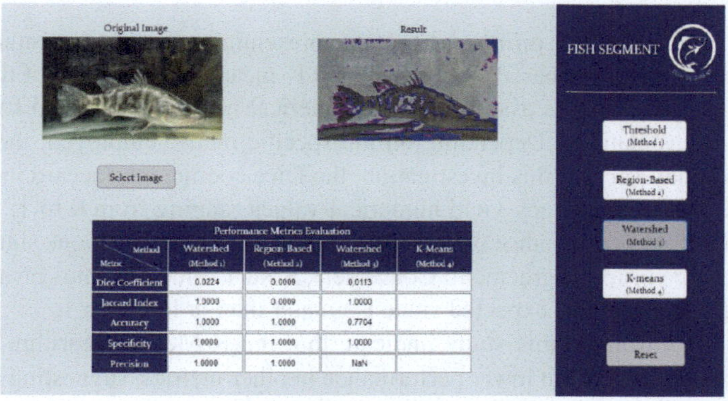

Fig. 6. Interface for Watershed technique button.

Fig. 7. Interface for K-means technique button.

technique to use depends on the certain image characteristics and the appropriateness of accuracy, precision and overlap. Table 1 shows the performance metric evaluation values for each technique tested in this study.

Table 1. Result of performance metric evaluation.

Heading level	Threshold	Region-based	Watershed	K-Means
Dice Coefficient	0.0224	0.0009	0.0113	0.5238
Jaccard Index	1.0000	0.0009	1.0000	0.3458
Accuracy	1.0000	1.0000	1.7704	0.9126
Specificity	1.0000	1.0000	1.0000	0.8814
Precision	1.0000	1.0000	NaN	0.7506

Table 1 comprises four primary columns representing various segmentation strategies: thresholding, region-based, watershed, and K-means. The efficacy of these image segmentation approaches is assessed using numerical measures obtained from performance metric evaluations. Depending on the specific metric employed, the values of these metrics may vary. In this investigation, the Dice coefficient, Jaccard index, precision, specificity, and accuracy yield numerical values ranging from 0 to 1. A score of zero indicates poor performance or no agreement, whereas a score of one signifies flawless performance or total agreement. Conversely, a NaN value indicates invalid results, irrespective of whether the expected value is greater than or less than 1.

The findings from in this study, indicate that the K-Means algorithm shows the highest Dice Coefficient but lower performance in other metrics, suggesting it excels in overlap but may face challenges in other aspects. However, the watershed method excels in accuracy and Jaccard Index but has issues with precision, making it highly accurate in overall classification but potentially unreliable for positive identifications. Although, the threshold method performs well in terms of accuracy, specificity, and precision but poorly in Dice Coefficient, suggesting it can correctly classify most instances but struggles with overlap. Furthermore, the Region-based method has perfect specificity and precision but performs poorly in other metrics, indicating it may be highly reliable in specific cases but not generally effective. Overall, each segmentation technique has its strengths and weaknesses, and the choice of method may depend on the specific requirements of the task at hand.

5 Conclusion and Future Work

In conclusion, this study demonstrates conclusively that the system can detect white spot disease on the bodies of seabass fish by utilizing four distinct segmentation techniques. The system's accuracy and efficiency were validated through standard performance evaluation metrics. This approach provides fish farmers and breeders with an early screening tool specifically designed to detect white spot disease in seabass. Additionally, it includes the development of an application that effectively assesses fish health at an early stage.

By integrating artificial intelligence-based technology, this method supports proactive fish health management, thereby ensuring a more sustainable, healthy, and transparent aquatic environment. Moreover, this technology has facilitated the development of an application for the early assessment of fish health, specifically targeting the detection of white spot disease in seabass. It offers fish farmers and breeders a valuable tool for early disease screening. This approach not only exemplifies the potential of social innovation and digital acceleration in enhancing aquaculture knowledge management but also underscores its significance for improving practices among fish breeders and farmers.

Future enhancements to this technique will focus on reducing background noise in underwater fish images, thereby increasing the accuracy of white spot disease detection on the fish's body. The system's broad scope, which encompasses a variety of underwater photographs, ensures its adaptability to diverse aquatic environments. Additionally, it provides fish breeders and farmers with actionable measures to improve the well-being of their livestock.

Acknowledgement. This study was funded by the Ministry of Higher Education (MoHE) under the Fundamental Research Grant Scheme (FRGS/1/2020/ICT02/UUM/02/2, S/O Code: 14856). The content of this article is solely the responsibility of the authors and does not necessarily reflect the official views of MoHE, Malaysia.

References

1. Moon, S.: What is image segmentation? - definition, types, use case, deep learning, trends. Datahunt. https://www.thedatahunt.com/en-insight/image-segmentation. Last accessed 17 April 2024 (2023)
2. Guide to image segmentation in computer vision: Best practices. Encord. https://encord.com/blog/image-segmentation-for-computer-vision-best-practice-guide/. Last accessed 20 April 2024
3. Schettini, R., Corchs, S.: Underwater image processing: state of the art of restoration and image enhancement methods. In: EURASIP Journal on Advances in Signal Processing, 2010, pp. 1–14 (2010)
4. White Spot Disease (ICH): Symptoms, causes, and Remedies-Pethelpful. https://pethelpful.com/fish-aquariums/white-spot-disease-ich-ick. Last accessed 20 March 2024
5. Li, D., Li, X., Wang, Q., Hao, Y.: Advanced techniques for the intelligent diagnosis of fish diseases: a review. Animals **12**(21), 2938 (2022). https://doi.org/10.3390/ani12212938
6. Azhar, A.S.B.M., Harun, N.H.B., Yusoff, N.B., Hassan, M.G.B., Chu, K.B.: Image enhancement on underwater images for protozoan white spot fish disease detection. In: 2022 International Conference on Intelligent Systems and Computer Vision (ISCV), pp. 1–4. Fez, Morocco (2022). https://doi.org/10.1109/ISCV54655.2022.9806095
7. Wong, S.L., Paramesran, R., Taguchi, A., Ong, S.H.: A review on the underwater image restoration and enhancement methods. ASM Sci. J. **14**, 1–13 (2021)
8. Chen, W., He, C., Ji, C., Zhang, M., Chen, S.: An improved K-means algorithm for underwater image background segmentation. Multimedia Tools Appl. **80**(14), 21059–21083 (2021). https://doi.org/10.1007/s11042-021-10693-7
9. Sankar, M.M., Rao, C.N., Sailaja, G., Bhuvaneswary, N., Gunasekhar, P.: White spot syndrome virus detection in shrimp images using image segmentation techniques. Int. J. Adv. Res. Comput. Sci. Softw. Eng. **3**, 9 (2013)

10. Kumar Rai, R., Gour, P., Singh, B.: Underwater image segmentation using clahe enhancement and thresholding. Int. J. Emerg. Technol. Adv. Eng. **2**(1), 118–123 (2012)
11. Ma, G.Q., Tian, Y.C., Li, X.L., Xing, K.Z., Xu, S.: Color image segmentation of live grouper fish with complex background in seawater. Appl. Mech. Mater. **743**, 293–302 (2015)
12. Eugene, P.: What is image acquisition in image processing? Easy Tech Junkie. https://www.easytechjunkie.com/what-is-image-acquisition-in-image-processing.htm. Last accessed 2 May 2024 (2023)
13. Image segmentation. MATLAB and Simulink. https://www.mathworks.com/discovery/image-segmentation.html. Last accessed 31 March 2024
14. Image segmentation: The basics and 5 key techniques. Datagen. https://datagen.tech/guides/image-annotation/image-segmentation/. Last accessed 20 March 2024 (2023)
15. Niu, Z., Li, H.: Research and analysis of threshold segmentation algorithms in image processing. J. Phys. Conf. Ser. IOP Publishing (2019)
16. Thresholding-based image segmentation: GeeksforGeeks. https://www.geeksforgeeks.org/thresholding-based-image-segmentation/. Last accessed 29 March 2024 (2023)
17. Region and edge-based segmentation: Online Courses and eBooks Library, https://www.tutorialspoint.com/region-and-edge-based-segmentation. Last accessed 10 May 2024
18. Watershed segmentation algorithm in image processing: Aegis Softtech, https://www.aegissofttech.com/articles/watershed-algorithm-and-limitations.html. Last accessed 10 May 2024 (2019)
19. The watershed transform: Strategies for Image Segmentation. MATLAB and Simulink. https://www.mathworks.com/company/newsletters/articles/the-watershed-transform-strategies-for-image-segmentation.html. Last accessed 11 May 2024
20. Introduction to image segmentation with K-means clustering: KDnuggets, https://www.kdnuggets.com/2019/08/introduction-image-segmentation-k-means-clustering.html. Last accessed 22 May 2024 (2019)
21. Monteiro, F.C., Campilho, A.C.: Performance evaluation of image segmentation. In: International Conference Image Analysis and Recognition, pp. 248–259. Springer, Berlin, Heidelberg (2006)

Malaysia Rice Yield Prediction Using Synthetic Data Generation of CTAB-GAN

Muhammad Khalifa Umana[1], Mohamad Farhan Mohamad Mohsin[1](✉), Mohamad Ghozali Hassan[2], Kamal Imran Mohd Sharif[2], Mohd Azril Ismail[2], and Maslina Ismail[3]

[1] School of Computing, Universiti Utara Malaysia, 06010 Sintok, Kedah, Malaysia
farhan@uum.edu.my

[2] School of Technology Management and Logistics, Universiti Utara Malaysia, 06010 Sintok, Kedah, Malaysia

[3] Department of Agriculture, Malaysia Putrajaya, Putrajaya Federal Territory, Sintok, Malaysia

Abstract. Machine learning and artificial intelligence advancement have significantly impacted various sectors, including agriculture. One prominent challenge researchers face in this field is reliable and accurate data availability. This paper explores the use of conditional tabular GAN (CTAB-GAN) to generate synthetic data to enhance agricultural yield prediction models, focusing on rice yield in Malaysia. The approach involves data collection, preparation for analysis, synthetic data generation at various scales, and model training. The study compares synthetic datasets of different sizes—25%, 50%, 75%, 100%, and an additional set of 250 rows—against the actual dataset. The results indicate that while synthetic data generally led to a performance decline in most models, decision tree-based models and the XGBoost Regressor improved performance with smaller synthetic datasets (25% of the original data). Notable improvements were observed in model performance when evaluated using four standard regression metrics. The findings suggest that CTAB-GAN can enhance model performance in specific agricultural contexts by generating synthetic data. It demonstrates how synthetic data can help resolve the problem of insufficient data, improve the precision of crop yield predictions, and contribute to more effective agricultural practices.

Keywords: Rice Yield · Generative Adversarial Networks · Precision Agriculture · Synthetic Data · Machine Learning

1 Introduction

Recent advancements in data-driven technologies have significantly improved various sectors, including agriculture. However, a critical challenge hindering progress in these fields is the insufficient supply of quality data [1–3]. Oftentimes, real-world data is constrained in quantity, expensive to get, or vulnerable to privacy concerns [4, 5]. To overcome these limitations, researchers have increasingly turned to synthetic data generation [6–9]. Artificially generated synthetic data may accurately mimic the attributes and patterns seen in actual data, providing a viable choice for training and assessing machine learning models [10, 11].

Z. Mat Aji et al. (Eds.): KMICe 2024, ISEM 49, pp. 473–484, 2025.
https://doi.org/10.1007/978-3-031-91485-0_37

Technological advancements have vastly improved agriculture, a vital business guaranteeing worldwide food security [12–14]. Data-driven approaches can revolutionize agriculture by optimizing resource use and improving plant production. However, collecting comprehensive agricultural data is occasionally hindered by seasonal oscillations, geographical variances, and the costs associated with large-scale data collection. [15–18]. This is accurate in nations such as Malaysia, where various meteorological and topographical factors affect agricultural output [19]. Synthetic data may efficiently overcome these challenges by providing ample data for training and evaluating models, accelerating agricultural technology research and development.

The Conditional Tabular GAN (CTAB-GAN) is a sophisticated generative adversarial network designed exclusively to generate high-quality tabular data [20]. CTAB-GAN is tailored explicitly for tabular datasets, which are often encountered in several practical domains, including agriculture [21–25]. Unlike traditional GANs primarily used for generating images or text, CTAB-GAN specializes in creating synthetic tabular data. It accurately models the intricate structures inherent in real-world datasets.

This study specifically examines the utilization of CTAB-GAN to provide artificial data to predict rice production in Malaysia. The data utilized in this study is exclusive to Malaysia, which has distinctive meteorological and topographical circumstances that impact rice cultivation in the area. The aim is to assess the efficacy of CTAB-GAN in generating authentic and valuable synthetic data that can improve rice yield prediction models. Our objective is to showcase the capabilities of synthetic data production in driving progress in agricultural research and technology in Malaysia.

2 Related Works

Synthetic data has gained significant attention in machine learning, especially in domains with a scarcity or sensitivity around high-quality data. GANs, a synthetic data generation technology, have demonstrated their efficacy across diverse sectors, including healthcare, finance, and autonomous driving [26–29].

The agricultural sector has investigated the utilization of synthetic data to tackle issues about the availability and reliability of data. Prior research has utilized GANs to produce artificial crop pictures and environmental data, resulting in enhanced accuracy of predictive models for tasks like disease identification and yield estimation [30–33]. Nevertheless, the existing research has concentrated chiefly on analyzing picture data, leaving the exploration of GANs in tabular agricultural data largely unexplored.

Predicting agricultural output is a crucial domain in agricultural investigation, carrying substantial consequences for sustenance safety and resource administration. [34, 35]. Conventional methods for predicting agricultural yields depend on collecting a large amount of field data and using advanced modeling techniques [16, 36, 37]. Nevertheless, these techniques are sometimes hampered by the scarcity and exorbitant expense of acquiring high-quality data [1–3]. Synthetic data is a viable option, as it offers an alternate method to create extensive and varied datasets that can be utilized for training and validating prediction models.

CTAB-GAN [20] is a technique specifically designed to create excellent-quality synthetic tabular data. CTAB-GAN is a specialized type of GAN that is mainly created to handle the distinct features of tabular datasets, in contrast to regular GANs that are commonly used for image and text data [25, 38]. The first research on CTAB-GAN has shown promise in creating authentic synthetic data for several purposes. However, its efficacy in the agricultural field for predicting crop production has not been thoroughly examined.

This work seeks to close this disparity by assessing the application of CTAB-GAN for producing artificial data to improve agricultural output predicting models, particularly for rice production in Malaysia.

3 Methodology

CTAB-GAN is a specific type of GAN framework specifically developed to produce synthetic tabular data of exceptional quality. The system employs a generative component that produces artificial samples and a discriminative component that distinguishes between real and synthetic data. The generative component aims to create indistinguishable samples by minimizing adversarial loss, while the discriminative component refines its ability to detect synthetic data. CTAB-GAN operates through a competitive learning process. One component generates synthetic data, while another evaluates its authenticity (Fig. 1).

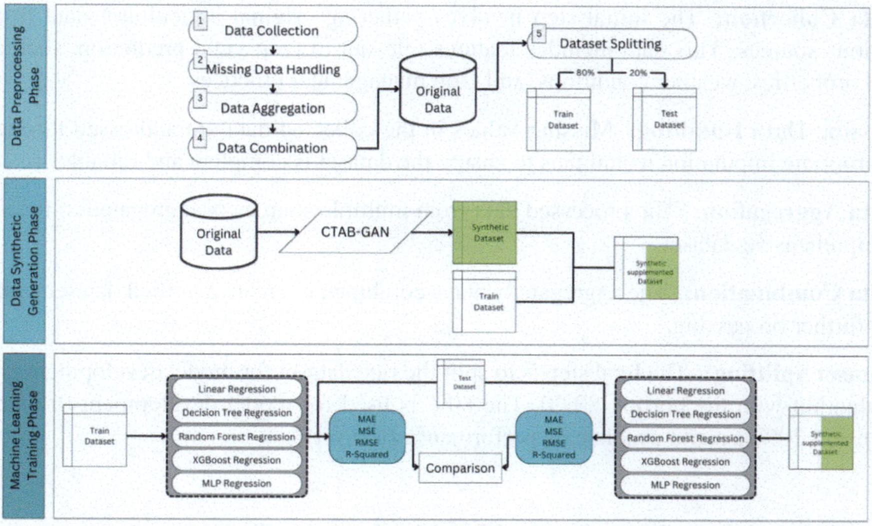

Fig. 1. Pipeline for synthetic data generation and crop yield prediction model development.

The methodology followed in this research is divided into three main phases: Data Preprocessing, Data Synthetic Generation, and Machine Learning Training. The diagram (see **Error! Reference source not found.**) illustrates the overall workflow.

3.1 Data Collection and Preprocessing

The study utilizes two datasets: the meteorological and rice production data. The datasets belong to two government agencies and were subsequently merged for prediction modeling.

The first dataset is the Meteorological or climate dataset. It was sourced from the Malaysian Meteorological Department and was a foundational dataset. This data encompassed four key variables: wind speed, temperature, humidity, and rainfall, collected monthly from 2010 to 2021 across all Malaysian states. These climatic factors recognized as critical determinants of plant growth and development, particularly for rice cultivation, have been extensively studied [13, 39].

The second dataset, rice production data, was obtained from the Department of Statistics Malaysia. This dataset provided essential insights into rice cultivation, including quantitative measures of yield, cultivated area, and planted area. Categorical variables such as location, year, and season complemented the dataset. Yield data spanned from 1.48 to 6.56 tons per hectare between 2011 and 2021, with variations across Malaysian states. These elements are crucial for analyzing rice production trends and are significant yield indicators.

The first stage of the technique centers on preparing the data for later synthetic generation and machine learning operations. This stage guarantees that the data is dependable, thorough, and appropriate for training the model.

Data Collection: The initial step involves gathering original agricultural data from various sources. This data includes features relevant to crop yield prediction, such as soil properties, weather conditions, and crop management practices.

Missing Data Handling: Missing values in the collected data are addressed through appropriate imputation techniques to ensure the dataset is complete and reliable.

Data Aggregation: The processed data from multiple sources is aggregated to form a comprehensive dataset.

Data Combination: The aggregated data is combined to create a unified dataset ready for further processing.

Dataset Splitting: The final step is to split the rice dataset for model development and evaluation with the portion 80:20. The 80% is used for model development (training set), and 20% is set to evaluate its performance (test set).

3.2 Data Synthetic Generation and Machine Learning Training

This study employed a systematic strategy that involved model training, synthetic data production, and performance evaluation to assess the efficacy of CTAB-GAN in producing synthetic data for agricultural yield prediction.

The CTAB-GAN model was trained on the original dataset to learn its underlying data distribution, enabling the generation of highly realistic synthetic samples. After training, CTAB-GAN produces synthetic datasets of various sizes, namely 25%, 50%, 75%, and 100% of the initial data size (66, 132, 198, and 264 rows, respectively). A synthetic dataset consisting of 250 rows is generated, which does not correspond to any specific percentage of the original data size. The synthetic datasets are merged with the original training dataset to create synthetic augmented datasets, augmenting the data's diversity and amount.

The training phase involved developing five regression models: Linear Regression, Lasso, Multilayer Perceptron (MLP), XGBoost, and two tree-based ensemble models – Decision Tree and Random Forest. This is done using both the original training dataset and additional synthetic datasets. Model performance was assessed using four standard regression error metrics to quantify prediction accuracy as follows:

- Mean Absolute Error (MAE): Calculate average prediction error
- Mean Squared Error (MSE): Averages squared prediction errors, amplifying more significant errors.
- Root Mean Squared Error (RMSE): Calculate estimation error by square rooting the MSE.
- R-squared: Percentage of variance explained by the model.

This configuration enables a thorough evaluation of model performance by comparing it with and without the incorporation of synthetic data. This analysis provides valuable insights into the efficacy of CTAB-GAN in creating beneficial synthetic data for predicting crop production.

4 Result

The predictive capabilities of the developed models were assessed using the four standard error metrics for the regression-based model. The impact of synthetic data on model performance is quantified by calculating the percentage change in evaluation metrics. Table 1 presents the capabilities of the experimented regression models trained solely on the original dataset. This serves as a standard for assessing the efficiency of the artificial data produced by CTAB-GAN.

Table 1. Prediction models performances using only original data.

Model	MAE	MSE	RMSE	R-squared
LinearRegression	12310,97	325585025	18043,97	0,9824
Lasso	12261,09	324807318	18022,41	0,9825
DecisionTreeRegressor	10102,09	532858257	23083,72	0,9712
RandomForestRegressor	7661,04	229086520	15135,60	0,9876
XGBRegressor	10416,57	950969944	30837,80	0,9486
MLPRegressor	15683,76	525259447	22918,54	0,9716

Table 2 compares the performance changes seen while utilizing synthetic data about the baseline performance documented in Table 1. The models are trained using datasets that incorporate both the actual and synthetic data produced by CTAB-GAN at various sizes of synthetic data. The 'Synthetic Data Size' column indicates the synthetic data generated. The performance metric scores are expressed as percentages, where a negative value indicates a decrease, and a positive value indicates an increase.

The figures provided depict the fluctuation in performance for the MAE, MSE, RMSE, and R-squared metrics across various synthetic data sizes and models. These figures are based on the data shown in Table 2. Figure 2 displays the variability of the MAE metric performance, Fig. 3 portrays the variability of the MSE metric performance, Fig. 4 demonstrates the variability of the RMSE metric performance, and Fig. 5 presents the variability of the R-squared metric performance.

By utilizing synthetic data produced by CTAB-GAN, it is evident that there is a consistent decline in performance as the amount of synthetic data increases for most models tested in this study. Two models - Linear and Lasso Regression exhibited substantial decreases in performance when trained using synthetic data. This was evident from the decline in model performance, as indicated by standard error metrics. The persistent decline can be clearly shown in Figs. 2, 3 and 4. These models are suggested to be very responsive to the synthetic data's quality and quantity.

In contrast, the Decision Tree Regressor and Random Forest Regressor showed significant improvements when trained with smaller synthetic data sets, namely 25% of the original data or 66 rows. The related figures demonstrate that these models exhibited positive MAE, MSE, RMSE, and R-squared improvements at this specific synthetic data size. This indicates that CTAB-GAN can generate valuable synthetic data for tree-based models, given particular circumstances.

The XGBoost Regressor had varied outcomes but demonstrated enhancements for most synthetic data sizes overall, except in the case where it was augmented with 75% synthetic data (198 rows). Table 2 and the figures illustrate the contrasting results, with specific synthetic data sizes (66, 132, and 264) demonstrating significant enhancements, while the size of 198 leads to a decrease in performance.

The performance of the MLP Regressor noticeably declined while using synthetic data, as seen by substantial increases in error metrics and drops in R-squared values. The trend of the MLP Regressor's sensitivity to the quality and amount of synthetic data is consistently seen across all data sizes, as demonstrated in Table 2 and the accompanying figures.

To summarize, incorporating synthetic data generally led to a decline in model performance for most models evaluated. However, smaller synthetic data sizes (25%) were shown to be advantageous for decision tree-based models. The improvement in performance for the XGBoost Regressor varied across different synthetic data sizes. These findings indicate that CTAB-GAN can produce valuable artificial data in some circumstances. However, its efficacy is contingent upon the model and the amount of the synthetic data.

Table 2. Comparison for synthetic data usage with base model.

Synthetic Data Size	Model	MAE	MSE	RMSE	R-squared
66	LinearRegression	−71%	−117%	−47%	−2%
66	Lasso	−72%	−117%	−47%	−2%
66	DecisionTreeRegressor	64%	81%	56%	2%
66	RandomForestRegressor	21%	42%	24%	1%
66	XGBRegressor	63%	92%	71%	5%
66	MLPRegressor	−33%	−89%	−38%	−3%
132	LinearRegression	−134%	−286%	−97%	−5%
132	Lasso	−135%	−287%	−97%	−5%
132	DecisionTreeRegressor	−24%	−107%	−44%	−3%
132	RandomForestRegressor	−7%	−21%	−10%	−0%
132	XGBRegressor	13%	44%	25%	2%
132	MLPRegressor	−81%	−290%	−97%	−8%
198	LinearRegression	−254%	−711%	−185%	−13%
198	Lasso	−255%	−713%	−185%	−13%
198	DecisionTreeRegressor	−27%	−440%	−132%	−13%
198	RandomForestRegressor	−97%	−232%	−82%	−3%
198	XGBRegressor	−56%	−4%	−2%	−0%
198	MLPRegressor	−100%	−372%	−117%	−11%
250	LinearRegression	−196%	−619%	−168%	−11%
250	Lasso	−197%	−621%	−168%	−11%
250	DecisionTreeRegressor	−32%	−105%	−43%	−3%
250	RandomForestRegressor	−30%	−36%	−16%	−0%
250	XGBRegressor	22%	78%	53%	4%
250	MLPRegressor	−77%	−272%	−93%	−8%
264	LinearRegression	−190%	−678%	−179%	−12%
264	Lasso	−192%	−680%	−179%	−12%
264	DecisionTreeRegressor	13%	−7%	−3%	−0%
264	RandomForestRegressor	−19%	11%	6%	0%
264	XGBRegressor	0%	60%	36%	3%
264	MLPRegressor	−58%	−192%	−71%	−6%

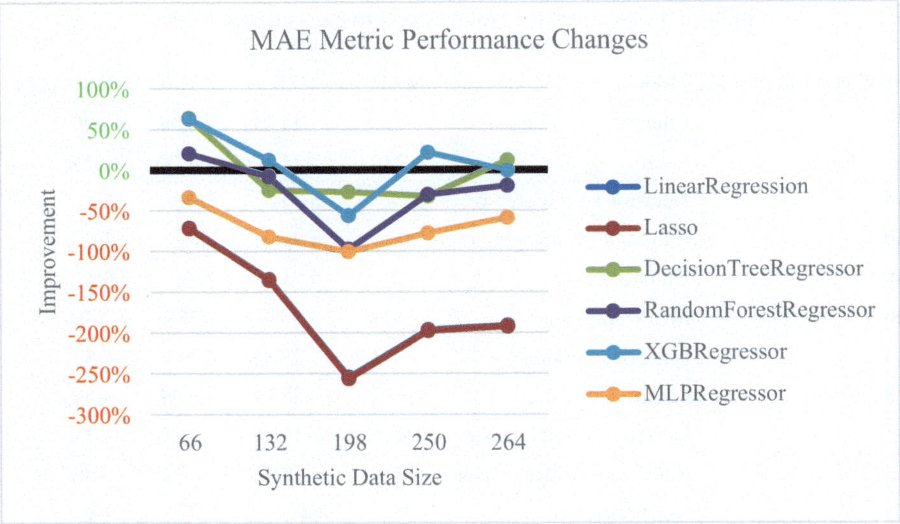

Fig. 2. MAE metric performance changes across different synthetic data sizes and models.

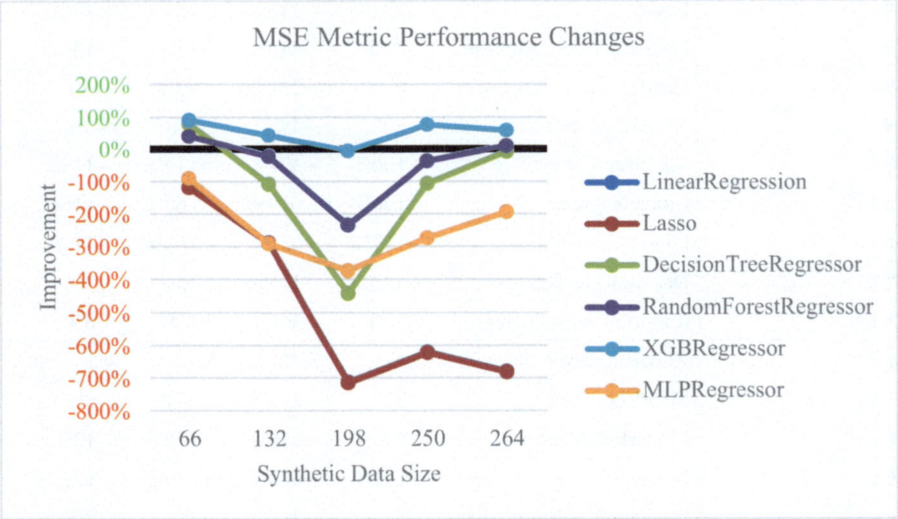

Fig. 3. MSE metric performance changes across different synthetic data sizes and models.

5 Discussion

5.1 Conclusion

The experimental findings provide valuable insights into the efficacy of utilizing synthetic data generated by CTAB-GAN to predict agricultural production. Utilizing synthetic data often resulted in a decline in performance for most models since more significant amounts of synthetic data consistently led to poorer model performance. The

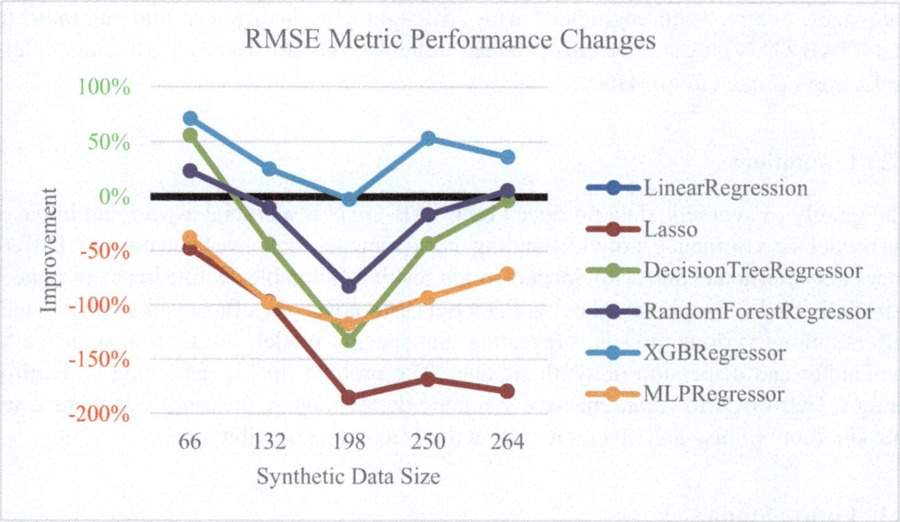

Fig. 4. RMSE metric performance changes across different synthetic data sizes and models.

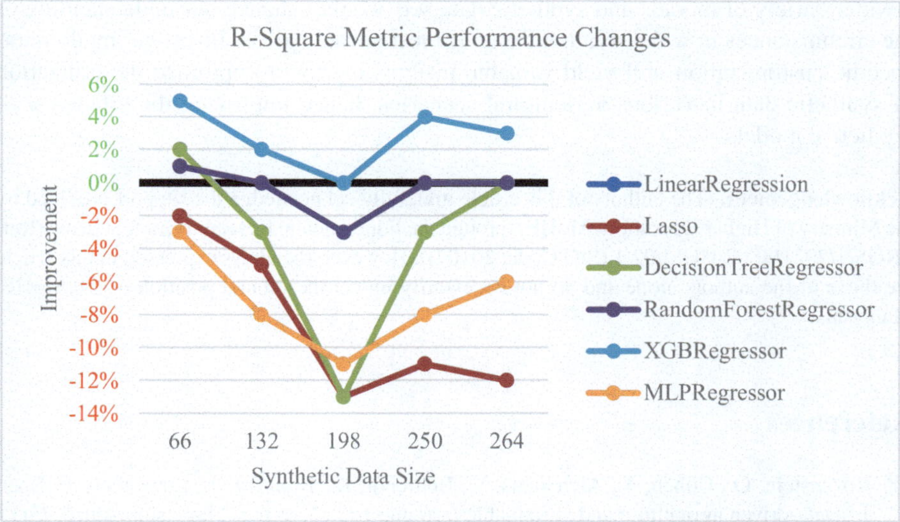

Fig. 5. R-squared metric performance changes across different synthetic data sizes and models.

addition of synthetic data resulted in notable performance degradation for Linear Regression and Lasso Regression, as seen by the rise of prediction error MAE, MSE, and RMSE. In conjunction with that, the R-squared indicates a drop result.

Nevertheless, several models showed enhancements under specific circumstances. When trained with reduced synthetic data sets (25% of the original data), the Decision Tree Regressor and Random Forest Regressor showed significant performance enhancements. Similarly, the XGBoost Regressor exhibited improvements for most synthetic data sizes, except when augmented with 75% synthetic data. These findings indicate that CTAB-GAN may successfully produce valuable synthetic data for particular models under appropriate circumstances.

5.2 Limitations

The quality of synthetic data produced by CTAB-GAN is a crucial aspect that impacts the model's performance, notwithstanding the advantages that have been noticed. Differences between actual and artificial data might result in a notable decline in performance, especially for linear and neural network models. Moreover, the efficacy of synthetic data differs among various models, suggesting that specific models are more responsive to the caliber and dispersion of synthetic data. The problem lies in achieving scalability using CTAB-GAN to create massive synthetic datasets while ensuring that there is no introduction of noise and divergence from the actual data distribution.

5.3 Future Studies

Future research should prioritize customizing synthetic data generation to align with specific agricultural applications and datasets. Furthermore, doing extensive trials using a wider variety of models and synthetic data sets would enhance our understanding of the circumstances in which synthetic data is most advantageous. Investigating domain-specific customization will yield valuable insights on how to optimize the utilization of synthetic data in various agricultural scenarios, hence improving the efficiency of predictive models.

Acknowledgement. The authors of this article gratefully acknowledge the support provided by the Ministry of Higher Education (MoHE) through the Fundamental Research Grant Scheme (Ref: FRGS/1/2021/SS02/UUM/02/1 (S/O Code: 20107)). However, the views expressed in this article are those of the authors alone and do not necessarily reflect the official position of the MoHE, Malaysia.

References

1. Rozenstein, O., Cohen, Y., Alchanatis, V., Behrendt, K., Bonfil, D.J., Lowenberg-DeBoer, J.: Data-driven agriculture and sustainable farming: friends or foes? Precision Agric. **25**(1), 520–531 (2024)
2. Kganyago, M., Adjorlolo, C., Mhangara, P., Tsoeleng, L.: Optical remote sensing of crop biophysical and biochemical parameters: an overview of advances in sensor technologies and machine learning algorithms for precision agriculture. Comput. Electron. Agric. **218**, 108730 (2024)
3. Zhu, F., Han, M., Sun, Y., Zeng, Y., Zhao, L., Zhong, P.: A machine learning framework for multi-step-ahead prediction of groundwater levels in agricultural regions with high reliance on groundwater irrigation. Environ. Model. Softw. **180**, 106146 (2024)
4. Wu, H., Zhao, Z., Chen, L.Y., Van Moorsel, A.: Federated learning for tabular data: exploring potential risk to privacy. In: Proceedings of 2022 IEEE 33rd International Symposium on Software Reliability Engineering (ISSRE), pp. 193–204 (2022)

5. Sun, C., Van Soest, J., Dumontier, M.: Generating synthetic personal health data using conditional generative adversarial networks combining with differential privacy. J. Biomed. Inform. **143**, 104404 (2023)
6. Gonzales, A., Guruswamy, G., Smith, S.R.: Synthetic data in health care: a narrative review. PLoS Digital Health **2**(1), e0000082 (2023)
7. Livieris, I.E., Alimpertis, N., Domalis, G., Tsakalidis, D.: An evaluation framework for synthetic data generation models. Artificial Intelligence Application and Innovation, pp. 320–335, Springer (2024)
8. Razghandi, M., Zhou, H., Erol-Kantarci, M., Turgut, D.: Variational autoencoder generative adversarial network for synthetic data generation in smart home. In: Proceedings of ICC 2022 - IEEE International Conference on Communications. IEEE, pp. 4781–4786 (2022)
9. Patrikar, A.M., Mahenthiran, A., Said, A.: Leveraging synthetic data for AI bias mitigation. In: Manser, K. E., Rao, R. M., Howell, C. L. (eds). Synthetic Data for Artificial Intelligence and Machine Learning: Tools, Techniques, and Applications. SPIE, p. 21 (2023)
10. Zellinger, M.J., Bühlmann, P.: Repliclust: synthetic data for cluster analysis (2023)
11. Drechsler, J., Haensch, A.C.: 30 years of synthetic data. Project Euclid (2023)
12. Zibaee, A.: Rice: importance and future. J. Rice Res. (2013). https://doi.org/10.4172/jrr.1000e102
13. Vaghefi, N., Shamsudin, M.N., Radam, A., Rahim, K.A.: Impact of climate change on rice yield in the main rice growing areas of Peninsular Malaysia. Res. J. Environ. Sci. **7**(2), 59–67 (2013)
14. Guo, Y., Fu, Y., Hao, F., Zhang, X., Wu, W., Senthilnath, J.: Integrated phenology and climate in rice yields prediction using machine learning methods. Ecol. Indic. **120**, 106935 (2021)
15. Grigorieva, E., Livenets, A., Stelmakh, E.: Adaptation of agriculture to climate change: a scoping review. Climate **11**(10), 202 (2023)
16. Vijai, C., Wisetsri, W., Elayaraja, M.: Climate change and its impact on agriculture. Int. J. Agric. Sci. Vet. Med. **11**, 1–8 (2023)
17. Alotaibi, M.: Climate change, its impact on crop production, challenges, and possible solutions. Notulae Botanicae Horti Agrobotanici Cluj-Napoca **51**(1), 13020 (2023)
18. Morales-García, J., Bueno-Crespo, A., Terroso-Sáenz, F., et al.: Evaluation of synthetic data generation for intelligent climate control in greenhouses. Appl. Intell. **53**, 24765–24781 (2023)
19. Ho, Y.B.: Exposures of pesticides and health impact on Malaysia's agricultural (paddy) community. Malays. J. Med. Health Sci. **19**, 1–4 (2023)
20. Zhao, Z., Kunar, A., Van der Scheer, H., et al.: CTAB-GAN: effective table data synthesizing. Proc. Mach. Learn. Res. (2021)
21. Sauber-Cole, R., Khoshgoftaar, T.M.: The use of generative adversarial networks to alleviate class imbalance in tabular data: a survey. J. Big Data **9**(1), 1–37 (2022)
22. Alkhalifah, T., Wang, H., Ovcharenko, O.: MLReal: Bridging the gap between training on synthetic data and real data applications in machine learning. Artif. Intell. Geosci. **3**, 101–114 (2022)
23. Wang, J., Yang, Y., Xia, B.: A simplified Cohen's Kappa for use in binary classification data annotation tasks. IEEE Access **7**, 164386–164397 (2019)
24. Yuan, S., Wu, P., Chen, Y.: Data-level hybrid strategy selection for disk fault prediction model based on multivariate GAN. Int. J. Cybern. Inform. **12**(6), 53–68 (2023)
25. Kawade, R.R., Jagtap, S.K.: Optimal trained ensemble of classification model for speech emotion recognition: considering cross-lingual and multi-lingual scenarios. Multimedia Tools Appl. **83**, 54331–54365 (2023)
26. Mansourifar, H., Moskovitz, A., Klingensmith, B., Mintas, D., Simske, S.J.: GAN-based satellite imaging: a survey on techniques and applications. IEEE Access **10**, 118123–118140 (2022)

27. Tan, H., Chen, R., Qin, M., Tang, L., Wu, Z., Luo, Q.: Tabular GAN-based oversampling of imbalanced time-to-event data for survival prediction. In: Proceedings of 2023 8th International Conference on Cloud Computing and Big Data Analytics (ICCCBDA). IEEE, pp 376–380 (2023)
28. Fallahian, M., Dorodchi, M., Kreth, K.: GAN-based tabular data generator for constructing synopsis in approximate query processing: challenges and solutions. Mach. Learn. Knowl. Extr. **6**, 171–198 (2024)
29. Bourou, S., El Saer, A., Velivassaki, T.H., Voulkidis, A., Zahariadis, T.: A review of tabular data synthesis using GANs on an IDS dataset. Information **12**, 375 (2021)
30. Anuradha, D., Kuchipudi, R., Ashreetha, B., Ramesh, J.V.N., Rami, A.: Enhancing agricultural yield forecasting with deep convolutional generative adversarial networks and satellite data. Int. J. Adv. Comput. Sci. Appl. **15**(2), 661–674 (2024)
31. Alam, M.S., Faria, F.T.J., Moin, M., Wase, A.A., Sani, M.R., Hasib, K.M.: PotatoGANs: utilizing generative adversarial networks, instance segmentation, and explainable AI for enhanced potato disease identification and classification (2024)
32. Alshammari, K., Alshammari, R., Alshammari, A., Alkhudaydi, T.: An improved pear disease classification approach using cycle generative adversarial network. Sci. Rep. (2024). https://doi.org/10.1038/s41598-024-57143-6
33. Stephen, A., Punitha, A., Chandrasekar, A.: Optimal deep generative adversarial network and convolutional neural network for rice leaf disease prediction. Vis. Comput. **40**, 919–936 (2024)
34. Sarr, A.B., Sultan, B.: Predicting crop yields in Senegal using machine learning methods. Int. J. Climatol. **43**, 1817–1838 (2023)
35. Baltagi, B.H., Bresson, G., Chaturvedi, A., Lacroix, G.: Robust dynamic space-time panel data models using ε-contamination: an application to crop yields and climate change (2023)
36. Barth, R., IJsselmuiden, J., Hemming, J., Henten, E.J.V.: Data synthesis methods for agricultural semantic segmentation: a capsicum annuum dataset. Comput. Electron. Agric. **144**, 284–296 (2018)
37. Fassò, A., Rodeschini, J., Moro, A.F., Shaboviq, Q., Maranzano, P., Otto, P.: Agrimonia: a dataset on livestock, meteorology and air quality in the Lombardy region, Italy. Sci. Data (2023)
38. Tran, T.K., Nguyen, C.T., Phung, N.H., Truong, T.T.H., Tran, D.C., Nguyen, T.T.: Evaluation of synthetic data generating methods in down syndrome prediction during the first trimester screening in Vietnam. In: Proceedings of 2022 International Conference on Multimedia Analysis and Pattern Recognition, MAPR 2022, pp. 3–8 (2022)
39. Tan, B.T., Fam, P.S., Firdaus, R.B.R., Tan, M.L., Gunaratne, M.S.: Impact of climate change on rice yield in Malaysia: a panel data analysis. Agric. MDPI **11**(6), 1–17 (2021)

Boosting QR Code Storage Data Text Capacity with Advanced Multiplexing Model Techniques

Azizi Abas(✉) and Fazli Azzali

Universiti Utara Malaysia, 06010 Sintok, Kedah, Malaysia
azizia@uum.edu.my

Abstract. This paper explores innovative strategies to enhance the data storage capacity of QR codes using advanced multiplexing techniques. QR codes have become ubiquitous in the digital age due to their simplicity and versatility, yet their data storage limitations pose challenges for applications requiring high information density. We propose a novel multiplexing model that leverages advanced encoding algorithms and multi-layered data storage to significantly increase the capacity of QR codes without compromising their readability or error correction capabilities. Our approach involves the integration of color multiplexing, layered data encoding, and adaptive error correction to optimize storage efficiency. Through a series of simulations and practical implementations, we demonstrate that our model can effectively double the storage capacity of standard QR codes while maintaining a high degree of scanning accuracy. This advancement can transform various industries, including logistics, advertising, and mobile payments, by embedding more comprehensive information within a single QR code.

Keywords: QR code · Multiplexing · Demultiplexing

1 Introduction

The QR code, a type of two-dimensional barcode, was originally developed by the Japanese company Denso Wave in 1994 [1]. Developed to store digital information, it is easily readable by any smartphone camera. In comparison, traditional one-dimensional barcodes have a limited storage capacity and can only encode alphabets and numbers [2]. QR codes, however, can hold a broader range of data formats and languages, providing much greater storage capacity and faster data retrieval [3]. This versatility has led to the widespread use of QR codes in industries such as retail, healthcare, and marketing, where they serve to store detailed product information, URLs, and contact details [4].

1.1 Advantages and Versatility of QR Codes

QR codes provide several advantages over traditional barcodes. Unlike one-dimensional barcodes, QR codes encode data both horizontally and vertically, allowing them to store substantially more information within a compact space [5]. The technology supporting QR codes also includes error correction mechanisms, enabling accurate data retrieval

Z. Mat Aji et al. (Eds.): KMICe 2024, ISEM 49, pp. 485–500, 2025.
https://doi.org/10.1007/978-3-031-91485-0_38

even if the code is partially damaged, which enhances their reliability and durability [5]. Additionally, QR codes can be scanned from any angle, offering 360-degree recognition [6]. These capabilities, along with their capacity to store various data types, have made QR codes a highly versatile tool for information storage and sharing [7].

1.2 Data Capacity of QR Code

QR codes have a storage limit, holding up to 2,953 bytes of data, which isn't much [3, 6], as reported by Denso Wave Incorporated [8]. Even the high-capacity coloured QR code (HiQ), which has three layers, only reaches about 89 kilobytes, still considered a relatively small storage capacity [9]. QR codes are primarily used for storing certain types of information, like text or web addresses [10], and they struggle to hold larger data types, such as high-resolution images [11] or secure data that requires more substantial storage [10].

When more data needs to be stored, the number of QR codes required increases [12–14], making it difficult to display them effectively in limited media space. Although some apps use coloured 2D barcodes like Cobra, Strata, and Focus [11], smartphone colour displays vary, and not all can accurately reproduce coloured QR codes.

There's also research on linking several QR codes together into one or breaking one QR code into multiple sections [15]. This method, called the structured append feature [8, 9] combines multiple QR codes into a single unit. However, this approach can slow down processing times for decoding and encoding, especially when updating, deleting, or adding data, as it may require regenerating the entire QR code. Additional QR codes may also be needed when the storage capacity is maxed out.

1.3 Research Aim

This research aims to increase the data capacity of a RGB mono-color QR code by designing and developing multiplexing algorithms. The objectives of this study are as follows:

a. To design and develop an encoding algorithm for an RGB mono-color QR code that enables larger data storage.
b. To design and build a decoding algorithm for an RGB mono-color QR code that reliably extracts all the stored data.
c. To propose and evaluate the proposed RGB mono-color QR code by comparing its performance against that of existing QR codes.

2 Literature Review

The rapid rise in QR code popularity can be attributed to its widely recognized convenience. As its use expanded, market demands shifted towards codes with greater storage capacity, the ability to support more diverse character sets, and compatibility with smaller print spaces. QR codes, also called matrix or 2D barcodes, consist of distinctive geometric patterns, including squares, dots, hexagons, and other shapes.

2.1 Introduction to QR Code

The QR code has gained popularity due to its fast scanning speed, high accuracy, and versatile applications [16]. Rapid and reliable access has become essential as digital data, ranging from emails to images and videos, continues to grow. This information may be stored on personal devices, like computers and smartphones, or in global data centres [17]. Recent advancements in storage system extensions have provided valuable insights into the objectives of these systems [8, 9].

The QR code is one method of storing information on paper through encoded symbols. A comparison of data density between different 2D barcodes printed at 600 dpi, such as QR code, High-Capacity Colored Code (HCCB), High-Capacity Colored Two-Dimensional Code (HCCB2D), and CQR Code-5, highlights key differences in their storage capabilities (see Table 1).

Table 1. A comparison of data density across selected two-dimensional barcodes printed at 600 dots per inch.

Two Dimensional Barcode	Data Density [Bits per in^2]
QR Code	5,136
HCCB	16,384
HCCB2D	15,409
CQR Code-5	17,163

The current QR code standard, version 40, establishes the framework for ongoing data storage and processing improvements. Algorithms are essential for encoding, decoding, and partially extracting data from QR codes, and numerous approaches have been developed and shared online, each targeting different QR code creation and processing aspects. One effective method for enhancing data capacity involves incorporating red, green, and blue (RGB) colors, which significantly expands the data storage potential of QR codes [9, 16, 17].

When creating a mono-colored QR code using RGB colors, it's crucial to consider data capacity techniques, encoding and decoding algorithms, and standard benchmarks. Research should focus on refining and integrating these techniques to develop an efficient RGB mono-colored QR code. Establishing a benchmark—such as the traditional black and white QR code—provides a baseline for assessing improvements in storage capacity. After defining input parameters, encoding algorithms are designed, followed by decoding algorithms that facilitate data extraction. QR codes are composed of multiple segments with specific functions; in QR code version 2, for example, there are eight distinct sections (see Fig. 1).

2.2 Multiplexing

Multiplexing is a technique that combines multiple similar items for transmission over a single line, helping to reduce line costs and simplify the tracking of individual QR codes

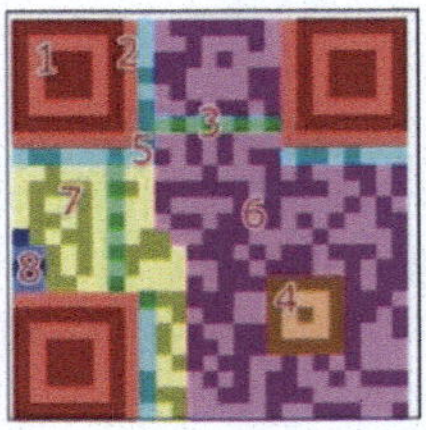

Fig. 1. The QR code version 2 is structured into eight distinct segments.

[18]. In contrast, demultiplexing reverses this process, separating the combined items back into their original forms [19]. Vongpradhip [20] has introduced an algorithm aimed at increasing QR code capacity by leveraging both multiplexing and demultiplexing. This process involves encoding data, breaking it into smaller parts, and generating QR codes with distinctive symbols which still be read by traditional QR code readers [21].

Pillai and Naresh [22] have proposed enhancing QR code capacity by incorporating multiple colors alongside the multiplexing technique, achieving up to three times the expended data capacity of traditional QR codes [23]. Similarly, Galiyawala and Pandya [18] have claimed that there is a method which can multiplexed QR codes by using color to increase storage capacity up to twenty four times. This approach even allows for embedding speech signals, although for optimal speed, it is advised to multiplex no more than twelve QR codes [23].

Encoding. Sartid Vongpradhip has developed a new approach to boost QR code data capacity by using multiplexing and demultiplexing techniques (see Fig. 2) [20]. This approach begins by dividing the original data into smaller segments, each of which is converted into a unique QR code pattern. Through the multiplexing process, the modules in each segment are modified into specialized symbols, allowing for the regeneration of

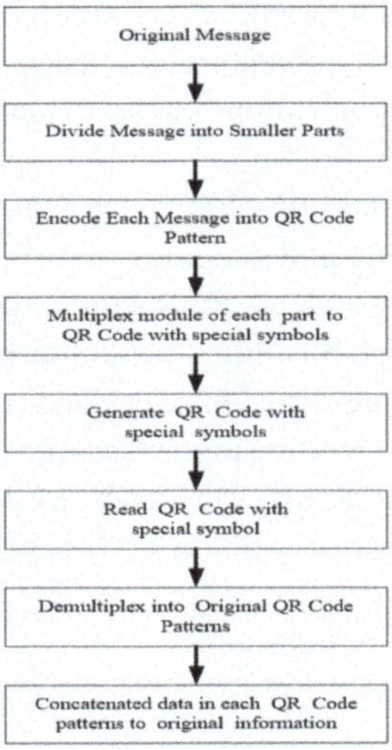

Fig. 2. An algorithm flow that explains the processes of multiplexing and demultiplexing.

the QR code with these customized symbols. Each module, represented by black and white special symbols, is then encoded according to this design [1].

Galiyawala and Pandya proposed a method that divides the original information into n segments, where n represents the number of QR codes involved in the multiplexing process [18]. This value also determines the distinct colors needed, calculated using the formula 2^n equation (see Table 3). This technique leverages multiple QR codes and colors to efficiently encode larger data sets (Table 2).

Table 2. Overview of various multiplexing techniques that relate to QR code technology that use Reed-Solomon error correction.

No.	Researchers and year	Colors	Increments	Transmission Capacity	Processing Time	Others
1	Kris Antoni Hadiputra Nurwono and Raymondus Kosala (2009)	• Red • Green • Blue	• 3 times based on layers combined			
2	M. Ramya and M. Jayasheela (2014)	• Cyan • Magenta • Yellow	• 3 times based on layers combined			
3	Henryk Blasinski, Orhan Bulan, and Gaurav Sharma (2013)	• Cyan • Magenta • Yellow	• 3 times			• Otsu Algorithm - the reduction of a grey level image to a binary image
4	Max E. Vizcarra Melgar, Alexandre Zaghetto, Bruno Macchiavello, and Anderson C. A. Nascimento (2012)	• Red • Green • Blue • Black • White	1,024 bits of information	• Size of 1.3 cm × 1.3 cm, resulting in a transmission capacity of approximately 605 bits/cm2		
5	Antonio Grillo, Alessandro Lentini, Marco Querini, and Giuseppe F. Italiano (2010)	• 4 and 16 colours	• 1.881 kilobytes/inch • (600 dpi)		• The average overhead introduced by HCC2D with 16 colors is about 25%	

QR codes are generated based on dividing data into n segments, with the multiplexing process beginning by identifying unique colors through a coloring code technique. Each pattern within the combined QR code is assigned to a distinctive color, generated from RGB combinations using MATLAB's color map matrix. The normalized RGB values for the colored QR code are detailed in Table 4. Once each distinct color is assigned, a vibrant QR code is produced as the result of the multiplexing process (see Fig. 3).

Table 3. An example of specific color requirements needed for multiplexing QR codes.

Number of QR codes for multiplexing (*n*)	Required distinct colours
2	4
3	8
4	16
5	32
6	64
...	...
14	16384

Table 4. The color combinations in the colored QR code based on normalized RGB values.

QR Code 1	QR Code 2	QR Code 3	Index	Red	Green	Blue	Color
0	0	0	0	0	0	0	Black
0	0	1	1	0	0	1	Blue
0	1	0	2	0	1	0	Green
0	1	1	3	1	0	0	Red
1	0	0	4	0	1	1	Cyan
1	0	1	5	1	0	1	Magenta
1	1	0	6	1	1	0	Yellow
1	1	1	7	1	1	1	White

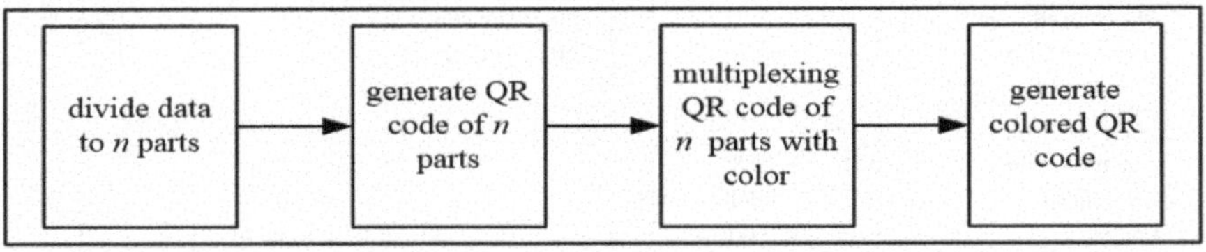

Fig. 3. The multiplexing process works for a colored QR code.

In the decoding phase described by Vongpradhip [20], the process begins by reading a QR code with special symbols, followed by demultiplexing to reconstruct the original pattern. Each module, or code word, is extracted and placed back into its respective layer, where 0 and 1 correspond to black and white modules, respectively. After restoring the modules, the QR code returns to its original form and can be read as usual (see Fig. 4).

Pillai and Naresh [22] have initiated the decoding process by reading the coloured QR code image and separating it into individual RGB colour layers. Each RGB layer contains binary information, which is then converted into decimal values for each byte.

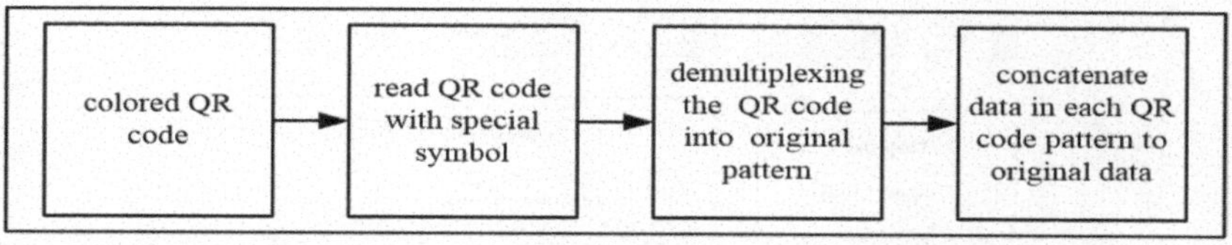

Fig. 4. The sequence of steps in the demultiplexing process for a QR code with special symbols outlines.

These decimal values serve as inputs to the Reed-Solomon decoder, which outputs ASCII equivalents that are converted back into the original data (see Fig. 5).

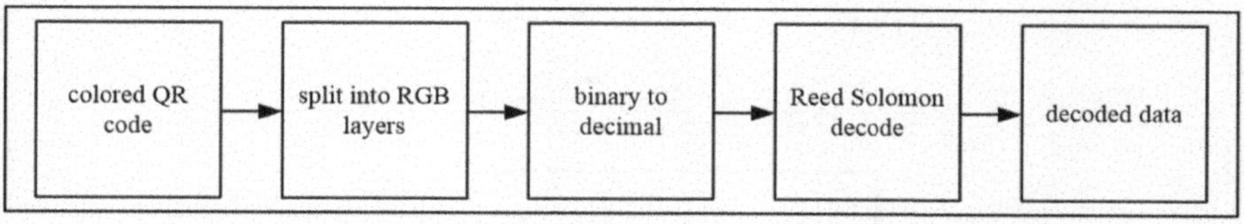

Fig. 5. The sequence of steps in the decoding process.

The demultiplexing process, as described by Galiyawala and Pandya [18], essentially reverses the multiplexing method to decode a coloured QR code. This involves using a reference table that assigns RGB colour combinations to data bits. By comparing each pixel in the QR code with the values in this table, the original pixel values can be recovered. After the QR code is restored to its original state, it can be scanned by standard reader devices. The decoded data is then merged to reconstruct the complete original message (see Fig. 6).

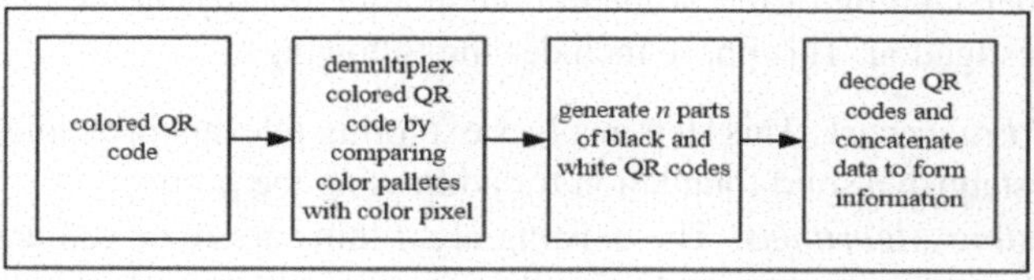

Fig. 6. The sequence flow of the demultiplexing and decoding processes.

3 Methodology

This research adopts a deductive approach [24], starting from general concepts—such as theories, principles, and core ideas—and building upon them through each chapter. Each phase in this approach has its own distinct methods and outcomes, forming a structured progression throughout the research process (see Fig. 7).

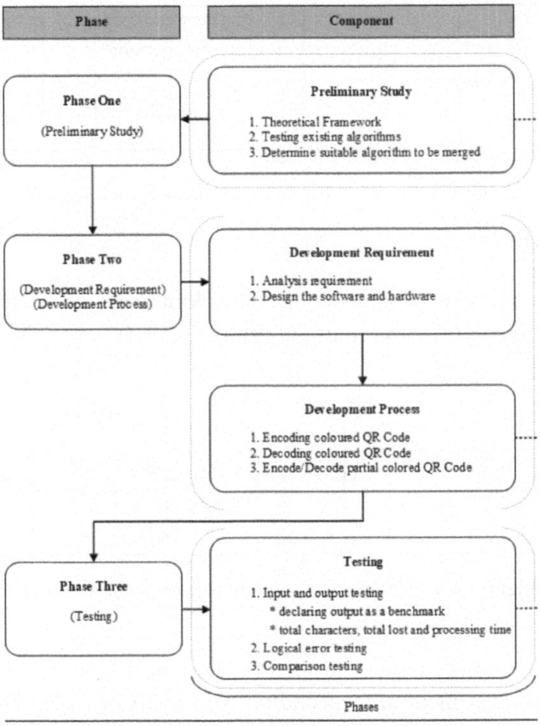

Fig. 7. The flow of methodology framework.

3.1 Phase 1: Preliminary Study

The first phase, known as the preliminary study (see Fig. 7), centres on gaining a comprehensive understanding of the problem domain and identifying the best approaches for further investigation. This phase includes the following.

a. *Theoretical Framework*: This step involves exploring relevant principles, theories, or models to establish a solid foundation for addressing the problem.
b. *Testing Existing Algorithms:* The existing algorithms are reviewed and assessed to determine their effectiveness and suitability for the task at hand.
c. *Determining the Suitable Algorithm for Integration:* Based on the evaluation, the most appropriate algorithms are chosen for integration in the subsequent development process.

The tasks in this phase focus on gathering information from various sources and selecting algorithms that can perform effectively before any modifications are applied to the chosen algorithm. This phase involves three main components:

a. Setting up the theoretical framework.
b. Evaluating existing algorithms.
c. Choosing the most appropriate one for integration.

3.2 Phase 2: Development Requirement (Development Process)

This phase focuses on designing and developing a mono RGB-colored QR code, where the actual planning and development take place. It is divided into two main components:

- **Development Requirements**:

  Analysis Requirement: This step involves identifying both the needs of the problem being addressed and the specifications for the system to be developed.
  Designing Software and Hardware: This includes planning the architecture, components, and workflows required for software and hardware.

- **Development Process**:

  a. *Encoding Colored QR Code:* This step entails creating a process for generating QR codes that can represent colors.
  b. *Decoding Colored QR Code:* Methods are developed to read and interpret these colored QR codes effectively.
  c. *Encoding/Decoding Partial Colored QR Code:* This involves managing situations where the QR code is incomplete or only partially available, ensuring the system can still process it.

3.3 Phase 3

Once the development process is completed, the system undergoes a testing phase to ensure it functions as intended. This phase includes:

Input and Output Testing: This step checks that the system accurately processes inputs and produces the expected outputs.

Declaring Output as a Benchmark: Comparing the system's output to established benchmarks.

Measuring performance metrics, including error rates, processing time, and overall efficiency.

Logical Error Testing: This involves identifying and addressing any logical errors or bugs within the code.

Comparison Testing: Here, the results are compared to expected outcomes or other systems to confirm the application's accuracy and performance.

4 Result

The multiplexing process involves several preparatory steps: converting binary data to text, transforming the American National Standards Institute (ANSI) code into Unicode Transformation Format 8-bit (UTF-8), and performing the multiplexing itself (see Fig. 8).

4.1 The Multiplexing Model

A sequential multiplexing process for converting ANSI text into a mono-colored QR code was proposed. The process begins with converting ANSI text to UTF-8 encoding. Next, the UTF-8 encoded text is transformed into a standard black and white QR code. Finally, the black and white QR code is converted into a single-color QR code, completing the process (see Fig. 8).

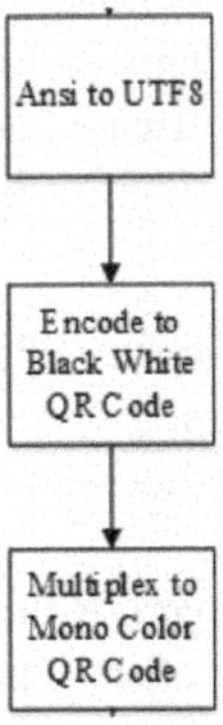

Fig. 8. The flow of multiplex processes.

Converting ANSI to UTF-8 Module. The conversion was necessary because the Base64 encoder generates the ANSI format when converting binary data to text. To create the black and white QR code, the text data needed to be in UTF-8 format. UTF-8 was chosen because most scanner devices use the JIS8 (QR 2000) format to scan data, which is compatible with UTF-8. The JIS (Japanese Industrial Standards) encoding system is designed for the Japanese language and offers support for code set conversion to UTF-8 [25].

Creating Blank Files Module. Once the encoding process was complete, the next step involved creating blank files, which served as a preprocessing task before executing the actual multiplexing process. These blank files were necessary because the multiplexing process required a specific number of empty files to embed the input data. In this experiment, *N* blank files were created and divided into three groups, with each group containing eight files.

The following paragraph will explain the rationale for creating eight files in each group. When initializing the blank files, three factors were considered: the maximum number of characters, the total number of files, and the location where the files would be stored. The total number of blank files used in this experiment was *N*, and they were all located in the current directory of the program's execution. When creating these blank files, an instruction was given to embed the 'space' symbol in each file, resulting in 2,952 'space' symbols embedded within every file.

Dividing Characters Module. In the next step, the characters from the input files were divided, with a specific number of characters allocated to each file. In this case, 2,952 characters were assigned to each file (see Table 5).

Table 5. The maximum number of characters, categorized by the error correction level.

Error Correction Level	Total Characters for each QR Code
L	2952
M	2330
Q	1662
H	1270

Several conditions were implemented to prevent logical errors during execution. If the number of characters embedded exceeded the limit of n files, the program would automatically terminate, displaying an error message. On the other hand, if the number of embedded characters were fewer than n files, the program would continue running but ensure that the 'space' symbols remained balanced within the files where the characters had been previously embedded. All 'space' symbols were replaced when inserting the encoded characters.

Table 6. The allocation of characters across the files.

	A File	Eight Files	Twenty-Four Files	N Files
Characters	2,952	8 x 2,952 23,616	24 x 2,952 70,848	*N* x 2,952 2,952*N*

Before the execution process began, two calculations were performed to determine the exact number of files to be used. The first calculation is as follows:

$$Total\ Files = Total\ Characters/Maximum\ Characters\ in\ Each\ File$$

If the value resulted in a remainder, only the integer part would be used. The following calculation involved determining the mod value, which is as follows:

$$Total\ Mod\ Value = Total\ Characters\ \%Maximum\ Characters\ in\ Each\ File$$

After completing the calculations, the total number of files would increase by one if there was a remainder in the mod value. This remainder would account for the remaining characters in the last file. If the character count was less than 2,952, the remaining 'space' symbols would remain in the final file, along with the remaining characters, ensuring each file reached 2,952 characters (see Table 6).

Once the files were ready to receive the characters, the characters were split into groups of 2,952 and assigned to the corresponding files until all the data was allocated. If the final file didn't reach 2,952 characters, the remaining space would be filled with 'space' symbols and kept in that file. After all the files were set up, the process would move forward by sending the completed files to the next stage.

Creating Black and White QR Code Module. This section details the steps involved in producing a set of N black and white QR codes using input data files that contain separated characters. The process begins with setting up the graphic output and text input files and assigning specific file names to the graphic output files for more effortless execution. Once the initialization is complete, the data from the input files is read and stored in an array to create the black and white QR codes. As there are multiple input files, the creation module is called repeatedly in an object-oriented manner to generate the necessary black and white QR codes.

Before generating the black and white QR codes, various essential parameters such as character set, error correction level, color, margin, height, and width need to be defined. Once these parameters are established, the QR code is created using the *ZXing* library, which utilizes the *QRCodeWriter* class to produce a byte matrix representing the graphic pixels of the QR code. This byte matrix is then utilized to generate the QR code image by processing both the graphic and buffer images. The creation of the black and white QR code involves assembling the design pixel by pixel, with each pixel's position determined along the x-axis and y-axis.

Development of Red, Green, and Blue (RGB) QR Code Module. This method marks the final step in the multiplexing process, broken down into three sub-processes: reading the image and converting it into an 8-bit structure, collecting the data, and generating the red, green, and blue QR codes. The first task involves capturing the color pixel information from the previously created black and white QR codes.

After initializing image sizes (height and width), the total image files, the image file names, and the images are read. In this experiment, *N* black and white QR codes are used as the basis to create red, green, and blue RGB QR codes. An array of buffered images is assigned to store the pixel color information for all the black and white QR codes. This array holds the color data based on the RGB scheme, using the x- and y-axis coordinates of each black and white QR code. Only two colors are used in the RGB scheme: black and white.

Once the pixel color information is collected, the next step is to convert the colors into digital values. Black is represented as 1 and white as 0. In the RGB scheme, black is coded as *RGB(000,000,000)* and white as *RGB(255,255,255)*. For example, eight black and white QR codes are selected and combined based on their x and y-axis coordinates to generate the red QR code. The pixel values at location *(0,0)* from these eight QR codes are merged into an 8-bit binary number, then converted to a decimal value between 0 and 254. This decimal number is placed in the first column of the RGB scheme, with the second and third columns set to 0. This process continues for all pixel locations, from *(0,0)* to *(width-1, height-1)*. The same procedure is followed to create the blue and green QR codes.

For multiplexing, the process generates three QR codes: red, green, and blue. Figure 4.4 illustrates the first step in converting binary to decimal at the index location *(0,0)* for each black and white QR code and assigning the value to the corresponding location in the red QR code. The red, green, and blue QR codes are created from the array of graphic images, each containing pixel color information. The multiplexing process combines eight images into one by converting binary numbers into 8-bit values, resulting in the final three QR codes.

4.2 Demultiplexing

Demultiplexing is the process of dividing a combined data stream back into its original individual channels or signals. It is essentially the opposite of multiplexing, where multiple signals or data streams are combined into a single stream.

Decoding Demultiplexing Module. The next phase is demultiplexing, which involves separating the red, green, and blue QR codes into N black and white QR codes. This process extracts the decimal values of each pixel from the red, green, and blue QR codes in the RGB color model and converts them into 8-bit binary numbers. Once the binary numbers are determined, they are split into eight separate 8-bit units. Each binary digits represents a pixel, where 0 indicates a white pixel and 1 indicates a black pixel in the corresponding black and white QR code.For example, if the pixel at location *(0,0)* in the red QR code has the RGB value of *(45, 0, 0)*, the 8-bit binary equivalent would be 00101101. The first binary digit, 0, corresponds to a white pixel in the first black and white QR code at location *(0,0)*, and this method is applied to all the other QR codes. This illustrates the "one to many" concept used in the demultiplexing process.

The flow of the process begins by identifying the image data of the red, green, and blue QR codes and grouping them into Mn black and white QR codes for each color channel. Demultiplexing starts by extracting the decimal value of the red element from the RGB value of the red QR code at each pixel location, beginning with *(0,0)* and continuing until *(177,177)*, which is determined by the width and height of the image. The process then continues with the green and blue QR codes, following a conditional sequence to ensure each is processed. The decimal values are assigned to temporary storage locations corresponding to the relevant red, green, and blue elements. For example, if the RGB value for a pixel in the red QR code is *RGB(46,0,0)*, the decimal value 46 is temporarily stored. The decimal values for all three color channels are collected and stored in temporary memory.

Once all the decimal values are gathered, they are converted into binary numbers. This conversion begins by assigning the decimal values as integers and converting them to string format for easier manipulation. The conversion itself is done through a standard mathematical process: dividing the decimal value by two, recording the remainder, and repeating the division until the value can no longer be divided by two (see Table 7).

Next, the remainders, when read from bottom to top, formed the binary representation, which in this instance was 10011101.

Once the binary values were extracted from the red, green, and blue QR codes, the subsequent step was to decompose them into individual bits and identify the index position of each bit. With each bit isolated, its corresponding pixel color was then determined: a bit value of 0 indicated a white pixel, while a bit value of 1 indicated a black pixel. This process considered three main factors to assign pixel colors accurately: the specific QR code type (red, green, or blue), the pixel's location, and the bit's position within its 8-bit binary sequence. The combination of these elements is detailed in the following process.

$$\textit{Type of QR code}\; x = red;\; y = green;\; z = blue$$

$$\textit{Pixel location} = a(0,0) \sim a(177,177)$$

$$\textit{Break out index}\; 8\; \textit{bits location into single bit} = p(0) \sim p(7)$$

Table 7. The calculation of converting decimal values into binary.

Divide value (÷)	Modulus value (%)
$157 \div 2 = 78$	with a remainder of 1
$78 \div 2 = 39$	with a remainder of 0
$39 \div 2 = 19$	with a remainder of 1
$19 \div 2 = 9$	with a remainder of 1
$9 \div 2 = 4$	with a remainder of 1
$4 \div 2 = 2$	with a remainder of 0
$2 \div 2 = 1$	with a remainder of 0
$1 \div 2 = 0$	with a remainder of 1

The determination of whether a pixel is black or white can be represented as a set of *{QR Code Type}{Pixel Location}{Single Bit}*, which is more specifically referred to as $\{x_i\}\{a_{ij}\}\{p_i\}$.

Decoding Black and White QR Codes. After all 24 black and white QR codes were generated, the next step was to decode them into a text file. This process involved three key stages: decoding each QR code, combining the decoded segments, and saving the final content to a file. The decoding began with the first QR code from the red set, extracting the text it contained. Each QR code held part of the full text, which had been encoded earlier. This process continued in order, decoding each black and white QR code from the red and green sets. The *MultiFormatReader* module was used to decode the QR codes.

Following the text extraction, the next task was to merge the decoded text segments. This was accomplished using a string class capable of handling up to 2,147,483,647 (231 - 1) characters. Each text segment extracted from the individual black QR codes was concatenated using the ' +' operator. In the final step, the entire merged text, stored in the buffer string class, was saved to the file directory using the *File class*. The resulting file was in *Base64* format, so it needed to be decoded with the appropriate decoder to restore the original content.

Experiment and Result of Decoding Demultiplexing Process. An experiment was conducted to measure the elapsed time for different error correction levels during the demultiplexing process, as presented in Table 8. The results indicate that error correction level Q had the shortest decoding time, with a total of 1 s and 614 ms. On the other hand, the error correction level L showed the longest decoding time, taking 1 s and 656 ms. The results indicate that the time needed to decode the colored QR code into a Base64 text file was not a major issue, as the variation in decoding times across different error correction levels was minimal, with the entire process averaging around 1.65 s.

Table 8. The elapsed time for the decoding and demultiplexing process.

Error Correction Level	RGB Demultiplexing	Black and White Demultiplexing	Total Multiplexing
L	0 s 402 ms	0 s 470 ms	1 s 656 ms
M	0 s 437 ms	0 s 465 ms	1 s 650 ms
Q	0 s 411 ms	0 s 450 ms	1 s 614 ms
H	0 s 421 ms	0 s 473 ms	1 s 655 ms

5 Conclusion

In conclusion, the proposed model for enhancing data capacity in QR codes constitutes a significant advancement in this field of research. By utilizing innovative multiplexing techniques, this model substantially improves the data capacity of a single QR code.

This enhancement is particularly advantageous for optimizing processing time efficiency and data storage across error correction levels. Furthermore, the model notably reduces the necessity for multiple QR codes, simplifying the user experience. The model's effectiveness was validated through a meticulously selected set of test data, which allowed for a dynamic evaluation of its data storage capabilities within the QR code framework. The experimental results demonstrated a remarkable increase in data capacity compared to existing QR code implementations. To ensure a thorough assessment, automated techniques were used to measure throughput time, total character capacity, and to evaluate various error correction levels.

These findings confirm that our model achieves a significant enhancement in data capacity, establishing its superiority over traditional methods. Additionally, results from partial extraction techniques indicate that QR code content can be customized more swiftly than previous approaches. The outstanding flexibility of this model enables it to accommodate a diverse range of inputs, including images, different character sets, and public/private keys. This adaptability makes the model well-suited for various applications, such as process control, product descriptions, and offline e-books, positioning it as a valuable resource for contemporary users.

Acknowledgement. This research was funded by the Ministry of Higher Education (MOHE) of Malaysia under the Fundamental Research Grant Scheme for Research Acculturation of Early Career Researchers (FRGS-RACER) (RACER/1/2019/ICT03/UUM//2).

References

1. Denso Wave Incorporated, http://www.qrcode.com. The Inventor of QR Cod. last accessed 14 August 2024
2. Smith, J.T., Peters, M.B.: State highway administration research report utilizing automatic identification tracking systems to compile operational field and structure data. University of Maryland Final Report (2014)

3. Grillo, A., Lentini, A., Querini, M., Italiano, G.F.: High capacity colored two dimensional codes. In: Proceedings of the International Multiconference on Computer Science and Information Technology, pp. 709–716 (2010)
4. Yusuf, B., et al.: Multicolor and multiple QR Code based information support system during Disaster for Elderly People (n.d.)
5. Galiyawala, H.J., Pandya, K.H.: To increase data capacity of QR code using multiplexing with color coding: An example of embedding speech signal in QR code. In: 11th IEEE India Conference: Emerging Trends and Innovation in Technology, INDICON 2014 (2015)
6. Boob, A., Shinde, R.D., Gaikwad, A.: QR code based mobile app and business process integration. Int. J. Multidiscipl. Current Res. 2 (2014)
7. You, L.Z., Swee, S.K., Kiong, W.E.: High capacity multi colored code system. In: Proceeding of 2017 International Conference on Robotics, Automation and Sciences, ICORAS 2017 (2017)
8. Yang, X., Xu, Z., Deng, L., Loy, C.C., Lau, R.Y.: High capacity coloured QR code. IEEE Trans. Multimedia **19**(3), 500–513 (2017)
9. Grillo, A., Lentini, A., Querini, M., Italiano, G.F.: High capacity colored QR code. In: Proceeding International Conference Data Storage Retrieval, pp. 77–85 (2010)
10. Victor, N.: Enhancing the data capacity of QR codes by compressing the data before generation. Int. J. Comp. Appl. **60**(2), 17–21 (2012)
11. Querini, M., Grillo, A., Lentini, A., Italiano, G.F.: 2D color barcodes for mobile phones. Int. J. Comput. Sci. Appl. **8**(1), 136–155 (2011)
12. Tiwari, S.: An introduction to QR code technology. In: 15th International Conference on Information Technology, pp. 39–44 (2017)
13. Galiyawala, H.J., Pandya, D.: Enhancement of QR code storage capacity using RGB color model. Int. J. Adv. Res. Comp. Sci. **5**(3) (2014)
14. Zhang, Y., Yao, X., Zhou, L.: Metric column dependency in QR codes. Int. J. Comp. Appl. **23**(4), 112–119 (2023)
15. Majumdar, P., Maiti, A., Bhattacharyya, K., Nath, N.: Challenges in QR code data storage. Journal of Secure Data Storage **9**(4), 223–229 (2015)
16. Denso Wave Incorporated, http://www.qrcode.com. Introduction of QR Code. Last accessed 14 August 2024
17. Feng, Z., Zheng, Y.: Differences between 1D and 2D barcodes. Journal of Barcode Technology **23**(2), 45–52 (2010)
18. Galiyawala, H.J., Pandya, K.H.: To increase data capacity of QR code using multiplexing with color coding : an example of embedding speech signal in QR code. In: 2014 Annual IEEE India Conference, pp. 1–6 (2014)
19. Toh, S.R., Goh, W., Yeo, C.K.: Data exchange via multiplexed color QR codes on mobile devices. In: Wireless Telecommunications Symposium (2016)
20. Vongpradhip, S.: Use multiplexing to increase information in QR Code. In: The 8th International Conference on Computer Science & Education, pp. 361–364 (2013)
21. Ferreira, R.A.S., André, P.S.: Colour multiplexing of quick-response (QR) codes. Electron Lett. **50**(24), 1828–1830 (2014)
22. Pillai, P.N., Naresh, K.: Improving the capacity of QR code by using color technique. Int. J. Adv. Res. Electr. Electron. Instrument. Eng. **3**(7), 10561–10568 (2014)
23. Galiyawala, H.J., Pandya, K.H.: To increase data capacity of QR code using multiplexing with color coding: An example of embedding speech signal in QR code. In: 11th IEEE India Conference: Emerging Trends and Innovation in Technology, pp. 1–6 (2015)
24. Trochim, W.M.K.: Research methods: knowledge base. Atomic Dog Publishing, Cincinnati (2015)
25. Stinner, V.: Programming with Unicode documentation (2017)

Binary-Based Classification Models for Malaysian Graduates Employment Analysis

Yuhanis Yusof[1(✉)], Fauziah Baharom[1], Nuratikah Jamaludin[1], and Salwati Badroddin[2]

[1] Universiti Utara Malaysia, Sintok, 06010 Kedah, Malaysia
yuhanis@uum.edu.my
[2] Ministry of Higher Education, Putrajaya, 62200 Putrajaya, Malaysia

Abstract. Graduate employability is a crucial facet within this unemployment context. Traditionally, an individual's employability has primarily hinged on their academic performance, typically measured by their Cumulative Grade Point Average (CGPA). However, recent research has unveiled that CGPA alone does not exclusively determine graduate employability; various other factors influence an individual's job prospects. The primary objective is determining suitable machine learning algorithms to predict graduates' employment status based on factors. This study used data spanning 10 years (2012 to 2022) obtained from the Malaysia Ministry of Higher Education graduate's tracer study. Four classification algorithms—Decision Tree, Naïve Bayes, Random Forest and Logistic Regression were employed and compared to determine the most effective model. The findings indicate that the Random Forest classifier yields the highest accuracy, initially at an average of 75%, while Logistic Regression was the 2nd best model. This study has managed to analyze 13 factors that impact graduate employment status. Important factors include Types of Higher Learning Institution and language proficiency (Bahasa Melayu and English). In addition, it is learned ICT skills is also relevant in determining whether a graduate will be working or not six months after graduating. Consequently, the insights from this study can assist higher learning institutions in Malaysia in better preparing their graduates with the requisite skills for a successful transition into the job market. Hence, data analytics enables the development of adaptive knowledge management systems that enhance the efficiency of knowledge dissemination.

Keywords: Explicit Knowledge · Graduate Employability · Data Classification

1 Introduction

Malaysia has a growing higher education sector with numerous universities and colleges. Historically, graduate employability has been a concern, with some graduates facing challenges finding employment matching their qualifications. Furthermore, employability rates often vary by field of study. Graduates from fields such as engineering, information technology, and healthcare tend to have relatively higher employability rates due to industry demand. On the other hand, graduates from some humanities and social

Z. Mat Aji et al. (Eds.): KMICe 2024, ISEM 49, pp. 501–513, 2025.
https://doi.org/10.1007/978-3-031-91485-0_39

science programs might face greater challenges in securing jobs directly related to their majors. Some graduates explore entrepreneurship opportunities to create employment for themselves and others. Malaysia has been supportive of entrepreneurial initiatives, with programs and funding available for start-ups.

Based on the outcome of Sistem Kajian Pengesanan Graduan (SKPG) or also known as National Tracer Study, Malaysia's graduate employability rate in 2021 has shown an increase of 1.1%, which is 85.5% compared to 84.4% in 2020. Likewise, for TVET graduates, the graduate employability rate recorded an increase in 2021 to 88.0% compared to 87.0% in the previous year [1]. Nevertheless, the findings of 2020 graduates show a decrease of 1.8% to 84.4% in 2020 compared to 86.2% in 2019. From the employability rate, 60.0% of graduates are employed, 17.8% continue their studies, followed by 4.6% waiting for job placement and 2.0% following up-skilling programs [2] Although the percentage of unemployed in 2020 increased compared to 2019, the number of unemployed graduates for both years is almost the same, around 41,000.

Various studies have been reported predicting graduates' employment based on selected factors. Predicting graduates' employment status is essential in today's dynamic job market. It is a valuable tool for educational institutions, policymakers, and students. Students can make informed choices about their career paths, majors, and additional training needs by understanding the likely outcomes of their educational investments. For universities and colleges, predictive data can help them tailor their programs to meet industry demands, ensuring graduates are better prepared for the workforce. Policymakers can also use this information to address skill gaps and labour market needs strategically.

This study focuses on the use of a data mining approach to predict the employment status of graduates in Malaysia. Data analytics plays a pivotal role in knowledge management by transforming vast amounts of raw data into actionable insights that drive decision-making and innovation. By leveraging advanced analytical techniques, organizations can uncover patterns, trends, and relationships within their data, enhancing their knowledge repositories and improving organizational learning. These insights help optimize processes, predict future outcomes, and foster a culture of evidence-based decision-making. Furthermore, data analytics supports the creation of more dynamic and responsive knowledge management systems that adapt to changing information landscapes, ultimately leading to more effective and efficient knowledge dissemination.

This work begins with data collection and the gathering and storage of extensive datasets. These data are obtained from the Sistem Kebolehpasaran Graduan (SKPG), managed by the Ministry of Higher Education. The SKPG is a big pool of data as all local graduates in Malaysia are required to complete a survey on employment before their convocation. To date, there are 1,072,868 graduates who have participated in the study, which was first introduced in 2006. Upon gathering the data, it undergoes preprocessing, which includes cleaning, transforming, and organizing it into a suitable format for analysis. As the scope of this study is on STEM graduates, the cleaning process filters the irrelevant data. STEM is an educational program that prepares students and graduates in science, technology, engineering, and mathematics. Suitable data mining algorithms are then deployed to uncover meaningful associations or correlations between the data. This is achieved through the binary classification task that learns the pattern of employment status (i.e. employed or unemployed). Using Decision Tree, Naïve Bayes, Random Forest and Logistic Regression learning models, factors affecting graduates' employability such as academic achievement, area of study, family's background and many more can be determined. These patterns can range from simple relationships to intricate, multi-dimensional insights. Finally, the discovered knowledge can be used for predictive modelling and/or decision-making.

2 Related Work

Existing research on employability utilizes data mining techniques to identify factors influencing graduates' employability status. Studies explore various factors, including academic performance, extracurricular activities, field of study, language or ICT skills and internship experiences. Machine learning algorithms like decision trees and support vector machines are employed to build models predicting a graduate's employment likelihood. However, the accuracy of these models can be limited by the size and scope of the data used for training. This section discusses several studies involving employability factors and machine learning models used in predicting employability status.

2.1 Factors Affecting Graduate's Employability

Various studies worldwide have identified factors that affect graduates' employability. An example is the work in Sri Lanka [3] where the variables under consideration were gender, degree classification, professional qualifications, English proficiency, IT skills, training and experience, and soft skills. It was reported that IT skills partially influence, and degree classification did not reflect a significant impact on graduate employability. Recently, in China, a study deploys exploratory factor analysis to test the employability factors of college students and utilizes multiple linear regression analysis to test the factors that influence employability [4]. The results of the research show that individual traits, social experience, and workplace training have a significant impact on college students' employability while knowledge understanding and learning ability, self-management ability, emotional intelligence, generic skills, professional ability, and career planning capability are the important factors of the employability [4].

To strengthen the body of knowledge, Clarke [5] proposes a framework that incorporates six dimensions–human capital, social capital, individual attributes, individual behaviours, perceived employability, and labour market factors to understand factors attributing to graduates' employability. The work supported an earlier study undertaken in Australia where technical expertise, generic skill mastery, and a successfully formed graduate identity are the specific details of graduates' employability. Apart from graduates' identity and background, research also reveals that university branding is also important in graduates' employment [6]. Universities with strong brand recognition and a reputation for academic excellence often have an advantage. Employers may have a positive bias toward graduates from prestigious universities, assuming they received a high-quality education. Some well-known universities have established relationships with top employers, leading to on-campus recruiting events, job fairs, and direct partnerships. While a prestigious university can offer advantages, it's the graduate's abilities and achievements that ultimately determine their success in the job market.

2.2 Data Mining in Analyzing Graduate's Employability

Researchers have increasingly used data mining models, such as machine learning algorithms, to forecast graduates' employability. These models leverage historical data on graduates, including their academic performance, skills, demographic profile, and other factors, to predict their chances of finding employment. In 2018, Othman, Shan, Yusoff and Chang reported the success of using three data mining algorithms in predicting graduates' employability at one of the public universities in Malaysia [7]. The three methods used were J48, Multi-Layer Perceptron, and Self-Organizing Map. J48 was the best classifier, producing 66.18% accuracy. The findings also revealed that age, industrial internship, and faculty contain the most information and affect the final class, i.e. employment status.

There is also a report on predicting the employability of IT graduates by determining whether the IT graduates would land in an IT-related career or not [8]. The study aims to examine the accuracy of data mining techniques used in supervised learning to discover the main descriptive features for prediction. Naïve Bayes, J48, Simple Cart, Logistic Regression, and Chaid Algorithms were among the supervised learning algorithms deployed in the study. The researcher gathered historical data from the University Job Placement Office Tracer Study of BSIT graduates from 2011 to 2015 and combined it with their academic records. The findings reveal three key criteria that have a direct impact on IT employability which are IT_Core, IT_Professional, and Gender. A similar study on IT graduates' employability was performed in Egypt [9]. The dataset used in the study was classified into Training, Soft skills, Hard skills, and In-demand skills. The Decision Tree model predicts the highest number of true positives among the five classification models. Furthermore, Logistic Regression and Support Vector Machine models predict the highest number of true negatives. It was noted the best classifier for the undertaken task was the Decision Tree while the least was Gaussian Naïve Bayes.

Besides IT graduates, data mining approaches have also been deployed in classifying the employability status of engineering graduates [10–12]. The aim of the studies was two-fold which are to determine 1) factors that influence graduates' employability, and 2) classifier that best learned from graduates' historical data. In considering the importance of statistics graduates to the market labour, a study was performed recently in Thailand [13]. It was reported that employability was influenced by graduation year, education level, major of study, type of university, ranking, and region of the university. The four models compared were Logistic Regression, Decision Tree, Random Forest, and k-Nearest Neighbour (kNN) models. The comparative result showed that the Random Forest and kNN models are suitable for employability prediction with 70.554% and 70.158% accuracy.

Nevertheless, all of the discussed work were on small dataset. The findings based on small datasets might not be generalizable to a wider population. This is because the limited data may not capture the full range of existing variations. Additionally, machine learning algorithms trained on such datasets may not perform well on unseen data, raising concerns about their real-world effectiveness. Hence, this study evaluates the effectiveness of machine learning algorithms on a larger dataset which is the Malaysian graduates (indicating the population instead of sample).

3 Methods and Materials

The undertaken study comprises four phases: preprocessing, data sampling, model development, and model evaluation. Connections between the phases are illustrated in Fig. 1.

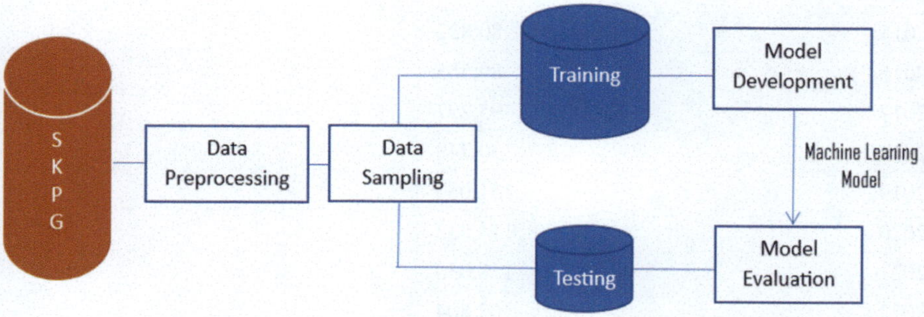

Fig. 1. Dataflow for graduates' employability status classification.

3.1 Data Preprocessing

This study utilizes graduates' records from 10 years of data (2012–2022) obtained from the Sistem Kajian Pengesanan Graduan (SKPG, also known as the Malaysia Tracer Study). As the population data is huge, this study focuses on graduates from Science, Technology, Engineering, and Mathematics (STEM) programs who are below the age of 30. Table 1 depicts the distribution of the filtered data, while Table 2 shows all features deployed in this study, including the target.

Data in Tables 2a, 2b, and 2c depict the 12 descriptive features while the "Employment_Status" was utilized as the class or target. Upon obtaining the filtered STEM data, it was then pre-processed by representing the actual data into unique codes, for example 0 for female and 1 for male (as for Gender).

3.2 Data Sampling

In data mining, before building the classification model, the pre-processed data need to be divided into training and testing subsets. The training subset is the more extensive set compared to the testing as discovering patterns is performed in the training set. This study investigates three data proportions for graduates' employability forecasting, and this includes 70:30, 80:20, and 90:10. Relevant analysis on the outcome of the classification model are then discussed based on these data proportions (Table 3).

3.3 Model Development and Evaluation

Machine learning models for binary-based classification can be built using various algorithms. In this study, four binary-based classification models were developed. And they

Table 1. Number of records based on year.

Year	No. of Graduates
2012	85,623
2013	84,107
2014	78,368
2015	80,822
2016	89,555
2017	95,297
2018	130,749
2019	131,680
2020	114,833
2021	118,030
2022	63,804
Total	1,072,868

are Decision Tree (i.e. J48), Naïve Bayes, Random Forest, and Logistic Regression. A comparison of which model is best suited to predict Malaysia graduate's employability status is performed based on three evaluation metrics: accuracy, precision, and recall. Accuracy provides an overall measure of a model's performance, while precision and recall focus on specific aspects of classification performance, especially in cases of imbalanced datasets or when false positives or false negatives have different implications. The accuracy calculates the ratio of correctly predicted instances to the total number of cases in the dataset, while precision measures the accuracy of optimistic predictions made by a model. It calculates the ratio of accurate positive predictions to all optimistic predictions (true and false positives). On the other hand, Recall quantifies a model's ability to identify all positive instances in the dataset correctly.

Data in Table 4 depicts the parameter setting used by the deployed classifiers: Decision Tree, Random Forest, Naïve Bayes and Linear Regression. The importance of parameter settings in classifiers cannot be overstated, as they directly influence the model's performance, accuracy, generalization ability, and computational efficiency. The proper parameter settings can distinguish between an underperforming model and one that excels at predicting unseen data. Nevertheless, the settings deployed in this study are the default settings commonly used in R studio.

4 Results and Discussion

The model training phase is crucial for building a classification model, while the testing phase is essential for evaluating how well the developed model will perform in practice. Tables 5, 6 and 7 depict the outcome of model testing, where values of three evaluation metrics are presented. On the other hand, Table 8 describes the overall result of the classifiers.

Table 2a. List of features.

Features	Value	Preprocessing Remark
Employment_Status	0, 1	0 = unemployed
		1 = employed
Gender	0, 1	0 = female
		1 = male
Age	0, 1, 2, 3	0 = 18 and below
		1 = 19 to 20
		2 = 21 to 25
		3 = 26 to 30
Ethnic	0, 1, 2, 3, 4, 5	0 = Melayu
		1 = Cina
		2 = India
		3 = Bumiputera Sabah/ Sarawak/ Lain-lain
		4 = Orang Asli
		5 = Lain-lain
Household Income	0, 1, 2, 3, 4, 5	0 = RM1000 and below
		1 = RM1001 – RM1500
		2 = RM1501 – RM2000
		3 = RM2001 – RM3000
		4 = RM3001 – RM5000
		5 = RM5000 and above
Entry Qualification	0,1,2,3,4,5,6,7,8,9	0 = not mentioned
		1 = SPM and equivalent
		2 = STPM and equivalent
		3 = Certificate and equivalent
		4 = Diploma and equivalent
		5 = Foundation
		6 = Matriculation
		7 = Bachelor Degree
		8 = Master Degree
		9 = Others

Based on data in Tables 5, 6 and 7, it is learned that the Random Forest is the best binary-based classifier. While logistic regression and Naive Bayes may find it challenging to predict non-linear relationships between characteristics and the target variable, Random Forest is highly adept. In addition, Random Forests are less susceptible to noise

Table 2b. List of features.

Features	Value	Preprocessing Remark
Types of HLI	0,1,2,3,4,5	0 = Private HLIs
		1 = Public HLIs
		2 = Polytechnic
		3 = Community College
		4 = ILKA
		5 = Vocational College
Field of Study	0,1,2,3,4,5,6,7	0 = Pendidikan
		1 = Sastera & Kemanusiaan
		2 = Sains Sosial, Perniagaan, Pentadbiran & Perundangan
		3 = Sains Semulajadi, Sains Matematik & Komputer, Teknologi Maklumat & Komunikasi
		4 = Kejuruteraan, Pembuatan dan Pembinaan
		5 = Pertanian, Perhutanan, Perikanan & Veterinar
		6 = Kesihatan & Kebajikan
		7 = Perkhidmatan
Features	**Value**	**Preprocessing Remark**
Program Level	0, 1, 2, 3	0 = Certification
		1 = Diploma
		2 = Bachelor Degree
		3 = Master Degree
		4 = PhD
CGPA	0, 1, 2, 3, 4	0 = 2.00 – 2.49
		1 = 2.50 – 2.99
		2 = 3.00 – 3.49
		3 = 3.50 – 3.69
		4 = 3.70 – 4.00
MUET	0,1,2,3,4,5,6	0 = not stated
		1 = Band 1
		2 = Band 2
		3 = Band 3

(*continued*)

Table 2b. (*continued*)

Features	Value	Preprocessing Remark
		4 = Band 4
		5 = Band 5
		6 = Band 6

Table 2c. List of features.

Features	Value	Preprocessing Remark
Bahasa_Melayu Proficiency	0,1,2,3,4,5	0 = not stated
		1 = Not satisfied
		2 = Slightly satisfied
		3 = Neutral
		4 = Satisfied
		5 = Very satisfied
English Proficiency	0,1,2,3,4,5	0 = not stated
		1 = Not satisfied
		2 = Slightly satisfied
		3 = Neutral
		4 = Satisfied
		5 = Very satisfied
ICT skills	0,1,2,3,4,5	0 = not stated
		1 = Not satisfied
		2 = Slightly satisfied
		3 = Neutral
		4 = Satisfied
		5 = Very satisfied

Table 3. Data proportions.

	Training	Testing
70:30	751008	321860
80:20	858294	214574
90:10	965581	107287

Table 4. Parameter settings for deployed classifiers.

Decision Tree	Random Forest	Naïve Bayes	Linear Regression
Induced binary tree	# of trees = 10	var_smoothing = 1e-9	No Regularization
Min. # of instances in leaves = 2	Min. # of instances in leaves = 2	priors = none	
Split subsets > 5	Split subsets > 5		
Max tree depth = 100	Max tree depth = 100		

Table 5. Binary-based classification testing results (70:30).

Classifier	70:30		
	Acc	Prec	Recall
Decision Tree	**0.7500**	0.5226	0.0175
Naïve Bayes	0.6314	0.5491	0.2184
Random Forest	0.7500	0.0894	0.5110
Logistic Regression	**0.7500**	0.5243	0.0458

Table 6. Binary-based classification testing results (80:20).

Classifier	80:20		
	Acc	Prec	Recall
Decision Tree	0.7503	0.5286	0.0180
Naïve Bayes	0.6308	0.5500	0.2190
Random Forest	**0.7504**	0.0874	0.5177
Logistic Regression	0.7499	0.5237	0.0451

Table 7. Binary-based classification testing results (90:10).

Classifier	90:10		
	Acc	Prec	Recall
Decision Tree	0.7508	0.5346	0.0199
Naïve Bayes	0.6277	0.5472	0.2189
Random Forest	**0.7509**	0.0942	0.5217
Logistic Regression	0.7501	0.5285	0.0450

Table 8. Average accuracy between classifiers.

Classifier	Average
Decision Tree	**0.7504**
Naïve Bayes	0.6299
Random Forest	**0.7504**
Logistic Regression	0.7500

and outliers than single-decision trees since they average predictions from several trees. Able to handle high-dimensional data: Random forests outperform decision trees, which may experience overfitting when faced with a high number of characteristics. Nevertheless, on average, the Decision Tree is at par with Random Forest. This can be seen in the data depicted in Table 8.

Even though the accuracy obtained is above 70% (except for Naïve Bayes), there is still room for improvement. Since the dataset under analysis is imbalanced, it has impacted the performance of classifiers, leading to biased models that fail to generalize well to unseen data, particularly in cases of minority class (i.e. unemployed). Out of the 1,072,868 records in the dataset, 75% are employed graduates while only a quarter are unemployed. When one class is overrepresented, many machine learning algorithms favour the majority class, leading to high accuracy but poor performance on the minority class.

In addition to determining the best classifier for binary-based classification of graduates' employability, it is essential to learn which descriptive features impact employability status most. Data in Table 9 shows the correlation between the features and the target (i.e. employability status). Even though most correlations are weak, as indicated by having only a slight linear relationship with the target variable, correlation does not imply causation. Just because two variables are correlated does not mean one causes the other. There might be a third influencing factor. The strongest correlations are with "Types of HLI" (0.1505) (positive) and "Bahasa Melayu Proficiency" (−0.1857) (negative). For the "Types of HLI" (0.1505), it has a moderate positive correlation. Positive correlations suggest that the target variable tends to increase as these features increase. Negative correlations indicate that the target variable decreases as these features increase.

Table 9. Correlation values between descriptive features and target.

Features	Correlation
Gender	0.0687
Age	−0.0535
Ethnic	−0.0358
Household Income	0.045
Entry Qualification	−0.071
Types of HLI	0.1505
Field of Study	−0.0539
Program Level	−0.0996
CGPA	0.0818
MUET	−0.0483
Bahasa Melayu Proficiency	−0.1857
English Proficiency	−0.1652
ICT	−0.1282

5 Conclusion

Graduate employability is crucial to a country's economic, social, and educational development. It contributes to economic growth, social stability, and international competitiveness while allowing graduates to use their education and skills to build meaningful careers. As a result, governments, educational institutions, and employers often work together to enhance graduate employability as a critical driver of national prosperity and well-being. Forecasting graduate employability is essential for ensuring that education and workforce development efforts are aligned with the ever-changing needs of the job market. It helps individuals make informed choices, educational institutions stay relevant, employers find the needed talent, and governments create policies that foster economic growth and stability. Nevertheless, graduate employability is also influenced by broader economic conditions. Economic growth, industry trends, and market demand for specific skills can impact employment opportunities for graduates. In addition, networking and soft skills, such as communication and problem-solving abilities, can be crucial in employability. These are among the factors to be considered in forecasting graduates' employability. Besides utilising a single classifier, future work can also be put into deploying hybrid models in forecasting employability. On top of that, there is also a need to investigate trend analysis and temporal patterns in predicting graduates' employability. As the study involves population data, innovative data handling would be much more helpful before deploying machine learning classifiers.

Acknowledgement. The Malaysia Ministry of Higher Education has financially supported this study under the grant scheme P58/614. It was also supported by the Data Management and Software Solutions Research Lab at the School of Computing, Universiti Utara Malaysia.

References

1. Kementerian Pengajian Tinggi: Laporan Kajian Pengesanan Graduan 2021 (2022)
2. Kementerian Pengajian Tinggi: Laporan Kajian Pengesanan Graduan 2020 (2021)
3. Jayasingha, D.G.M.S., Suraweera, S.M.B.L.: An analysis of the factors affecting the graduates' employability in case of Rajarata University of Sri Lanka. IRE Journals. 3 (2020)
4. Jiang, L., Chen, Z., Lei, C.: Current college graduates' employability factors based on university graduates in Shaanxi Province, China. Front Psychol. 13 (2023)
5. Clarke, M.: Rethinking graduate employability: the role of capital, individual attributes and context. Studies in Higher Education. 43 (2018)
6. Blanco, M., Bares, L., Hrynevych, O.: University brand as a key factor of graduates employment. Marketing and Management of Innovations (2019)
7. Othman, Z., Shan, S.W., Yusoff, I., Kee, C.P.: Classification techniques for predicting graduate employability. Int. J. Adv. Sci. Eng. Info. Technol. 8 (2018)
8. Piad, K.C.: Determining the dominant attributes of information technology graduates employability prediction using data mining classification techniques. J. Theoret. Appl. Info. Technol. 96 (2018)
9. ElSharkawy, G., Helmy, Y., Yehia, E.: Employability prediction of information technology graduates using machine learning algorithms. Int. J. Adv. Comp. Sci. Appl. 13 (2022)
10. Karim, Z.I.A., Maat, S.M.: Employability skills model for engineering technology students. J. Techn. Edu. Train. 11 (2019)
11. Maaliw, R.R., et al.: Employability prediction of engineering graduates using ensemble classification modelling. In: 2022 IEEE 12th Annual Computing and Communication Workshop and Conference, CCWC 2022 (2022)
12. Vinutha, K., Yogisha, H.K.: Prediction of employability of engineering graduates using machine learning techniques. In: Proceedings of the 2021 8th International Conference on Computing for Sustainable Global Development, INDIACom 2021 (2021)
13. Panityakul, T., Suriyaamorn, W., Chinram, R., Kongchouy, N.: Classification models for employability of statistics and related fields graduates from Thailand universities. Int. J. Math. Comp. Sci. 17 (2022)

Human-Centric Approach to Enhancing Charisma Detection in Speech Through Adaptive Deep Learning Model

Lee Qi Zun, Mohamad Sabri Sinal Zainal(✉), and Yuhanis Yusof

Universiti Utara Malaysia, Sintok, 06010 Kedah, Malaysia
msabri@uum.edu.my

Abstract. Charisma detection in human speech is challenging but significant for improving human-computer interactions. Traditional methods of charisma assessment are often subjective and inconsistent. Many existing studies on charisma detection using computational approaches focus on numerous features, leading to performance and model evaluation complexity. This study introduces a novel solution by employing a deep learning-based approach to identify charismatic speech in humans objectively. The proposed model leverages Long Short-Term Memory (LSTM) networks to analyze audio features contributing to charisma detection. Using 32 self-labelled speech samples, the model was built and trained to identify these features. Precision, recall, and F1 score metrics validated the model's effectiveness and efficiency. Achieving an overall accuracy of 87.5%, the results indicate that the model reliably identifies charismatic speech, successfully fulfilling its intended purpose. This study contributes to developing an objective method for charisma detection, offering valuable applications in various domains requiring charisma assessment.

Keywords: Charisma Detection · Adaptive Modeling · Artificial Intelligence

1 Introduction

In today's digital age, where human-computer interaction and daily communication intertwine, charisma has emerged as a crucial factor in shaping leadership personality traits [1]. This elusive quality, characterized by a distinctive fusion of compelling language, emotive expression, and persuasive rhetoric, wields considerable influence across various spheres, including oratory, leadership, and the entertainment industry. Studies indicate that effective leadership hinges not only on vision but also on charismatic appeal [2]. Charismatic individuals are frequently seen as more persuasive, inspiring, and influential, enhancing personal and professional opportunities. Moreover, charisma can remarkably effect dramatic transformations in individuals, communities, and even populations, whether consciously exercised or unwittingly practiced [3].

Generally, evaluating charisma has always been subjective and heavily reliant on personal impressions and judgments. Although it seems easy for us to identify charismatic

Z. Mat Aji et al. (Eds.): KMICe 2024, ISEM 49, pp. 514–526, 2025.
https://doi.org/10.1007/978-3-031-91485-0_40

and non-charismatic individuals, it is unclear what it is about an individual influencing our judgement [4]. This inherent subjectivity generates obstacles in achieving consistent and impartial assessments. However, recent advancements in data analysis and artificial intelligence have paved the way for innovative approaches to charisma detection [5]. Detection models are a powerful subset of data analysis techniques designed to identify patterns, anomalies, or specific characteristics within datasets. These models helped analyze various dimensions of verbal communication, such as tone, pitch, cadence, and word choice, to evaluate the presence and strength of charismatic traits when applied to speech [4].

Existing studies on charisma detection can be broadly categorized into three: audio-based [6] and text-based [7]. Audio-based detection often faces challenges related to the audio input quality, which can significantly influence the detection of charismatic traits [8]. Poor audio quality can affect the analysis of tone, pitch, and cadence, compromising the integrity of the detection model. On the other hand, text-based detection requires large datasets to analyze linguistic features of charisma. This approach faces difficulties in data collecting and the variability of language used among individuals with different linguistic backgrounds and levels of mastery [7]; [9].

A novel Adaptive Deep Learning-based Charisma Detection System (ADLCDS) is proposed to address the limitations of existing approaches. This model uses LSTM networks to analyze key features contributing to charisma detection, such as speech melody, fluency, and loudness. The proposed model is designed to adaptively manage the dynamic nature of audio inputs and utilize the proposed model to detect charisma in speech with high accuracy. The model was built and trained with 32 self-labelled speech samples to identify these features. Then, the performance of the proposed model was tested and validated through precision, recall, and F1 score metrics. The remainder of this paper is organized as follows: Literature Review, which provides an overview of existing research on charisma and its assessment; Methodology, which details the proposed model; Results and Discussion, which presents the findings based on the model's performance; and Conclusion, which discusses further improvements and potential directions for future research.

2 Literature Review

Interest in detecting charisma has undoubtedly grown over time. However, there's still a notable lack of research focused on accurately identifying charisma using a single type of data, especially in a way that adapts to the data's nature. This literature review highlights the unique contribution of this paper, which aims to address this gap through the development of the ADLCDS. The review is organized into two main sections: the first looks at emotion detection techniques that use machine learning, and the second explores the traits that characterize a charismatic speaker.

2.1 Emotion Detection Techniques

The literature on emotion detection was reviewed to draw relevant parallels due to the shortage of studies directly related to charisma detection. Speech Emotion Recognition

(SER) has garnered significant research interest, either as a Machine Learning (ML) problem or in the domain of affective computing. This field's increasing potential, advancements in models, and real-world applications have been explored in depth. In emotion detection, the para-linguistic information that can be quantitatively represented involves features such as pitch, intensity, and Mel-Frequency Cepstral Coefficients (MFCC) [10]. Speech data are then digitized using signal processing methods, and speech features are obtained. These features vary depending on the emotions such as sadness, fear, anger, happiness, boredom, confusion, etc. [11]. In this case, SER is typically achieved through three core steps: data processing, feature selection/extraction, and classification based on emotional features. On top of that, feature selection must be applied to seek the most relevant feature subsets that contribute to the main topic: emotion recognition [12].

Researchers have recently employed various ML and Deep Learning (DL) techniques in SER, but some weaknesses exist, especially in capturing the techniques and methods that effectively facilitate these steps [10]. Multiple deep learning paradigms were used and tested throughout all the related works, which concentrated on speech-emotion recognition. For example, the Recurrent Neural Network (RNN) classifier, Multivariate Linear Regression (MLR), and Support Vector Deep (SVM) are a few of the most popular deep learning models to be utilized in classifying speaker emotions [12]. These deep learning models were then tested and evaluated using Various datasets such as IEMOCAP, Emo-DB, and RAVDESS to examine the accuracy of different models in classifying the emotion [13]. Compared to studies related to speech emotion recognition, there is no prepared dataset or databases available on the Internet for this study, which focused on determining and assessing speaker charisma.

2.2 Features for Charismatic Communication

The review further explores several features that are relevant to charismatic communication. The quality of speech transmission has an impact on the perception of charisma, which proves that the phonetics aspect of a speech plays a significant role in determining whether a speech is charismatic or not [14]. An important example was the analysis of Steve Jobs, a former Apple CEO renowned for his charismatic speaking style. The first detailed acoustic profile of Jobs' speech was during two of his most famous product presentations, during the iPhone 4 and iPad 2 launches [15]. The study suggested that melody, loudness, tempo, and fluency are the specific acoustic features that were instrumental in defining a charismatic speaker's impact [15]. The study aimed to provide actionable advice for improving one's charismatic presence by analyzing these features. It highlighted how technological advancements could further bridge the gap between behavioural characteristics and computational analysis of human speech [15].

On top of that, research on speech melody offered valuable insights into the acoustic features of charismatic communication. One study has proposed a musical approach to speech melody that utilized musical notation to represent pitch contours and intervallic variations [16]. This approach addressed limitations in traditional phonetic and phonological methods by offering a more comprehensive representation of pitch dynamics across entire sentences. The study has successfully quantified pitch changes associated with different sentence types by converting fundamental-frequency trajectories from Hertz to semitones and averaging them across speakers. This method demonstrated that speech prosody could be analyzed through a musical framework, which combined speech rhythm and melody into a unified representation of prosody [16]. The study's observations on the atonal nature of speech and the potential for using musical scores to represent pitch dynamics provided additional tools for analyzing charismatic communication.

Table 1. Speech emotion recognition vs charisma detection.

Domain	Dataset	Related studies	Modality	Categories
SER	IEMOCAP	Niu et al.[17], Pandharipande et al.[18]	Audio-visual	Anger, happiness, excitement, sadness, frustration, fear, surprise, neutral and other
SER	SAVEE	Yogesh et al.[19], Sivanagaraja et al. [20]	Audio-visual	Angry, disgust, sad, fear, happy, surprise and neutral
SER	RAVDESS	Darekar and Dhande [21]	Audio-visual	Happy, sad, angry, fearful, surprise, disgust, calm, and neutral
SER	Emo-DB	Niu et al.[17], Pandharipande et al.[18]	Speech	Angry, boredom, disgusting, fear, happy, sad and neutral
SER	CASIA	Wen et al. [22], Xie et al. [23]	Speech	Anger, fear, happy, neutral, sad and surprise
Charisma detection	-	-	Audio	Non-charismatic, neutral, charismatic

In summary, the literature review has highlighted the significant research gaps in detecting charisma and emotion through Table 1, establishing the novel nature of the ADLCDS. This review laid the groundwork for developing a deep learning-based approach to analyze and assess charisma by examining current emotion detection methods and exploring features relevant to charismatic communication. Finally, the design of the ADLCDS framework is built through the insights gained from emotion recognition technologies and the study of critical charismatic traits, such as melody, loudness, tempo, and fluency, to determine the three target categories, which include Non-Charismatic, Neutral, and Charismatic. The following section will detail the methodology employed to develop the ADLCDS.

3 Methodology

In this section, the proposed model of the ADLCDS is elaborated in detail. The model has several vital layers, including audio preprocessing, feature extraction, model training, and testing. Each of the layers contributes to the different functionalities of the system.

3.1 Proposed Models for Charisma Detection

Figure 1 presents a detailed flowchart illustrating the sequential layers of the proposed model. The layers include audio preprocessing, feature extraction, training, testing using a Long Short-Term Memory (LSTM) network, and classification into different charismatic categories. The details for each section are elaborated in the subsections below.

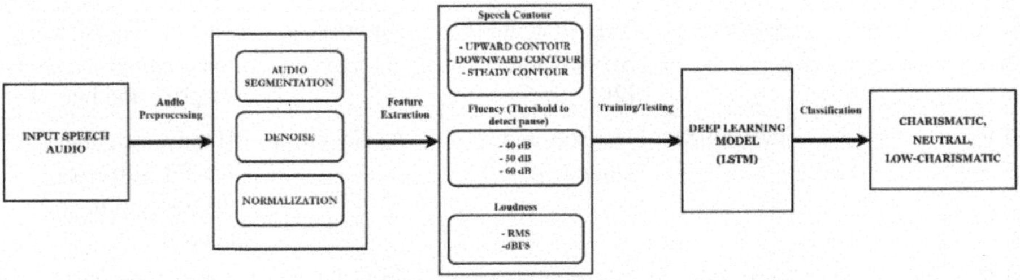

Fig. 1. Flowchart of ADLCDS.

Audio Preprocessing. Audio preprocessing was divided into three main steps. First, the raw audio files were segmented into 30-s intervals to ensure uniformity in the data input length and help in subsequent consistent processing. Second, background noise was removed from the audio segments to enhance the clarity of the speech data. Finally, the volume levels of the audio segments were normalized to ensure consistency across all samples.

Feature Extraction. This stage involves extracting attributes such as fluency and speech melody and applying transformations to optimize these features for the model. The specific features considered for this proposed model are detailed below.

Speech Contour. Speech contour refers to the melodic pattern of speech, capturing the variations in pitch across time [13]. This feature is essential for understanding the intonational characteristics of a speaker, which significantly influence the perception of charisma [16]. It helps recognize familiar melodies usually conveyed via pitch [24]. The speech contour is strongly related to charisma since the pitch variations can represent a speaker's emotion, emphasis, and engagement.

Fluency. Speech fluency is crucial in charismatic communication, representing a temporal phenomenon and spoken language competence [25]. In analyzing charisma, fluency is assessed by examining the presence and length of silent pauses in the speech. One research on Second Language (L2) acquisition [26] has extensively analyzed acoustic measures of fluency, focusing mainly on the role of silent pauses.

Loudness. Loudness is a fundamental acoustic feature in speech that significantly impacts the perception of a speaker's charisma [15]. It refers to the perceived intensity or volume of the speaker's voice, influencing audience engagement. In charismatic communication, a speaker's ability to effectively modulate loudness enhances their presence, commands attention, and conveys authority [27]. In the ADLCDS, loudness is measured using Root Mean Square (RMS) Loudness and dBFS Loudness [28]. RMS loudness indicates the average power of the audio signal, while dBFS (Decibels relative to Full Scale) measures loudness on a logarithmic scale relative to the maximum possible level. Both RMS and dBFS loudness are calculated and recorded for each audio segment.

Training/Testing. This section outlined selecting, training, and optimizing the deep learning model used in the ADLCDS. Each process is described in detail below.

Model Selection. The extracted features during feature extraction were compiled into a dataset consisting of 32 self-labelled speech audios from diverse speakers. This dataset was then trained and tested with different machine learning paradigms, including SVM, Random Forest, Decision Tree, and LSTM networks. The models are being trained with a ratio of 7:3 and split randomly to ensure the competency of these initial models.

Feature Engineering. Eight features were extracted from the speech audio during feature extraction, which includes upward contour, steady contour, downward contour, pause detection at −40 dB, −50 dB, −60 dB, RMS loudness, and dBFS loudness. The process of identifying the most relevant features for accurately detecting speech-based charisma is presented in detail in Table 4. This involves feature engineering, which encompasses raw data selection, manipulation, and transformation to enhance the model's performance in charisma detection.

Model Tuning and Optimization. This process involved experimenting with different training and testing sizes and varying the number of epochs to identify the model's most influential parameters to improve the accuracy of the proposed model.

4 Results and Discussion

This section provided a comprehensive overview of the ADLCDS's construction and its functionality in classifying speech samples into various charisma categories using the proposed input. Detailed information on the dataset and evaluation strategy is presented in Sects. 4.1 and 4.2. Section 4.3 elaborates on the development of the adaptive model and presents the results from testing the proposed model with the specified input.

4.1 Dataset for Evaluation

This section discusses the dataset used to evaluate the proposed model. Table 2 presents the total size of the labelled datasets, which includes 32 audio files from diverse speakers categorized into Low-Charismatic, Neutral, and Charismatic. Several steps were taken to determine the target category for each audio sample. First, a list of charismatic individuals was compiled from articles featuring notable speakers such as Steve Jobs, Barack Obama, and Donald Trump, frequently cited in the literature on charismatic communication

[15][29]. Next, the relevant literature on the characteristics of non-charismatic, neutral, and charismatic individuals was reviewed. According to Stephanie Berger [30], the charismatic effect arises from distinct elements, including message content and delivery. Since the ADLCDS focuses on audio-based charisma detection, phonetic features such as pitch range and speech rate are considered for categorization. The audio quality for each data bitrate is set at 1536 kbps, lasting 10 min.

Table 2. The sample size for labelled datasets.

Category	Sample size
Low charismatic	6
Neutral	9
Charismatic	17
Total	**32**

4.2 Evaluation Strategy

Four performance metrics were used to evaluate the effectiveness of the proposed model: precision, recall, F1-score, and accuracy. Precision indicates the proportion of accurate positive predictions compared to all optimistic predictions made by the model. At the same time, recall measures the proportion of true positives among the total actual positives, reflecting the model's ability to identify relevant instances. The F1-score is the harmonic mean of precision and recall, providing a balanced assessment of overall effectiveness. Accuracy measures the proportion of correctly classified instances among all evaluations. These metrics support a comprehensive assessment of the model's performance across various charisma categories, as detailed below:

$$Precision = \frac{TP}{TP + FP} \tag{1}$$

$$Recall = \frac{TP}{TP + FN} \tag{2}$$

$$F1\,Score = 2 \times \frac{Precision \times Recall}{Precision + Recall} \tag{3}$$

$$Accuracy = \frac{TP + TN}{TP + TN + FP + FN} \tag{4}$$

where

True Positive (TP): The number of correctly predicted positive instances.
True Negative (TN): The number of correctly predicted negative instances.
False Positive (FP): The number of incorrectly predicted positive instances.
False Negative (FN): The number of incorrectly predicted negative instances.

4.3 Discussion

This section presents the results and discussion of the ADLCDS. It includes an analysis of the initial performances of various deep learning models, specifically focusing on the chosen LSTM model regarding training accuracy and model loss. Furthermore, the overall effectiveness of the ADLCDS in detecting speech charisma is evaluated, supported by detailed performance metrics and evaluation results.

Performance Evaluation. Initially, the accuracy of various machine learning models, including SVM, Random Forest, Decision Tree, and LSTM networks, was compared using the proposed input. The LSTM networks were selected for charisma detection in speech because they effectively process sequential data and capture long-term dependencies in speech patterns. As shown in Table 3, the results indicated that the SVM, Random Forest, and Decision Tree models each achieved an initial accuracy of 0.5. In contrast, the LSTM model demonstrated a higher initial accuracy of 0.65625, suggesting it performed better during the initial testing. This finding supports the conclusion that the LSTM model is the most effective for this study. As a result, the LSTM model was chosen for further tuning and optimization to improve accuracy in assessing charisma.

Table 3. Comparison of initial accuracy for initial models.

Models	Technique	Accuracy
Machine learning	SVM	0.5
	Random forest	0.5
	Decision tree	0.5
Deep learning	LSTM	0.65625

Next, the accuracy and performance of the LSTM model were analyzed with different parameters, such as the number of epochs and the training/testing size ratio. Table 4 illustrates the model's performance under various configurations. The optimal configuration was achieved after several trials using a reduced feature set, a specific number of epochs, and an appropriate training/testing size ratio. This configuration showed consistent improvement in model performance and reliability. The comparison with other parameters highlighted that these settings were optimal for achieving the best accuracy and minimizing overfitting. In this model, using a lower number of epochs with a higher testing size proved beneficial because it helped prevent overfitting, which is the strategy of early-stopping [31]. It can ensure that the model is generalized better to future unseen data. Fewer epochs in this model pointed out that the model was trained for a shorter period, reducing the risk of becoming too tailored to the training data. A larger testing size provided a more accurate evaluation of the model's performance. This approach balanced training and testing, enhancing the model's ability to perform well on new data.

Table 4 shows that the ADLCDS model was evaluated using the original feature set of eight features and a reduced set of five features. The eight features included upward

Table 4. Model performance with varying parameters.

Features	Epoch	Testing size	Accuracy
5	5	0.4	0.6923
	5	0.35	0.5833
	5	**0.3**	**0.9**
	5	0.25	0.625
	5	0.2	0.7143
	5	0.15	0.8
	5	0.1	0.75
	10	0.3	0.8
	50	0.3	0.5
	100	0.3	0.6
	500	0.3	0.7
8	5	0.4	0.4615
	5	0.35	0.5
	5	0.3	0.5
	5	0.25	0.625
	5	0.2	0.5714
	5	0.15	0.8
	5	0.1	0.75
	10	0.15	0.6
	50	0.15	0.8
	100	0.15	0.4
	500	0.15	0.6–0.8

contour, steady contour, downward contour, pause detection at −40 dB, −50 dB, −60 dB, RMS loudness, and dBFS loudness, while the reduced set included downward contour, steady contour, dBFS loudness, and pause detection at −40 dB and −50 dB.

Besides, Table 4 indicates that the model's accuracy varied significantly based on the features and testing size. With the reduced feature set, the model achieved a peak accuracy of 0.9 when trained for five epochs with a testing size of 0.3, showing consistent performance across epochs ranging from 0.8 to 0.9. In contrast, the original eight features yielded lower accuracy, peaking at 0.8 with a testing size of 0.15 and a higher number of epochs. Although accuracy improved with more epochs, it did not reach the levels achieved with the reduced feature set. A smaller testing size also limited the model's ability to generalize, as seen with an accuracy of 0.4615 when using five epochs and a testing size of 0.4, suggesting that the additional features did not positively contribute to performance. The analysis revealed that the reduced feature set significantly outperformed the original's accuracy. The optimal configuration comprised the five key features and a testing size of 0.3, providing the best balance between model complexity and performance for the ADLCDS.

Subsequently, the analysis of the proposed model's accuracy and loss metrics based on the training phase is presented and discussed. Figure 2 illustrates the model's accuracy and loss during the training phase of the LSTM model.

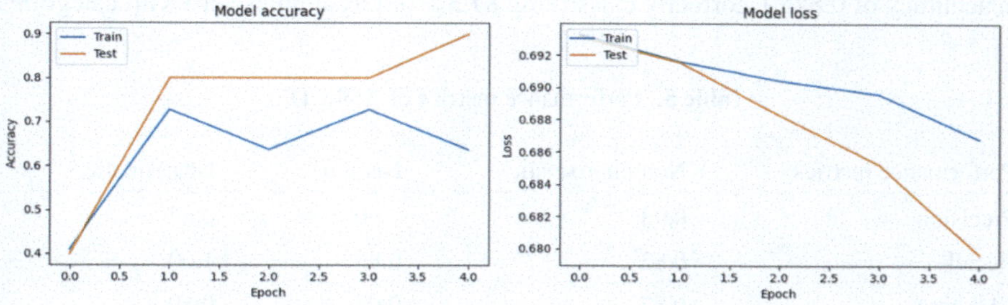

Fig. 2. Model accuracy and model loss.

The ADLCDS achieved an accuracy of 0.9 within just five epochs, indicating the effective classification of speech samples based on the selected features. At the same time, the model showed a consistent decrease in loss during training, reflecting its ability to learn and refine its parameters. This high accuracy with limited epochs and loss reduction demonstrates the model's efficiency and successful optimization.

Results are presented in a confusion matrix along with four performance metrics to evaluate overall performance. Figure 3 displays the confusion matrix for the ADLCDS, detailing the model's classification of speech samples into three categories: Low Charismatic, Neutral, and Charismatic. Each cell in the matrix represents the number of samples for specific combinations of actual and predicted classifications.

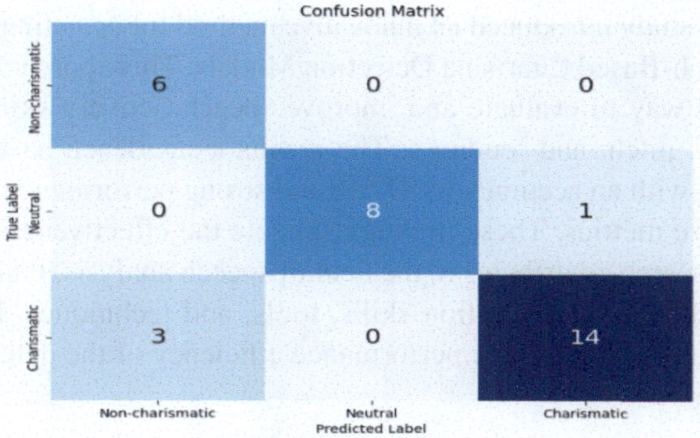

Fig. 3. Confusion matrix for ADLCDS.

Due to the imbalanced dataset, evaluation metrics beyond accuracy were used, including precision, recall, and the F1-score, to assess the model's performance. This approach helped reduce bias toward the majority class. Table 5 shows that for Non-Charismatic samples, the model achieved a precision of 0.9333 and a recall of 0.823, resulting in an F1-score of 0.8750, indicating predictive solid accuracy. Neutral samples had a perfect precision of 1.0000, a recall of 0.8889, and an F1-score of 0.9412, showing excellent classification performance. In the Charismatic category, the precision was 0.6667, the recall was 1.0000, and the F1-score was 0.8000, suggesting effective identification but indicating room for improved precision. Overall, the model achieved an accuracy of 0.8750, correctly classifying 87.5% of the samples across all categories.

Table 5. Performance metrics of ADLCDS.

Performance metrics	Non-charismatic	Neutral	Charismatic
Precision	0.93	1.00	0.67
Recall	0.82	0.89	1.00
F1-Score	0.87	0.94	0.80
Accuracy	0.88	0.88	0.88

Although deep learning models typically require large datasets, the dataset used in this study is carefully curated and processed for quality. Raw audio files were segmented into 30-s intervals, with background noise removal and volume normalization to maintain consistency. These preprocessing steps improve the clarity of the extracted features, allowing the model to focus on crucial voice characteristics. Data augmentation and transfer learning techniques also help the model learn effectively from this dataset and achieve high accuracy despite the smaller sample size.

5 Conclusion

In conclusion, this study introduced an innovative method for detecting speech charisma through new Speech-Based Charisma Detection Models. This approach offers a distinctive and advanced way to evaluate and improve speech delivery skills by objectively measuring fluency, pitch, and loudness. The results as evidence have proven that the model is effective, with an accuracy of 87.5% and strong performance across precision, recall, and F1-score metrics. These findings validate the effectiveness of the proposed model, and this research contributes to the field of speech analysis. It also has the potential to enhance future communication skills, tools, and techniques. Moving forward, efforts should focus on refining the performance efficiency of the detection model.

References

1. Denrich, D.: Charismatic leadership and vulnerability: a comprehensive study of cult dynamics. https://digitalcommons.ursinus.edu/cgi/viewcontent.cgi?article=1010&context=psych_pres. Last accessed: 1 May 2024

2. Khatri, N., Templer, K.J., Budhwar, P.S.: Great (transformational) leadership = charisma + vision. South Asian Journal of Global Business Research **1**, 38–62 (2012)
3. Bhide, A.: Charisma: the phenomenon and its psychology: a mental health perspective. Indian Journal of Psychiatry **50**, 274 (2008)
4. Rosenberg, A., Hirschberg, J.: Charisma perception from text and speech. Speech Commun. **51**, 640–655 (2009)
5. Feng, X., Zhang, S., Liu, X., Srinivasan, K., Lamberton, C.P.: An AI method to score celebrity visual potential from human faces. SSRN Electronic Journal (2021)
6. Yang, Z., et al.: What makes a speaker charismatic? producing and perceiving charismatic speech. In: Proceeding10th International Conference on Speech Prosody, pp. 685–689 (2020)
7. Mohamaddoust, R., Mohammadzadeh, J., Khalilian, M., Nikravanshalmani, A.: Measuring and analyzing charisma on Twitter using combination weighting and TOPSIS method. Int. J. Nonlinear Analysis and Appl. **12**, 1143–1158 (2021)
8. Batliner, A., Möbius, B.: Prosody in automatic speech processing, https://opus.bibliothek.uni-augsburg.de/opus4/frontdoor/deliver/index/docId/71287/file/Batliner_71287.pdf. Last accessed: 1 May 2024
9. Amiriparian, S., et al.: Synchronization in interpersonal speech. Frontiers in Robotics and AI 6 (2019)
10. Madanian, S., et al.: Speech emotion recognition using machine learning — A systematic review. Intelligent Systems with Applications **20**, 200266 (2023)
11. Ülgen Sönmez, Y., Varol, A.: In-depth investigation of speech emotion recognition studies from past to present – The importance of emotion recognition from speech signal for AI. Intelligent Systems with Applications **22**, 200351 (2024)
12. Kerkeni, L., et al.: Automatic speech emotion recognition using machine learning. In: Social Media and Machine Learning, IntechOpen (2019)
13. Rathi, T., Tripathy, M.: Analyzing the influence of different speech data corpora and speech features on speech emotion recognition: A review. Speech Commun. **162**, 103102 (2024)
14. Schuller, B.W., et al.: Computational charisma—A brick by brick blueprint for building charismatic artificial intelligence. Frontiers in Computer Science 5 (2023)
15. Niebuhr, O., Voße, J., Brem, A.: What makes a charismatic speaker? a computer-based acoustic-prosodic analysis of Steve Jobs' tone of voice. Comput. Hum. Behav. **64**, 366–382 (2016)
16. Chow, I., Brown, S.: A musical approach to speech melody. Front. Psychol. **9**, 247 (2018)
17. Niu, Y., Zou, D., Niu, Y., He, Z., Tan, H.: A breakthrough in speech emotion recognition using deep retinal convolution neural networks, https://arxiv.org/pdf/1707.09917. Last accessed: 1 May 2024
18. Pandharipande, M., Chakraborty, R., Panda, A., Kopparapu, S.K.: An Unsupervised Frame Selection Technique for Robust Emotion Recognition in Noisy Speech. In: 26th European Signal Processing Conference (EUSIPCO), pp. 2055–2059 (2018)
19. Yogesh, C.K., et al.: A new hybrid PSO assisted biogeography-based optimization for emotion and stress recognition from the speech signal. Expert Syst. Appl. **69**, 149–158 (2017)
20. Tatinati, S., Mun Kit, H., Wai Hoong Khong, A., Yubo, W.: End-to-end speech emotion recognition using multi-scale convolution networks. In: Asia-Pacific Signal and Information Processing Association Annual Summit and Conference (APSIPA ASC), pp. 189–192 (2017)
21. Darekar, R.V., Dhande, A.P.: Emotion recognition from Marathi speech database using adaptive artificial neural network. Biologically Inspired Cognitive Architectures **23**, 35–42 (2018)
22. Wen, G., Li, H., Huang, J., Li, D., Xun, E.: Random deep belief networks for recognizing emotions from speech signals. Comput. Intell. Neurosci. **2017**, 1–9 (2017)
23. Xie, Y., et al.: Speech emotion classification using attention-based LSTM. In: IEEE/ACM Transactions on Audio, Speech, and Language Processing, pp. 1675–1685 (2019)

24. McDermott, J.H., Lehr, A.J., Oxenham, A.J.: Is relative pitch specific to pitch? Psychol. Sci. **19**, 1263–1271 (2008)
25. Kormos, J., Dénes, M.: Exploring measures and perceptions of fluency in the speech of second language learners. System **32**, 145–164 (2004)
26. De Jong, N.H., Bosker, H.R.: Choosing a threshold for silent pauses to measure second language fluency. In: Eklund, R. (ed.) The 6th Workshop on Disfluency in Spontaneous Speech (DiSS), pp. 17–20. Universitetsservice US-AB, Stockholm, Sweden (2013)
27. Seshadri, G., Yegnanarayana, B.: Perceived loudness of speech based on the characteristics of the glottal excitation source. J. Acoust. Soc. Am. **126**, 2061–2071 (2009)
28. Wendl, M., Lee, H.: The effect of dynamic range compression on loudness and quality perception about crest factor. J. Audio Eng. Soc. 86–93 (2014)
29. McMahon, T.P.: Comparing charismatic leaders' communication styles. Taylor & Francis (2024)
30. Berger, S., Niebuhr, O., Peters, B.: Winning over an audience - a perception-based analysis of prosodic features of charismatic speech. In: Jahrestagung für Akustik, pp. 1454–1457 (2017)
31. Ying, X.: An overview of overfitting and its solutions. J. Phys: Conf. Ser. **1168**, 022022 (2019)

Leveraging Elitist Genetic Algorithms Strategy in Automated Test Case Generation for Improving Convergence Rate

Nor Iadah Yusop(✉) and Li Kai

School of Computing, Universiti Utara Malaysia, 06010 Sintok, Kedah, Malaysia
noriadah@uum.edu.my

Abstract. Efficient test case generation is essential for ensuring software quality. Genetic algorithms offer a faster alternative to manual methods for creating test cases. However, traditional genetic algorithms can be slow due to crossover and mutation probabilities. This paper presents an improved genetic algorithm framework to address these limitations and accelerate test case generation. An elite genetic algorithm with an elitist retention strategy is proposed. A pseudocode is provided to illustrate the algorithm's steps. Experimental results show that the improved algorithm outperforms traditional and random methods in terms of speed and iteration count, offering valuable insights for future software testing research. The demonstrated superiority of elite genetic algorithms in this new approach, combined with the shared knowledge of automation tools and scripting languages, has the potential to significantly advance software testing methodologies and contribute to overall enhancements in software quality.

Keywords: Elite Genetic Algorithm · Automatic Test Case Generation · Convergence Rate Improvement

1 Introduction

Genetic algorithms have proven to be effective and efficient in automating test case generation, significantly reducing development time compared to manual methods [1]. Their ability to create and solve objective functions without extensive prior knowledge makes them versatile for various applications [2]. However, traditional genetic algorithms can suffer from slow convergence, which is often attributed to crossover and mutation probabilities ([3, 4]). High crossover rates can diminish genetic diversity, hindering the search for optimal solutions and slowing convergence. This has contributed to the limited adoption of automatic test case generation technologies among test engineers [5]. This study seeks to address the slow convergence issue in genetic algorithms for test case generation, aiming to improve their speed and resource efficiency, thereby increasing their acceptance and practical use.

This paper presents the findings from a study that investigates the challenge of slow convergence rates in traditional genetic algorithms for test case generation. To address

Z. Mat Aji et al. (Eds.): KMICe 2024, ISEM 49, pp. 527–541, 2025.
https://doi.org/10.1007/978-3-031-91485-0_41

this issue, an enhanced genetic algorithm is developed and implemented, aiming to accelerate convergence while preserving test case quality. Through comparative analysis with traditional and random algorithms, the effectiveness of this optimized approach is demonstrated, highlighting its practical value and potential for broader application.

A systematic approach is adopted to enhance the convergence rate of genetic algorithms in test case generation. This involves the design of an Elite Genetic Algorithm (EGA) that incorporates an elitist retention strategy to safeguard the most promising solutions across generations. The performance of the EGA is rigorously evaluated through experiments, comparing it to traditional genetic and random algorithms. Key metrics such as convergence speed and test case quality are assessed to provide empirical evidence of the EGA's improvements.

The paper is organized as follows: Sect. 2 provides a literature review that establishes theoretical foundations and highlights existing challenges in genetic algorithms for test case generation. Section 3 presents the design and implementation of the EGA. Section 4 describes the experimental setup, methodology, and results of the comparative analysis. Lastly, Sect. 5 discusses the implications of the findings, draws conclusions, and outlines potential avenues for future research in automatic test case generation.

2 Literature Review

This section covers key concepts in test cases, including their classification, generation methods, and writing standards, such as black-box and white-box classifications. It also explains genetic algorithm basics, reviews their development and applications, addresses challenges in using genetic algorithms for test case generation, and suggests possible solutions.

2.1 Test Case

Test cases are essential for software testing, encompassing input data, expected output, preconditions, postconditions, and specific instructions for testers. Effective test cases enhances testing efficiency and reduces costs. Test case generation primarily involves two approaches: functional and structural testing. Functional testing generates cases based on requirements, while structural testing examines the program's internal logic. Structural testing includes branch, statement, and path coverage. Manual test case generation is time-consuming, inefficient, and prone to errors. Automatic test case generation offers a more efficient and effective solution.

White-box test cases, based on a program's internal logic, focus on error-prone areas. These test cases assess logical paths by comparing actual and expected states, emphasizing program logic coverage. Structural test case generation methods include static and dynamic techniques. Static methods, like symbolic execution and interval analysis, examine the program without running it. Dynamic methods involve running the program with selected inputs to test specific paths. Techniques such as random testing, Korel's method, iterative relaxation, and exploration methods are used to drive program execution and test various paths [6].

The random method generates test data randomly. Korel's method randomly selects data and incrementally drives program execution. The iterative relaxation method is similar to Korel's method. The trial-and-error method involves running a program with initial input values and evolving them based on results. Exploration methods include genetic algorithms, simulated annealing, ant colony optimization, and particle swarm optimization.

Recent research on automatic test case generation has focused on optimizing scaling functions through intelligent algorithms. Integrating intelligent algorithms with software testing is crucial for addressing practical challenges. However, not all intelligent algorithms consistently produce optimal results for test case generation. As software systems become more complex, generating test data remains a challenging task. Genetic algorithms are particularly effective for handling unpredictable and stochastic search problems.

2.2 Genetic Algorithms

Given the increasing scale and complexity of software systems, traditional testing methods have become increasingly time-consuming, particularly in test case design. To address this, researchers have explored metaheuristic search algorithms, such as genetic algorithms, to generate optimal test cases. This section will discuss the principles, advantages, and limitations of genetic algorithms, along with their application in test case generation.

Genetic algorithms are computational models inspired by biological evolution. They select the fittest individuals from a population and apply genetic operations to evolve superior solutions over generations. These algorithms have been used in various fields, including automatic test case generation.

Before proceeding, it is essential to understand the key terms and concepts. Chromosomes, the carriers of genetic information, are represented as encodings of solutions in genetic algorithms. Genes, the fundamental units of heredity, are depicted as elements within these solutions. In the computational domain, a genotype represents a population as a solution understandable by a computational system, while a phenotype represents a population as real-world solutions. Similar to biological processes, genetic algorithms map decodings to specific mathematical model vectors (encoding) and vice versa (decoding). An individual, a complete solution, is expressed dominantly by chromosomes. Moreover, a population, a collection of individuals, signifies a collection of solutions in genetic algorithms. Fitness, a measure of an organism's suitability, is used to evaluate the quality of individuals. Selection, based on fitness, implements the principle of survival of the fittest [7]. Recombination and hybridization of genes, known as crossover, involves the exchange of genes between two parent individuals to create a new individual [8]. Mutation, the creation of new chromosomes during copying, is represented in genetic algorithms as the alteration of a gene within an individual, leading to increased population diversity [7].

2.3 Application Traditional Genetic Algorithm in Automatic Test Case Generation

Genetic algorithms are used to create highly effective test data sets [9]. In each cycle of the evolutionary process, the automated test data generation tool uses the existing test data to run the program being tested, tracking and recording the execution paths for each data set. The effectiveness of each individual data set is evaluated using a fitness function, which aims to maximize the coverage of the program's execution paths. The best-performing data sets are selected to create the next generation of test data. After several generations of this process, the algorithm produces an optimal set of test cases. The basic steps are as follows:

Step 1. Randomly generate an initial set of test cases based on the population parameter rule.
Step 2. Check if the population meets the test goal. If it does, go to **Step 7**; if not, proceed to **Step 3**.
Step 3. Determine if the maximum iteration limit has been reached. If it has, go to **Step 7**; if not, proceed to **Step 4**.
Step 4. Calculate the fitness value for each individual in the population.
Step 5. Apply genetic operations - selection, crossover, and mutation - to create a new population.
Step 6. Return to **Step 2**.
Step 7. End of the algorithm and the test case is produced.

2.4 Preliminary Work

The concept of software testing emerged in 1957 as an essential aspect of software development. While software testing is a crucial method, the design of test cases can be time-consuming and labor-intensive [10]. This has led to extensive research on automated test case generation. Experts have proposed various methods, categorized into static and dynamic approaches. The random static method, widely used in automated testing tools [11], generates test cases randomly within specified ranges. While this method offers objectivity and simplicity, it often produces redundant test cases, leading to inefficient execution.

Dynamic test case generation, a widely studied academic topic, enhances the test case generation process through heuristic search and utilizes feedback from the tested program to determine test case adequacy [12]. The dynamic method, starting from the program's actual state, iteratively adjusts the solution based on feedback, ensuring optimal outcomes. Research has validated the effectiveness of dynamic testing frameworks combined with heuristic search algorithms for generating test cases for software structures. Intelligent algorithms, such as ant colony algorithms, random algorithms, and genetic algorithms, excel in solving these problems, with genetic algorithms being particularly popular ([11, 13–15]). Studies have consistently demonstrated the superiority of genetic algorithms over random algorithms in terms of iteration count and test data generation for various testing scenarios ([16, 17]).

Genetic algorithms have faced challenges in efficiently addressing multi-path program coverage [15]. To overcome this, researchers have proposed various modifications to the algorithm. One approach involves replacing the individual fitness function with a clustering convergence graph fitness function, which has shown significant improvements in multi-path program coverage but at the cost of increased time consumption.

Ghiduk [18] redefined chromosomes, crossover operations, and mutation operations in genetic algorithms to enhance their adaptability to path size and chromosome length changes during iterations. Another approach focuses on addressing the premature convergence issue of traditional genetic algorithms. Cheng and Zhong [19] proposed a hybrid optimization algorithm that incorporates immune factors to avoid local convergence by reducing the number of similar individuals in the population. However, this method may lead to slow convergence in later stages of evolution.

Efforts to improve the fitness function in genetic algorithms, such as those by Gao et al. [20], have led to increased convergence speed but at the expense of population diversity and coverage. Yuan and Liu [21] combined genetic and simulated annealing algorithms to create a hybrid genetic algorithm that demonstrated advantages in running efficiency and solution quality for benchmark programs. However, its applicability to complex programs remains limited.

Recent research has explored further enhancements to genetic algorithms. Li et al. applied adaptive random testing to the initialization population and proposed a method to improve test data generation [22]. While this approach effectively reduces average evolution generations, it neglects the combined effects of selection, crossover, and mutation operations. Fan et al. [23] focused on designing a fitness function related to the balance of the program, enhancing test data generation efficiency. Aburas [24] used static analysis to guide mutation operators in the elicitation genetic algorithm, achieving higher branch coverage but compromising convergence speed.

Despite these advancements, a common limitation of genetic algorithms for automatic test case generation remains: the difficulty in reducing iteration numbers while increasing convergence rates. This research aims to address this challenge by exploring methods to enhance the convergence speed of genetic algorithms, ultimately contributing to faster and more balanced automatic test case generation.

2.5 Identify Solution

Previous studies have often overlooked the potential for reduced convergence rates in genetic algorithms. This article addresses the slow convergence speed of genetic algorithms by proposing a solution to minimize iterations and enhance test case generation efficiency.

The primary cause of slow convergence is the loss of promising individuals during the evolutionary process [25]. This leads to a decline in the population's quality and an increased number of generations required to reach the optimal solution. The elite retention strategy preserves the best individuals encountered so far, directly replicating them to the next generation without genetic operations. During selection, the fittest individual is stored in the optimal solution set, while others undergo selection, crossover, and mutation. In each generation, the best offspring individual is compared to the best parent individual. If the offspring is superior, it directly enters the next generation. If the parent is superior, it randomly replaces an offspring individual. This ensures that promising individuals are not lost due to genetic operations, protecting the best individuals and maintaining global optimality [26]. This approach effectively improves convergence ability and speed, leading to faster discovery of optimal solutions and more rapid test case generation.

With these assumptions, the proposed Elite Genetic Algorithm (EGA) integrates an elite retention strategy. Theoretically, this approach preserves promising individuals, increases the proportion of high-quality candidates in the population, minimizes unnecessary genetic operations, and steers the evolution process more accurately, thereby reducing iteration count.

3 Methodology

This section presents the research design, covering the entire process from research planning to algorithm development, experimentation, data collection, analysis, and evaluation.

3.1 Research Design

The research is divided into three main phases: (1) algorithm design and implementation, (2) experimental testing, and (3) analysis and evaluation, as shown in Fig. 1. The classical random algorithm, traditional genetic algorithm, and optimized genetic algorithm are each used to generate test cases for four different programs, highlighting the improvements in test case generation speed achieved by the optimized algorithm.

3.2 Improved Design and Implementation

The first stage involves developing a new algorithm (the EGA algorithm). This stage builds on the problem analysis from the Literature Review Section. To preserve elite individuals, several enhancements have been identified, leading to the design of a new algorithm. Simulations will be conducted to assess whether the optimized algorithm achieves faster performance and fewer iterations in automatic test case generation.

Build the Algorithm. The elite retention strategy ensures that the best individuals from the current generation are preserved without undergoing crossover or mutation, allowing them to pass directly to the next generation. The elite retention process by [27] suggests that the strategy safeguards superior genes, preventing their loss during crossover and mutation operations.

The strategy to the *i*th 1 generation in the fitness value of the best individual as I generation threshold value, as shown in formula $\underset{I-1}{\overset{M}{f}}$. If the fitness value of individuals in generation I is below the threshold set by generation I − 1, genetic retention is applied; otherwise, crossover and mutation are performed. If a higher-fitness individual

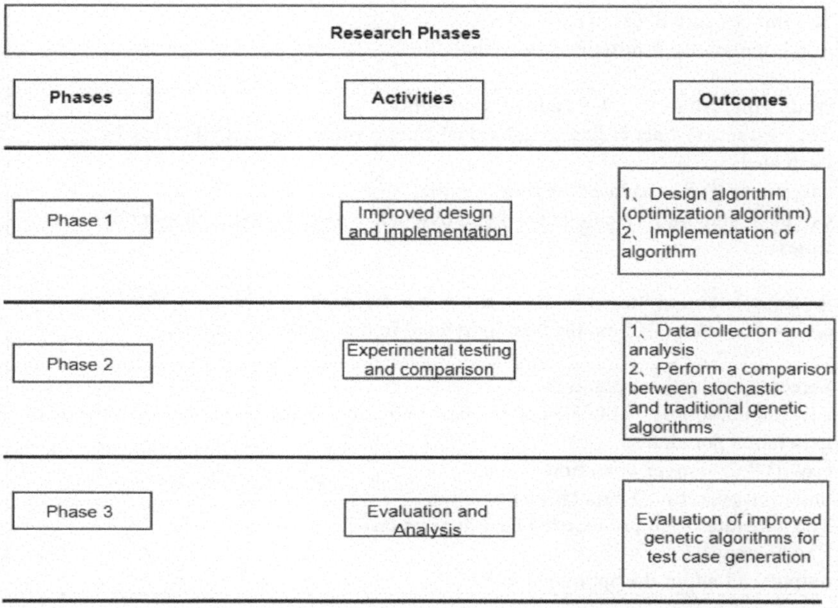

Fig. 1. Research design.

appears in the next generation, it will replace the current optimal individual. This strategy effectively enhances the algorithm's convergence ability [28].

$$K = \{\frac{1, f_I^j < f_{I-1}^m}{0, f_{I'} \geq f_{I-1}^m} \tag{1}$$

A is the first generation of I first J individuals in the process of evolution. K = 1, the individual is kept elite; K = 0, then it is not preserved. The elite retention strategy minimizes redundant calculations of high-quality genes and provides greater protection for them, enhancing algorithm efficiency [29]. This design optimization approach effectively safeguards elite genes. Next, we will optimize the elite retention strategy within the traditional genetic algorithm.

Pseudocode of the Improved Algorithm. The sample pseudocode in Fig. 2 outlines the steps for measuring processing time and iteration count during automatic test case generation. The evolution process will terminate upon meeting one of two conditions: (1) identifying the best individual, or (2) reaching the maximum specified number of generations.

3.3 Experimental Testing and Data Collection

To assess the effectiveness of the improved algorithm, three standard programs were used to automatically generate test cases. For all the three programs, both the number of iterations and the time taken for test case generation were recorded.

```
The init () # initialization parameter
init_Population() # Initializes the population

while Final Num >     0: # Judge the condition
The evaluate () # according to individual fitness value, the individual can be
evaluated
cal_average() # calculates the fitness value
cal_max_and_min() # calculates the individuals with the maximum and minimum
fitness

# Elitist Retention Strategy - Keep the best individual
elite = find_elite() # Find the best individual in the current population

Select () # selection operation
new population[0] = elite # Keep the best individual in the first position of the next
generation population
cross() # Crossover operation
mutation operation() # mutation operation
The generate (#) to produce the next generation
Final Num -= 1
output() # Output the optimal solution
```

Fig. 2. Sample of pseudocode.

Experimental Testing. A comparative analysis was performed to assess the effectiveness of enhancements to the traditional genetic algorithm through the inclusion of an elite retention strategy and differential mutation operation. The new algorithm was evaluated against both the conventional and random genetic algorithms. These improvements were implemented in Python, and performance tests were conducted on three benchmark programs: triangle classification, day-of-the-week calculation, and complex number functions. Test cases were generated for branch coverage using each algorithm.

The analysis focused on comparing the improved genetic algorithm's performance with the standard version, specifically regarding convergence quality and efficiency. The iteration count and time required for each test case generation were recorded.

The experiments were conducted on a Windows 10 system using PyCharm. To ensure reliability, each experiment was independently repeated 100 times. The experiments concluded when either full target path coverage was achieved or the set number of iterations was reached.

Data Collection. To address the branch coverage principle, which ensures that each branch is executed at least once [30], the following data were collected:

- **Average Coverage**: The average coverage of the generated test cases across multiple runs is assessed. Higher coverage suggests better software exploration and a greater chance of detecting faults.
- **Average Number of Iterations**: The average number of iterations required for each optimization run was compared to assess the resource efficiency of the new and traditional genetic algorithms in generating branch coverage test cases.

- **Average Execution Time per Generation**: This metric assesses the computational efficiency of the genetic algorithm. A lower average time per generation indicates a more efficient algorithm, which is essential for addressing complex problems or when rapid solutions are required.
- **Minimum Number of Iterations**: The minimum number of iterations necessary to attain a satisfactory level of performance or solution. Algorithms that converge more rapidly are generally preferred as they efficiently find high-quality solutions.

4 Evaluation and Analysis

The final phase of this study involves evaluation and analysis. By comparing the performance of the three algorithms through experiments, we will assess the proposed method's effectiveness and feasibility in comparison to existing methods. The experiments aim to address the following questions:

Q1: Does the algorithm with differential mutation require fewer iterations?
Q2: Can it reduce time consumption?

These comparative experiments will evaluate the effectiveness of the improved genetic algorithm model in engineering applications. The methodology involves three primary steps: (1) refining the design using the flowchart and pseudocode, (2) conducting experiments to collect data parameters, and (3) evaluating and analyzing the experimental results. This section provides an overview of the improved algorithm's concept and design. The subsequent section will delve into the specific implementation and experimentation using Python code.

4.1 Experiments Analysis and Results

In the methodology section, we presented a refined genetic algorithm incorporating an elite retention strategy and examined its advantages. To validate the effectiveness of this enhanced algorithm in automatic test case generation, experiments were conducted on three benchmark programs: triangle classification, day-of-the-week calculation, and Bessel function. The standard genetic algorithm, random algorithm, and the EGA were applied to these programs. Comparisons between the EGA, standard GA, and random algorithm provide a clear benchmark for evaluating the performance of the algorithm and highlight the contribution of the elite retention strategy to test case generation efficiency.

4.2 Experiments Design

To evaluate the improved EGA, it was compared with traditional GA and random algorithms in generating branch coverage test cases for triangle classification and date calculation. Metrics assessed included the minimum number of iterations, average number of iterations, average coverage rate, and average execution time per generation.

The experiments were conducted using Python and executed in PyCharm on a Windows 10, 64-bit system equipped with a 12th Gen Intel(R) Core (TM) i5-12450H processor operating at 2.00 GHz. The three-branch coverage test case generation methods shared common parameters: a population size of 50 and a maximum of 100 evolutionary iterations. To ensure consistency, each program underwent 50 test runs under each method. Test data generation continued until the target path was covered or the maximum iteration limit was reached. The three benchmark programs namely Triangle, Week, and Bessel are described next.

The Triangle program takes three integer inputs and classifies them as equilateral, isosceles, scalene, or non-triangular. Experimental parameters included a value range for triangle sides of [−10, 10], a population size of 100, a chromosome length of 21, 50 iterations, a crossover probability of 0.7, and a mutation probability of 0.01. The experiment aimed to compare the efficiency of traditional GA and EGA in generating test cases for different triangle types. The results demonstrate the effectiveness of genetic algorithms in improving test case generation speed and reducing iteration numbers, showcasing the benefits of the elite retention strategy.

The Week program calculates the day of the week for a given date using three integers. The program parameters were defined as follows: year range of [−10, 50], month range of [−10, 30], and day range of [−10, 50]. The population size was set to 100, chromosome length to 52, iterations to 50, crossover probability to 0.7, and mutation probability to 0.01. The experiment used random algorithms, traditional GA, and EGA to generate test cases that cover various date conditions, including leap year exceptions. Performance was evaluated based on the efficiency of test case generation across different algorithms.

The Bessel program calculates the Bessel function for specified parameters nnn and xxx, addressing up to 12 distinct scenarios. In the experiments, random algorithms, traditional genetic algorithms (GA), and the improved Elite Genetic Algorithm (EGA) were employed to automatically generate test cases for these scenarios. The focus was on evaluating the algorithms' effectiveness in handling various combinations of *xxx* and *nnn* for generating test cases.

5 Results and Analysis

To assess the effectiveness and feasibility of this method for generating branch coverage test cases, we compared the enhanced EGA with the traditional genetic algorithm (GA) and a random algorithm. This comparison, detailed in Fig. 3, was conducted using the triangle classification, Bessel function, and Week calculation programs. We recorded the minimum number of iterations, average number of iterations, average coverage, and average execution time per generation for each algorithm. Figure 3 reveals that, in terms of average coverage for test case generation across the three programs, the EGA outperforms the traditional genetic algorithm. This indicates that the EGA strategy effectively ensures higher coverage efficiency.

The subsequent analysis will focus on examining the speed improvements of the EGA compared to the other two algorithms for each program.

5.1 Triangle Experiment

The results of the triangle experiment are shown in Fig. 4. The EGA achieves an average of 17.44 iterations, compared to 18.94 iterations for the traditional Genetic Algorithm

Programs		Randomized algorithm	GA	EGA
triangle	Minimum number of iterations	-	11	5
	Average number of iterations	-	18.94	17.44
	Average coverage rate	0.379886	0.9142857	0.9314285
	Average execution time per generation	-	0.0289558	0.0067983
week	Minimum number of iterations	-	27	15
	Average number of iterations		47.1	40.72
	Average coverage rate	0.1213714	0.9414285	0.9442857
	Average execution time per generation	-	0.496375	0.4884387
Bessel	Minimum number of iterations	-	12	11
	Average number of iterations	-	23.26	19.3
	Average coverage rate	0.365	0.925	0.96166666
	Average execution time per generation	-	0.16825438	0.16842347

Fig. 3. Results of experiments.

(GA), representing an optimization of 7.92%. This indicates a significant improvement in the average number of iterations with the Elite Genetic Algorithm. Additionally, the minimum number of iterations required by the EGA is 5, whereas the GA involves a minimum of 11 iterations, demonstrating greater efficiency in reaching the optimal solution. The average execution time per generation for the EGA is 0.0067983 s, compared to 0.0289558 s for the GA, highlighting the enhanced time efficiency of the EGA.

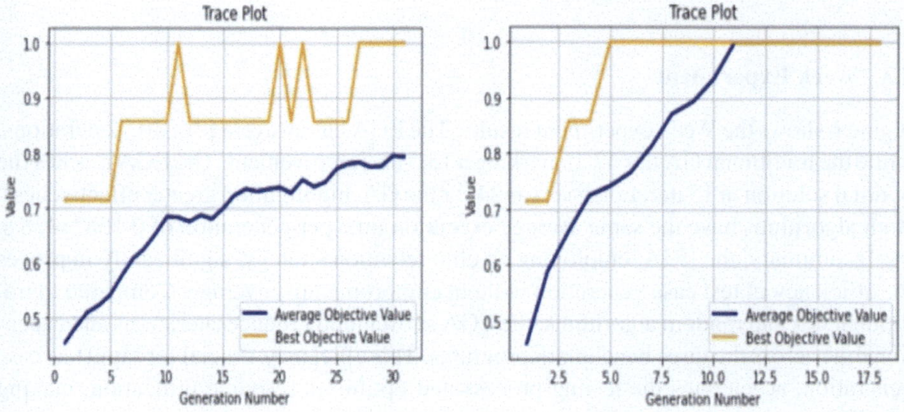

Fig. 4. Triangle experiment.

5.2 Bessel Experiment

The results of the Bessel experiment, presented in Fig. 5, show that the Elite Genetic Algorithm (EGA) achieves an average of 19.3 iterations, compared to 23.26 for the traditional Genetic Algorithm (GA), marking a 17.02% improvement in the average number of iterations. This indicates a significant enhancement in efficiency with the EGA. In terms of minimum iterations, the EGA requires 11 iterations, while the GA requires 12, demonstrating the EGA's superior efficiency in finding the optimal solution. The average execution time per generation for the EGA is 0.16825438 s, compared to 0.16842347 s for the GA, indicating a slight improvement in time efficiency with the EGA.

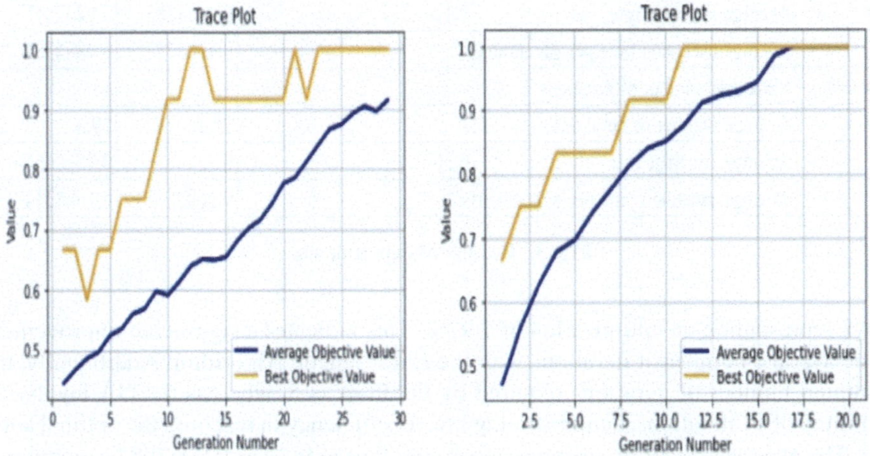

Fig. 5. Bessel experiment.

5.3 Week Experiment

Figure 6 shows the Week experiment results. The EGA has an average of 40.72 iterations, while the traditional GA has 47.1, marking a 13.54% improvement. The EGA reaches the optimal solution in 15 iterations, versus 24 for the GA, highlighting greater effectiveness. Both algorithms have the same average execution time per generation of 0.16825438 s.

In summary, the EGA, employing an elite retention strategy, significantly improves the efficiency of test case generation without compromising coverage. Compared to traditional GA and random algorithms, the EGA substantially reduces iterations and execution time across the three benchmark programs. This efficiency, crucial for rapid test case generation, accelerates the testing process and optimizes resource utilization, making the EGA particularly valuable in resource-constrained or time-sensitive scenarios.

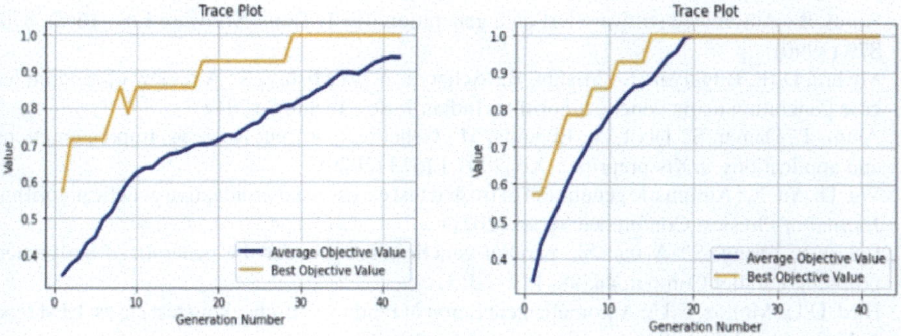

Fig. 6. Week experiment.

6 Conclusion

An automated approach utilising genetic algorithms was proposed to mitigate the inefficiencies, subjectivity, and workload inherent in manual test case generation. An elite retention strategy was introduced to expedite the convergence speed of the standard genetic algorithm. This strategy preserves high-quality solutions, thereby accelerating the identification of optimal results. The performance of traditional, elite, and random genetic algorithms using three benchmark programs was evaluated via experiments.

Key performance indicators, including the optimal number of iterations, runtime, and average number of iterations, are analyzed to assess the effectiveness of the improved algorithm in automatic test case generation. The findings unequivocally demonstrate that the EGA, incorporating the elite retention strategy, significantly surpasses both traditional genetic and random algorithms in terms of time efficiency and effectiveness. The innovative approach presented in this paper, employing elite genetic algorithms in conjunction with shared knowledge of automation tools and scripting languages, has the potential to revolutionize software testing methodologies. This breakthrough could significantly enhance software quality and efficiency. To foster innovation and uphold high standards, continued exploration of new testing techniques should be supported by knowledge sharing and collaboration.

References

1. He, Y.W., Liu, Z.Y., Li, J.Y., Peng, X.Y.: A survey of test case automatic generation methods based on genetic algorithm. Journal of Langfang Normal University, Natural Science Edition (2020)
2. Ji, F.: Research on software testing automation mechanism based on improved genetic algorithms. Information Technology 10 (2019)
3. Bao, X., Xiong, Z., Zhang, W., Wu, B., Zhang, N.: An improved genetic algorithm based path test case generation method. Computer Science **45**(8), 174–178 (2018)
4. Luo, F., Lyu, F., Hou, Z.: An Improved Genetic Algorithm based on Elite Retention Strategy and Explosion Operators. Journal of Xihua University (Natural Science Edition), 3 (2018)
5. Garousi, V., Felderer, M., Kuhrmann, M., Herkiloğlu, K.: What industry wants from academia in software testing? Hearing practitioners' opinions. Proceedings of the 21st International Conference on Evaluation and Assessment in Software Engineering (2017)

6. Korel, B.: Automated software test data generation. IEEE Trans. Software Eng. **16**(8), 870–879 (1990)
7. Mishra, D.B., Bilgaiyan, S., Mishra, R., Acharya, A.A., Mishra, S.: A review of random test case generation using genetic algorithm. Indian J. Sci. Technol. (2017)
8. Alam, T., Qamar, S., Dixit, A., Benaida, M.: Genetic algorithm: Reviews, implementations, and applications. arXiv preprint arXiv:2007.12673 (2020)
9. Wu, D., Yu, X.: Automatic generation of trusted test cases based on adaptive genetic algorithm. Journal of Physics: Conference Series (2021)
10. Liu, Y.Y., Wang, S.: A scalable parallel genetic algorithm for the generalized assignment problem. Parallel Comput. **46**, 98–119 (2015)
11. Bird, D.L., Munoz, C.U.: Automatic generation of random self-checking test cases. IBM Syst. J. **22**(3), 229–245 (1983)
12. Wang, J., Cai, Y.: Research on testing data generation algorithms based on maintaining population diversity. Comp. Res. Develop. **49**(5), 1039–1048 (2012)
13. Srivastava, P.R., Baby, K.: Automated software testing using metaheuristic technique based on an ant colony optimization. 2010 international symposium on electronic system design (2010)
14. Bo, F.: Automated software test data generation based on simulated annealing genetic algorithms. Computer Engineering and Applications 12 (2005)
15. Harman, M., McMinn, P.: A theoretical and empirical study of search-based testing: Local, global, and hybrid search. IEEE Trans. Software Eng. **36**(2), 226–247 (2009)
16. Jones, B.F., Sthamer, H.-H., Eyres, D.E.: Automatic structural testing using genetic algorithms. Softw. Eng. J. **11**(5), 299–306 (1996)
17. Jones, B.F., Eyres, D.E., Sthamer, H.-H.: A strategy for using genetic algorithms to automate branch and fault-based testing. Comput. J. **41**(2), 98–107 (1998)
18. Ghiduk, A.S.: Automatic generation of basis test paths using variable length genetic algorithm. Inf. Process. Lett. **114**(6), 304–316 (2014)
19. Cheng, L., Zhong, L.: A parallel immune genetic algorithm for a multimodal function optimization problem. Microelectronics and Computer **32**, 117–121 (2015)
20. Gao, X.D., Zhou, L.J., Zhang, S.D., Liu, H.M.: Research on test data automatic generation based on improved genetic algorithm. Comput. Sci. **44**(3), 209–213 (2017)
21. Yuan, M.F., et al.: Path planning of express vehicle based on adaptive elite genetic algorithm. Journal of Navigation and Positioning (2021)
22. Li, Z.B., Li, Q.B., Zhang, J.G.: Research on the initial population distribution of genetic algorithm for test data generation. J. Info. Eng. Univ. **21**(2), 236–241 (2020)
23. Fan, S.P., Zhang, Y., Ma, B.Y., Wan, L., Yao, N.M., Song, Y.: Evolutionary generation of test data for paths coverage based on balance optimization theory. ACTA Electronica SINICA **48**(7), 1303–1310 (2020)
24. Aburas, A.: An Improved Genetic Algorithm with Initial Population Strategy and Guided Mutation. Proceedings of Sixth International Congress on Information and Communication Technology: ICICT 2021, London, Volume 2 (2022)
25. Shen, J.X., Lu, H.F.: Adaptive fitting of AGV steering ladder model based on improved genetic algorithm. Lifting and transport machinery (2023)
26. Zhang, Y.X., Qi, Y.X.: Improved genetic algorithm for solving TSP. Shanxi Electronic Technology **1**, 28–30 (2016)
27. Wang, L., Luo, X.H., Yu, M., Xia, S.X.: Genetic Algorithm Used in Functional Verification Based on Elite Strategy. Journal of East China University of Science and Technology: Natural Science Edition **5**, 676–681 (2016)
28. Xu, Y.F., Cao, L., Huang, J.W., Kan, X.: Extraction of bio-impedance characteristic parameters through genetic algorithm based on elitist preservation strategy. Chinese Journal of Medical Physics, 214–221 (2021)

29. Yuan, G.H., Liu, Z.C.: Research on test case generation based on hybrid genetic algorithm. J. Heilongjiang Univ. Technol. **19**(10), 33–38 (2019)
30. Myers, G.J., Badgett, T., Thomas, T.M., Sandler, C.: The art of software testing, Vol. 2. Wiley Online Library (2004)

Knowledge Engineering Approach in Early Detection of Oak Wilt Disease

Nur Azzah Abu Bakar(✉), Aniza Mohamed Din, Lam Hui Shan, Ho Yi Xin, and Kam Weng Kee

Universiti Utara Malaysia, 06010 Sintok, Kedah, Malaysia
nurazzah@uum.edu.my

Abstract. Oak wilt is a destructive fungal disease that affects oak trees, causing rapid defoliation and death if left untreated. Early detection is critical for controlling its spread, minimizing ecosystem damage, and reducing extensive economic consequences. Detection of oak wilt requires expertise as oak wilt symptoms can be confused with other diseases or environmental stressors; to date, the expertise remains limited. This study aims to develop an alternative tool for the early detection of oak wilt disease affecting three oak species, i.e. red oak, white oak, and live oak. The study follows a standard knowledge engineering methodology involving problem assessment, knowledge acquisition, knowledge representation, design, development, and evaluation phases. The domain knowledge was acquired from the secondary knowledge source via knowledge mining activities, analyzed and represented into conceptual maps, flowcharts, rules and inference networks. A total of 26 rules are built in the knowledge base, with 32 primitive premises. The prototype was designed to work in a forward chaining fashion and was developed using JavaScript. The evaluation of the prototype revealed that it successfully detects the presence of oak wilt from the symptoms. The findings also demonstrate the prototype's usability, especially its ease of use and explanation facility. Early detection will allow timely interventions, reducing oak wilt's ecological and economic impact on forests and communities.

Keywords: Oak Wilt Disease · Diagnostic Expert System · Forest Management Tools

1 Introduction

Oak trees can be taxonomically categorized into three main groups: red oaks, white oaks, and live oaks. The most distinctive feature differentiating them is the leaf shape, where the red oaks have fan-shaped leaves with sharply pointed tips. The leaves of the white oaks are also fan-shaped but with rounded or blunt tips. The live oaks have oval leaves with pointed to rounded tips. Meanwhile, oak wilt is an aggressive disease affecting many oak species. It is one of the most serious tree diseases in the United States and Canada, killing thousands of oaks yearly in forests, woodlots, and home landscapes [1]. The disease is caused by a fungal pathogen named *Ceratocystis fagacearum*. It has become much more apparent since the 1980s due to home construction in oak wood,

Z. Mat Aji et al. (Eds.): KMICe 2024, ISEM 49, pp. 542–553, 2025.
https://doi.org/10.1007/978-3-031-91485-0_42

which caused wounds to the oak trees. The white oaks are the most resistant to the wilt disease; they can be killed over a period of one to several years after being infected. The red oaks are the most critical hosts with mortality, which can occur within three weeks after infection. The live oaks are also considered an essential host with sustained high mortality levels.

Wilt disease spreads in two ways, i.e. via local spread (also known as root grafts) and overland spread. The root of an infected tree can graft to the root of the other trees in the same group, forming a continuous underground network. The fungus spreads through the connected root system to create an infection centre. Root grafts do not commonly occur between trees in different species groups. Thus, mixing the various species groups is one of the strategies to retard the local spread and limit the impact of the disease. On the other hand, overland spread involves insect vectors, and the disease can spread if the fungus is carried by insects from an infected tree to the wound on a healthy tree. Oak trees often sustain wounds from construction equipment, storms, pruning tools, or vandalism. The scars usually attract insects, including those that have visited the infected trees. Like root grafts, this overland movement of the fungus creates a new infection centre.

The symptoms of oak wilt disease vary from one species group to another. For example, an infected red oak usually shows rapid leaf discolouration and wilting. In the white oak, the branches die slowly, one after another, for one to several years. Leaf wilting occurs similarly to the red oak, but the progress is much slower in the white oak. The death of live oak is influenced by the timing of infection and weather conditions; generally, it will succumb within 1 to 6 months of the infection.

Oak wilt disease often leads to extensive economic and ecological consequences. Early detection is crucial for effectively managing and mitigating its impacts, as symptoms can manifest subtly before the disease progresses. The financial implications are substantial. In the United States, removing urban oak trees due to oak wilt can cost between 18 million and 60 million dollars [2]. In Canada, similar economic impacts are anticipated, with urban centres facing significant costs related to tree management and replacement as oak wilt spreads. The financial burden of managing oak wilt extends beyond immediate removal costs, encompassing long-term ecological restoration and the loss of ecosystem services that healthy oak forests provide. The environmental consequences of oak wilt are profound. The disease disrupts forest ecosystems, leading to biodiversity loss and altering habitat structures. The interaction between oak wilt and other environmental stressors, such as drought, further complicates management efforts [3].

This paper aims to present the design and development of a prototype of an expert system for early detection of oak wilt disease. Detection of oak wilt requires expertise, and this expertise remains limited. This gap is concerning, given that misdiagnosis can lead to inappropriate management actions, exacerbating the spread of the disease. Furthermore, the complexity of oak wilt symptoms, which can be confused with other diseases or environmental stresses, underscores the need for specialized training and knowledge among forestry professionals and land managers [4]. Additionally, the reliance on laboratory-based diagnostics can delay responses, making it imperative to enhance field-based expertise and rapid assessment tools [5].

2 Related Works

The importance of early disease detection cannot be overstated, as it directly correlates with the economic viability of agricultural practices. Ezhilarasan [6] noted that early identification of diseases can significantly reduce maintenance costs and improve the overall health of crops. Recent studies on oak wilt disease detection have increasingly incorporated artificial intelligence (AI) and non-AI techniques to enhance the accuracy and efficiency of identifying this significant threat to oak trees. Oak wilt poses a considerable risk to oak populations, necessitating the development of effective detection methods [2]. One of the most promising approaches involves using spectral reflectance data to distinguish oak wilt from other stressors, such as drought. Research by Sapes et al. [7] emphasizes the importance of integrating physiological and spectral data to improve pre-visual oak wilt detection, highlighting that spatial patterns can significantly reduce misclassification rates. This study builds upon previous work that demonstrated the capability of canopy spectral reflectance to detect oak wilt at a landscape scale, utilizing phylogenetic discrimination to enhance detection accuracy [7]. The findings indicate that physiological changes associated with oak wilt can be detected before visible symptoms appear, which is crucial for timely intervention [3].

Another significant advancement is the development of rapid, field-based diagnostic tools [8]. Researchers have created a handheld device that uses DNA amplification techniques to detect fungi from tree samples within 30 min. This tool has been shown to have high sensitivity and specificity, making it an efficient alternative to traditional lab testing, which can be slow and expensive. This rapid detection method enables faster intervention to prevent the spread of the disease.

AI techniques are being employed to automate and refine the detection process. For instance, machine learning algorithms have been utilized to classify plant diseases based on visual cues, demonstrating the potential for rapid and accurate oak wilt identification [6]. The application of deep learning frameworks, particularly convolutional neural networks (CNNs), has shown promise in classifying various plant diseases, although specific studies on oak wilt using CNNs are limited [9]. These AI-driven methods can process large datasets of images, allowing for more efficient monitoring of oak health across extensive areas.

Moreover, integrating AI with remote sensing technologies is paving the way for innovative solutions in oak wilt detection. By leveraging high-resolution imagery and machine learning, researchers can develop models that predict the spread of oak wilt based on environmental and biological factors. This approach not only enhances the understanding of oak wilt dynamics but also aids in formulating management strategies to mitigate its impact. However, the specific application of these advanced techniques to oak wilt detection remains an area that requires further exploration to optimize their effectiveness [10].

Expert systems is one of the AI approaches that is appropriate for the detection type of problem-solving paradigm. Expert systems that utilize advanced machine learning and deep learning techniques to analyze images of plant leaves and other parts allow for the timely identification and classification of diseases. Machine learning techniques, particularly those involving deep neural networks, have shown remarkable efficacy in recognizing plant diseases through image classification. For instance, Sladojević et al.

[11] demonstrated that deep neural networks could effectively classify plant diseases based on leaf images, outperforming traditional methods in accuracy. Similarly, Varshney et al. [12] highlighted the potential of a machine learning expert system in plant disease detection, emphasizing its role in improving agricultural productivity by facilitating early diagnosis. The application of expert systems is not limited to a single type of plant; various studies have successfully implemented it for different crops including guava [13], tomatoes [6], mango [14], papaya [15], passion fruit [16], banana [17, 18], teak tree [19], corn [20], tobacco [21], spinach [22], rice plant [23] and many others. The wide use of expert systems in various agriculture domains proved the feasibility of expert system technology in plant disease diagnosis. The integration of the above-mentioned advanced detection methods into expert systems adds to its significance in improving the management of oak wilt. For example, a comprehensive monitoring system can be developed by combining spectral analysis, molecular diagnostics, and remote sensing. Such a system would enable forest managers to detect oak wilt early, implement control measures effectively, and preserve oak populations.

In conclusion, applying expert systems and innovative detection technologies represents a significant advancement in the fight against oak wilt disease. These methods not only facilitate early diagnosis but also empower forest management strategies to mitigate the impact of this devastating disease. The ongoing research and development in this field promise to enhance the capabilities of these systems further, ultimately contributing to improved agricultural resilience.

3 Methodology

This study was carried out according to the knowledge engineering methodology, which consists of the phases described in Sub-sects. 3.1 to 3.4.

3.1 Problem Assessment

This phase involves the assessment of the various criteria that make oak wilt disease detection a suitable problem requiring an expert system solution. The problem meets several criteria, including the scarcity of expertise and the severity of the consequences if detection is delayed. Disease detection is also one of the problem-solving paradigms of expert systems in which many successful systems have been developed in the past.

3.2 Knowledge Acquisition and Representation

The knowledge regarding oak wilt disease was acquired through a series of knowledge-mining activities from multiple sources, but the primary source of knowledge is from [1]. Domain concepts and objects extracted from the knowledge source are represented into a cognitive map (Fig. 1). Wilt disease identification domain consists of many domain-specific terms; the figure partially represents the domain knowledge. Similar terms are used throughout the prototype when asking questions and displaying or explaining the results.

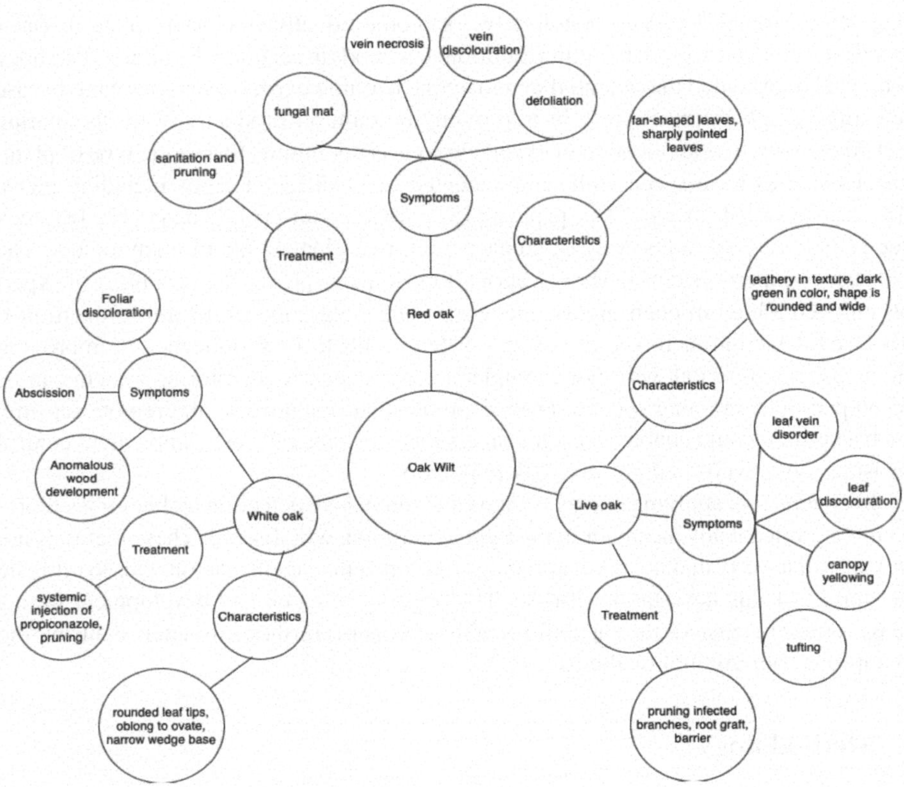

Fig. 1. Partial representation of oak wilt disease domain knowledge.

The detection of the disease follows the steps as shown in Fig. 2. It starts with determining the oak group whether it is red, white or live oak. Once determined, several questions about the observable symptoms are asked before the results are displayed. The flowchart illustrates that the prototype works in a forward chaining fashion. No hypotheses were developed beforehand, meaning that the detection is solely based on symptoms provided by the user. If enough symptoms are evidenced, the prototype concludes with the presence of the disease or vice versa.

A total of 26 rules have been incorporated into the prototype's knowledge base. The inference network in Fig. 3 shows the rules for detecting oak wilt disease in red oaks. The two boxes labeled with dots represent the rules for detecting the disease in the other two types of oak.

3.3 Design and Development

Expert systems commonly interact with users through a series of questions and answers. This interaction mimics the process when a human expert asks his client several questions to solve the client's problem. These questions are observable, which means the client can directly tell the expert the answers in the form of yes or no or give values, symptoms, or

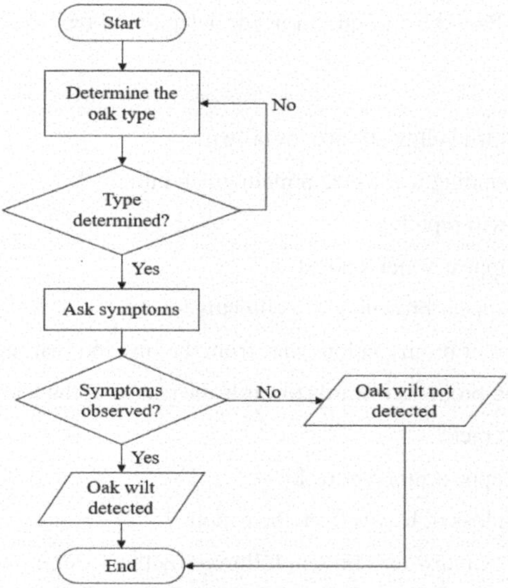

Fig. 2. The oak wilt disease detection steps.

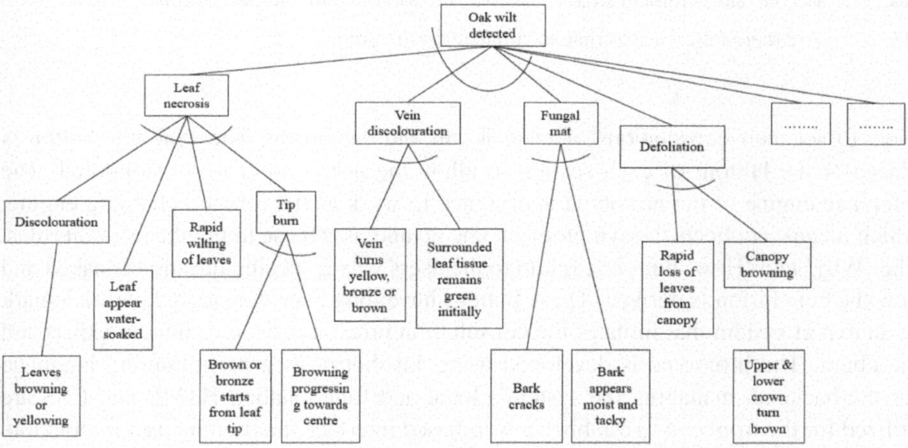

Fig. 3. An inference network for oak wilt disease detection rules in red oaks.

any other type of answers. In an expert system context, these questions are represented as primitive premises or conditions of rules. Thirty-two primitive premises were selected in this study, as there were 32 questions to detect oak wilt disease. However, the questions prompted on the screen are less than 32, depending on the selected type of oak tree under examination, i.e., whether it is red, white, or live oak. Some of these questions are listed in Table 1. As mentioned earlier, many domain-specific terms are used in the questions as laypeople are not usually involved in detecting oak wilt disease.

The interface of the prototype consists of 5 buttons, i.e. 'Start', 'Submit', 'Exit', 'Why', and 'How'. The first page contains information about the prototype, allowing

Table 1. Partial list of questions for detection of oak wilt disease.

No.	Question
1.	Does the vein turn yellow, bronze or brown?
2.	Does the surrounding leaf tissue remain green initially?
3.	Do the leaves wilt rapidly?
4.	Does the leaf appear water-soaked?
5.	Do the leaves appear browning or yellowing?
6.	Does the brown or bronze colour start from the tips and margin?
7.	Is the browning progress inwards towards the centre of the leaf?
8.	Does the bark crack?
9.	Does the bark appear moist or tacky?
10.	Is there a rapid loss of leaves from the canopy?
11.	Does the upper canopy turn brown, followed by the lower crown in late summer?
12.	Do the leaves turn drab green?
13.	Are there any signs of yellowing of leaves along the vein?
14.	Do the leaves remain attached with discolouration only at the margins?
15.	Are there fallen leaves that accumulate on the ground?

users to set their expectations of what it can and cannot do. The 'Submit' button is placed at the bottom of each section to allow the user's input to be submitted. The inference engine of the prototype is designed to work as the forward chaining engine, which means it collects the symptoms in one go and uses these to fire the relevant rules. The 'Why' and 'How' buttons explain to the users why a specific question is asked and how the conclusion is derived. These buttons have long been considered the trademark of an expert system that imitates the consultation process between a human expert and the client. The prototype is developed using JavaScript as a programming language for the backend, managing the system's logic and functionality. HTML and CSS are utilized for the front end to establish a web-based interface, facilitating user interaction. This integration between the backend and frontend is crucial in ensuring a user-friendly experience.

3.4 Evaluation

The prototype was evaluated with only nine respondents; 6 of them tested the functionality, while the other 3 tested the prototype's usability. The small number of respondents involved in the evaluation is partly because this study is conducted out of the region where the oak trees grow. Two types of instruments are used: the test plan for the functionality test and the questionnaire for usability test. Usability testing provides invaluable insights into the user's experience with the expert system, essential for ensuring the system's effectiveness, efficiency, and user satisfaction. It allows us to evaluate how effectively the system interacts with and supports the intended users in accomplishing their tasks [24]. The test plan consists of 16 test cases, including the following functions: start the application; choose the leaf shape; diagnosis of leaf necrosis; diagnosis of vein discolouration; detection of fungal mat; diagnosis of defoliation; diagnosis of foliar discolouration; diagnosis of abscission; diagnosis of anomalous wood development; detection of canopy yellowing; detection of leaf vein disorders; detection of tufting; detection of leaf discolouration; WHY explanation; HOW explanation and EXIT.

Meanwhile, the questionnaire measures the ease of use, clarity of the questions, clarity of the explanation, presentation of the results, and general consideration on five scales: 1 = strongly Disagree; 2 = disagree; 3 = neutral; 4 = agree; and 5 = strongly *Agree*.

4 Expert System for Early Detection of Oak Wilt Disease

The main page (Fig. 4) displays brief but adequate information about the prototype, including its purpose and scope. This introduction is essential for users to set realistic expectations towards the prototype.

Fig. 4. The main page of expert system for early detection of oak wilt disease.

Upon clicking the 'Start' button, users will be prompted with the first question (Fig. 5(a)), asking them to choose the type of oak tree. If users click *Not Sure*, the images of different oak leaves will appear on the screen, and the user can click on any of the pictures to indicate the type of oak tree (Fig. 5(b)).

The symptoms differ between the different oak groups; therefore, only questions relevant to the group are displayed next. For example, the first 11 questions will be asked in the red oak case. Images are used to add clarity to the questions. Figure 6(a) to (d) show the questions displayed to detect the presence of oak wilt disease in red oaks. Figure 6(b) shows an explanation when the user clicks on the WHY button, as questions 1 and 2 are to detect the presence of *vein discolouration*. Depending on the answers to these questions, several other questions will be asked, and several intermediate conclusions will be made. Figure 6(c) shows the final result after the user clicks *Yes* for all questions. In this case, oak wilt is detected. The lower section displays the reasoning that has taken

(a) (b)

Fig. 5. The *chosen type of oak tree* interface.

place to conclude that *Oak wilt was detected.* Should the user answer *No* to any of these questions, the Result section will display *No oak wilt detected*, as seen in Fig. 6(d).

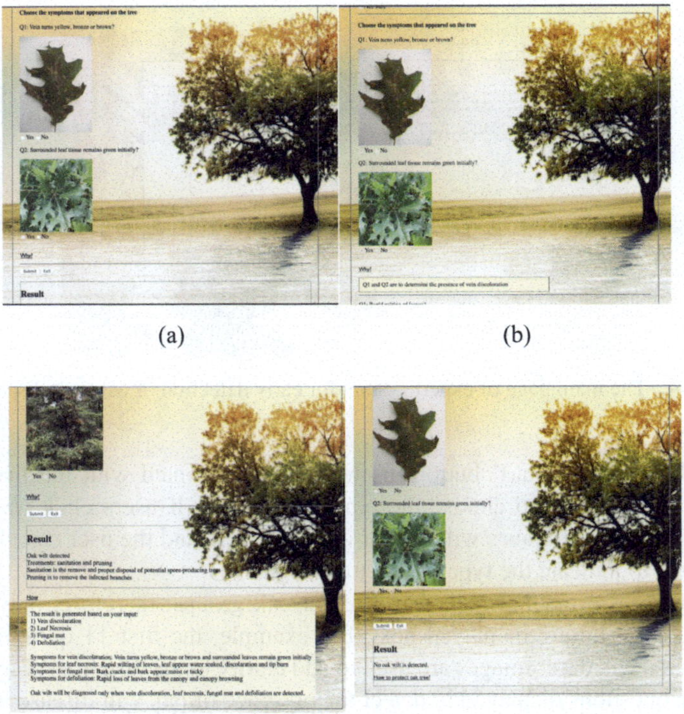

(a) (b)

(c) (d)

Fig. 6. Screenshots of oak wilt disease detection in red oak.

5 Findings

The functionality test's results show that every test case was successful, indicating that every function operates as it should. As for the usability test, the result is summarized in Table 2.

Table 2. Summary of usability test results.

No.	Question	Score in %				
		1	2	3	4	5
1	The prototype is easy to use					100.00
2	The prototype is easy to obtain information					100.00
3	The prototype is easy to navigate					100.00
4	The interface is easy to understand					100.00
5	The prototype provides an easy Exit					100.00
6	All terms used in the questions are clear				66.67	33.33
7	All questions are clear.				66.67	33.33
8	All questions have adequate options for answers			33.3	33.33	33.33
9	The system provides a clear explanation of the WHY question					100.00
10	The system provides a clear explanation of the HOW question				33.33	66.67
11	The result or output of the system is easy to understand					100.00
12	The results or output meet expectations (display what the system intends to display as its final result: diagnosis, advice, approval, etc.)				33.33	66.67
13	The system is sound				33.33	66.67

The scores in Table 2 show the positive feedback from the respondents. Only Question 8 receives *Neutral* feedback from 1 respondent. The prototype only provides two answer options, i.e. *Yes or No* as users can easily observe the symptoms; hence, the answers should be very straightforward. Responses for the other questions are positive, i.e., either Agree or Strongly Agree to all the evaluated aspects. The highest score for both *Agree* and *Strongly Agree* feedback is 100%, derived from ease of use (Questions 1 to 5) and explanation facility (Questions 9 and 10) constructs. The *explanation facility* construct has one respondent who just *Agree* with the HOW explanation. This could be due to the missing information that occurred during the testing, i.e., when the prototype failed to display its reasoning, which led to the conclusion. The mistake, however, has been fixed during the modification phase. Feedback on the *clarity of the questions* is also positive, but the percentage of *Agree* is more significant than *Strongly Agree*. The underlying reason is that the respondents who tested the prototype are not the real intended users due to difficulties in getting their participation. Thus, it is unsurprising that they find clarity issues as they are unfamiliar with the domain-specific terms used in the questions.

6 Conclusion

In general, this study addresses issues related to oak wilt disease, which must be detected as early as possible to mitigate substantial economic loss and disruption to the oak tree forest ecosystem. The prototype demonstrates the potential of the knowledge engineering approach to be applied in this domain and to replicate it in detecting other types of plant or tree disease. The current version of the prototype has several shortcomings which can be improved. Thus, future work should focus on getting involvement from the authorities, real experts, and users to strengthen the results. The current study does not incorporate machine learning or any other advanced techniques; incorporating these would lead to better early detection results. Furthermore, the prototype developed in this study could be used as a knowledge-acquisition tool in future studies.

References

1. O'Brien, J.G., Mielke, M.E., Starkey, D., Juzwik, J.: How to identify, prevent and control oak wilt. NA-FR-01–11, U.S. Department of Agriculture, Pennsylvania (2011)
2. Pedlar, J.H., McKenney, D.W., Hope, E.S., Reed, S.E., Sweeney, J.: Assessing the climate suitability and potential economic impacts of oak wilt in Canada. Scientific Reports **10**(1) (2020)
3. Fallon, B., et al.: Spectral differentiation of oak wilt from foliar fungal disease and drought is correlated with physiological changes. Tree Physiol. **40**(3), 377–390 (2020)
4. Moore, M., Juzwik, J., Saiapina, O., Ahmed, S., Yang, A., Abbas, A.: Use of sodium hydroxide DNA extraction methods for nested PCR detection of bretziella fagacearum in the sapwood of oak species in Minnesota. Plant Health Progress **23**(2), 132–139 (2022)
5. Bourgault, É., et al.: Benchmarking a fast and simple on-site detection assay for the oak wilt pathogen bretziella fagacearum. Frontiers in Forests and Global Change (2022)
6. Ezhilarasan, G., Ranganathan, S.R., Mani, L.S., Kadry, S.: An intelligent deep residual learning framework for tomato plant leaf disease classification. Int. J. Electr. Comp. Eng. (IJECE) **14**(3), 3168 (2024). Sapes, G., Schroeder, L., Scott, A., Clark, I., Juzwik, J., Montgomery, R. A. and Cavender-Bares, J.: Mechanistic links between physiology and spectral reflectance enable pre-visual detection of oak wilt and drought stress. In: Proceedings of the National Academy of Sciences 121(7) (2024)

7. Sapes, G., et al.: Mechanistic links between physiology and spectral reflectance enable pre-visual detection of oak wilt and drought stress. In: Proceedings of the National Academy of Sciences 121(7) (2024)
8. Novi, V.T., Aboubakr, H.A., Moore, M.J., Zarouri, A., Juzwik, J., Abbas, A.: A rapid LAMP assay for the diagnosis of oak wilt with the naked eye. Plant Methods **20**, 119 (2024)
9. Hassan, S., Maji, A., Jasiński, M., Leonowicz, Z., Jasińska, E.: Identification of plant-leaf diseases using CNN and transfer-learning approach. Electronics **10**(12), 1388 (2021)
10. Weingarten, E., et al.: Early detection of a tree pathogen using airborne remote sensing. Ecological Applications **32**(2) (2020)
11. Sladojević, S., Arsenović, M., Anderla, A., Ćulibrk, D., Stefanović, D.: Deep neural networks based recognition of plant diseases by leaf image classification. Computational Intelligence and Neuroscience **2016**, 1–11 (2016)
12. Varshney, D., Babukhanwala, B., Khan, J., Saxena, D. and Singh, A.: Machine learning techniques for plant disease detection. In: 5th International Conference on Trends in Electronics and Informatics (ICOEI) (2021)
13. Dheir, I., Abu-Naser, S.S.: Knowledge based system for diagnosing guava problems. Int. J. Acad. Info. Sys. Res. (IJAISR) **3**(3), 9–15 (2019)
14. Elqassas, R., Abu Naser, S.S.: Expert system for the diagnosis of mango diseases. Int. J. Acad. Eng. Res. (IJAER) **2**(8), 10–18 (2018)
15. Abu-Saqer, M.M., Abu-Naser, S.S.: Developing an expert System for Papaya Plant Disease Diagnosis. International Journal of Academic Engineering Research (IJAER) ISSN: 2000-001X 3(4), 14–21 (2019)
16. Radwan, H.I.A., El-Hamarnah, H.A., Al-Saloul, N.J.H, Lafi, O.I.A., Abu-Naser, S.S.: A proposed expert system for passion fruit diseases. Int. J. Acad. Eng. Res. (IJAER) **6**(5), 24–33 (2022). ISSN: 2643-9085
17. Budiyanto, G., et al.: Web based expert system for diagnosing disease pest on banana plant. Int. J. Eng. Technol. **7**(4), 4715–4721 (2018)
18. Almadhoun, H.R., Abu-Naser, S.S.: Banana knowledge based system diagnosis and treatment. Int. J. Acad. Pedagog. Res. (IJAPR) **2**(7), 1–11 (2018)
19. Imamah, W.A., Eka, M.R., Nuris, F.: Mobile expert system using forward chaining for diagnosing teak tree disease. Journal of Physics: Conference Series **1569**, 022072 (2020)
20. Budianto, I., Winarsih, F.: Expert System for early detection of disease in corn plant using Naive Bayes method. Jurnal Mantik **3**(4), 308–317 (2020)
21. Aldaour, A.F., Abu-Naser, S.S.: An expert system for diagnosing tobacco diseases using CLIPS. Int. J. Acad. Eng. Res. (IJAER) **3**(3), 2–18 (2019)
22. Abu Al-Qumboz, M.N., Abu-Naser, S.S.: Spinach expert system: diseases and symptoms. Int. J. Acad. Info. Sys. Res. (IJAISR) **3**(3), 16–22 (2019)
23. Habib, A., Sunan, M., Koesdijarto, R., Sajiyo, S., Ronando, E.: The empirical study on algorithm optimization in expert systems for diagnosing rice plant diseases. Jurnal Ilmiah Penelitian dan Penerapan Teknologi Sistem Informasi **8**(1) (2024)
24. Preece, A.D.: Towards a methodology for evaluating expert systems. Expert. Syst. **7**(4), 215–223 (1990)

Correction to: Leveraging Deep Learning for Enhancing Blockchain Smart Contract Security: A Systematic Literature Review

Zhyar Yassin Abdalla, Mohammed Nawzad M. Ramzi, Mohammed Alsamman, and Nur Haryani Zakaria

Correction to:
Chapter 12 in: Z. Mat Aji et al. (Eds.): *Digital Innovation in Knowledge Management*, ISEM 49, https://doi.org/10.1007/978-3-031-91485-0_12

In the originally published version of the chapter 12, the region name (Kurdistan) in the affiliation of the author (Mohammed Nawzad M. Ramzi) was removed. Now, it is corrected as "Information Technology Department, Tishk International University, Sulaymaniyah, Iraq".

The updated version of this chapter can be found at
https://doi.org/10.1007/978-3-031-91485-0_12

Z. Mat Aji et al. (Eds.): KMICe 2024, ISEM 49, p. C1, 2026.
https://doi.org/10.1007/978-3-031-91485-0_43

Author Index

Z. Mat Aji et al. (Eds.): KMICe 2024, ISEM 49, pp. 555–557, 2025.
https://doi.org/10.1007/978-3-031-91485-0

The manufacturer's authorised representative in the EU is Springer Nature Customer Service Centre GmbH, Europaplatz 3, 69115 Heidelberg, Germany. If you have any concerns regarding our products, please contact ProductSafety@springernature.com

Printed and bound by CPI Group (UK) Ltd, Croydon, CR0 4YY
30/09/2026
02225569-0018